Laminar Flow Analysis

LAMINAR FLOW ANALYSIS

David F. Rogers

Professor of Aerospace Engineering
United States Naval Academy
Annapolis, Maryland

Published by the Press Syndicate of the University of Cambridge
The Pitt Building, Trumpington Street, Cambridge CB2 1RP
40 West 20th Street, New York, NY 10011-4211, USA
10 Stamford Road, Oakleigh, Victoria 3166, Australia

©Cambridge University Press 1992

First published 1992

Printed in the United States of America

Library of Congress Cataloging-in-Publication Data
Rogers, David F., 1937-
  Laminar Flow Analysis / David F. Rogers.
    p.   cm.
  Includes bibliographical references.
  ISBN 0-521-41152-1
  1. Laminar flow.   2. Boundary layer.   3. Navier-Stokes equations.
  I. Title.
  TL574.L3R64   1992
  629.132'32—dc20   92-9137
                                                                  CIP

A catalog record for this book is available from the British Library.

ISBN 0-521-41152-1 hardback

*Dedicated to
Professor Henry T. Nagamatsu
Teacher, colleague, friend.
You taught me more than you know.*

CONTENTS

	Preface	xi
Chapter 1	Derivation of the Navier-Stokes Equations	1
1-1	Introduction	1
1-2	Stress-Rate-of-Strain Relation	7
1-3	Angular Velocity and Fluid Deformation	8
1-4	The Viscosity Coefficients of Fluids	12
1-5	The Momentum Equation	12
1-6	Derivation of the Energy Equation	13
1-7	Boundary Conditions	20
1-8	Equation of State and the Relation of the Transport Properties to the State Variables	21
1-9	Special Forms of the Navier-Stokes Equations	21
1-10	Classification	22
Chapter 2	Exact Solutions of the Navier-Stokes Equations	23
2-1	Introduction	23
2-2	Analytical Solutions	23
2-3	Parallel Flow Through a Straight Channel	24
2-4	Couette Flow	26
2-5	The Suddenly Accelerated Plane Wall	28
2-6	Similarity Analysis	31
2-7	Two-dimensional Stagnation Point Flow	35
2-8	Iteration Scheme	40
2-9	Numerical Solution	41
2-10	Axisymmetric Stagnation Point Flow	44
Chapter 3	Boundary Layer Theory	53
3-1	Concept of a Boundary Layer	53
3-2	Derivation of the Boundary Layer Equations	54
3-3	The Flow Past a Flat Plate — Blasius Solution	60

3–4	The General Two-dimensional Incompressible Boundary Layer — The Falkner-Skan Equations	65
3–5	Group Properties of the Blasius Equation	69
3–6	Numerical Integration of the Falkner-Skan Equation	70
3–7	Skin Friction, Displacement and Momentum Thickness	77
3–8	Physical Interpretation of the Falkner-Skan Equation	81
3–9	Axisymmetric Boundary Layers — The Mangler Transformation	84
3–10	Boundary Layers with Mass Transfer	90
3–11	Nonsimilar Boundary Layers	93
3–12	Locally Nonsimilar Boundary Layer Method	96

Chapter 4 Thermal Layers and Forced Convection Boundary Layers 105

4–1	Introduction	105
4–2	General Characteristics of Thermal Layers	105
4–3	Simple Couette Flow	110
4–4	Nonsimple Couette Flow	115
4–5	Flow in a Channel with Straight Parallel Walls — Poiseuille Flow	119
4–6	The Energy Equation for the Boundary Layer	122
4–7	Forced Convection Boundary Layer — Parallel Flow Past a Flat Plate	127
4–8	Forced Convection Boundary Layer Flows with Pressure Gradient and Nonisothermal Surface Conditions	136
4–9	Numerical Integration of the Energy Equation for a Forced Convection Boundary Layer	139
4–10	Forced Convection Boundary Layer Flows with Mass Transfer	147
4–11	Nonsimilar Forced Convection Boundary Layer	154

Chapter 5 Free Convection Boundary Layers 159

5–1	Introduction	159
5–2	Free Convection Boundary Layers	159
5–3	Numerical Integration of the Free Convection Boundary Layer Equations	164
5–4	Results for a Free Convection Boundary Layer on an Isothermal Vertical Flat Plate	164
5–5	Results for a Free Convection Boundary Layer on a Nonisothermal Vertical Flat Plate	170

5–6	Free Convection Boundary Layer on a Nonisothermal Vertical Flat Plate with Mass Transfer	173
5–7	Nonsimilar Free Convection Boundary Layer	178
5–8	Nonsimilar Free Convection Boundary Layer with Mass Transfer	182
5–9	Combined Forced and Free Convection Boundary Layer Flows on a Nonisothermal Surface	183
5–10	Numerical Integration of the Governing Equations for Combined Forced and Free Convection	186
5–11	Nonsimilar Combined Forced and Free Convection Boundary Layers	190

Chapter 6 The Compressible Boundary Layer 195

6–1	Introduction	195
6–2	Variation of Transport Properties	195
6–3	Analytical Solutions of the Compressible Navier-Stokes Equations — Couette Flow	199
6–4	Compressible Boundary Layer	208
6–5	Transformation of the Compressible Boundary Layer Equations	210
6–6	The Low Speed Compressible Boundary Layer	223
6–7	The Compressible Boundary Layer on a Flat Plate	225
6–8	The Similar Compressible Boundary Layer with Unit Prandtl Number	230
6–9	The Similar Hypersonic Compressible Boundary Layer with Nonunit Prandtl Number	232
6–10	Displacement and Momentum Thickness in a Compressible Boundary Layer	234
6–11	Numerical Integration of the Compressible Boundary Layer Equations	236
6–12	Results for the Low Speed Compressible Boundary Layer	237
6–13	Results for the Compressible Boundary Layer with Unit Prandtl Number	241
6–14	Results for the Similar Hypersonic Compressible Boundary Layer with Nonunit Prandtl Numbers	257
6–15	The Similar Compressible Boundary Layer with Mass Transfer	261
6–16	Results for the Similar Compressible Boundary Layer with Mass Transfer	262
6–17	Hypersonic Shock Wave-Boundary Layer Interaction	270
6–18	The Nonsimilar Compressible Boundary Layer	276

References		283
Appendix A	Runge-Kutta Integration Scheme	289
Appendix B	Satisfaction of Asymptotic Boundary Conditions	299
Appendix C	Numerical Results for the Boundary Layer Flows	321
Appendix D	General Boundary Layer Program	353
Appendix E	Problems	381
Index		411

PREFACE

For many years the author has been engaged in teaching fluid dynamics. At the undergraduate level, one of the most neglected subjects in fluid dynamics is viscous flow. At this level, viscous flow is frequently presented as a qualitative discussion of the characteristics of laminar and turbulent boundary layers and of the effects of separation and reverse flow. Analytical results are generally confined to the simpler exact solutions of the Navier-Stokes equations, e.g., flow in a channel. Blasius's solution of the flat plate laminar boundary layer and perhaps the von Karman momentum integral technique are briefly discussed and the results stated. Yet undergraduates are quite capable of understanding and appreciating both the underlying mathematics and the physics of classical viscous flows, *provided* they can actually *obtain* the solutions themselves.

Today, almost every undergraduate engineering student either owns or has ready access to a personal computer capable of numerically solving the governing equations for classical laminar boundary layers. This capability leads to both a new approach to teaching viscous flow and a great danger that the fundamental mathematics, modeling and analytical skills developed from a study of viscous flow will be neglected in favor of developing numerical computational skills. Just presenting a numerical technique or providing a 'canned' program to solve a problem is not enough. The student must understand the underlying physical, mathematical and modeling concepts inherent in any solution, be it analytic or numerical.

In this book the major emphasis is to present a technique of analysis which aids the formulation, the understanding and the solution of the defined class of problems. The student must first choose a mathematical model and derive the governing equations based on realistic assumptions, or he must become thoroughly aware of the limitations and assumptions associated with existing models. An appropriate solution technique is then selected. The solution technique can be either analytical or numerical. Analysis does not end once a solution is obtained; in a sense it only begins. The investigator must now interpret the results in light of the previous assumptions and the physical world. In light of these results, he may decide to change the analysis technique, remove some limiting assumptions, or even change the mathematical model or the specified boundary conditions.

Selecting the material coverage for the present volume proved difficult. Since the study of turbulent flows continues to undergo intensive investigation, the material is restricted to the analysis of laminar flows. The study of even laminar viscous flow encompasses such a wide area that it is impossible to cover it in any depth in a single volume. The laminar flows which illustrate the fundamental mathematics and physics with the least complexity are the similar flows. In discussing nonsimilar flows it was decided to use the locally nonsimilar method, since it represents a logical extension of the similar analyses, rather than introduce the approximate momentum integral techniques, singular perturbation methods, direct finite difference integration or other computational fluid dynamics techniques. These techniques are adequately covered in other texts. Both the breadth and depth of the subject matter is increased by supplementing the classical analyses with computer-aided analysis algorithms. These algorithms contribute to the understanding of many of the problems of laminar flow that are not amenable to analytical solutions. In this way problems which previously have only been formulated in the classroom can now be completely solved and analyzed.

The computer algorithms allow the reader to follow each step in the numerical solution. This is an important part of the analysis. A thorough understanding of a computer algorithm leads to a better understanding of the mathematical model and its limitations, and hence to a better physical understanding. The student can make improvements and modifications to the algorithms. Each algorithm is fully documented. Each algorithm was implemented in a higher level language and the pseudocode algorithm derived from that implementation.[†] The fully annotated computer runs increase student understanding. They can also be used by students to check their own implementations.

The book begins by deriving the Navier-Stokes equation for a viscous compressible variable property fluid. The second chapter considers exact solutions of the incompressible Navier-Stokes equations. In Chapter 3, the boundary layer equations are derived and the incompressible hydrodynamic boundary layer equations solved with and without mass transfer at the wall. Next, in Chapter 4 forced convection is considered. In Chapter 5 free convection is discussed, and in Chapter 6 the compressible laminar boundary layer is analyzed. Appendix A develops the fourth-order Runge-Kutta numerical integration scheme. Appendix B discusses the Nachtsheim-Swigert iteration scheme for asymptotic boundary conditions. Appendix C is an attempt to present in a unified manner the important numerical results for similar laminar boundary layers. Appendix D presents, in pseudocode suitable for implementation in any programming language, a program for solving the general similar laminar boundary layer equations. Finally, Appendix E presents a number of problems.

[†]The author recommends True Basic, an implementation of ANSI Basic, as an implementation tool. Experience has shown that ANSI Basic provides an excellent working environment for students with limited computer experience. True Basic and the True Basic implementations of the pseudocode are available from True Basic Inc., 1-800-TR-BASIC (see also Appendix D.)

A unifying mechanism among the various topics is the logical progression from simple to complex governing differential equations and boundary conditions. The similar hydrodynamic boundary layer is governed by a single nonlinear ordinary differential equation and a single asymptotic boundary condition. Forced convection is governed by a pair of uncoupled ordinary differential equations which can be solved successively. Each equation is subject to a single asymptotic boundary condition. Free convection and the similar compressible boundary layer are both governed by a pair of coupled ordinary differential equations which must be solved simultaneously. They are subject to a pair of asymptotic boundary conditions (see Section 1–10). Analogous situations occur for the locally non-similar method.

The problems in Appendix E are generally of three types: numerical, parametric and directed analysis. The numerical problems are generally of the form — 'Given the following specific information, calculate a required result'. This is the type of problem that is generally found in textbooks. Examples are Problems 1–4 and 4–6. The parametric problems are of the form — 'investigate the effect of the variation of a parameter on the numerical solution'. These problems are generally amenable to solutions using modifications of programs derived from the pseudocode algorithms discussed in the text and given in Appendix D. They allow the student to ask the question 'What if ——— ?' and lead to a better physical understanding of the mathematical model. The author has found this type of problem very useful in teaching the concept of parametric analysis. Examples are Problems 3–4 and 4–8. The directed analysis problems seek to guide the student through an original analysis of a given problem. They are frequently skeletons of the analyses in classical papers existing in the literature. They serve to extend and diversify the material covered in any specific course. For example, there are several directed analysis problems on rotational flows which serve as an introduction to that topic. Frequently the result of a directed analysis problem is a differential equation or system of differential equations that require numerical solution. In this case, the student is shown how to put these equations in a form similar to that of the equations solved by one of the analysis algorithms. In this way, the directed analysis problems serve as a unifying mechanism for the various topics in fluid dynamics. These problems generally require considerable effort on the part of the student. It has therefore been the policy to assign various directed analysis problems to individual students at the rate of one per week. In this way it is possible, while still covering the basic material, to tailor the course to individual interests and needs, e.g., rotational flow. Examples of directed analysis problems are Problems 2–5, 3–10 and 4–11.

The prerequisites for understanding the material are a course in classical potential flow theory, gasdynamics or compressible, nonviscous, heat conducting flow, college level mathematics through classical differential equations, skill in a higher level computer language, e.g. ANSI Basic, Fortran, C or Pascal and the availability of a computer. The book is suitable for a senior-level elective course or a first year graduate course on viscous flows.

As with any book, a number of people have contributed to its existence, form and substance. First let me thank my partner and wife Nancy, who typeset the book using TeX from rough handwritten notes, roughly marked up typewritten copy and incomplete computer files. She made multiple changes first one way and then back to the original way, and frequently to still a third way, while I made up my mind about what I wanted. More than that no author nor husband can expect. Professor John Anderson deserves a note of thanks for encouraging me to finally put the book into publishable form. Special thanks are due my colleague, office mate and co-author, Professor J. Alan Adams, for first introducing me to forced and free convection flows. Finally, my very special thanks to Professor Gabriel (Gabby) Karpouzian, who read the entire manuscript, checked the equations and the cross-references and made many valuable contributions. Still, as in any book, I am sure that typos, incorrect cross-references, numerical errors, etc. remain. Those are my responsibility. When you find them please bring them to my attention.

The pedagogical approach taken in this text is such that it requires innovation and involvement on the part of both the student and professor. I hope that it is as much 'fun' for you as it is and was for me.

David F. Rogers
Annapolis, Maryland
USA

ONE

DERIVATION OF THE COMPRESSIBLE NAVIER-STOKES EQUATIONS

1-1 Introduction

The Navier-Stokes equations, together with the energy and continuity equations, are the cornerstones of continuum fluid dynamics. The equations were first derived for incompressible flow by Navier in 1822 and subsequently placed on a more rigorous basis by St. Venant, Poisson and Stokes.

Although the final form of the equations does not seem to be in dispute, the route to their formulation is a matter of some difference in opinion. It is not evident that all procedures are mutually consistent.

The stimulus for formulation of the laws of friction in a fluid was the success of the equations of elasticity, i.e., the generalized Hooke's law, which postulates that the components of the stress and strain tensors are linearly related by a number of constants whose values depend on the material under consideration. If, however, the material is *homogeneous* and *isotropic*, and if it is also stipulated that the equations be *invariant* under a rotation of coordinates, i.e., their form be the same for an arbitrarily-oriented set of orthogonal coordinates, the number of independent material constants reduces to two. In the analogous fluid dynamic situation, it is assumed that the component stresses are linearly dependent on the *rate-of-strain* rather than the strain. This assumption is fundamental to the derivation of the Navier-Stokes equations.

Three independent postulates furnish the fundamental relations of fluid dynamics. They are conservation of mass, Newton's second law of motion and conservation of energy. The fundamental relations of fluid dynamics are stated in terms of particle properties, e.g., the principle of conservation of mass states that *particles* of matter can neither be created nor destroyed. In order to apply these relations to a fluid we introduce the concept of a field. A field is a region whose properties are defined as functions of space and time. A one-to-one correspondence between particle properties and field properties in a fluid is achieved by assuming that as particles move through the field they take on the properties of

2 Laminar Flow Analysis

the field point that they occupy. Using this concept we can apply the fundamental relations to a continuous fluid, i.e., a fluid not composed of discrete particles.

We first consider the principle of conservation of mass. Without loss of generality we consider a cubical control volume as shown in Figure 1–1. We assume that the properties of the flow are known at the center of the element. Some of these properties are the density ρ, the pressure P, and the velocity components u, v, w in the x, y, z directions, respectively. Further, we assume that the values of these quantities on the faces of the element are adequately represented by a first-order Taylor series expansion about the center of the element. This assumption fundamentally depends on the fluid being continuous and homogeneous, i.e., the composition is the same throughout.

Considering the mass flow into and out of the control volume, and neglecting the possibility of the production of matter inside the control volume, i.e., sources or sinks of matter, the principle of conservation of mass stated as a rate equation is

$$\begin{pmatrix} \text{The rate at which} \\ \text{mass leaves the} \\ \text{control volume} \\ \text{through the faces} \end{pmatrix} - \begin{pmatrix} \text{The rate at which} \\ \text{mass enters the} \\ \text{control volume} \\ \text{through the faces} \end{pmatrix} = \begin{pmatrix} \text{The rate at which} \\ \text{mass decreases} \\ \text{within the} \\ \text{control volume} \end{pmatrix}$$

If, for convenience, we consider that a fluid flows diagonally through the field represented by the elemental volume from the lower rear left corner to the upper front right corner, we see that mass leaves the volume through the right hand,

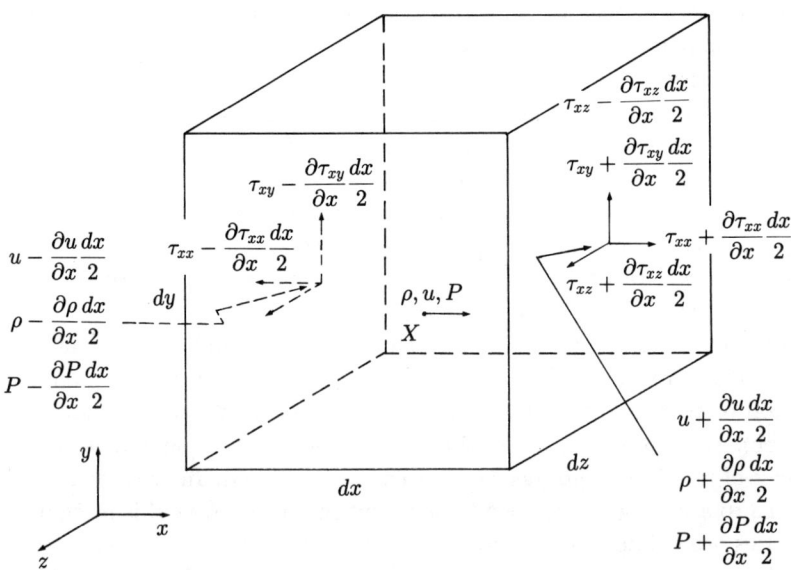

Figure 1–1. Elemental control volume.

front and upper faces and enters the volume through the left hand, rear and lower faces. The rate at which mass leaves the control volume is then

$$\begin{pmatrix} \text{The rate at which} \\ \text{mass leaves the} \\ \text{control volume} \\ \text{through the faces} \end{pmatrix} = \left(\rho + \frac{\partial \rho}{\partial x}\frac{dx}{2}\right)\left(u + \frac{\partial u}{\partial x}\frac{dx}{2}\right)dydz$$
$$+ \left(\rho + \frac{\partial \rho}{\partial y}\frac{dy}{2}\right)\left(v + \frac{\partial v}{\partial y}\frac{dy}{2}\right)dxdz$$
$$+ \left(\rho + \frac{\partial \rho}{\partial z}\frac{dz}{2}\right)\left(w + \frac{\partial w}{\partial z}\frac{dz}{2}\right)dxdy$$

Similarly, the rate at which mass enters the control volume is

$$\begin{pmatrix} \text{The rate at which} \\ \text{mass enters the} \\ \text{control volume} \\ \text{through the faces} \end{pmatrix} = \left(\rho - \frac{\partial \rho}{\partial x}\frac{dx}{2}\right)\left(u - \frac{\partial u}{\partial x}\frac{dx}{2}\right)dydz$$
$$+ \left(\rho - \frac{\partial \rho}{\partial y}\frac{dy}{2}\right)\left(v - \frac{\partial v}{\partial y}\frac{dy}{2}\right)dxdz$$
$$+ \left(\rho - \frac{\partial \rho}{\partial z}\frac{dz}{2}\right)\left(w - \frac{\partial w}{\partial z}\frac{dz}{2}\right)dxdy$$

Without loss of generality, we assume that the rate at which mass leaves the control volume is larger than the rate at which mass enters the control volume. The rate at which mass decreases within the control volume is then

$$\begin{pmatrix} \text{The rate at which} \\ \text{mass decreases} \\ \text{within the} \\ \text{control volume} \end{pmatrix} = -\frac{\partial \rho}{\partial t}dxdydz$$

Combining these results and neglecting the higher-order terms, i.e., terms of order $(dx)^2$, $(dy)^2$, $(dz)^2$, yields the continuity equation

$$\frac{\partial \rho}{\partial t} + \frac{\partial}{\partial x}(\rho u) + \frac{\partial}{\partial y}(\rho v) + \frac{\partial}{\partial z}(\rho w) = 0 \qquad (1-1)$$

Here note that no restrictions are placed on the observer's frame of reference and that the only assumptions with respect to the fluid are that it be continuous and homogeneous.

From Newton's second law, which is restricted to an inertial frame of reference, we have

$$m\frac{D\vec{V}}{Dt} = \sum \vec{F}_{\text{ext}} \qquad (1-2)$$

Here, m is the mass of a particle and $D\vec{V}/Dt$ is the total acceleration of the particle. Again we introduce the field concept in order to apply Newton's second law to a continuous fluid. The total acceleration of a fluid particle moving through the velocity field is composed of a convective contribution due to movement of the particle through the velocity field, and a local contribution due to the change in the velocity field with time at any position within the field, i.e., unsteady field effects.

Let us now determine the x component of the convective acceleration of a particle moving through the velocity field. If the x component of the velocity of the particle at some point in the field is u_1, then the x component of its velocity at some other neighboring point is u_2 and can be determined using a first-order Taylor series expansion. Thus

$$u_2 = u_1 + \frac{\partial u}{\partial x}\bigg|_1 dx + \frac{\partial u}{\partial y}\bigg|_1 dy + \frac{\partial u}{\partial z}\bigg|_1 dz + \text{h.o.t.}^{\dagger} \qquad (1-3)$$

Due to the particle's movement through the field its velocity is changing. Thus, the particle is accelerating. Hence we have

$$\begin{pmatrix} \text{The } x \text{ component} \\ \text{of the convective} \\ \text{acceleration} \end{pmatrix} = \lim_{(t_2-t_1) \to 0} \frac{u_2 - u_1}{t_2 - t_1} = u\frac{\partial u}{\partial x} + v\frac{\partial u}{\partial y} + w\frac{\partial u}{\partial z}$$

since $u = dx/dt$, $v = dy/dt$ and $w = dz/dt$.

The x component of the local acceleration is due to unsteady field effects, i.e., the field velocity at a fixed point in the field is a function of time. Hence, the x component of the local acceleration is $\partial u/\partial t$. The total or substantial acceleration is the sum of the local and convective accelerations. Thus, the x component of the total acceleration is

$$\frac{Du}{Dt} = \frac{\partial u}{\partial t} + u\frac{\partial u}{\partial x} + v\frac{\partial u}{\partial y} + w\frac{\partial u}{\partial z} \qquad (1-4a)$$

Similarly, the y and z components of the total acceleration are, respectively

$$\frac{Dv}{Dt} = \frac{\partial v}{\partial t} + u\frac{\partial v}{\partial x} + v\frac{\partial v}{\partial y} + w\frac{\partial v}{\partial z} \qquad (1-4b)$$

and

$$\frac{Dw}{Dt} = \frac{\partial w}{\partial t} + u\frac{\partial w}{\partial x} + v\frac{\partial w}{\partial y} + w\frac{\partial w}{\partial z} \qquad (1-4c)$$

Here D/Dt is the substantial or Eulerian derivative. In Cartesian coordinates

$$\frac{D}{Dt} = \frac{\partial}{\partial t} + u\left(\frac{\partial}{\partial x}\right) + v\left(\frac{\partial}{\partial y}\right) + w\left(\frac{\partial}{\partial z}\right)$$

---

†h.o.t. means higher-order terms.

The external forces acting on the control volume $\vec{F}_{\text{ext}}$ are divided into body or volume forces and into surface forces. A body force is one which acts on the entire mass of the element. Examples are gravity and electromagnetic forces. We use X, Y, Z to represent the x, y and z components of the body force per unit mass (see Figure 1–1). Surface forces act only at the surface of the control volume. Examples are viscous shearing forces and normal pressure forces. Figure 1–1 shows the stresses (force/unit area) acting on the elemental volume. Here, the notation is such that the first subscript indicates the face on which the stress is acting and the second the direction in which it is acting. Thus, for example, τ_{xy} is the stress acting on the face perpendicular to the x-axis in the y direction. The components of the stress on the faces of the elemental control volume form a shearing stress tensor, which describes the state of stress in the fluid. The matrix of the shearing stress tensor is

$$[\tau] = \begin{bmatrix} \tau_{xx} & \tau_{yx} & \tau_{zx} \\ \tau_{xy} & \tau_{yy} & \tau_{zy} \\ \tau_{xz} & \tau_{yz} & \tau_{zz} \end{bmatrix} \tag{1-5}$$

From Figure 1–1 we see that the acceleration in the x direction is equal to the sum of the body force term, a normal force term involving τ_{xx} (which includes the pressure), and two shearing force terms involving τ_{yx} and τ_{zx}. Thus, the x component of Newton's second law is

$$\frac{Du}{Dt} = X + \frac{1}{\rho}\left(\frac{\partial \tau_{xx}}{\partial x} + \frac{\partial \tau_{yx}}{\partial y} + \frac{\partial \tau_{zx}}{\partial z}\right) \tag{1-6a}$$

Similarly, the equations in the y and z directions are

$$\frac{Dv}{Dt} = Y + \frac{1}{\rho}\left(\frac{\partial \tau_{xy}}{\partial x} + \frac{\partial \tau_{yy}}{\partial y} + \frac{\partial \tau_{zy}}{\partial z}\right) \tag{1-6b}$$

$$\frac{Dw}{Dt} = Z + \frac{1}{\rho}\left(\frac{\partial \tau_{xz}}{\partial x} + \frac{\partial \tau_{yz}}{\partial y} + \frac{\partial \tau_{zz}}{\partial z}\right) \tag{1-6c}$$

Equations (1–6) are known as Cauchy's equations of motion.

The shearing stress tensor is symmetrical, i.e., $\tau_{yx} = \tau_{xy}$, $\tau_{xz} = \tau_{zx}$ and $\tau_{yz} = \tau_{zy}$. In order to show this, consider one of the faces of the elemental volume as illustrated in Figure 1–2. Conservation of angular momentum requires that the torque acting on the element be equal to the time rate-of-change of the angular momentum on the element. Thus

$$\tau_{yx}\frac{h}{2}h - \tau_{xy}\frac{h}{2}h = \dot{\omega}I_{zz} \tag{1-7}$$

6   Laminar Flow Analysis

where I_{zz} is the polar moment of inertia and $\dot{\omega}$ is the angular acceleration. For a square of unit depth we have

$$\tau_{yx}\frac{h^2}{2} - \tau_{xy}\frac{h^2}{2} = \dot{\omega}\frac{h^4}{6}$$

or

$$\tau_{yx} - \tau_{xy} = \dot{\omega}\frac{h^2}{3}$$

The angular velocity due to the shearing stresses is

$$\omega = 3\int \frac{(\tau_{yx} - \tau_{xy})}{h^2} dt \qquad (1-8)$$

As the volume (in this two-dimensional case the area) of the element approaches zero, i.e., $h \to 0$, the angular velocity must approach zero. This is only possible if the integrand in Eq. (1–8) is zero. Hence,

$$\tau_{yx} = \tau_{xy} \qquad (1-9)$$

The normal stress τ_{xx} includes the pressure which is assumed to be a scalar and hence independent of direction. Therefore, for convenience we separate out the pressure by writing

$$\tau_{xx} = \tau'_{xx} - P \qquad (1-10)$$

where the minus sign indicates that the pressure exerts a force acting on the elemental area in a direction opposite to the positive outer normal. Equation (1–6a) thus becomes

$$\frac{Du}{Dt} = X - \frac{1}{\rho}\frac{\partial P}{\partial x} + \frac{1}{\rho}\left(\frac{\partial \tau'_{xx}}{\partial x} + \frac{\partial \tau_{yx}}{\partial y} + \frac{\partial \tau_{zx}}{\partial z}\right) \qquad (1-11a)$$

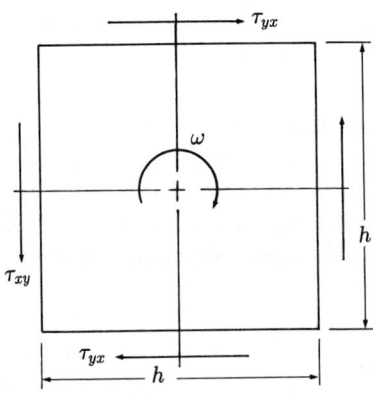

Figure 1–2. Two-dimensional angular momentum.

and similarly Eqs. (1–6b) and (1–6c) become

$$\frac{Dv}{Dt} = Y - \frac{1}{\rho}\frac{\partial P}{\partial y} + \frac{1}{\rho}\left(\frac{\partial \tau_{xy}}{\partial x} + \frac{\partial \tau'_{yy}}{\partial y} + \frac{\partial \tau_{zy}}{\partial z}\right) \qquad (1-11b)$$

$$\frac{Dw}{Dt} = Z - \frac{1}{\rho}\frac{\partial P}{\partial z} + \frac{1}{\rho}\left(\frac{\partial \tau_{xz}}{\partial x} + \frac{\partial \tau_{yz}}{\partial y} + \frac{\partial \tau'_{zz}}{\partial z}\right) \qquad (1-11c)$$

The matrix of the stress tensor becomes

$$[\tau'] = \begin{bmatrix} \tau'_{xx} & \tau_{yx} & \tau_{zx} \\ \tau_{xy} & \tau'_{yy} & \tau_{zy} \\ \tau_{xz} & \tau_{yz} & \tau'_{zz} \end{bmatrix} \qquad (1-12)$$

1–2 Stress–Rate-of-Strain Relation

We are now at a critical point in the discussion: the form of the stress–rate-of-strain relation for a viscous flow. The most general assumption for the stress–rate-of-strain relationship, which is still linear, is that an arbitrary stress component is a linear function of all the rates-of-strain; e.g.

$$\tau'_{xx} = \alpha_1 \dot{\epsilon}_{xx} + \alpha_2 \dot{\epsilon}_{xy} + \alpha_3 \dot{\epsilon}_{xz} + \cdots \alpha_9 \dot{\epsilon}_{zz} \qquad (1-13)$$

where the α_n's are independent material properties, at most functions of the thermodynamic state variables, and independent of the rates-of-strain, i.e., the $\dot{\epsilon}$'s. Here, the same subscripting convention used for the stresses is used for the rates-of-strain. As there are nine stresses and nine rates-of-strain, a maximum of 81 independent material properties is possible.

It has already been established that the stress tensor is symmetric. In addition, it is subsequently shown that the rate-of-strain tensor is also symmetric. Further, if it is assumed that the fluid is isotropic and homogeneous, together with the stipulation that the stress–rate-of-strain relation be invariant under a rotation of orthogonal coordinates, the number of independent properties reduces to two. The resulting equations then take the form (see Liepmann and Roshko [Liep57])

$$\tau'_{xx} = 2\mu\dot{\epsilon}_{xx} + \lambda(\dot{\epsilon}_{xx} + \dot{\epsilon}_{yy} + \dot{\epsilon}_{zz}) \qquad (1-14a)$$

$$\tau'_{yy} = 2\mu\dot{\epsilon}_{yy} + \lambda(\dot{\epsilon}_{xx} + \dot{\epsilon}_{yy} + \dot{\epsilon}_{zz}) \qquad (1-14b)$$

$$\tau'_{zz} = 2\mu\dot{\epsilon}_{zz} + \lambda(\dot{\epsilon}_{xx} + \dot{\epsilon}_{yy} + \dot{\epsilon}_{zz}) \qquad (1-14c)$$

and

$$\tau_{xy} = \tau_{yx} = 2\mu\dot{\epsilon}_{yx} \qquad (1-15a)$$

$$\tau_{yz} = \tau_{zy} = 2\mu\dot{\epsilon}_{zy} \qquad (1-15b)$$

$$\tau_{xz} = \tau_{zx} = 2\mu\dot{\epsilon}_{xz} \qquad (1-15c)$$

8   Laminar Flow Analysis

Here μ is the well known dynamic viscosity, and λ is the second coefficient of viscosity, or the dilatational viscosity.

To obtain the components of the rate-of-strain tensor and hence the components of the stress tensor in terms of more convenient and physically meaningful quantities, we consider the angular velocity and deformation of a fluid element.

1-3 Angular Velocity and Fluid Deformation

A continuous fluid is assumed to be composed of arbitrarily small particles (mathematical points). Since these points are infinitely small it is not reasonable to consider the rotation of a single particle. However, by considering the behavior of the intersection of two originally orthogonal fluid lines, i.e., two mathematical lines always connecting the same particles—two fluid curves, during a short interval of time the local rotation and distortion of the fluid is determined.

Consider two lines lying in the xy plane (see Figure 1-3). The z component of angular velocity is defined as the average rotation of the tangents to these fluid lines at their intersection divided by the time interval. Taking the limit as $\Delta t \to 0$

$$\omega_z = \lim_{\Delta t \to 0} \frac{1}{2}\left(\frac{\Delta\alpha + \Delta\beta}{\Delta t}\right) = \frac{1}{2}\left(\frac{d\alpha}{dt} + \frac{d\beta}{dt}\right) \qquad (1-16)$$

We now wish to determine the angular velocity in terms of the translational

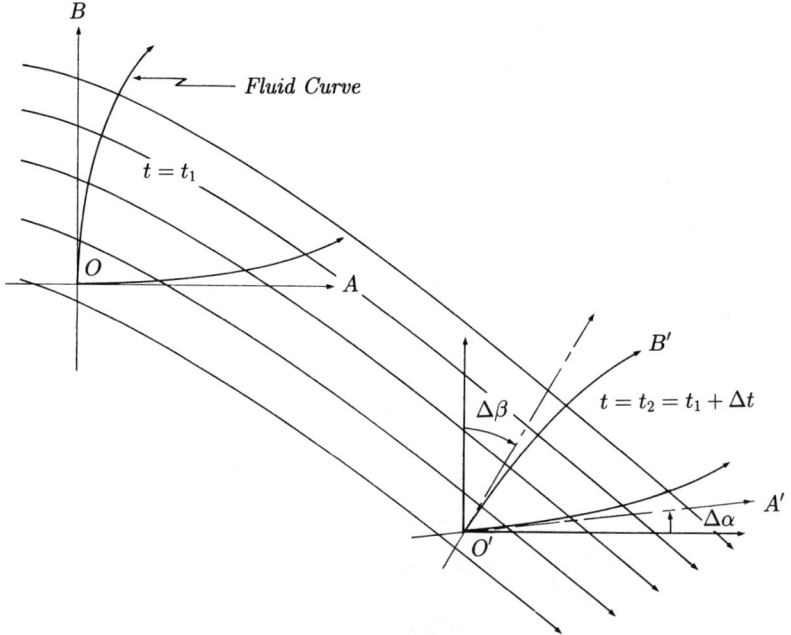

Figure 1-3. Angular rotation and distortion in a fluid.

velocity gradients in the flow. For simplicity consider a two-dimensional fluid element at some initial time t_0 and subsequently at some time $t_0 + \Delta t$. At this subsequent time the fluid element has *in general* undergone translation, rotation and deformation. Figure 1–4 shows the shape of the element at the initial time t_0 and at the subsequent time $t_0 + \Delta t$. From Figure 1–4 the angular displacement of one side of the element is

$$\tan(d\alpha) = \frac{\left(v_0 + \frac{\partial v}{\partial x}\Big|_0 dx\right)dt - v_0 dt}{\left(u_0 + \frac{\partial u}{\partial x}\Big|_0 dx\right)dt + dx - u_0 dt}$$

which can be written as

$$\tan(d\alpha) = \frac{\frac{\partial v}{\partial x}\Big|_0 dt}{\left(1 + \frac{\partial u}{\partial x}\Big|_0 dt\right)} \tag{1-17}$$

Assuming the gradient $\partial u/\partial x|_0$ is at most of order one, then $\partial u/\partial x|_0\, dt \ll 1$ and is negligible with respect to one. Hence, Eq. (1–17) becomes

$$\tan(d\alpha) = \frac{\partial v}{\partial x}\Big|_0 dt$$

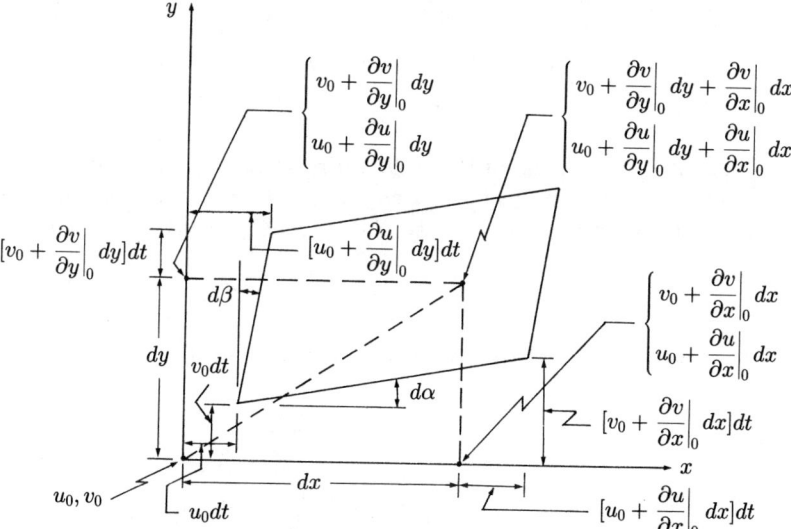

Figure 1–4. Translation, rotation, and deformation of a fluid element.

10   Laminar Flow Analysis

Further, for small angles $\tan(d\alpha)$ is approximately equal to $d\alpha$. Thus

$$d\alpha = \frac{\partial v}{\partial x}\bigg|_0 dt \qquad (1-18)$$

Similarly, from Figure 1–4 the angular displacement of the other side of the element is

$$\tan(d\beta) \approx d\beta = -\frac{\partial u}{\partial y}\bigg|_0 dt \qquad (1-19)$$

where the negative sign results if a counterclockwise angular displacement is considered positive.

Using these results and Eq. (1–16), we see that the angular velocity depends on the two cross-gradients of the x and y components of velocity, i.e.

$$\omega_z = \frac{1}{2}\left(\frac{\partial v}{\partial x} - \frac{\partial u}{\partial y}\right) \qquad (1-20a)$$

The angular velocity is a vector quantity in the same sense as the translational velocity. Thus, Eq. (1–20a) is taken as the z component of the angular velocity vector. The x and y components are obtained in a similar manner. They are

$$\omega_x = \frac{1}{2}\left(\frac{\partial w}{\partial y} - \frac{\partial v}{\partial z}\right) \qquad (1-20b)$$

and

$$\omega_y = \frac{1}{2}\left(\frac{\partial u}{\partial z} - \frac{\partial w}{\partial x}\right) \qquad (1-20c)$$

The angular velocity vector is written as

$$\vec{\omega} = \frac{1}{2}\left[\left(\frac{\partial w}{\partial y} - \frac{\partial v}{\partial z}\right)\vec{i} + \left(\frac{\partial u}{\partial z} - \frac{\partial w}{\partial x}\right)\vec{j} + \left(\frac{\partial v}{\partial x} - \frac{\partial u}{\partial y}\right)\vec{k}\right] \qquad (1-21)$$

where $\vec{i}, \vec{j}, \vec{k}$ are unit vectors in the x, y, z directions, respectively.

For reasons of mathematical simplicity it is preferable to use the vorticity function, ζ, instead of angular velocity. The vorticity vector, $\vec{\zeta}$, is defined as

$$\vec{\zeta} = 2\vec{\omega} \qquad (1-22)$$

The other components of the vorticity vector are ξ and η

$$\xi = 2\omega_x = \frac{\partial w}{\partial y} - \frac{\partial v}{\partial z} \qquad (1-23a)$$

$$\eta = 2\omega_y = \frac{\partial u}{\partial z} - \frac{\partial w}{\partial x} \qquad (1-23b)$$

$$\zeta = 2\omega_z = \frac{\partial v}{\partial x} - \frac{\partial u}{\partial y} \qquad (1-23c)$$

Angular Velocity and Fluid Deformation 11

We now consider the deformation of the fluid element. In particular we consider the rate-of-shear deformation, i.e., the rate at which the shape of the element is distorted by the shearing stresses acting on the surfaces of the element. In a fluid the *rate-of-shearing-deformation* (strain) is defined as the rate of change of the angle between two orthogonal fluid curves. For the two-dimensional element shown in Figure 1–3, the rate-of-shear deformation is given by the decrement of the angle AOB at the intersection of the two fluid lines. Thus

$$\dot{\gamma}_{xy} = \lim_{\Delta t \to 0} \frac{\pi/2 - \angle A'O'B'}{\Delta t} = \lim_{\Delta t \to 0} \frac{\Delta\alpha - \Delta\beta}{\Delta t} = \frac{d\alpha}{dt} - \frac{d\beta}{dt}$$

Using Eqs. (1–18) and (1–19) we have

$$\dot{\gamma}_{xy} = \frac{\partial v}{\partial x} + \frac{\partial u}{\partial y} \tag{1-24}$$

where the quantity $\dot{\gamma}_{xy}$ is the rate-of-shear deformation. However, for mathematical reasons it is again preferable to work with the shear rate-of-strain defined as $\dot{\epsilon}_{xy} = (\dot{\gamma}_{xy})/2$. There are six shearing rates-of-strain occurring in pairs; e.g.

$$\dot{\epsilon}_{xy} = \frac{1}{2}\left(\frac{\partial v}{\partial x} + \frac{\partial u}{\partial y}\right) = \dot{\epsilon}_{yx} \tag{1-25}$$

There are three elongation terms, of which the x component is

$$\dot{\epsilon}_{xx} = \frac{\partial u}{\partial x} \tag{1-26}$$

These nine quantities form the rate-of-strain tensor. The matrix of the rate-of-strain tensor is

$$[\dot{\epsilon}] = \begin{bmatrix} \dot{\epsilon}_{xx} & \dot{\epsilon}_{yx} & \dot{\epsilon}_{zx} \\ \dot{\epsilon}_{xy} & \dot{\epsilon}_{yy} & \dot{\epsilon}_{zy} \\ \dot{\epsilon}_{xz} & \dot{\epsilon}_{yz} & \dot{\epsilon}_{zz} \end{bmatrix} \tag{1-27}$$

In terms of the velocity gradients we have

$$[\dot{\epsilon}] = \begin{bmatrix} \dfrac{\partial u}{\partial x} & \dfrac{1}{2}\left(\dfrac{\partial v}{\partial x} + \dfrac{\partial u}{\partial y}\right) & \dfrac{1}{2}\left(\dfrac{\partial w}{\partial x} + \dfrac{\partial u}{\partial z}\right) \\ \dfrac{1}{2}\left(\dfrac{\partial v}{\partial x} + \dfrac{\partial u}{\partial y}\right) & \dfrac{\partial v}{\partial y} & \dfrac{1}{2}\left(\dfrac{\partial w}{\partial y} + \dfrac{\partial v}{\partial z}\right) \\ \dfrac{1}{2}\left(\dfrac{\partial w}{\partial x} + \dfrac{\partial u}{\partial z}\right) & \dfrac{1}{2}\left(\dfrac{\partial w}{\partial y} + \dfrac{\partial v}{\partial z}\right) & \dfrac{\partial w}{\partial z} \end{bmatrix} \tag{1-28}$$

which we note is symmetric about the diagonal. The sum of the diagonal terms in the matrix of the rate-of-strain tensor is called the trace, $\dot{\theta}$

$$\dot{\theta} = \dot{\epsilon}_{xx} + \dot{\epsilon}_{yy} + \dot{\epsilon}_{zz} = \text{div}\vec{V} \tag{1-29a}$$

12   Laminar Flow Analysis

The trace is invariant under a transformation of coordinates and is called the *volume rate-of-dilatation* of the fluid. In terms of the velocity components we have

$$\dot{\theta} = \frac{\partial u}{\partial x} + \frac{\partial v}{\partial y} + \frac{\partial w}{\partial z} \tag{1-29b}$$

Thus, we have been able to write the rate-of-strain tensor and ultimately the dynamic equation of motion in terms of the velocity gradients in the fluid. Returning to the stress tensor, we note that the trace of the matrix of the stress tensor is

$$\tau'_{xx} + \tau'_{yy} + \tau'_{zz} = (3\lambda + 2\mu)\dot{\theta}$$

and
$$\tau_{xx} + \tau_{yy} + \tau_{zz} = -3P + (3\lambda + 2\mu)\dot{\theta} \tag{1-30}$$

1-4 The Viscosity Coefficients of Fluids

Since the average of the three normal stresses in a fluid is generally taken to be the thermodynamic pressure, Eq. (1–30) implies that, unless the second term on the right vanishes, the pressure depends not only on the density and the temperature but also on the dilatation. In an incompressible fluid, the dilatation, as is subsequently shown, is zero; and the pressure depends only on the temperature and density as in a gas at rest. However, in a compressible fluid the dilatation is not in general zero. Hence $(3\lambda + 2\mu)$ must be zero in order for the pressure to be independent of the dilatation. This is Stokes' approximation.

The best theoretical justification of Stokes' approximation is derived from the kinetic theory of gases. For a monatomic gas, kinetic theory shows that the ratio $(3\lambda + 2\mu)/3\mu$ must be of the order of the square of the ratio of the volume occupied by the molecules to the volume of the fluid. Except for very dense fluids or fluids under high pressure this ratio is quite small.

Although complete justification does not exist, we assume that Stokes' approximation is valid for all liquids and gases with ultimate justification coming from comparison with experimental results. Thus, we assume

$$(3\lambda + 2\mu) = 0 \tag{1-31}$$

As found in the simplest frictional law of Newton, the material property μ is the familiar *dynamic coefficient of viscosity*. λ is called the *dilatational viscosity*, or the second coefficient of viscosity. The combination $(3\lambda + 2\mu)/3 = \kappa$ is called the *bulk viscosity*.

1-5 The Momentum Equation

Adopting Eq. (1–31), a typical stress component becomes

$$\tau'_{xx} = 2\mu \frac{\partial u}{\partial x} - \frac{2}{3}\mu \operatorname{div}\vec{V} \tag{1-32}$$

Substitution of Eqs. (1–15) and (1–30) and the corresponding relations for the other stresses into Eq. (1–6) yields the Navier-Stokes equations for compressible flow. Since the dynamic viscosity depends on temperature (for gases μ increases with temperature but is essentially independent of pressure), it is necessary to allow for its variation in the equations of motion. Thus the x component becomes

$$\frac{Du}{Dt} = X - \frac{1}{\rho}\frac{\partial P}{\partial x} + \frac{1}{\rho}\frac{\partial}{\partial x}\left[\mu\left(2\frac{\partial u}{\partial x} - \frac{2}{3}\operatorname{div}\vec{V}\right)\right]$$

$$+ \frac{1}{\rho}\frac{\partial}{\partial y}\left[\mu\left(\frac{\partial u}{\partial y} + \frac{\partial v}{\partial x}\right)\right] + \frac{1}{\rho}\frac{\partial}{\partial z}\left[\mu\left(\frac{\partial w}{\partial x} + \frac{\partial u}{\partial z}\right)\right] \quad (1-33a)$$

Similarly, the equations for the y and z components are

$$\frac{Dv}{Dt} = Y - \frac{1}{\rho}\frac{\partial P}{\partial y} + \frac{1}{\rho}\frac{\partial}{\partial y}\left[\mu\left(2\frac{\partial v}{\partial y} - \frac{2}{3}\operatorname{div}\vec{V}\right)\right]$$

$$+ \frac{1}{\rho}\frac{\partial}{\partial z}\left[\mu\left(\frac{\partial v}{\partial z} + \frac{\partial w}{\partial y}\right)\right] + \frac{1}{\rho}\frac{\partial}{\partial x}\left[\mu\left(\frac{\partial u}{\partial y} + \frac{\partial v}{\partial x}\right)\right] \quad (1-33b)$$

$$\frac{Dw}{Dt} = Z - \frac{1}{\rho}\frac{\partial P}{\partial z} + \frac{1}{\rho}\frac{\partial}{\partial z}\left[\mu\left(2\frac{\partial w}{\partial z} - \frac{2}{3}\operatorname{div}\vec{V}\right)\right]$$

$$+ \frac{1}{\rho}\frac{\partial}{\partial x}\left[\mu\left(\frac{\partial w}{\partial x} + \frac{\partial u}{\partial z}\right)\right] + \frac{1}{\rho}\frac{\partial}{\partial y}\left[\mu\left(\frac{\partial v}{\partial z} + \frac{\partial w}{\partial y}\right)\right] \quad (1-33c)$$

1–6 Derivation of the Energy Equation

Before deriving the energy equation, a few remarks on the relationship between classical thermodynamics and fluid dynamics are in order. Classical thermodynamics considers very general systems in equilibrium. The initial and final equilibrium states of a thermodynamic system are governed by the laws of thermodynamics even if the intermediate states are not in equilibrium. Continuum fluid dynamics (and the theory of elasticity) attempts to determine the details of the intermediate nonequilibrium states. In so doing, it is necessary to assume that the equilibrium thermodynamic concept of state variables, such as pressure, density, temperature, etc., has meaning in the intermediate nonequilibrium states. Hence, continuum fluid dynamics is an extension and a generalization of classical thermodynamics to nonequilibrium states. Thus, the energy equation that is subsequently derived is considered an extension of and *not* a result of the first law of thermodynamics.

Again we consider an elemental control volume, $dxdydz$. The principle of conservation of energy describes how the total energy in the control volume

14 Laminar Flow Analysis

changes due to transport, heat addition and work done on the volume. Thus, the principle of conservation of energy stated as a rate equation is

$$\begin{pmatrix} \text{Rate of increase} \\ \text{of total energy} \\ \text{in the control} \\ \text{volume} \end{pmatrix} = \begin{pmatrix} \text{Rate of heat} \\ \text{addition to} \\ \text{the control} \\ \text{volume} \end{pmatrix} + \begin{pmatrix} \text{Rate at which work} \\ \text{is done on the} \\ \text{fluid in the control} \\ \text{volume} \end{pmatrix}$$

The conservation principle now contains energy terms associated with the flow crossing the control volume boundary and work terms due to pressure and viscous forces, in addition to the energy terms which occur in the analysis of a closed system.

The total energy of the fluid in the volume is the sum of the internal energy and the kinetic energy. Here we neglect the potential energy contribution. Hence, the rate of increase of the total energy in the control volume is

$$\begin{pmatrix} \text{Rate of increase} \\ \text{of total energy in} \\ \text{the control volume} \end{pmatrix} = \frac{D}{Dt}(\rho e + \frac{1}{2}\rho U^2)$$

where here e is the internal energy/unit mass and $U^2 = u^2 + v^2 + w^2$.

Heat can be added as a bulk rate of heat addition, as for example from radiation, combustion or chemical reaction. Heat can also be transferred across the surface of the elemental volume by conduction. The net rate of heat transferred across the faces perpendicular to the x, y, z directions, respectively, is

$$\frac{\partial (q/A)_x}{\partial x}dx \qquad \frac{\partial (q/A)_y}{\partial y}dy \qquad \frac{\partial (q/A)_z}{\partial z}dz$$

Since the heat transfer from the elemental volume decreases the total energy in the volume, we have

$$\begin{pmatrix} \text{Rate of heat addition} \\ \text{to the control volume} \end{pmatrix} = \rho Q - \left[\frac{\partial (q/A)_x}{\partial x} + \frac{\partial (q/A)_y}{\partial y} + \frac{\partial (q/A)_z}{\partial z} \right]$$

where Q is the bulk rate of heat transfer per unit mass and q/A is the heat transfer rate due to conduction through the surfaces of the elemental volume. The bulk rate of heat addition is a volume term and includes heat addition from *external* sources only, e.g., from radiation. Conceptually this is similar to the internal generation term used in conduction heat transfer.

The rate at which work is done on the control volume by the volume or body forces is

$$\rho X u + \rho Y v + \rho Z w$$

where X, Y and Z are the components of the body force per unit mass.

Recall that we have previously considered the normal stresses acting on the surfaces of the element to be composed of the normal pressure, a scalar independent of direction, plus a component dependent on direction (see Eq. 1–10). We first consider the rate at which the scalar normal pressure forces do work on the elemental volume. Considering only the x direction (see Figure 1–1), where we again assume that ρ, u, v, w, P, etc. are known at the center of the element and the pressure, velocity, etc. on the faces are given by a first-order Taylor series expansion about the center, the rate at which work is done on the left hand face by the normal pressure force is

$$\left[Pu - \frac{\partial(Pu)}{\partial x}\frac{dx}{2}\right]dy\,dz$$

The rate at which work is done on the right hand face by the normal pressure force is

$$-\left[Pu + \frac{\partial(Pu)}{\partial x}\frac{dx}{2}\right]dy\,dz$$

The total rate at which work is done on the element of volume (per unit volume) by the x component of the normal pressure forces is the sum of these two contributions

$$\frac{\left[Pu - \frac{\partial(Pu)}{\partial x}\frac{dx}{2}\right]dy\,dz - \left[Pu + \frac{\partial(Pu)}{\partial x}\frac{dx}{2}\right]dy\,dz}{dx\,dy\,dz}$$

which after reduction and neglecting higher-order terms is

$$-\frac{\partial(Pu)}{\partial x}$$

Similar results are obtained for the y and z directions. Thus, the rate at which the normal pressure forces do work on the elemental volume is

$$-\left[\frac{\partial(Pu)}{\partial x} + \frac{\partial(Pv)}{\partial y} + \frac{\partial(Pw)}{\partial z}\right]$$

This contribution, which is also present for inviscid compressible flow, represents the work done in compressing the elemental volume.

Using a similar analysis, the rate at which work is done on the elemental volume by the viscous shearing forces is obtained. For simplicity consider only the rate at which the shearing forces acting in the x direction do work on the elemental control volume. The shearing stresses on each face of the element contribute to the rate at which work is done.

The net rate at which work is done on the element by the shearing stresses acting in the x direction on the faces perpendicular to the x-axis (per unit volume) is (see Figure 1–1)

$$\frac{\left(\tau'_{xx} + \frac{\partial \tau'_{xx}}{\partial x}\frac{dx}{2}\right)\left(u + \frac{\partial u}{\partial x}\frac{dx}{2}\right)dy\,dz - \left(\tau'_{xx} - \frac{\partial \tau'_{xx}}{\partial x}\frac{dx}{2}\right)\left(u - \frac{\partial u}{\partial x}\frac{dx}{2}\right)dy\,dz}{dx\,dy\,dz}$$

16   Laminar Flow Analysis

Expanding and neglecting higher-order terms reduces this expression to

$$\frac{\partial}{\partial x}(u\tau'_{xx})$$

The net rate at which work is done on the element by the shearing stresses acting in the x direction on the faces perpendicular to the y-axis (per unit volume) is

$$\frac{\left(\tau_{yx} + \dfrac{\partial \tau_{yx}}{\partial y}\dfrac{dy}{2}\right)\left(u + \dfrac{\partial u}{\partial y}\dfrac{dy}{2}\right)dx\,dz - \left(\tau_{yx} - \dfrac{\partial \tau_{yx}}{\partial y}\dfrac{dy}{2}\right)\left(u - \dfrac{\partial u}{\partial y}\dfrac{dy}{2}\right)dx\,dz}{dx\,dy\,dz}$$

Expanding and neglecting higher-order terms yields

$$\frac{\partial}{\partial y}(u\tau_{yx})$$

The net rate at which work is done on the element by the shearing stresses acting in the x direction on the faces perpendicular to the z-axis (per unit volume) is

$$\frac{\left(\tau_{zx} + \dfrac{\partial \tau_{zx}}{\partial z}\dfrac{dz}{2}\right)\left(u + \dfrac{\partial u}{\partial z}\dfrac{dz}{2}\right)dx\,dy - \left(\tau_{zx} - \dfrac{\partial \tau_{zx}}{\partial z}\dfrac{dz}{2}\right)\left(u - \dfrac{\partial u}{\partial z}\dfrac{dz}{2}\right)dx\,dy}{dx\,dy\,dz}$$

After expanding and neglecting higher-order terms we have

$$\frac{\partial}{\partial z}(u\tau_{zx})$$

Thus, the net rate at which work is done on the element by the shearing stresses acting in the x direction is

$$\frac{\partial}{\partial x}(u\tau'_{xx}) + \frac{\partial}{\partial y}(u\tau_{yx}) + \frac{\partial}{\partial z}(u\tau_{zx})$$

Using our previous results for the stress–rate-of-strain relations allows writing this result in terms of the velocity gradients in the flow, i.e.

$$\frac{\partial}{\partial x}(u\tau'_{xx}) + \frac{\partial}{\partial y}(u\tau_{yx}) + \frac{\partial}{\partial z}(u\tau_{zx}) = \frac{\partial}{\partial x}\left\{u\left[2\mu\frac{\partial u}{\partial x} - \frac{2}{3}\mu\left(\frac{\partial u}{\partial x} + \frac{\partial v}{\partial y} + \frac{\partial w}{\partial z}\right)\right]\right\}$$

$$+ \frac{\partial}{\partial y}\left\{u\left[\mu\left(\frac{\partial u}{\partial y} + \frac{\partial v}{\partial x}\right)\right]\right\} + \frac{\partial}{\partial z}\left\{u\left[\mu\left(\frac{\partial u}{\partial z} + \frac{\partial w}{\partial x}\right)\right]\right\}$$

Derivation of the Energy Equation

The net rate at which work is done on the element by the shearing stresses acting in the y and z directions is obtained in a similar manner. Combining these results yields the net rate at which work is done by all of the shearing stresses acting on the element. Thus

$$\frac{\partial}{\partial x}\left\{u\left[2\mu\frac{\partial u}{\partial x} - \frac{2}{3}\mu\left(\frac{\partial u}{\partial x} + \frac{\partial v}{\partial y} + \frac{\partial w}{\partial z}\right)\right]\right.$$
$$\left. + v\left[\mu\left(\frac{\partial v}{\partial x} + \frac{\partial u}{\partial y}\right)\right] + w\left[\mu\left(\frac{\partial w}{\partial x} + \frac{\partial u}{\partial z}\right)\right]\right\}$$
$$+ \frac{\partial}{\partial y}\left\{v\left[2\mu\frac{\partial v}{\partial y} - \frac{2}{3}\mu\left(\frac{\partial u}{\partial x} + \frac{\partial v}{\partial y} + \frac{\partial w}{\partial z}\right)\right]\right.$$
$$\left. + u\left[\mu\left(\frac{\partial u}{\partial y} + \frac{\partial v}{\partial x}\right)\right] + w\left[\mu\left(\frac{\partial w}{\partial y} + \frac{\partial v}{\partial z}\right)\right]\right\}$$
$$+ \frac{\partial}{\partial z}\left\{w\left[2\mu\frac{\partial w}{\partial z} - \frac{2}{3}\mu\left(\frac{\partial u}{\partial x} + \frac{\partial v}{\partial y} + \frac{\partial w}{\partial z}\right)\right]\right.$$
$$\left. + u\left[\mu\left(\frac{\partial u}{\partial z} + \frac{\partial w}{\partial x}\right)\right] + v\left[\mu\left(\frac{\partial v}{\partial z} + \frac{\partial w}{\partial y}\right)\right]\right\} \qquad (1-34)$$

where Stokes' approximation (Eq. 1–31) is used in arriving at this result.

Before writing out the conservation of energy equation let us relate the conduction heat transfer rate to more general fluid quantities — in particular, to the temperature gradient. The most general linear relationship is

$$\left(\frac{q}{A}\right)_x = \beta_1 \frac{\partial T}{\partial x} + \beta_2 \frac{\partial T}{\partial y} + \beta_3 \frac{\partial T}{\partial z}$$

$$\left(\frac{q}{A}\right)_y = \beta_4 \frac{\partial T}{\partial x} + \beta_5 \frac{\partial T}{\partial y} + \beta_6 \frac{\partial T}{\partial z}$$

$$\left(\frac{q}{A}\right)_z = \beta_7 \frac{\partial T}{\partial x} + \beta_8 \frac{\partial T}{\partial y} + \beta_9 \frac{\partial T}{\partial z}$$

where the β_n's are generalized coefficients of heat conduction which depend on the material properties. However, our previous assumptions that the fluid is isotropic and that the heat transfer rate is invariant with respect to a coordinate rotation reduces the nine independent material properties to one. This is k, the coefficient of heat conduction for the fluid. Thus

$$\left(\frac{q}{A}\right)_x = -k\frac{\partial T}{\partial x} \qquad \left(\frac{q}{A}\right)_y = -k\frac{\partial T}{\partial y} \qquad \left(\frac{q}{A}\right)_z = -k\frac{\partial T}{\partial z} \qquad (1-35)$$

18   Laminar Flow Analysis

Note that these relationships are identical to Fourier's law, which is commonly used to relate the heat transfer to the temperature gradient. Combining these results allows writing the general conservation of energy equation as

$$\frac{D}{Dt}\left(\rho e + \frac{1}{2}\rho U^2\right) = \rho Q + \left[\frac{\partial}{\partial x}\left(k\frac{\partial T}{\partial x}\right) + \frac{\partial}{\partial y}\left(k\frac{\partial T}{\partial y}\right) + \frac{\partial}{\partial z}\left(k\frac{\partial T}{\partial z}\right)\right]$$

$$- \left[\frac{\partial}{\partial x}(Pu) + \frac{\partial}{\partial y}(Pv) + \frac{\partial}{\partial z}(Pw)\right]$$

$$+ \rho\left[Xu + Yv + Zw\right]$$

$$+ \frac{\partial}{\partial x}\left\{u\left[2\mu\frac{\partial u}{\partial x} - \frac{2}{3}\mu\dot\theta\right] + \mu v\dot\gamma_{xy} + \mu w\dot\gamma_{xz}\right\}$$

$$+ \frac{\partial}{\partial y}\left\{\mu u\dot\gamma_{xy} + v\left[2\mu\frac{\partial v}{\partial y} - \frac{2}{3}\mu\dot\theta\right] + \mu w\dot\gamma_{zy}\right\}$$

$$+ \frac{\partial}{\partial z}\left\{\mu u\dot\gamma_{xz} + \mu v\dot\gamma_{zy} + w\left[2\mu\frac{\partial w}{\partial z} - \frac{2}{3}\mu\dot\theta\right]\right\} \quad (1-36)$$

In order to put the energy equation in a more convenient form, the x, y, z components of the Navier-Stokes equations are multiplied by u, v, w, respectively, and substituted into the energy equation. The result is

$$\frac{D}{Dt}\rho\left(e + \frac{1}{2}U^2\right) = -\left[\frac{\partial}{\partial x}(Pu) + \frac{\partial}{\partial y}(Pv) + \frac{\partial}{\partial z}(Pw)\right] + \rho Q + \mu\Phi$$

$$+ \rho\frac{D}{Dt}\left(\frac{U^2}{2}\right) + \left(u\frac{\partial P}{\partial x} + v\frac{\partial P}{\partial y} + w\frac{\partial P}{\partial z}\right)$$

$$+ \frac{\partial}{\partial x}\left(k\frac{\partial T}{\partial x}\right) + \frac{\partial}{\partial y}\left(k\frac{\partial T}{\partial y}\right) + \frac{\partial}{\partial z}\left(k\frac{\partial T}{\partial z}\right) \quad (1-37)$$

where Φ is the viscous dissipation function given by

$$\Phi = 2\left[\left(\frac{\partial u}{\partial x}\right)^2 + \left(\frac{\partial v}{\partial y}\right)^2 + \left(\frac{\partial w}{\partial z}\right)^2\right] - \frac{2}{3}\left(\frac{\partial u}{\partial x} + \frac{\partial v}{\partial y} + \frac{\partial w}{\partial z}\right)^2$$

$$+ \left(\frac{\partial v}{\partial x} + \frac{\partial u}{\partial y}\right)^2 + \left(\frac{\partial w}{\partial x} + \frac{\partial u}{\partial z}\right)^2 + \left(\frac{\partial w}{\partial y} + \frac{\partial v}{\partial z}\right)^2 \quad (1-38)$$

Note that the viscous dissipation function is always positive and hence any viscous flow is dissipative, i.e., entropy increases.

Derivation of the Energy Equation

From the continuity equation we have

$$P\left(\frac{\partial u}{\partial x} + \frac{\partial v}{\partial y} + \frac{\partial w}{\partial z}\right) = -\frac{DP}{Dt} + \rho \frac{D}{Dt}\left(\frac{P}{\rho}\right)$$

Multiplying the continuity equation by $(e + U^2/2)$ and subtracting from the energy equation yields

$$\rho\left[\frac{De}{Dt} + \frac{D}{Dt}\left(\frac{P}{\rho}\right)\right] = \frac{DP}{Dt} + \rho Q + \mu \Phi$$

$$+ \frac{\partial}{\partial x}\left(k\frac{\partial T}{\partial x}\right) + \frac{\partial}{\partial y}\left(k\frac{\partial T}{\partial y}\right) + \frac{\partial}{\partial z}\left(k\frac{\partial T}{\partial z}\right) \quad (1-39)$$

Further, assuming the equilibrium thermodynamic enthalpy relation holds, i.e.

$$h = \frac{P}{\rho} + e \quad (1-40)$$

we have for the energy equation

$$\rho \frac{Dh}{Dt} = \frac{DP}{Dt} + \rho Q + \mu \Phi + \frac{\partial}{\partial x}\left(k\frac{\partial T}{\partial x}\right) + \frac{\partial}{\partial y}\left(k\frac{\partial T}{\partial y}\right) + \frac{\partial}{\partial z}\left(k\frac{\partial T}{\partial z}\right) \quad (1-41)$$

Collecting our previous results, we have for the continuity equation

$$\frac{\partial \rho}{\partial t} + \frac{\partial}{\partial x}(\rho u) + \frac{\partial}{\partial y}(\rho v) + \frac{\partial}{\partial z}(\rho w) = 0 \quad (1-1)$$

and for the Navier-Stokes equation

$$\frac{Du}{Dt} = X - \frac{1}{\rho}\frac{\partial P}{\partial x} + \frac{1}{\rho}\frac{\partial}{\partial x}\left[\mu\left(2\frac{\partial u}{\partial x} - \frac{2}{3}\text{div}\vec{V}\right)\right]$$

$$+ \frac{1}{\rho}\frac{\partial}{\partial y}\left[\mu\left(\frac{\partial u}{\partial y} + \frac{\partial v}{\partial x}\right)\right] + \frac{1}{\rho}\frac{\partial}{\partial z}\left[\mu\left(\frac{\partial w}{\partial x} + \frac{\partial u}{\partial z}\right)\right] \quad (1-33a)$$

$$\frac{Dv}{Dt} = Y - \frac{1}{\rho}\frac{\partial P}{\partial y} + \frac{1}{\rho}\frac{\partial}{\partial y}\left[\mu\left(2\frac{\partial v}{\partial y} - \frac{2}{3}\text{div}\vec{V}\right)\right]$$

$$+ \frac{1}{\rho}\frac{\partial}{\partial z}\left[\mu\left(\frac{\partial v}{\partial z} + \frac{\partial w}{\partial y}\right)\right] + \frac{1}{\rho}\frac{\partial}{\partial x}\left[\mu\left(\frac{\partial u}{\partial y} + \frac{\partial v}{\partial x}\right)\right] \quad (1-33b)$$

$$\frac{Dw}{Dt} = Z - \frac{1}{\rho}\frac{\partial P}{\partial z} + \frac{1}{\rho}\frac{\partial}{\partial z}\left[\mu\left(2\frac{\partial w}{\partial z} - \frac{2}{3}\text{div}\vec{V}\right)\right]$$

$$+ \frac{1}{\rho}\frac{\partial}{\partial x}\left[\mu\left(\frac{\partial w}{\partial x} + \frac{\partial u}{\partial z}\right)\right] + \frac{1}{\rho}\frac{\partial}{\partial y}\left[\mu\left(\frac{\partial v}{\partial z} + \frac{\partial w}{\partial y}\right)\right] \quad (1-33c)$$

20 Laminar Flow Analysis

Equations (1–1), (1–33) and (1–41) are the governing equations for the flow of a continuous, homogeneous, isotropic, compressible, viscous, heat conducting Stokesian fluid. They are a system of five coupled partial differential equations in eight dependent and four independent variables. Since there are more dependent variables than equations, the system is indeterminant. One of the additional required relations is provided by the equation of state

$$P = P(\rho, T) \tag{1-42}$$

which relates the equilibrium thermodynamic state variables. In addition, the transport properties are related to the thermodynamic state variables by equations of the form

$$k = k(T, \rho) \tag{1-43}$$

and

$$\mu = \mu(T, \rho) \tag{1-44}$$

With these additional relations the system is determinant.

1–7 Boundary Conditions

In order to complete the discussion we must specify the applicable boundary conditions. In general, boundary conditions are divided into three classes: initial conditions, surface boundary conditions and field boundary conditions.

Initial conditions are specified at the initial time or occasionally at an initial position on the surface, for example, at the leading edge of a semi-infinite flat plate. A complete set of initial conditions is generally specified if the velocity vector and two independent thermodynamic state variables such as P and T are given.

Surface boundary conditions are specified at the surface of the body. This usually involves specifying some condition on the velocity components and on the temperature or heat transfer rate at the surface. At ordinary densities, the velocity of the fluid at the surface is assumed equal to the velocity of the surface, i.e., the relative velocity between the fluid and the surface is zero. This is the so-called no-slip condition.

Similarly, the temperature of the fluid at the surface is assumed equal to the surface temperature. This is the no-temperature-jump condition.[†] Instead of specifying the temperature at the surface, the heat transfer rate can be given. Equation (1–35) then implies that the temperature gradient at the surface is specified. This serves to determine the temperature of the surface, and of the fluid adjacent to the surface, which will support the required heat transfer rate. Of particular interest is the condition specified by zero heat transfer rate from

---

[†] At sufficiently low densities these conditions must be relaxed to allow slip and temperature jump at the surface (see, e.g., [Scha58]).

the surface to the fluid. This is the so-called adiabatic wall, and the surface temperature is the adiabatic wall temperature.

In addition, the mass flow rate through the surface is generally specified. If the mass flow rate through the surface is zero, then the surface is a streamline of the flow, which implies that the velocity component normal to the surface is zero. Positive mass flow rates are generally referred to as blowing or injection and negative mass flow rates as suction.

Field boundary conditions are specified at some point in the flow field, generally at a large distance from the body. Typical examples are: the velocity components are required to approach zero; the velocity components and/or the thermodynamic variables are required to approach a constant value or some specified functional form.

1–8 Equation of State and the Relation of the Transport Properties to the State Variables

Since we are principally concerned with gases, we assume that the equation of state is that for a thermally perfect gas

$$P = \rho RT \qquad (1-45)$$

where R is the gas constant.

Experimental results show that for gases the viscosity μ and the heat conductivity k are largely dependent on the temperature, with only a very weak dependence on pressure (see, e.g., Weast [Weas87] and Grieser and Goldthwaite [Grie63]). Thus, we assume

$$\mu, k \sim T \qquad (1-46)$$

1–9 Special Forms of the Navier-Stokes Equations

Considerable simplification of the governing system of equations results if the fluid is assumed incompressible. This assumption does not preclude temperature variations. However, the temperature variations are generally small. Hence, the transport properties are assumed constant. For an incompressible fluid the variation of internal energy and enthalpy are

$$de = c\,dt \qquad (1-47)$$

and

$$dh = c\,dt + \frac{dP}{\rho} \qquad (1-48)$$

where c is the specific heat.

For incompressible flow the continuity equation becomes

$$\mathrm{div}\vec{V} = \frac{\partial u}{\partial x} + \frac{\partial v}{\partial y} + \frac{\partial w}{\partial z} = 0 \qquad (1-49)$$

22 Laminar Flow Analysis

The Navier-Stokes equations (Eqs. 1–33a,b,c) become

$$\frac{\partial u}{\partial t} + u\frac{\partial u}{\partial x} + v\frac{\partial u}{\partial y} + w\frac{\partial u}{\partial z} = X - \frac{1}{\rho}\frac{\partial P}{\partial x} + \nu\left(\frac{\partial^2 u}{\partial x^2} + \frac{\partial^2 u}{\partial y^2} + \frac{\partial^2 u}{\partial z^2}\right) \quad (1-50a)$$

$$\frac{\partial v}{\partial t} + u\frac{\partial v}{\partial x} + v\frac{\partial v}{\partial y} + w\frac{\partial v}{\partial z} = Y - \frac{1}{\rho}\frac{\partial P}{\partial y} + \nu\left(\frac{\partial^2 v}{\partial x^2} + \frac{\partial^2 v}{\partial y^2} + \frac{\partial^2 v}{\partial z^2}\right) \quad (1-50b)$$

$$\frac{\partial w}{\partial t} + u\frac{\partial w}{\partial x} + v\frac{\partial w}{\partial y} + w\frac{\partial w}{\partial z} = Z - \frac{1}{\rho}\frac{\partial P}{\partial z} + \nu\left(\frac{\partial^2 w}{\partial x^2} + \frac{\partial^2 w}{\partial y^2} + \frac{\partial^2 w}{\partial z^2}\right) \quad (1-50c)$$

The energy equation (Eq. 1–41) becomes

$$c\left(\frac{\partial T}{\partial t} + u\frac{\partial T}{\partial x} + v\frac{\partial T}{\partial y} + w\frac{\partial T}{\partial z}\right) = Q + \nu\Phi + \frac{k}{\rho}\left(\frac{\partial^2 T}{\partial x^2} + \frac{\partial^2 T}{\partial y^2} + \frac{\partial^2 T}{\partial z^2}\right) \quad (1-51)$$

where $\nu = \mu/\rho$ is the kinematic viscosity. Here Φ reduces to

$$\Phi = 2\left[\left(\frac{\partial u}{\partial x}\right)^2 + \left(\frac{\partial v}{\partial y}\right)^2 + \left(\frac{\partial w}{\partial z}\right)^2\right] + \left(\frac{\partial v}{\partial x} + \frac{\partial u}{\partial y}\right)^2 + \left(\frac{\partial w}{\partial x} + \frac{\partial u}{\partial z}\right)^2 + \left(\frac{\partial w}{\partial y} + \frac{\partial v}{\partial z}\right)^2$$

$$(1-52)$$

1–10 Classification

The boundary-value problems associated with the Navier-Stokes equations are broadly classified into three types:

1. Uncoupled—dynamics of the flow.
2. Uncoupled—dynamics and thermodynamics of the flow.
3. Coupled—dynamics and thermodynamics of the flow.

In order to illustrate the first of these, consider the boundary value problem described by Eqs. (1–49) to (1–51) along with appropriate boundary conditions. Looking at the continuity equation and the three components of the momentum equation, we see that these four equations are independent of the solution of the energy equation, Eq. (1–51). The momentum and energy equations are uncoupled. Hence, the dynamics of the flow can be considered separately from the thermodynamics of the flow. Because this is possible, many analyses consider only the dynamics of the flow.

Once the solution for the velocity field is obtained from the still coupled continuity and momentum equations, it is used in the energy equation to obtain the solution for the temperature field. This is the second case mentioned above, i.e., uncoupled dynamics and thermodynamics of the flow.

Referring back to the compressible flow case, Eqs. (1–1), (1–33) and (1–41), we see that the energy and momentum equations are coupled, and the entire system must be solved simultaneously.

Each of the above cases is separately discussed in the succeeding chapters.

TWO

EXACT SOLUTIONS OF THE NAVIER-STOKES EQUATIONS

2–1 Introduction

Because of the great complexity of the full compressible Navier-Stokes equations, no known general analytical solution exists. Hence, it is necessary to simplify the equations either by making assumptions about the fluid, about the flow or about the geometry of the problem in order to obtain analytical solutions. Typical assumptions are that the flow is laminar, steady, two-dimensional, the fluid incompressible with constant properties and the flow is between parallel plates. By so doing it is possible to obtain analytical, exact and approximate solutions to the Navier-Stokes equations.

Before proceeding let us clearly define what is meant by analytical, exact and approximate solutions. An analytical solution is obtained when the governing boundary value problem is integrated using the methods of classical differential equations. The result is an algebraic expression giving the dependent variable(s) as a function(s) of the independent variable(s). An exact solution is obtained by integrating the governing boundary value problem numerically. The result is a tabulation of the dependent variable(s) as a function(s) of the independent variables(s). An approximate solution results when methods such as series expansion and the von Karman-Pohlhausen technique are used to solve the governing boundary value problem (see Schlichting [Schl60], p. 239).

2–2 Analytical Solutions

Finding analytical solutions of the Navier-Stokes equations, even in the uncoupled case (see Section 1–10), presents almost insurmountable mathematical difficulties due to the nonlinear character of the equations. However, it is possible to find analytical solutions in certain particular cases, generally when the nonlinear convective terms vanish naturally. Parallel flows, in which only one velocity component is different from zero, of a two-dimensional, incompressible fluid have this characteristic. Examples for which analytical solutions exist are parallel flow through a straight channel, Couette flow and Hagen-Poiseuille flow, i.e., flow in a cylindrical pipe. Here we discuss parallel flow through a straight channel and Couette flow.

2–3 Parallel Flow Through A Straight Channel

A flow is considered parallel if only one component of the velocity is different from zero. In order to illustrate this concept, consider two-dimensional steady flow in a channel with straight parallel sides (see Figure 2–1). This flow is two-dimensional since the velocity is in the x direction and, as we shall see, its variation is in the y direction. We consider the fluid to be incompressible and to have constant properties. Under these circumstances the momentum and energy equations are uncoupled (see Section 1–10). Thus, we consider only the continuity and the momentum equations, i.e., the velocity field, and reserve our discussion of the energy equation, i.e., the temperature field, until Chapter 4. The continuity and momentum equations for two-dimensional, steady, incompressible, constant property flow are

continuity
$$\frac{\partial u}{\partial x} + \frac{\partial v}{\partial y} = 0 \tag{1-49}$$

x momentum
$$u\frac{\partial u}{\partial x} + v\frac{\partial u}{\partial y} = -\frac{1}{\rho}\frac{\partial P}{\partial x} + \nu\left(\frac{\partial^2 u}{\partial x^2} + \frac{\partial^2 u}{\partial y^2}\right) \tag{1-50a}$$

y momentum
$$u\frac{\partial v}{\partial x} + v\frac{\partial v}{\partial y} = -\frac{1}{\rho}\frac{\partial P}{\partial y} + \nu\left(\frac{\partial^2 v}{\partial x^2} + \frac{\partial^2 v}{\partial y^2}\right) \tag{1-50b}$$

Since the flow is constrained by the flat parallel walls of the channel, no component of the velocity in the y direction is possible, i.e., $v = 0$. This implies

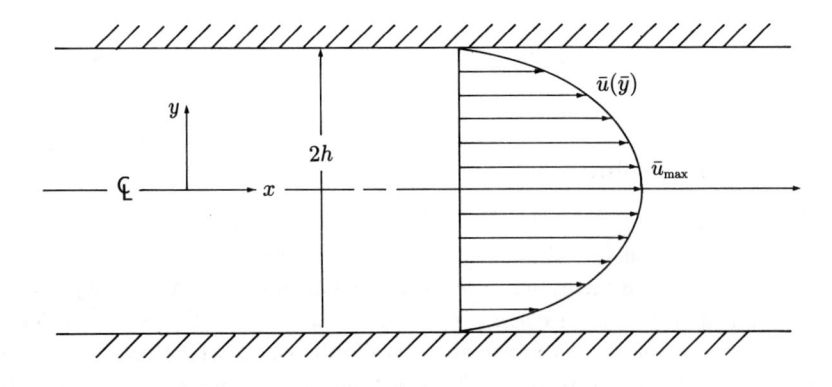

Figure 2–1. Parallel flow through a straight channel.

that $v = 0$ everywhere. Hence, the gradients in v are also equal to zero, i.e., $\partial v/\partial y = \partial v/\partial x = \partial^2 v/\partial y^2 = \partial^2 v/\partial x^2 = 0$.

From the continuity equation we have

$$\frac{\partial u}{\partial x} = -\frac{\partial v}{\partial y} = 0$$

which implies that $\partial u^2/\partial x^2 = 0$. The momentum equations are thus reduced to

$$\frac{dP}{dx} = \mu \frac{d^2 u}{dy^2} \qquad (2-1)$$

and

$$\frac{\partial P}{\partial y} = 0$$

Since u is not a function of x, the distance along the axis, there is no physical mechanism to provide a change in the pressure gradient. Thus, the pressure gradient is considered constant, i.e., $dP/dx = $ constant. Note that under these conditions all of the nonlinear convective terms in the momentum equations are eliminated. Equation (2-1) is a linear second-order ordinary differential equation. It is easily integrated twice to yield

$$u(y) = \frac{1}{2}\frac{1}{\mu}\frac{dP}{dx}y^2 + Ay + B \qquad (2-2)$$

where A and B are integration constants.

In order to evaluate the integration constants, we apply the no-slip boundary conditions at the channel walls. The boundary conditions at the walls are

$$y = \pm h \qquad u = 0 \qquad (2-3)$$

Using these boundary conditions to evaluate the integration constants, we have

$$A = 0$$
$$B = -\frac{1}{2}\frac{h^2}{\mu}\frac{dP}{dx}$$

Substitution into Eq. (2–2) yields

$$u(y) = -\frac{1}{2}\frac{h^2}{\mu}\frac{dP}{dx}\left[1 - \left(\frac{y}{h}\right)^2\right] \qquad (2-4)$$

Here we see that the velocity distribution in the channel is parabolic and symmetrical about the axis. The maximum velocity, which occurs at the center of the channel, is given by

$$u_m = -\frac{1}{2}\frac{h^2}{\mu}\frac{dP}{dx} \qquad (2-5)$$

26 Laminar Flow Analysis

Introducing nondimensional variables

$$\bar{u} = \frac{u}{u_m} \qquad \bar{y} = \frac{y}{h}$$

yields
$$\bar{u} = 1 - \bar{y}^2 \qquad (2-6)$$

This nondimensional velocity distribution is shown in Figure 2–1. Using Newton's law of friction, which is obtained with the help of Eqs. (1–15a) and (1–25), the shearing stress at the channel walls is given by

$$\tau_{(y=\pm h)} = \mu \frac{du}{dy}\bigg|_{(y=\pm h)} \qquad (2-7)$$

or using nondimensional variables

$$\tau_{(\bar{y}=\pm 1)} = \mu \frac{u_m}{h} \frac{d\bar{u}}{d\bar{y}}\bigg|_{(\bar{y}=\pm 1)} \qquad (2-8)$$

Thus, using Eqs. (2–5) and (2–6), the shearing stress at the channel walls is

$$\tau_{(\bar{y}=\pm 1)} = \pm h \frac{dP}{dx} \qquad (2-9)$$

2–4 Couette Flow

Continuing our discussion of the analytical solutions, consider the flow between two parallel infinite flat surfaces, one of which is moving in its plane with a velocity U (see Figure 2–2). The flow is considered steady, two-dimensional and incompressible with constant properties. Using the same physical and geometric arguments presented in the previous discussion of channel flow, the governing equations again reduce to

$$\frac{dP}{dx} = \mu \frac{d^2 u}{dy^2} \qquad (2-1)$$

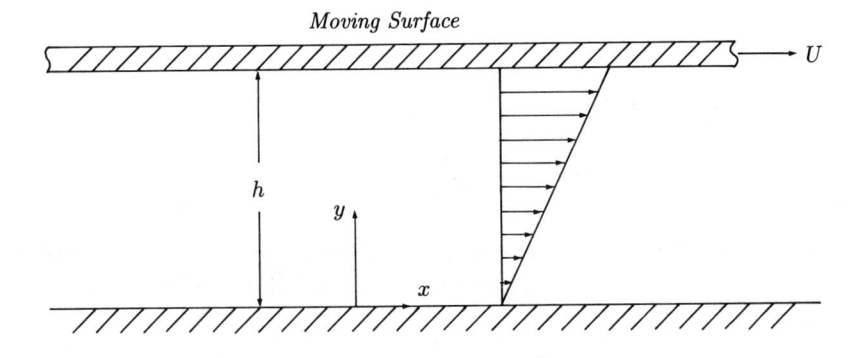

Figure 2–2. Couette flow.

No-slip boundary conditions are assumed to apply at both the moving and the stationary surfaces, i.e.,

$$y = 0 \quad u = 0 \qquad y = h \quad u = U \qquad (2-10a,b)$$

For simple Couette flow the pressure gradient is assumed zero. Thus, the governing equation becomes

$$\frac{d^2u}{dy^2} = 0$$

Integrating twice and evaluating the constants of integration from the boundary conditions yields the linear velocity distribution shown in Figure 2–2.

$$\frac{u}{U} = \frac{y}{h} \quad \text{or} \quad \bar{u} = \bar{y} \qquad (2-11)$$

Turning now to nonsimple Couette flow, i.e., when the pressure gradient is nonzero, the governing differential equation is now Eq. (2–1) with the boundary conditions given in Eqs. (2–10a, b). Integrating Eq. (2–1) twice again yields

$$u(y) = \frac{1}{2}\frac{1}{\mu}\frac{dP}{dx}y^2 + Ay + B \qquad (2-2)$$

where A and B are constants of integration. Again, the pressure gradient is constant. Using the boundary conditions to evaluate A and B yields

$$B = 0$$

$$A = \frac{1}{h}\left(U - \frac{1}{2}\frac{1}{\mu}\frac{dP}{dx}h^2\right)$$

Substituting into Eq. (2–2), introducing the nondimensional variables

$$\bar{u} = \frac{u}{U} \qquad \bar{y} = \frac{y}{h} \qquad \bar{P} = -\frac{h^2}{2\mu U}\frac{dP}{dx} \qquad (2-12a,b,c)$$

and rearranging yields

$$\bar{u} = \bar{y}\,[1 + \bar{P}(1-\bar{y})] \qquad (2-13)$$

From Eq. (2–13) we see that the shape of the nondimensional velocity distribution is determined by the nondimensional pressure gradient, $\bar{P}$. Nondimensional velocity profiles for several values of $\bar{P}$ are shown in Figure 2–3.

The results shown in Figure 2–3 indicate that the slope of the velocity profile is zero for some particular value of the nondimensional pressure gradient $\bar{P}$. This implies that the shearing stress at the surface is zero (see Eq. 2–7). Differentiating Eq. (2–13) and setting the result equal to zero yields

$$1 + \bar{P} - 2\bar{y}\bar{P} = 0 \qquad (2-14)$$

28 Laminar Flow Analysis

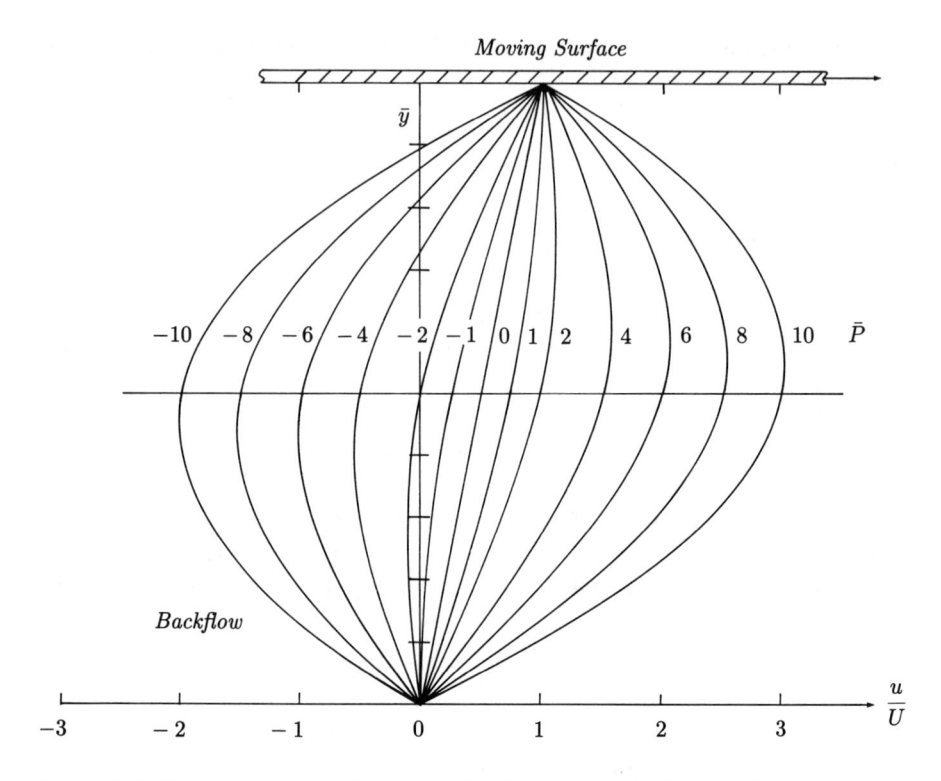

Figure 2–3. Nondimensional velocity profiles for nonsimple Couette flow.

Hence, the shearing stress at the stationary surface is zero for $\bar{P} = -1$ and at the moving surface for $\bar{P} = +1$. Using Eqs. (2–13) and (2–14) we see that for $\bar{P} < -1$, regions of backflow, i.e., $u \leq 0$, exist near the stationary surface. For $\bar{P} > 1$, the velocity in the flow exceeds the velocity of the moving plate. Physically, a region of backflow exists when the force due to the momentum of the fluid in the flow direction is overcome by the adverse pressure gradient[†] in the flow direction. Similarly, velocities greater than that of the moving plate occur when a favorable pressure gradient in the flow direction adds to the momentum of the fluid in that direction.

One further analytical solution, the suddenly accelerated plane wall, is presented below in order to illustrate several solution techniques.

2–5 The Suddenly Accelerated Plane Wall

Consider the two-dimensional parallel flow of an incompressible fluid near a flat plate which is suddenly accelerated from rest and moves in its own plane with

---

[†]An adverse pressure gradient means the pressure *increases* in the flow direction.

a constant velocity U (see Figure 2–4). Recall that the Navier-Stokes equations for two-dimensional incompressible, constant property flow are

$$\frac{\partial u}{\partial t} + u\frac{\partial u}{\partial x} + v\frac{\partial u}{\partial y} = -\frac{1}{\rho}\frac{\partial P}{\partial x} + \nu\left(\frac{\partial^2 u}{\partial x^2} + \frac{\partial^2 u}{\partial y^2}\right) \qquad (1-50a)$$

$$\frac{\partial v}{\partial t} + u\frac{\partial v}{\partial x} + v\frac{\partial v}{\partial y} = -\frac{1}{\rho}\frac{\partial P}{\partial y} + \nu\left(\frac{\partial^2 v}{\partial x^2} + \frac{\partial^2 v}{\partial y^2}\right) \qquad (1-50b)$$

and the continuity equation is

$$\frac{\partial u}{\partial x} + \frac{\partial v}{\partial y} = 0 \qquad (1-49)$$

Since the plate is accelerated only in the x direction the v component of the velocity is always and everywhere zero. Hence

$$v = \frac{\partial v}{\partial t} = \frac{\partial v}{\partial x} = \frac{\partial v}{\partial y} = \frac{\partial^2 v}{\partial x^2} = \frac{\partial^2 v}{\partial y^2} = 0 \qquad (2-15)$$

Thus, the y-momentum equation yields

$$\frac{\partial P}{\partial y} = 0$$

and $P = \text{constant} + f(x)$. Further, since the pressure is initially constant everywhere and the sudden acceleration does not allow sufficient time for a pressure gradient to develop in the direction along the plate, we take

$$\frac{\partial P}{\partial x} = 0$$

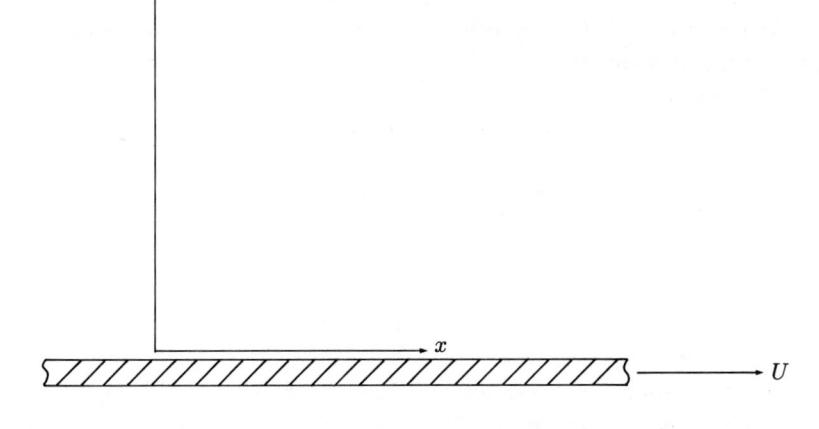

Figure 2–4. Suddenly accelerated flat plate.

30 Laminar Flow Analysis

which requires $f(x) = $ constant. The mathematical model is then based upon a uniform pressure field for all x, y and t equal to or greater than zero.

From the continuity equation we have

$$\frac{\partial u}{\partial x} = -\frac{\partial v}{\partial y} = 0$$

and the governing differential equations reduce to

$$\frac{\partial u}{\partial t} = \nu \frac{\partial^2 u}{\partial y^2} \qquad t > 0 \qquad (2-16)$$

The initial and boundary conditions are

$$u = 0 \quad \text{for all } y \qquad t \leq 0$$

$$\left. \begin{array}{l} u = U \quad y = 0 \\ u \to 0 \quad y \to \infty \end{array} \right\} \quad t > 0 \qquad (2-17a,b)$$

where for $t > 0$ the first boundary condition expresses the fact that there is no slip at the surface. The second boundary condition is read: In the limit as y approaches infinity the velocity approaches zero. It represents the physical condition that the influence of the wall, i.e., the effect of viscosity, decreases asymptotically to zero as the distance above the plate increases.

The governing differential equation, Eq. (2–16), has the same form as the equation describing the diffusion of heat by conduction in the space $y > 0$ when at $t = 0$ the wall temperature is suddenly changed. The kinematic viscosity, ν, which appears in the governing equation is sometimes called the momentum diffusivity. It plays the same role in the momentum equation that the thermal diffusivity, α, does in the energy equation. Equation (2–16) is generally called the diffusion equation. The diffusion equation is first encountered in a classical course in differential equations. There, an analytical solution is obtained by the classical separation of variables technique. Thus, it seems reasonable to look for a solution in the same manner. In an attempt to separate the variables we assume

$$u = F(t)G(y) \qquad (2-18)$$

Substituting into the differential equation yields

$$F'G = \nu F G''$$

or

$$\frac{F'}{F} = \nu \frac{G''}{G} = i\gamma^2 \qquad (2-19)$$

where here the prime denotes differentiation with respect to the appropriate argument. Since each side of Eq. (2–19) is a function of only one variable, they can individually be satisfied only if they are each equal to a constant, in this case $i\gamma^2$.[†] Hence, we have

$$\frac{F'}{F} = i\gamma^2 \qquad (2-20a)$$

$$\frac{G''}{G} = \frac{i\gamma^2}{\nu} \qquad (2-20b)$$

The solution of Eq. (2–20a) is

$$F = C_1 e^{i\gamma^2 t} \qquad (2-21a)$$

and to Eq. (2–20b)

$$G = C_2 e^{\frac{i\gamma^2}{\nu} y} + C_3 e^{-\frac{i\gamma^2}{\nu} y} \qquad (2-21b)$$

The initial and boundary conditions are given by Eq. (2–17). There are three possible cases, $i = 0, \pm 1$. The case of $i = 0$ yields the trivial solution $u = 0$ everywhere and hence is discarded. For $i = \pm 1$ the particular solutions of interest are obtained by using Eqs. (2–17) to determine C_1, C_2 and C_3. However, attempts to obtain a solution in this manner *fail*. The reason is that, unlike the boundary value problem for heat conduction in a plate where both boundary conditions are given at finite values of the independent variable, here we have an asymptotic boundary condition imposed at infinity. Hence, there is no characteristic length associated with the problem, and it is not possible to obtain the eigenvalues which satisfy the boundary conditions. However, a closed-form analytical solution of this problem is possible using a similarity analysis.

2–6 Similarity Analysis

A similarity analysis is used to investigate the conditions under which the solutions of a particular boundary value problem have similar forms for different values of the independent variables. If similarity exists, then by proper selection of the dependent and independent variables the solution for all values of the independent variable collapses into a single curve or function. One very important result of the search for similar solutions is that in those cases in which 'similarity' exists the governing partial differential equations are reduced to ordinary differential equations. This is a considerable mathematical simplification. Reduction to ordinary differential equations allows use of the generalized techniques developed for solving ordinary differential equations. To illustrate this technique reconsider the suddenly accelerated flat plate. Recalling Figure 2–4 and Eqs. (2–16) and (2–17) we have

$$\frac{\partial u}{\partial t} = \nu \frac{\partial^2 u}{\partial y^2} \qquad t > 0 \qquad (2-16)$$

$$\left. \begin{array}{ll} y = 0 & u = U \\ y \to \infty & u \to 0 \end{array} \right\} \quad t > 0 \qquad (2-17a,b)$$

[†]The values of γ that satisfy the equations are called the eigenvalues of the solution.

We now seek a similarity solution. In particular, we seek a transformation of variables which reduces the governing partial differential equation to an ordinary differential equation. Since a partial differential equation involves more than one independent variable and an ordinary differential equation only one, it is reasonable to assume an independent variable transformation which attempts to combine the two independent variables. Thus, we assume

$$\eta = By^m t^n \qquad (2-22)$$

where B, m, n are, as yet, undetermined constants and η is a transformed independent variable. In addition, to nondimensionalize the equations we assume a dependent variable transformation of the form

$$u = Af(\eta) \qquad (2-23)$$

where A is again an as yet undetermined constant. Using these transformations yields

$$\frac{\partial u}{\partial t} = A\frac{\partial \eta}{\partial t}\frac{\partial f}{\partial \eta} = ABny^m t^{n-1} f' \qquad (2-24)$$

where here the prime denotes differentiation with respect to η. Further

$$\frac{\partial u}{\partial y} = A\frac{\partial \eta}{\partial y}f' = ABmy^{m-1}t^n f' \qquad (2-25a)$$

and

$$\frac{\partial^2 u}{\partial y^2} = ABm(m-1)y^{m-2}t^n f' + AB^2 m^2 y^{2(m-1)} t^{2n} f'' \qquad (2-25b)$$

Substituting Eqs. (2-24) and (2-25) into the differential equation, (Eq. 2-16), yields

$$ABny^m t^{n-1} f' = \nu ABm(m-1)y^{m-2}t^n f' + \nu AB^2 m^2 y^{2(m-1)} t^{2n} f'' \qquad (2-26)$$

Since A appears in each term it can be eliminated. Hence, its value is arbitrary *with respect to the differential equation.*

We now seek to select values of m, n, A, B such that reduction to a nondimensional ordinary differential equation is achieved. Inspection of Eq. (2-26) shows that for $m = 1$ the first term on the right and the y dependence of the second term are eliminated. Thus, we have

$$Bnyt^{(n-1)} f' = \nu B^2 t^{2n} f'' \qquad (2-27)$$

and

$$\eta = Byt^n \qquad (2-28)$$

Using Eq. (2-28) allows Eq. (2-27) to be rewritten as

$$n\eta t^{-1} f' = \nu B^2 t^{2n} f'' \qquad (2-29)$$

If $n = -1/2$ the time dependence is eliminated. Finally, choosing $B^2 = 1/4\nu$, where the 4 is introduced for later convenience, yields a nondimensional ordinary differential equation

$$f'' + 2\eta f' = 0 \qquad (2-30)$$

where
$$\eta = \frac{1}{2}\frac{y}{\sqrt{\nu t}} \qquad (2-31)$$

$$u = Af(\eta) \qquad (2-32)$$

Now looking at the boundary conditions we have for $t > 0$, $\eta = 0$ when $y = 0$ and $\eta \to \infty$ when $y \to \infty$. Hence

$$\eta = 0 \qquad f(0) = \frac{U}{A} \qquad (2-33)$$

$$\eta \to \infty \qquad f(\eta) \to 0 \qquad (2-34)$$

The first of these boundary conditions, Eq. (2–33), is inconvenient in its present form. Thus, we take $A = U$ and have

$$\eta = 0 \qquad f(0) = 1 \qquad (2-35)$$

Note that the arbitrariness of A as revealed by the differential equation is used to achieve a simplified boundary condition.

A closed-form analytical solution to the boundary value problem given by Eqs. (2–30), (2–34) and (2–35) is obtained by letting

$$\phi = \frac{df}{d\eta} = f' \qquad (2-36)$$

Upon substitution into Eq. (2–30) we have

$$\phi' + 2\eta\phi = 0 \qquad (2-37)$$

The solution of this ordinary differential equation is (see [Murp60], Eq. 173, p. 327)

$$\phi = f' = C_1 e^{-\eta^2} \qquad (2-38)$$

Hence, after integrating

$$f = C_1 \int_0^\eta e^{-\eta^2} d\eta + C_2 \qquad (2-39)$$

The boundary condition at $\eta = 0$ yields $C_2 = 1$. The boundary condition as $\eta \to \infty$ yields

$$C_1 = \frac{-1}{\int_0^\infty e^{-\eta^2} d\eta} \qquad (2-40)$$

34 Laminar Flow Analysis

The solution is then

$$f(\eta) = 1 - \frac{\int_0^\eta e^{-\eta^2}\,d\eta}{\int_0^\infty e^{-\eta^2}\,d\eta} \qquad (2-41)$$

Evaluation of the integral from zero to infinity yields

$$f(\eta) = 1 - \frac{2}{\sqrt{\pi}} \int_0^\eta e^{-\eta^2}\,d\eta \qquad (2-42)$$

The second term on the right is the error function and $1 - \text{erf}(\eta)$ is the complementary error function. Thus,

$$f(\eta) = \frac{u}{U} = \text{erfc}(\eta) \qquad (2-43)$$

Here, a similarity analysis has been used to obtain a closed-form analytical solution to a problem which previously did not yield to any solution using separation of variables.

Figure 2–5a gives y vs u/U for various values of time t. Note that the extent of the viscous zone increases with increasing time. Due to the action of fluid viscosity, after an infinite time the entire flow field above the plate is moving with the velocity of the plate. Figure 2–5b shows η vs u/U. This figure illustrates that the similarity variables collapse the solutions given in Figure 2–5a into a single solution.

An exact solution to this boundary value problem can be obtained by numerical integration and compared with the analytical solution to illustrate the

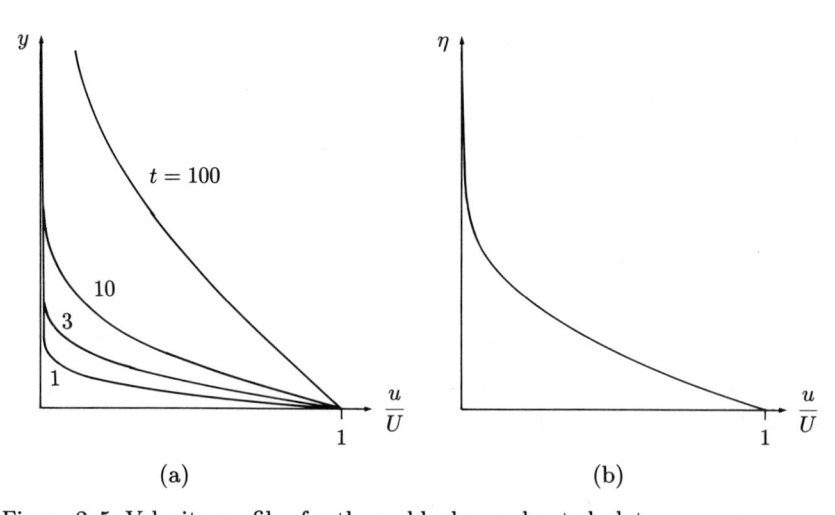

Figure 2–5. Velocity profiles for the suddenly accelerated plate.

accuracy of numerical integration techniques. Having demonstrated the utility of similarity analysis, we proceed to apply this technique to problems which generally do not yield closed-form analytical solutions (see Hansen [Hans64]).

2–7 Two-dimensional Stagnation Point Flow

Consider the two-dimensional steady flow of an incompressible viscous nonheat-conducting fluid impinging on a plate perpendicular to the flow direction (see Figure 2–6). Assume that the flow at a large distance above the plate is given by the corresponding inviscid (potential) flow and that no-slip conditions prevail at the plate surface.

First, consider the inviscid flow solution. We take the plate to be at $y = 0$ and the stagnation point at $x = 0$, $y = 0$ (see Figure 2–6). The flow is impinging on the plate from the positive y direction. Under these circumstances the potential flow solution yields the following expressions for the stream and velocity potential functions (see [Rose63], p. 155)

$$\psi_i = K\,xy \qquad (2-44)$$

$$\phi_i = \frac{K}{2}(x^2 - y^2) \qquad (2-45)$$

where the i subscript indicates the inviscid flow solution. Differentiation yields U_i and V_i, the x and y components of the inviscid flow velocity, i.e.,

$$U_i = \frac{\partial \psi_i}{\partial y} = K\,x \qquad (2-46)$$

$$V_i = -\frac{\partial \psi_i}{\partial x} = -K\,y \qquad (2-47)$$

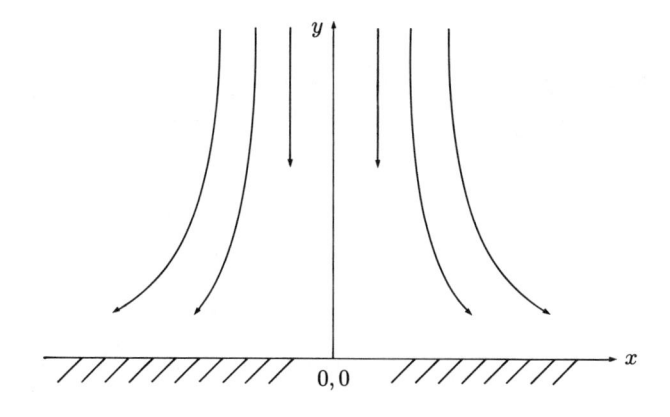

Figure 2–6. Two-dimensional stagnation point flow.

In a potential flow the incompressible Bernoulli equation is applicable and is used to obtain a relationship between the pressure at any point in the flow and the stagnation point. Thus, we have

$$P_0 - P_i = \frac{1}{2}\rho(U_i^2 + V_i^2) = \frac{1}{2}K^2(x^2 + y^2) \qquad (2-48)$$

where the zero subscript indicates the stagnation point.

Returning to the viscous flow case, we have that the governing equations for the dynamics of the two-dimensional steady flow of a viscous incompressible constant property fluid are

continuity

$$\frac{\partial u}{\partial x} + \frac{\partial v}{\partial y} = 0 \qquad (1-49)$$

momentum

$$u\frac{\partial u}{\partial x} + v\frac{\partial u}{\partial y} = -\frac{1}{\rho}\frac{\partial P}{\partial x} + \nu\left(\frac{\partial^2 u}{\partial x^2} + \frac{\partial^2 u}{\partial y^2}\right) \qquad (1-50a)$$

$$u\frac{\partial v}{\partial x} + v\frac{\partial v}{\partial y} = -\frac{1}{\rho}\frac{\partial P}{\partial y} + \nu\left(\frac{\partial^2 v}{\partial x^2} + \frac{\partial^2 v}{\partial y^2}\right) \qquad (1-50b)$$

The continuity equation is integrated by introducing a stream function such that

$$u = \frac{\partial \psi}{\partial y} \qquad v = -\frac{\partial \psi}{\partial x} \qquad (2-49)$$

Hence, the momentum equations become

$$\psi_y \psi_{xy} - \psi_x \psi_{yy} = -\frac{1}{\rho}\frac{\partial P}{\partial x} + \nu(\psi_{xxy} + \psi_{yyy}) \qquad (2-50)$$

and

$$\psi_y \psi_{xx} - \psi_x \psi_{xy} = \frac{1}{\rho}\frac{\partial P}{\partial y} + \nu(\psi_{xxx} + \psi_{xyy}) \qquad (2-51)$$

The appropriate boundary conditions at the plate surface, i.e., at $y = 0$, are $u = v = 0$ or $\psi_y = \psi_x = 0$. At the stagnation point, i.e., at $x = 0$, $y = 0$, we have $P = P_0$. Recall that at a large distance above the plate it was assumed that the inviscid flow is recovered. Hence, as $y \to \infty$, $u \to U_i$, $v \to V_i$. Thus, $u \to Kx$ and $v \to -Ky$, and the pressure field approaches the inviscid pressure field given by Eq. (2–48).

Notwithstanding our experience with the suddenly accelerated flat plate, we seek a solution by separation of variables. Thus, we assume

$$\psi(x, y) = F(x)G(y) = KxG(y) \qquad (2-52)$$

where K is a constant. Assuming a solution of this form is essentially equivalent to predetermining the solution as a function of x. This form of the solution is justified on the basis of the required behavior at infinity, i.e., as $y \to \infty$, $u = \psi_y = KxG'(y) \to Kx$. Here the prime denotes differentiation with respect to the appropriate argument. Our result also implies that as $y \to \infty$, $G'(y) \to 1$. Since Eqs. (2–50) and (2–51) represent two equations in the two dependent variables ψ and P, we must also make an assumption about the form of the viscous pressure field. Assuming that the functional form of the viscous pressure field is closely approximated by that for the inviscid pressure field, we take[†]

$$P_0 - P = \frac{1}{2}\rho K^2 (x^2 + F(y)) \qquad (2-53)$$

Substitution of Eqs. (2–52) and (2–53) into the governing differential equations Eqs. (2–50) and (2–51) yields

$$K^2 x G' G' - K^2 x GG'' = K^2 x + \nu K x G'''$$

and
$$-K^2 GG' = -\frac{K^2}{2} F' + \nu K G''$$

where here the prime denotes differentiation with respect to the argument y. Dividing the first of these equations through by $K^2 x$ and the second by $-K^2$ yields

$$G'G' - GG'' = 1 + \frac{\nu}{K} G''' \qquad (2-54)$$

and
$$GG' = \frac{1}{2} F' - \frac{\nu}{K} G''' \qquad (2-55)$$

Immediately we note that the governing partial differential equations have been reduced to ordinary differential equations. Since no equation which is a function of x results, we know that we have correctly chosen the functional form with respect to x in Eqs. (2–52) and (2–53).

From Eq. (2–52) the boundary conditions at $y = 0$ are

$$G(0) = G'(0) = 0 \qquad (2-56)$$

[†]The rationale for this assumption is illustrated by assuming

$$P_0 - P = \frac{1}{2}\rho(u^2 + v^2)$$

Substituting for u and v as obtained from Eq. (2–52), we have

$$P_0 - P = \frac{1}{2}\rho K^2 [(xG')^2 + G^2]$$

Realizing that as $y \to \infty$ $G' \to 1$, we finally assume for simplicity that the pressure field has the form given by Eq. (2–53). However, note that $F(y)$ is taken as an arbitrary function of y for greater generality.

38   Laminar Flow Analysis

At $x = 0$, $y = 0$, i.e., at the stagnation point, from Eq. (2–53) we have

$$P_0 - P = \frac{1}{2}\rho K\bigl(F(0)\bigr) = 0$$

Since at $x = 0, y = 0$, $P = P_0$

$$F(0) = 0 \qquad (2-57)$$

Recalling that at a large distance above the plate it is assumed that the inviscid flow solution is recovered, we write

$$u = \psi_y = KxG' \to U \to Kx$$

Hence as $y \to \infty$

$$G'(y) \to 1 \qquad (2-58)$$

Although an analytical solution of the two-point asymptotic boundary value problem given by Eqs. (2–54) to (2–58) does not presently exist, exact numerical solutions can be obtained. However, to obtain any insight into the results, it is necessary to parameterize the solutions with respect to ν/K. This is inconvenient and also expensive in terms of computation time. Hence, before proceeding further we look for affine stretching transformations for the dependent and independent variables which remove the factor ν/K, i.e., we nondimensionalize the equations. Thus, we let

$$G(y) = \alpha f(\eta) \qquad (2-59a)$$

$$F(y) = \gamma g(\eta) \qquad (2-59b)$$

$$\eta = \beta y \qquad (2-59c)$$

where α, γ, and β are nonzero constants.

Substituting into Eqs. (2–54) and (2–55) yields

$$f'^2 - ff'' = \frac{1}{(\alpha\beta)^2} + \frac{\nu}{K}\frac{\beta}{\alpha}f''' \qquad (2-60)$$

$$ff' = \frac{1}{2}\frac{\gamma}{\alpha^2}g' - \frac{\nu}{K}\frac{\beta}{\alpha}f'' \qquad (2-61)$$

Before choosing α, γ and β we look at the boundary conditions. The reason is that a particular choice of α, γ and β which eliminates ν/K from the differential equation may result in boundary conditions involving ν/K. Under these circumstances, the solution requires parameterization with respect to the boundary conditions. Transformation of Eqs. (2–56) to (2–58) at $\eta = 0$ yields

$$f(0) = f'(0) = g(0) = 0 \qquad (2-62)$$

and as $\eta \to \infty$

$$f'(\eta) = \frac{1}{\alpha\beta} \qquad (2-63)$$

From an examination of Eqs. (2–60), (2–61) and (2–63) we see that an appropriate choice is $\alpha\beta = 1$, $\beta/\alpha = K/\nu$ and $\gamma = 2\alpha^2$. After solving for α, β, γ we have

$$\beta = \left(\frac{K}{\nu}\right)^{1/2} \qquad \alpha = \left(\frac{\nu}{K}\right)^{1/2} \qquad \gamma = 2\frac{\nu}{K} \qquad (2-64)$$

and the governing two-point asymptotic boundary value problem is

$$f''' + f'' + (1 - f'^2) = 0 \qquad (2-65)$$
$$g' = f'' + ff' \qquad (2-66)$$

with boundary conditions

$$\eta = 0 \quad f(0) = f'(0) = g(0) = 0 \qquad (2-67)$$
$$\eta \to \infty \quad f'(\eta) \to 1 \qquad (2-68)$$

Here, we see that not only are the momentum equations uncoupled from the energy equation but in addition the x and y momentum equations, Eqs. (2–65) and (2–66), respectively, are also uncoupled. Thus, we can solve the x momentum equation, Eq. (2–65), independently of the y momentum equation, Eq. (2–66), and subsequently use the solution of Eq. (2–65) in obtaining that of Eq. (2–66). That is, we use the values of $f'(\eta)$ and $f''(\eta)$ obtained from a solution of Eq. (2–65) in Eq. (2–66) to solve for $g(\eta)$.

Equation (2–66) can be directly integrated to yield

$$g = \frac{f^2}{2} + f' + \text{constant}$$

From the boundary conditions given in Eq. (2–56) we see that the constant of integration is zero, hence

$$g = \frac{f^2}{2} + f' \qquad (2-69)$$

No known closed-form analytical solution of the remaining two-point asymptotic boundary value problem is available. Thus, we look for an exact numerical solution. The procedure is to seek an exact solution of Eq. (2-65) and to subsequently use that solution to obtain $g(\eta)$.

Equation (2–65) is a third-order nonlinear[†] ordinary differential equation. Numerical integration of this equation requires a knowledge of $f(0)$, $f'(0)$ and $f''(0)$

[†]A differential equation is nonlinear if powers and/or products of the dependent variable and/or its derivatives occur; e.g., $y(dy/dx) + y = 0$ is nonlinear, while $x(dy/dx) + y = 0$ is linear.

to start the integration. However, from the given boundary conditions only $f(0)$ and $f'(0)$ are known. The third required boundary condition is specified at infinity. The procedure is to estimate the unknown value of $f''(0)$ and then perform the numerical integration out to some large value of η which we call $\eta_{\max}$. $\eta_{\max}$ is taken to be equivalent to infinity. When the integration has proceeded to $\eta_{\max}$ the value of $f'(\eta_{\max})$ is compared with the required asymptotic value of one. If $f'(\eta_{\max})$ is within some specified small value of one then the outer boundary condition is said to be satisfied and we have a solution of the governing two-point asymptotic boundary value problem. If not, we estimate a new value of $f''(0)$ and repeat the procedure. Since $\eta_{\max} \neq \infty$, $f'(\eta_{\max})$ cannot equal one precisely. Thus, we consider the outer boundary condition to be satisfied if

$$f'(\eta_{\max}) = 1 \pm \epsilon_1 \qquad (2-70)$$

where ϵ_1 is some small quantity, say 5×10^{-7}. This is the so-called 'shooting' method.

2-8 Iteration Scheme

In general the first estimate for $f''(0)$ does not yield a solution. Arbitrary guessing of subsequent estimates of $f''(0)$, of course, proves to be quite inefficient. Hence, a logical method of determining the new estimates for $f''(0)$ must be used. The Newton-Raphson method (see Appendix B) is frequently used for estimating the unknown gradients needed to obtain numerical solutions to linear and nonlinear differential equations. Although the Newton-Raphson iteration scheme assures convergence to the required outer boundary condition at $\eta_{\max}$, it does not insure asymptotic convergence to the specified outer boundary condition required for this two-point asymptotic boundary value problem. Before continuing the discussion, recall that asymptotic convergence implies that as $f'(\eta) \to 1$ its first derivative approaches zero, i.e., $f''(\eta) \to 0$ as $\eta \to \eta_{\max}$. Considering the Newton-Raphson iteration scheme in this context reveals that it does not insure asymptotic convergence; in fact, the Newton-Raphson iteration scheme might be satisfied by $f'(\eta_{\max}) = 1 + \epsilon$ and (say) $f''(\eta_{\max}) = 0.5$. Hence, at $\eta_{\max} + \Delta\eta$, $f'(\eta)$ would not satisfy the outer boundary condition. In addition, if the initial estimate of $f''(0)$ is very far from the correct value the solutions tend to diverge. Under these circumstances the derivatives required in the Newton-Raphson method cannot be calculated in any meaningful manner.

In order to eliminate the problems associated with the Newton-Raphson iteration scheme, the Nachtsheim-Swigert [Nach65] iteration scheme is used. This technique is fully discussed in Appendix B. In summary the Nachtsheim-Swigert iteration scheme is so structured that asymptotic convergence to the correct outer boundary conditions is assured. This is accomplished by imposing the additional condition that

$$f''(\eta_{\max}) = \epsilon_2 \qquad (2-71)$$

2-9 Numerical Solution

The results of a typical run of the stag2d program described in Section D-12 are shown below.

```
f"(0) = 1.25
η       f"          f'          f
0       1.25        0           0
6.      0.222783    1.57637     6.42524

f"(0) = 1.251
0       1.251       0           0
6.      0.237166    1.61137     6.48878

f"(0) = 1.23368
0       1.23368     0           0
6.      1.24093e-2  1.03421     5.41737

f"(0) = 1.23272
0       1.23272     0           0
6.      1.45136e-3  1.00402     5.35978

f"(0) = 1.2326
0       1.2326      0           0
6.      1.75541e-4  1.00049     5.35303

f"(0) = 1.23259
0       1.23259     0           0
6.      2.1323e-5   1.00006     5.35221

f"(0) = 1.23259
0       1.23259     0           0
6.      2.60013e-6  1.00001     5.35211

f"(0) = 1.23259
0       1.23259     0           0
6.      3.25885e-7  1.          5.3521

Convergence achieved
```

Table 2-1. Solutions for two-dimensional stagnation point flow.

η	f''	f'	f	g
0	1.232588	0	0	0
0.2	1.03445	0.226612	2.33223e−2	0.226884
0.4	0.846325	0.414456	8.80566e−2	0.418333
0.6	0.675171	0.566281	0.186701	0.583709
0.8	0.525131	0.685937	0.312423	0.734742
1.	0.398013	0.777865	0.459227	0.88331
1.2	0.293776	0.846671	0.622028	1.04013
1.4	0.211003	0.896809	0.796652	1.21414
1.6	0.147351	0.932348	0.979779	1.41233
1.8	9.99638e−2	0.956834	1.16886	1.63995
2.	6.58254e−2	0.973217	1.36197	1.9007
2.2	4.20396e−2	0.983853	1.55776	2.19716
2.4	2.60203e−2	0.990549	1.75525	2.53101
2.6	1.55973e−2	0.994634	1.95381	2.90331
2.8	9.04887e−3	0.997046	2.153	3.31474
3.	5.07797e−3	0.998424	2.35256	3.76569
3.2	2.7549e−3	0.999186	2.55233	4.25637
3.4	1.44421e−3	0.999593	2.75221	4.78692
3.6	7.31269e−4	0.999803	2.95215	5.3574
3.8	3.57497e−4	0.999908	3.15212	5.96784
4.	1.6868e−4	0.999958	3.35211	6.61828
4.2	7.67926e−5	0.999982	3.5521	7.3087
4.4	3.37239e−5	0.999992	3.7521	8.03912
4.6	1.42847e−5	0.999997	3.9521	8.80954
4.8	5.83719e−6	0.999999	4.1521	9.61997
5.	2.30344e−6	1.	4.3521	10.4704
5.2	8.80753e−7	1.	4.5521	11.3608
5.4	3.29701e−7	1.	4.7521	12.2912
5.6	1.24507e−7	1.	4.9521	13.2616
5.8	5.12017e−8	1.	5.1521	14.2721
6.	2.623e−8	1.	5.3521	15.3225

Here the initial guess for the unknown initial condition $f''(0)$ for the x momentum equation (Eq. 2–65) is taken as 1.25.[†] After perturbing the initial guess for $f''(0)$ in order to calculate the Nachtsheim-Swigert iteration derivatives at the edge of the boundary layer, convergence within the required error values of $f''(\eta_{max}) = \pm 1 \times 10^{-6}$ and $f'(\eta_{max}) - 1 = \pm 1 \times 10^{-6}$ occurs after six iterations.

Solution of the y momentum equation is obtained by using these results in Eq. (2–69). The results are tabulated in Table 2–1 and shown in Figure 2–7.

[†]Since no prior knowledge of the solution of equations of this type was available, the value of $f''(0) = 1.25$ was arrived at by initially limiting the value of η_{max} as discussed in Appendix B.

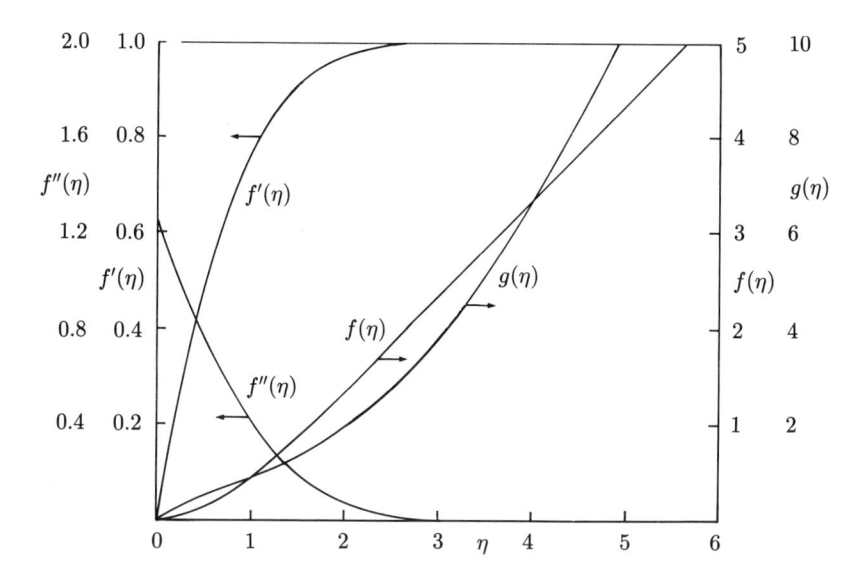

Figure 2-7. Nondimensional results for two-dimensional stagnation point flow.

The last two iterations are of interest. Notice that the same value is printed in each case. This implies that the correction to $f''(0)$ between iterations is less than 5×10^{-6}. Hence, convergence to the correct outer boundary condition is very sensitive to the value of $f''(0)$.

Figure 2-8 shows the behavior of the solutions, i.e., $f'(\eta)$ vs η, for various initial estimates of $f''(0)$ for a fixed value of $\eta_{\max} = 6$. Here we see that unless the initial estimate of $f''(0)$ is close to the actual value the solutions diverge radically. If an attempt is made to use these values in calculating the derivatives required in the Nachtsheim-Swigert iteration technique, the results are meaningless. Convergence to the correct boundary conditions does not occur, or at best occurs very slowly. However, Figure 2-8 suggests a method (see Appendix B and Nachtsheim and Swigert [Nach65]) which minimizes this problem. By initially limiting the value of $\eta_{\max}$ to, say, 2, the solutions do not diverge, and hence the calculated derivatives can then be used in the Nachtsheim-Swigert iteration scheme to generate more realistic initial estimates for $f''(0)$. However, asymptotic convergence to the required accuracy cannot, in general, be obtained for this small value of $\eta_{\max}$. Therefore, these initial iterations only serve to yield more reasonable estimates of $f''(0)$. When this is accomplished, usually within two or three iterations, the value of $\eta_{\max}$ is increased and new Nachtsheim-Swigert derivatives calculated. With these new initial estimates for $f''(0)$, convergence to the correct value of $f''(0)$ is quite rapid. Although the stag2d program assumes that the Nachtsheim-Swigert derivatives are constant, additional program sophistication allows calculation of the Nachtsheim-Swigert derivatives using the most recent values of $f'(\eta_{\max})$ and $f''(\eta_{\max})$. Convergence

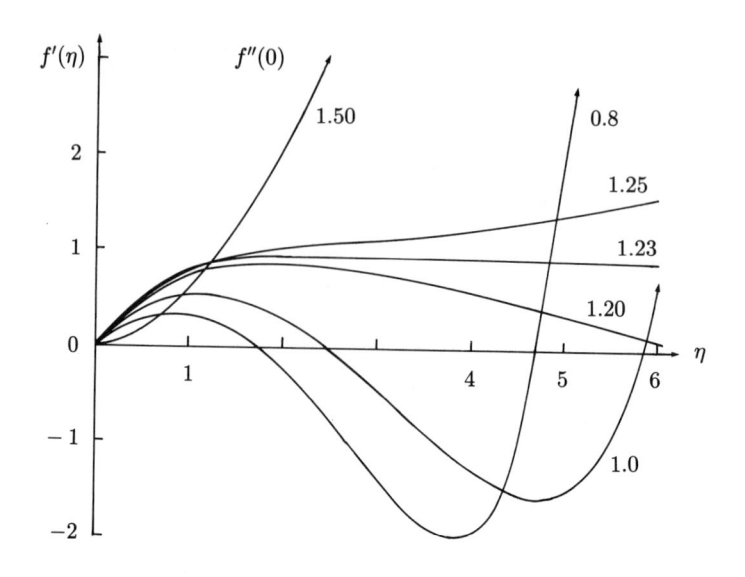

Figure 2–8. Effect of $f''(0)$ on $f'(\eta)$.

is somewhat improved in this case. These modifications are left as an exercise for the reader (see Problem 2–5).

2–10   Axisymmetric Stagnation Point Flow

The previously obtained solution for two-dimensional stagnation point flow can be extended to the case of an axisymmetric stream impinging on a plane wall. The solution obtained is representative of that near the forward stagnation point of an axisymmetric blunt body. Figure 2–9 illustrates the problem under discussion. Here x and y are the radial and axial directions, respectively, with u and v the velocity components in the x and y directions, respectively. The plane is assumed to be perpendicular to the flow direction, with the stagnation point as the center of the coordinate system. The momentum equations for steady axisymmetric incompressible flow with constant properties are obtained by transforming Eqs. (1–49) and (1–50) into cylindrical polar coordinates. They are then (see Goldstein [Gold38], p. 143)

$$u\frac{\partial u}{\partial x} + v\frac{\partial u}{\partial y} = -\frac{1}{\rho}\frac{\partial P}{\partial x} + \nu\left(\frac{\partial^2 u}{\partial x^2} + \frac{1}{x}\frac{\partial u}{\partial x} - \frac{u}{x^2} + \frac{\partial^2 u}{\partial y^2}\right) \quad (2-72)$$

$$u\frac{\partial v}{\partial x} + v\frac{\partial v}{\partial y} = -\frac{1}{\rho}\frac{\partial P}{\partial y} + \nu\left(\frac{\partial^2 v}{\partial x^2} + \frac{1}{x}\frac{\partial v}{\partial x} + \frac{\partial^2 v}{\partial y^2}\right) \quad (2-73)$$

The axisymmetric continuity equation is

$$\frac{\partial}{\partial x}(xu) + \frac{\partial}{\partial y}(xv) = 0 \quad (2-74)$$

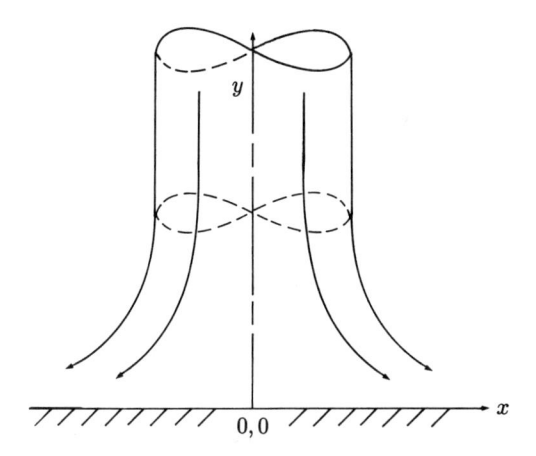

Figure 2-9. Axisymmetric stagnation point flow.

The boundary conditions are again taken to be no-slip at the plate surface and no-mass-transfer through the plate surface. Further, it is required that the inviscid solution be recovered at a large distance from the body. For axisymmetric inviscid stagnation point flow, the velocity components are given by (see Moore [Moor64], p. 79)

$$U_i = Kx \qquad (2-75)$$

$$V_i = -2Ky \qquad (2-76)$$

and the pressure distribution by

$$P_0 - P_i = \frac{1}{2}\rho K^2 (x^2 + 4y^2) \qquad (2-77)$$

where the zero subscript indicates the stagnation point at $x = y = 0$.[†]

The boundary conditions at the surface become $u = v = 0$ at $y = 0$. Further, at the stagnation point, $y = 0$, $x = 0$, $P = P_0$. At a large distance from the plate, i.e., as $y \to \infty$, $u \to U_i \to Kx$, $v \to V_i \to -2Ky$.

The continuity equation is automatically satisfied by introducing a stream function of the form

$$u = \frac{1}{x}\frac{\partial \psi}{\partial y} \qquad v = -\frac{1}{x}\frac{\partial \psi}{\partial x} \qquad (2-78)$$

[†]Alternatively, we may take $U_i = Kx/2$, $V_i = -Kx$, in which case the pressure distribution is

$$P_0 - P_i = \frac{1}{2}\rho K^2\left[\left(\frac{x}{2}\right)^2 + y^2\right]$$

46 Laminar Flow Analysis

Introducing the stream function into the momentum equations yields

$$\frac{1}{x^2}\psi_y\psi_{xy} - \frac{1}{x^2}\psi_x\psi_{yy} - \frac{1}{x^3}\psi_y^2 =$$
$$-\frac{1}{\rho}\frac{\partial P}{\partial x} + \nu\left(\frac{1}{x}\psi_{yyy} + \frac{1}{x}\psi_{xxy} - \frac{1}{x^2}\psi_{xy}\right) \quad (2-79)$$

and $\quad \dfrac{1}{x^2}\psi_x\psi_{xy} - \dfrac{1}{x^2}\psi_{xx}\psi_y + \dfrac{1}{x^3}\psi_x\psi_y =$
$$-\frac{1}{\rho}\frac{\partial P}{\partial y} - \nu\left(\frac{1}{x}\psi_{xxx} + \frac{1}{x}\psi_{xyy} + \frac{1}{x^3}\psi_x - \frac{1}{x^2}\psi_{xx}\right) \quad (2-80)$$

By analogy with two-dimensional stagnation point flow, we seek solutions of the viscous momentum equations by separation of variables. To this end we assume that

$$\psi(x,y) = Kx^a G(y) \quad (2-81)$$

where a is an as yet undetermined constant. Choosing this form for $\psi(x,y)$ presupposes that the x dependence is a power law. We again assume that

$$P_0 - P = \frac{1}{2}\rho K^2[x^2 + F(y)] \quad (2-82)$$

Substituting into the momentum equations, the x momentum equation is

$$(a-1)K^2 x^{2a-3}G'^2 - aK^2 x^{2a-3}GG'' =$$
$$K^2 x + \nu\left[Kx^{a-1}G''' + a(a-2)Kx^{a-3}G'\right] \quad (2-83)$$

and the y momentum equation becomes

$$2aK^2 x^{2a-4}GG' = \frac{K^2 F'}{2} - \nu\left[a(a-2)^2 Kx^{a-4}G + aKx^{a-2}G''\right] \quad (2-84)$$

where here the prime denotes differentiation with respect to the argument y. Inspection of Eqs. (2–83) and (2–84) shows that if $a = 2$ these partial differential equations reduce to ordinary differential equations. Thus, after dividing by K^2 and $4K^2$ in the x and y momentum equations, respectively, we have

$$\frac{\nu}{K}G''' + 2GG'' + (1 - G'^2) = 0 \quad (2-85)$$

and
$$GG' = -\frac{F'}{8} - \frac{\nu}{2K}G'' \quad (2-86)$$

Here, note that Eqs. (2–85) and (2–86) are very similar in form to Eqs. (2–54) and (2–55). Again, the equations are nondimensionalized by seeking affine stretching transformations of the form

$$G(y) = \alpha f(\eta) \qquad (2-87a)$$
$$F(y) = \gamma g(\eta) \qquad (2-87b)$$
$$\eta = \beta y \qquad (2-87c)$$

where α, γ, β are nonzero constants. Substituting into Eqs. (2–85) and (2–86) yields

$$f''' + \frac{2K}{\nu}\frac{\alpha}{\beta}ff'' + \frac{K}{\nu}\frac{1}{\alpha\beta^3}\left[1 - (\alpha\beta)^2 f'^2\right] = 0 \qquad (2-88)$$

and

$$ff' = \frac{\gamma}{8}\frac{1}{\alpha^2}g' - \frac{\nu}{2K}\frac{\beta}{\alpha}f'' \qquad (2-89)$$

Before choosing particular values for α, γ and β the boundary conditions must be investigated. The transformed surface boundary conditions are

$$\eta = 0 \qquad f(0) = f'(0) = g(0) = 0 \qquad (2-90a)$$

The transformed boundary condition at infinity is

$$\eta \to \infty \qquad f(\eta) \to \frac{1}{\alpha\beta} \qquad (2-90b)$$

Hence, in order to obtain a simplified boundary condition at infinity we take $\alpha\beta = 1$. Finally, we take

$$\frac{2K}{\nu}\frac{\alpha}{\beta} = 1$$

and

$$\gamma = 8\alpha^2$$

Hence

$$\beta = \left(\frac{2K}{\nu}\right)^{1/2} \qquad \alpha = \left(\frac{\nu}{2K}\right)^{1/2} \qquad \gamma = \frac{4\nu}{K} \qquad (2-91)$$

The governing two-point asymptotic boundary value problem is then given as

$$f''' + ff'' + \frac{1}{2}(1 - f'^2) = 0 \qquad (2-92)$$
$$g' = ff' + f'' \qquad (2-93)$$

with boundary conditions

$$\eta = 0 \qquad f(0) = f'(0) = g(0) = 0 \qquad (2-94)$$
$$\eta \to \infty \qquad f(\eta) \to 1 \qquad (2-95)$$

48   Laminar Flow Analysis

As was the case for two-dimensional stagnation point flow, the transformed y momentum equation is immediately integrable. Hence

$$g(\eta) = \frac{f^2}{2} + f' \qquad (2-96)$$

Comparing Eqs. (2–92) and (2–65), we see that they differ only by the factor of one-half multiplying the last term. The solution of Eq. (2–92) is obtained by the modifications to the **eqmot2d** routine used with **stag2d**. Specifically, eqmot2d becomes

equations of motion — two-dimensional axisymmetric stagnation point flow
subroutine eqmot(eta, x(), param(), f())
f(1) = −x(3)∗x(1) −0.5∗(1 − x(2)∗x(2))
f(2) = x(1)
f(3) = x(2)
return

A typical run using the above modified eqmot2d routine is

f″(0) = 1.23259

η	f″	f′	f
0	1.232588	0	0
6.	0.423879	3.29404	11.6197

f″(0) = 1.233588

| 0 | 1.233588 | 0 | 0 |
| 6. | 0.425169 | 3.30122 | 11.6401 |

f″(0) = 0.912905

| 0 | 0.912905 | 0 | 0 |
| 6. | −0.022935 | 0.881218 | 4.87142 |

f″(0) = 0.929481

| 0 | 0.929481 | 0 | 0 |
| 6. | 2.7727e−3 | 1.01441 | 5.23484 |

f″(0) = 0.92747

| 0 | 0.92747 | 0 | 0 |
| 6. | −3.22936e−4 | 0.998322 | 5.19087 |

f″(0) = 0.927705

| 0 | 0.927705 | 0 | 0 |
| 6. | 3.79112e−5 | 1.0002 | 5.19599 |

f″(0) = 1.23259

f"(0) = 0.927677

0	0.927677	0	0
6.	−4.31852e−6	0.999977	5.19539

f"(0) = 0.92768

0	0.92768	0	0
6.	6.21323e−7	1.	5.19546

f"(0) = 0.92768

0	0.92768	0	0
6.	4.34516e−8	1.	5.19545

Convergence achieved

As shown for two-dimensional stagnation point flow, arbitrary estimation of the value of $f''(0)$ can lead to results for the Nachtsheim-Swigert iteration derivatives, which do not yield convergence to the required outer boundary condition $f'(\eta_{max}) \to 1$. Hence, it is desirable to have some knowledge of an approximate range of values for $f''(0)$. If either an analytical or exact solution of a boundary value problem similar to that under investigation is known, then it is reasonable to use the value of $f''(0)$ obtained for the known solution as the first estimate for the problem under investigation. Considering the similarity of Eqs. (2–92) and (2–65), the first estimate of $f''(0)$ is taken to be that for two-dimensional stagnation point flow, i.e., 1.232588. Convergence to within $f''(\eta_{max}) = \pm 1 \times 10^{-6}$ and $f'(\eta_{max}) - 1 = \pm 1 \times 10^{-6}$ in seven iterations yields a value of $f''(0) = 0.927680$. Note that the correction value for the last iteration is less than 5×10^{-6}. The complete solution including the y momentum equation (see Eq. 2–96) is shown in Table 2–2. The value of $f''(0) = 0.927680$ agrees with that originally calculated by Homann [Homa36].

The nondimensional velocity and pressure function results for axisymmetric stagnation point flow are shown graphically in Figure 2–10. Also shown are the results for two-dimensional stagnation point flow.

From an analysis of the previous assumptions and transformations we see that for both two-dimensional and axisymmetric flow the local velocity components and the stream function are related to their respective inviscid values by

$$\frac{u}{U_i} = f'(\eta) \qquad \frac{v}{V_i} = -\frac{f'(\eta)}{\eta} \qquad \frac{\psi}{\psi_i} = \frac{f(\eta)}{\eta} \qquad (2-97a,b,c)$$

Further, the pressure field for two-dimensional viscous stagnation point flow is

$$P_0 - P = \frac{\rho}{2} K^2 \left(x^2 + \frac{2g(\eta)}{\eta^2} y^2 \right) \qquad (2-98)$$

and that for axisymmetric viscous stagnation point flow

$$P_0 - P = \frac{\rho K^2}{2} \left(x^2 + 4 \frac{2g(\eta)}{\eta^2} y^2 \right) \qquad (2-99)$$

Laminar Flow Analysis

Table 2–2. Solutions for axisymmetric stagnation point flow.

η	f''	f'	f	g
0	0.92768	0	0	0
0.2	0.82771	0.175537	0.017887	0.175697
0.4	0.728152	0.33111	6.88836e−2	0.333483
0.6	0.630002	0.466891	0.149011	0.477993
0.8	0.53477	0.583305	0.254348	0.615651
1.	0.444284	0.681115	0.381092	0.753731
1.2	0.360449	0.761462	0.525629	0.899605
1.4	0.284977	0.825853	0.684612	1.0602
1.6	0.219151	0.876098	0.855027	1.24163
1.8	0.163652	0.914204	1.03424	1.44903
2.	0.1185	0.94225	1.22004	1.6865
2.2	8.30998e−2	0.962255	1.41061	1.95716
2.4	0.056379	0.976069	1.60453	2.26332
2.6	3.69739e−2	0.985294	1.80073	2.60661
2.8	2.34217e−2	0.991248	1.99843	2.9881
3.	1.43227e−2	0.994959	2.19708	3.40854
3.2	8.4508e−3	0.997192	2.39631	3.86835
3.4	4.80892e−3	0.998488	2.59589	4.36782
3.6	2.63824e−3	0.999213	2.79567	4.9071
3.8	1.39496e−3	0.999605	2.99556	5.48628
4.	7.10674e−4	0.999808	3.1955	6.10542
4.2	3.48767e−4	0.99991	3.39547	6.76453
4.4	1.6484e−4	0.999959	3.59546	7.46363
4.6	7.50188e−5	0.999982	3.79546	8.20272
4.8	3.2869e−5	0.999993	3.99545	8.98181
5.	1.38626e−5	0.999997	4.19545	9.80091
5.2	5.62723e−6	0.999999	4.39545	10.66
5.4	2.19832e−6	1.	4.59545	11.5591
5.6	8.26459e−7	1.	4.79545	12.4982
5.8	2.99049e−7	1.	4.99545	13.4773
6.	1.04212e−7	1.	5.19545	14.4964

Referring to Eqs. (2–48) and (2–77) shows that the viscous pressure fields are given by similar modifications of the inviscid pressure fields. Looking at the tabulated results, we see that close to the plate surface $2g(\eta)/\eta^2 > 1$ and hence the local pressure P is larger than in the inviscid case, whereas far from the plate, i.e., for larger values of η, $2g(\eta)/\eta^2 < 1$. Hence, the local pressure P is less than in the inviscid case. Further, the effect is less for axisymmetric than for two-dimensional viscous stagnation point flow.

Returning to the stream function and velocity components, we see from the numerical results that the viscous stream function and the velocity components

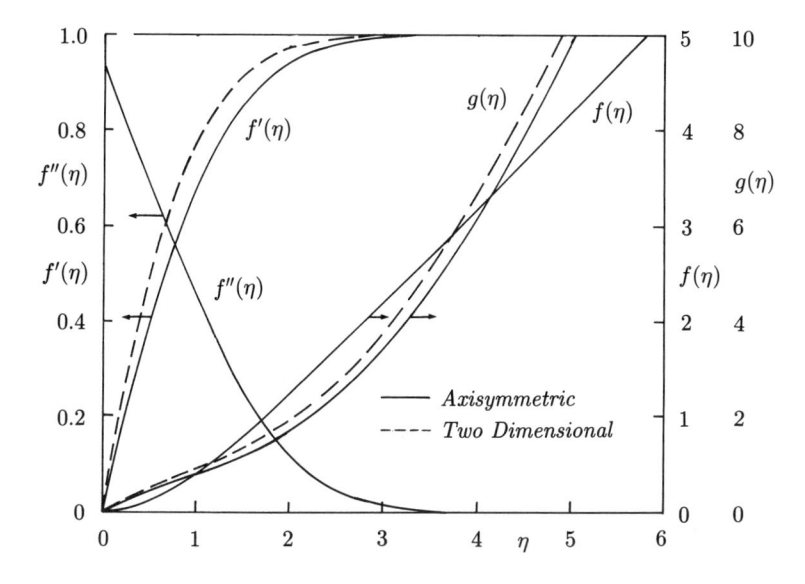

Figure 2–10. Nondimensional results for axisymmetric stagnation point flow.

are always less than the corresponding inviscid results. Further, the effect is greater for axisymmetric than for two-dimensional stagnation point flow.

Finally, we investigate the shearing stress at the wall. From the results given in Chapter 1, we see that the shearing stress at the wall for both axisymmetric and two-dimensional stagnation point flow is given by

$$\tau = \mu \frac{\partial u}{\partial y}\bigg|_{y=0} \qquad (2-100)$$

In transformed coordinates, the shearing stress at the wall for both two-dimensional and axisymmetric stagnation point flow is

$$\tau = \mu K x f''(0) \qquad (2-101)$$

Hence, the shearing stress at the wall is greater for two-dimensional stagnation point flow than for axisymmetric stagnation point flow.

The technique developed in this chapter for solving an asymptotic two-point boundary value problem is useful in succeeding chapters, where solutions for the velocity profiles in laminar boundary layer flows are obtained.

THREE

BOUNDARY LAYER THEORY

3-1 Concept of a Boundary Layer

Referring to the governing equations of a viscous, compressible, heat conducting fluid flow, i.e., the Navier-Stokes equations (see Eqs. 1-1, 1-33, 1-41),[†] we see that many of the terms are multiplied by the dynamic viscosity μ or the thermal conductivity k. For fluids of small viscosity and thermal conductivity, the approximation $\mu = k = 0$ would appear to be reasonable, i.e., we consider an inviscid, nonheat conducting fluid. This approximation reduces the Navier-Stokes equations to Euler's equations (see, for example, Kuethe and Schetzer [Kuet59]). However, Euler's equations are of lower order than the Navier-Stokes equations. Hence, all of the boundary conditions set forth for the Navier-Stokes equations (see Section 1-7) cannot be imposed. Generally the no-slip, no-temperature-jump boundary conditions at the surface are relaxed. The no-slip condition is replaced by the requirement that the surface of the body be a streamline of the flow. The temperature of the fluid at the surface and the surface temperature are also allowed to differ.

Since Euler's equations do not represent the complete problem, it follows that there must be a region of the flow where they do not apply. The boundary conditions relaxed in the Euler boundary value problem are those of no-slip and no-temperature-jump at the surface; thus, it is reasonable to expect that the region in which the Euler equations do not apply is near the surface. Experimental observation confirms this expectation. The layer of fluid in which the effects of viscosity and thermal conductivity are important is known as the boundary layer. The concept of a boundary layer is due to L. Prandtl [Pran28]. Thus, for the purpose of mathematical analysis the flow past a body is divided into two regions: a region 'far' from the surface of the body, in which the effects of viscosity

---

[†] Here, as in Chapter 1, Navier-Stokes equations implies the continuity, momentum and energy equations.

54   Laminar Flow Analysis

and thermal conductivity are negligible, and a region close to the surface, where they are not negligible.

In order to more fully understand the implications of this concept, we again refer to the Navier-Stokes equations and assume that the viscosity and thermal conductivity are small but not negligible. If the terms involving the viscosity and thermal conductivity are not negligible, then it follows that the velocity and thermal gradients in the boundary layer are large. The inclusion of the effects of viscosity and thermal conductivity allows the imposition of the no-slip and no-temperature-jump boundary conditions at the surface. This, along with the division of the flow into an inviscid and a viscous region and the requirement that the velocity and thermal gradients are large, implies that the boundary layer is thin with respect to the scale of the flow.

In general, there is both a velocity and a thermal boundary layer. This is true even if the fluid temperature in the inviscid region is equal to the surface temperature. Referring to the energy equation (Eq. 1–41) provides the necessary explanation. Here, we see that the large velocity gradients associated with the velocity boundary layer generate large thermal gradients through viscous dissipation. Hence, a thermal boundary layer exists. An argument similar to that above shows that, for consistency, the thermal boundary layer is also thin with respect to the scale of the flow.

The concept of a thin boundary layer results in considerable mathematical simplification of the full compressible Navier-Stokes equations.

3–2   Derivation of the Boundary Layer Equations

We now proceed to derive the boundary layer equations for a two-dimensional viscous, compressible, heat conducting fluid. The derivation given below is basically founded in physical arguments. For a derivation based on more mathematically elegant arguments, the reader is referred to Stewartson [Stew64].

Prandtl's boundary layer concept is fundamentally founded on the idea of a *thin* layer. An order-of-magnitude analysis is used to neglect certain terms in the compressible Navier-Stokes equations.[†] If δ, the boundary layer thickness, is small with respect to some characteristic length ℓ associated with the physical problem, then $\bar{\delta} = \delta/\ell << 1$, and terms of order $\bar{\delta}$, $O(\bar{\delta})$, can be neglected with respect to terms of order one, $O(1)$. If $\bar{\delta}$ is at most of $O(0.1)$ then the sequence of order-of-magnitudes is[‡]

$$O(\bar{\delta}^2),\; O(\bar{\delta}),\; O(1),\; O(1/\bar{\delta})\; O(1/\bar{\delta}^2)$$

[†]Prandtl's original derivation was based on an incompressible fluid.

[‡]An order-of-magnitude is an approximation of the size of a quantity or parameter. A parameter which is of the order of unity, $O(1)$, is generally considered to lie between 0.3 and 3.0. Similarly, a parameter of order-of-magnitude one tenth, $O(0.1)$, generally lies between 0.03 and 0.3. These order ranges are based on the fact that 10^n is the approximate center of the logarithmic decade lying between $3 \times 10^{n-1}$ and 3×10^n. For example, $\ln 0.3 \approx -1.12$, $\ln 1.0 = 0$ and $\ln 3.0 \approx 1.1$.

First, the compressible Navier-Stokes equations are nondimensionalized by introducing the following transformations

$$\bar{x} = \frac{x}{\ell} \qquad \bar{y} = \frac{y}{\ell} \qquad \bar{t} = \frac{tU_e}{\ell}$$

$$\bar{u} = \frac{u}{U_e} \qquad \bar{v} = \frac{v}{U_e} \qquad \bar{h} = \frac{h}{h_e} \qquad M_e = \frac{U_e}{a_e}$$

$$\bar{P} = \frac{P}{\rho_e U_e^2} \qquad \bar{\rho} = \frac{\rho}{\rho_e} \qquad \bar{\mu} = \frac{\mu}{\mu_e} \tag{3-1}$$

For convenience we consider two-dimensional flow and neglect surface curvature effects.[†] This implies that the centripetal forces are neglected.

The continuity equation becomes

$$\frac{\partial \bar{\rho}}{\partial \bar{t}} + \frac{\partial}{\partial \bar{x}}(\bar{\rho}\bar{u}) + \frac{\partial}{\partial \bar{y}}(\bar{\rho}\bar{v}) = 0 \tag{3-2}$$

The x and y momentum equations (Eqs. 1–33) become

$$\frac{\partial \bar{u}}{\partial \bar{t}} + \bar{u}\frac{\partial \bar{u}}{\partial \bar{x}} + \bar{v}\frac{\partial \bar{u}}{\partial \bar{y}} = \bar{X} - \frac{1}{\bar{\rho}}\frac{\partial \bar{P}}{\partial \bar{x}}$$

$$+ \frac{1}{Re_\ell}\frac{1}{\bar{\rho}}\frac{\partial}{\partial \bar{x}}\left[\bar{\mu}\left\{2\frac{\partial \bar{u}}{\partial \bar{x}} - \frac{2}{3}\left(\frac{\partial \bar{u}}{\partial \bar{x}} + \frac{\partial \bar{v}}{\partial \bar{y}}\right)\right\}\right]$$

$$+ \frac{1}{Re_\ell}\frac{1}{\bar{\rho}}\frac{\partial}{\partial \bar{y}}\left[\bar{\mu}\left(\frac{\partial \bar{u}}{\partial \bar{y}} + \frac{\partial \bar{v}}{\partial \bar{x}}\right)\right] \tag{3-3}$$

and

$$\frac{\partial \bar{v}}{\partial \bar{t}} + \bar{u}\frac{\partial \bar{v}}{\partial \bar{x}} + \bar{v}\frac{\partial \bar{v}}{\partial \bar{y}} = \bar{Y} - \frac{1}{\bar{\rho}}\frac{\partial \bar{P}}{\partial \bar{y}}$$

$$+ \frac{1}{Re_\ell}\frac{1}{\bar{\rho}}\frac{\partial}{\partial \bar{y}}\left[\bar{\mu}\left\{2\frac{\partial \bar{v}}{\partial \bar{y}} - \frac{2}{3}\left(\frac{\partial \bar{u}}{\partial \bar{x}} + \frac{\partial \bar{v}}{\partial \bar{y}}\right)\right\}\right]$$

$$+ \frac{1}{Re_\ell}\frac{1}{\bar{\rho}}\frac{\partial}{\partial \bar{x}}\left[\bar{\mu}\left(\frac{\partial \bar{u}}{\partial \bar{y}} + \frac{\partial \bar{v}}{\partial \bar{x}}\right)\right] \tag{3-4}$$

where

$$\bar{X} = \frac{X\ell}{U_e^2} \qquad \bar{Y} = \frac{Y\ell}{U_e^2} \tag{3-5}$$

An order-of-magnitude analysis seeks to estimate the size (order-of-magnitude) of the parameters involved in a problem, or the various terms in an equation. The parameters or terms of a certain order-of-magnitude and smaller are then neglected with respect to those of larger order-of-magnitude. This is one method of separating the primary and secondary effects in a physical problem. Such a procedure generally results in a simplification of the mathematical model used to analyze a physical problem.

[†] For a discussion of curvature effects see Rosenhead [Rose63], p. 201.

56   Laminar Flow Analysis

and
$$\text{Re}_\ell = \frac{\rho_e U_e \ell}{\mu_e}$$

The energy equation (Eq. 1–41) becomes

$$\frac{\partial \bar{h}}{\partial \bar{t}} + \bar{u}\frac{\partial \bar{h}}{\partial \bar{x}} + \bar{v}\frac{\partial \bar{h}}{\partial \bar{y}} = (\gamma-1)\text{M}_e^2\left(\frac{\partial \bar{P}}{\partial \bar{t}} + \bar{u}\frac{\partial \bar{P}}{\partial \bar{x}} + \bar{v}\frac{\partial \bar{P}}{\partial \bar{y}}\right) + \bar{Q} + \frac{(\gamma-1)}{\text{Re}_\ell}\text{M}_e^2\bar{\nu}\bar{\Phi}$$

$$+ \frac{1}{\text{Re}_\ell}\frac{1}{\text{Pr}}\frac{1}{\bar{\rho}}\left[\frac{\partial}{\partial \bar{x}}\left(\bar{\mu}\frac{\partial \bar{h}}{\partial \bar{x}}\right) + \frac{\partial}{\partial \bar{y}}\left(\bar{\mu}\frac{\partial \bar{h}}{\partial \bar{y}}\right)\right] \quad (3-6)$$

where
$$\bar{Q} = \frac{Q\ell}{U_e h_e} \quad (3-7)$$

$$\bar{\Phi} = 2\left[\left(\frac{\partial \bar{u}}{\partial \bar{x}}\right)^2 + \left(\frac{\partial \bar{v}}{\partial \bar{y}}\right)^2\right] - \frac{2}{3}\left(\frac{\partial \bar{u}}{\partial \bar{x}} + \frac{\partial \bar{v}}{\partial \bar{y}}\right)^2 + \left(\frac{\partial \bar{v}}{\partial \bar{x}} + \frac{\partial \bar{u}}{\partial \bar{y}}\right)^2 \quad (3-8)$$

and the Prandtl number, $\text{Pr} = \mu c_p/k$, is assumed constant in arriving at the above results.

We now proceed to estimate the order-of-magnitude of each term in Eqs. (3–2) to (3–4) and Eq. (3–6). Since u increases from a value of zero at the wall to U_e at the outer edge of the boundary layer, it is reasonable to consider $\bar{u} = \text{O}(1)$. A similar argument yields $\bar{T} = \text{O}(1)$ and $\bar{\rho} = \text{O}(1)$. Since the above arguments are valid at any station along the surface, it is reasonable to assume that

$$\frac{\partial \bar{u}}{\partial \bar{x}} = \text{O}(1) \quad \text{and} \quad \frac{\partial \bar{T}}{\partial \bar{x}} = \text{O}(1)$$

Thus, we see that differentiation with respect to $\bar{x}$ does not change the order-of-magnitude of the term. Alternatively, we say that the operator $\partial/\partial \bar{x} = \text{O}(1)$.

From the continuity equation (Eq. 3–2) we have

$$\frac{\partial}{\partial \bar{y}}(\bar{\rho}\bar{v}) = \text{O}(1)$$

or since $\bar{\rho} = \text{O}(1)$
$$\frac{\partial \bar{v}}{\partial \bar{y}} = \text{O}(1) \quad (3-10)$$

Integrating between the surface and the edge of the boundary layer yields

$$\bar{v} = \int_0^{\bar{y}} \frac{\partial \bar{v}}{\partial \bar{y}} d\bar{y} \approx \int_0^{\bar{\delta}} \text{O}(1) d\bar{y} = \text{O}(\bar{\delta}) \quad (3-11)$$

where in performing the integration $\partial \bar{v}/\partial \bar{y}$ was replaced by its average value which is of O(1). Equations (3–10) and (3–11) taken together show that the operator $\partial/\partial \bar{y} = \text{O}(1/\bar{\delta})$.

Derivation of the Boundary Layer Equations

The order-of-magnitude of the pressure gradient term in the x momentum equation is estimated from Euler's equations applied to the inviscid flow at the edge of the boundary layer. Since $\bar{v} = O(\bar{\delta})$, $\bar{v} \ll \bar{u}$ and Euler's equations in nondimensional form reduce to

$$\bar{u}\frac{\partial \bar{u}}{\partial \bar{x}} = -\frac{1}{\bar{\rho}}\frac{\partial \bar{P}}{\partial \bar{x}} = O(1) \qquad (3-12)$$

Hence, $\partial \bar{P}/\partial \bar{x} = O(1)$.

For many fluids and most gases, the dynamic viscosity is essentially independent of pressure and hence only a function of the temperature. For liquids the viscosity varies inversely as the temperature, whereas for gases it varies directly as the temperature. For gases, the maximum temperature in the boundary layer is proportional to the square of the Mach number, whereas for liquids the maximum temperature in the boundary layer is proportional to the external velocity squared (product of Prandtl number and Eckert number). Hence, except for large velocities or Mach numbers it is reasonable to assume that $\bar{\mu} = O(1)$.

Writing out the nondimensional x momentum equation (Eq. 3–3) and indicating the order-of-magnitude of each term yields

$$\underset{O(1)}{\frac{\partial \bar{u}}{\partial \bar{t}}} + \underset{O(1)O(1)}{\bar{u}\frac{\partial \bar{u}}{\partial \bar{x}}} + \underset{O(\bar{\delta})O(\frac{1}{\bar{\delta}})}{\bar{v}\frac{\partial \bar{u}}{\partial \bar{y}}} = \underset{O(1)}{\bar{X}} - \underset{O(1)O(1)}{\frac{1}{\bar{\rho}}\frac{\partial \bar{P}}{\partial \bar{x}}}$$

$$+ \frac{1}{\mathrm{Re}_\ell}\frac{1}{\bar{\rho}}\left[\underset{O(1)O(1)}{\bar{\mu}\left(\frac{4}{3}\frac{\partial^2 \bar{u}}{\partial \bar{x}^2}\right.} + \underset{O(1)}{\frac{1}{3}\frac{\partial^2 \bar{v}}{\partial \bar{x}\partial \bar{y}}} + \underset{O(\frac{1}{\bar{\delta}^2})}{\left.\frac{\partial^2 \bar{u}}{\partial \bar{y}^2}\right)}\right.$$

$$+ \frac{2}{3}\underset{O(1)}{\left(2\frac{\partial \bar{u}}{\partial \bar{x}}\right.} - \underset{O(1)\,O(1)}{\frac{\partial \bar{v}}{\partial \bar{y}}\left.\right)\frac{\partial \bar{\mu}}{\partial \bar{x}}} + \underset{O(\frac{1}{\bar{\delta}})}{\left(\frac{\partial \bar{u}}{\partial \bar{y}}\right.} + \underset{O(\bar{\delta})\,O(\frac{1}{\bar{\delta}})}{\left.\frac{\partial \bar{v}}{\partial \bar{x}}\right)\frac{\partial \bar{\mu}}{\partial \bar{y}}}\left.\right] \qquad (3-13)$$

Looking at the term in the square brackets on the right side of the equation we see that the terms $\bar{\mu}(\partial^2 \bar{u}/\partial \bar{y}^2)$ and $(\partial \bar{u}/\partial \bar{y})(\partial \bar{\mu}/\partial \bar{y})$ are of $O(1/\bar{\delta}^2)$. Hence, we may neglect all the other terms in the square brackets with respect to these terms. We further note that all the other terms in the equation, e.g., $\bar{u}(\partial \bar{u}/\partial \bar{x})$, are of order one, $O(1)$. For consistency, the product of $1/\mathrm{Re}_\ell$ and the remaining terms in the square bracket must be of order one. Thus

$$\mathrm{Re}_\ell = O\left(\frac{1}{\bar{\delta}^2}\right) \qquad (3-14)$$

or

$$\bar{\delta} = O\left(\frac{1}{\sqrt{\mathrm{Re}_\ell}}\right) \qquad (3-15)$$

This result implies that the boundary layer equations do not apply for sufficiently small Reynolds numbers.

58   Laminar Flow Analysis

Reverting to dimensional equations and neglecting terms of $O(\bar{\delta})$ with respect to those of $O(1)$, the x momentum equation becomes

$$\frac{\partial u}{\partial t} + u\frac{\partial u}{\partial x} + v\frac{\partial u}{\partial y} = X - \frac{1}{\rho}\frac{\partial P}{\partial x} + \frac{1}{\rho}\frac{\partial}{\partial y}\left(\mu\frac{\partial u}{\partial y}\right) \qquad (3-16)$$

Writing out the y momentum equation (Eq. 3–4) and indicating the order-of-magnitude of each term yields

$$\underset{O(\bar{\delta})}{\frac{\partial \bar{v}}{\partial \bar{t}}} + \underset{O(1)O(\bar{\delta})}{\bar{u}\frac{\partial \bar{v}}{\partial \bar{x}}} + \underset{O(\bar{\delta})O(1)}{\bar{v}\frac{\partial \bar{v}}{\partial \bar{y}}} = \bar{Y} - \frac{1}{\bar{\rho}}\frac{\partial \bar{P}}{\partial \bar{y}}$$

$$+ \underset{O(\bar{\delta}^2)}{\frac{1}{\mathrm{Re}_\ell}}\underset{O(1)}{\frac{1}{\bar{\rho}}}\left[\underset{O(1)}{\bar{\mu}}\left(\underset{O(\frac{1}{\delta})}{\frac{4}{3}\frac{\partial^2 \bar{v}}{\partial \bar{y}^2}} + \underset{O(\frac{1}{\delta})}{\frac{1}{3}\frac{\partial^2 \bar{u}}{\partial \bar{x}\partial \bar{y}}} + \underset{O(\bar{\delta})}{\frac{\partial^2 \bar{v}}{\partial \bar{x}^2}}\right)\right.$$

$$\left. + \underset{O(1)}{\left(\frac{4}{3}\frac{\partial \bar{v}}{\partial \bar{y}} - \frac{2}{3}\frac{\partial \bar{u}}{\partial \bar{x}}\right)}\underset{O(1)\,O(\frac{1}{\delta})}{\frac{\partial \bar{\mu}}{\partial \bar{y}}} + \underset{O(\frac{1}{\delta})}{\left(\frac{\partial \bar{u}}{\partial \bar{y}} + \frac{\partial \bar{v}}{\partial \bar{x}}\right)}\underset{O(\bar{\delta})\,O(1)}{\frac{\partial \bar{\mu}}{\partial \bar{x}}}\right] \qquad (3-17)$$

Inspection of Eq. (3–17) shows that the order-of-magnitude indicated for each term is $O(\bar{\delta})$ or less. Hence we may write

$$\frac{1}{\bar{\rho}}\frac{\partial \bar{P}}{\partial \bar{y}} = O(\bar{\delta}) \qquad (3-18)$$

Since, to our order of approximation, terms of $O(\bar{\delta})$ are neglected with respect to terms of $O(1)$, the y momentum equation (in dimensional variables) reduces to

$$\frac{\partial P}{\partial y} = 0 \qquad (3-19)$$

Consequently, the pressure is constant across the boundary layer and $P = P(x)$ only. Since the pressure gradient is constant across the boundary layer, the pressure variation in the inviscid flow at the edge of the boundary layer is imposed directly on the surface. This implies that, to a consistent order of approximation, $P = P(x)$ can be calculated from the inviscid flow solution *without regard to the presence of the thin boundary layer*.

Finally, turning our attention to the energy equation we note that for many liquids and gases of interest the Prandtl number is of order one, i.e., $\mathrm{Pr} = O(1)$. Further, we note that the specific heat at constant pressure c_p is also of $O(1)$, and hence $\bar{h} = O(1)$. Writing out the terms and indicating the order-of-magnitude for each term yields

$$\underset{O(1)}{\frac{\partial \bar{h}}{\partial \bar{t}}} + \underset{O(1)O(1)}{\bar{u}\frac{\partial \bar{h}}{\partial \bar{x}}} + \underset{O(\bar{\delta})O(\frac{1}{\bar{\delta}})}{\bar{v}\frac{\partial \bar{h}}{\partial \bar{y}}} = (\gamma-1)M_e^2 \left[\underset{O(1)}{\frac{\partial \bar{P}}{\partial \bar{t}}} + \underset{O(1)O(1)}{\bar{u}\frac{\partial \bar{P}}{\partial \bar{x}}} + \underset{O(\bar{\delta})O(\bar{\delta})}{\bar{v}\frac{\partial \bar{P}}{\partial \bar{y}}} \right]$$

$$+ \underset{O(\bar{\delta}^2)}{\bar{Q}} + \underset{O(1)}{\frac{(\gamma-1)}{\text{Re}_\ell}M_e^2 \bar{\nu}\bar{\Phi}} + \underset{O(\bar{\delta}^2)O(1)O(1)\ O(1)\ O(1)}{\frac{1}{\text{Re}_\ell}\frac{1}{\text{Pr}}\frac{1}{\bar{\rho}}\left[\bar{\mu}\left(\frac{\partial^2 \bar{h}}{\partial \bar{x}^2} + \underset{O(\frac{1}{\bar{\delta}^2})}{\frac{\partial^2 \bar{h}}{\partial \bar{y}^2}}\right)\right.}$$

$$\left. + \underset{O(1)\ O(1)}{\frac{\partial \bar{h}}{\partial \bar{x}}\frac{\partial \bar{\mu}}{\partial \bar{x}}} + \underset{O(\frac{1}{\bar{\delta}})O(\frac{1}{\bar{\delta}})}{\frac{\partial \bar{h}}{\partial \bar{y}}\frac{\partial \bar{\mu}}{\partial \bar{y}}}\right] \qquad (3-20)$$

and

$$\bar{\Phi} = 2\left[\underset{O(1)}{\left(\frac{\partial \bar{u}}{\partial \bar{x}}\right)^2} + \underset{O(1)}{\left(\frac{\partial \bar{v}}{\partial \bar{y}}\right)^2}\right] - \frac{2}{3}\underset{O(1)}{\left(\frac{\partial \bar{u}}{\partial \bar{x}} + \frac{\partial \bar{v}}{\partial \bar{y}}\right)^2} + \underset{O(\bar{\delta})}{\left(\frac{\partial \bar{v}}{\partial \bar{x}}} + \underset{O(\frac{1}{\bar{\delta}})}{\frac{\partial \bar{u}}{\partial \bar{y}}\right)^2} \qquad (3-21)$$

Neglecting terms of $O(\bar{\delta})$ with respect to those of $O(1)$ reduces the energy equation (in dimensional variables) to

$$\frac{\partial h}{\partial t} + u\frac{\partial h}{\partial x} + v\frac{\partial h}{\partial y} = \frac{1}{\rho}\left(\frac{\partial P}{\partial t} + u\frac{\partial P}{\partial x}\right) + Q$$

$$+ \nu\left(\frac{\partial u}{\partial y}\right)^2 + \frac{1}{\text{Pr}}\frac{1}{\rho}\frac{\partial}{\partial y}\left(\mu\frac{\partial h}{\partial y}\right) \qquad (3-22)$$

Thus, the thin boundary layer approximation results in the reduction of the compressible Navier-Stokes equations for a two-dimensional plane flow to

continuity
$$\frac{\partial \rho}{\partial t} + \frac{\partial}{\partial x}(\rho u) + \frac{\partial}{\partial y}(\rho v) = 0 \qquad (3-23)$$

x momentum
$$\frac{\partial u}{\partial t} + u\frac{\partial u}{\partial x} + v\frac{\partial u}{\partial y} = X - \frac{1}{\rho}\frac{\partial P}{\partial x} + \frac{1}{\rho}\frac{\partial}{\partial y}\left(\mu\frac{\partial u}{\partial y}\right) \qquad (3-16)$$

y momentum
$$\frac{\partial P}{\partial y} = 0 \qquad (3-19)$$

energy
$$\frac{\partial h}{\partial t} + u\frac{\partial h}{\partial x} + v\frac{\partial h}{\partial y} = \frac{1}{\rho}\left(\frac{\partial P}{\partial t} + u\frac{\partial P}{\partial x}\right) + Q$$

$$+ \nu\left(\frac{\partial u}{\partial y}\right)^2 + \frac{1}{\text{Pr}}\frac{1}{\rho}\frac{\partial}{\partial y}\left(\mu\frac{\partial h}{\partial y}\right) \qquad (3-22)$$

60 Laminar Flow Analysis

These equations are called the compressible boundary layer equations. As can easily be seen, they represent a considerable mathematical simplification compared to the compressible Navier-Stokes equations. First, we have only three rather than four coupled equations. In addition, the order of the system of equations is reduced. It is now second order in x and third order in y. Further analysis shows that these equations are all parabolic (in y) rather then elliptic partial differential equations, i.e., they are wave like in y. Looking at Eqs. (3–16), (3–22), (3–23), we see that for an incompressible fluid with constant properties the momentum and energy equations are again uncoupled, i.e., the classification previously discussed in Section 1–10 is applicable.

3–3 The Flow Past a Flat Plate — Blasius Solution

Now that we have obtained the boundary layer equations let us consider a typical problem. One of the fundamental problems of viscous flow is the steady, two-dimensional uniform flow of an incompressible viscous fluid past an infinitesimally thin semi-infinite flat plate parallel to the flow (see Figure 3–1). We take the x direction along the plate, with $x = 0$ at the leading edge of the plate, and y normal to the plate, with $y = 0$ at the plate surface. The corresponding inviscid flow solution shows that the plate creates no disturbance in the flow. Hence, the pressure gradient dP/dx along the plate is zero. Under these circumstances the boundary layer equations are uncoupled. Here we consider only the dynamics of the flow, i.e., the continuity and momentum equations. The thermodynamics of the flow is considered in Chapter 4. Under these conditions the governing boundary value problem is reduced to

$$\frac{\partial u}{\partial x} + \frac{\partial v}{\partial y} = 0 \qquad (3-24)$$

$$u\frac{\partial u}{\partial x} + v\frac{\partial u}{\partial y} = \nu\frac{\partial^2 u}{\partial y^2} \qquad (3-25)$$

with the no-slip and no-mass-transfer boundary conditions given by

$$y = 0 \qquad u = v = 0 \qquad (3-26)$$

The condition that the viscous flow approach the inviscid flow at large (on the scale of the boundary layer) distances above the plate is given by

$$y \to \infty \qquad u \to U \qquad (3-27)$$

Here, U is the inviscid flow velocity at the edge of the boundary layer.

Noting that Eqs. (3–24) and (3–25) are two equations in the unknown dependent variables, u and v, we introduce the stream function, ψ, such that the continuity equation is satisfied. Thus, we obtain one equation in one unknown

The Flow Past a Flat Plate — Blasius Solution

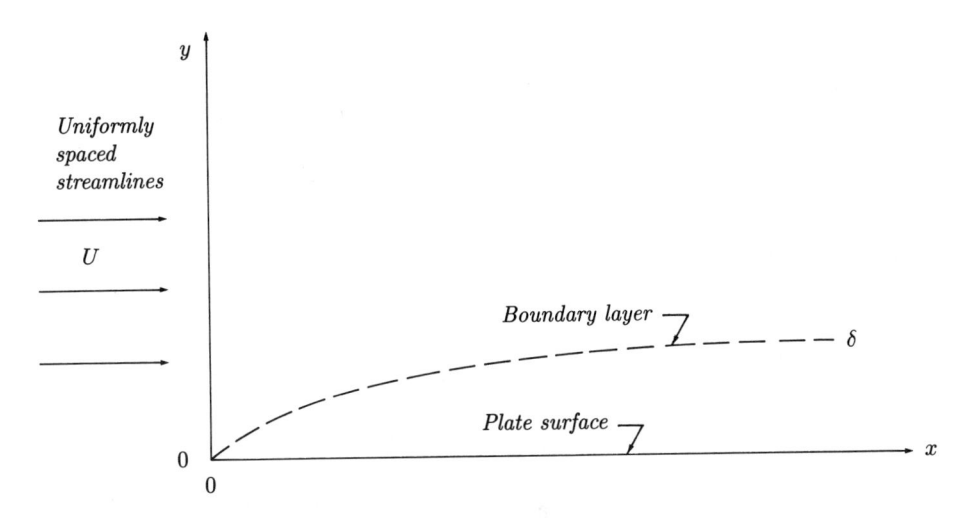

Figure 3–1. Flow past a semi-infinite flat plate.

dependent variable, i.e.[†]

$$u = \psi_y \qquad v = -\psi_x \qquad (3-28)$$

yields

$$\psi_y \psi_{xy} - \psi_x \psi_{yy} = \nu \psi_{yyy} \qquad (3-29)$$

with boundary conditions given by

$$y = 0 \qquad \psi_x = \psi_y = 0 \qquad (3-30)$$

$$y \to \infty \qquad \psi_y \to U \qquad (3-31)$$

Since the physical problem under consideration has no preferred or characteristic length it is reasonable to suppose that the velocity profiles at various distances from the leading edge are similar.[‡] This implies that the shape of the velocity profiles, $u(y)$, at various positions, x, can be made identical by selecting suitable scale factors for $u(y)$ and y. Further, if the velocity profiles are similar it implies that the governing partial differential equation can be reduced to an ordinary differential equation. Appropriate physical arguments often lead to the

---

[†] Here the subscripts indicate partial differentiation, e.g.

$$\psi_x = \frac{\partial \psi}{\partial x} \qquad \psi_y = \frac{\partial \psi}{\partial y}$$

[‡] A more detailed discussion of similarity analysis is given in Hansen [Hans64]. Hansen indicates that problems which have a characteristic length do not in general yield to similarity analysis. Such problems generally have boundary conditions specified at finite values of the independent variable. Hence, it is unlikely that a single transformation can be found which allows the solution to satisfy the boundary conditions.

correct similarity scaling factors for $u(y)$ and y as well as the correct functional form of the similarity transform. Recall that for the suddenly accelerated flat plate previously discussed in Section 2–5 the growth of the region of fluid affected by viscosity is inversely proportional to the square root of the time, i.e., it grew as $1/\sqrt{\nu t}$. Although the present problem is steady, a characteristic time is obtained by considering the time necessary for a fluid particle to travel from the leading edge of the plate to any point downstream. If we calculate this time based on conditions outside of the boundary layer, we have $t = x/U$. Thus, we expect the boundary layer, which represents the region of fluid affected by viscosity, to grow as $1/\sqrt{(\nu x/U)}$. By analogy with the suddenly accelerated flat plate, we expect the transformed similarity coordinate to be proportional to $y\sqrt{U/\nu x}$. Although, as we shall see, these physical arguments lead to the correct formulation for the similarity transformations, this is not necessarily the case. Hence, we take a more formal viewpoint.

In order to obtain the appropriate similarity transformations, consider the general affine transformations

$$\eta = A x^a y^b \tag{3-32}$$

$$\psi(x,y) = B x^c y^d f(\eta) \tag{3-33}$$

where the transformed stream function, f, is assumed to be only a function of the transformed independent variable, η, and A, B, a, b, c, d are as yet undetermined constants. The factors A and B are included to allow nondimensionalization of the resulting equation. The assumed power law form of the transformations is based on experience (see Sections 2–7 and 2–10) and the desire for generality.

Recalling Eq. (3–29), we note that ψ_x, ψ_y, ψ_{yy} and ψ_{yyy} are required. Performing the indicated partial differentiations yields

$$\psi_x = cBx^{c-1}y^d f + aABx^{a+c-1}y^{d+b} f' \tag{3-34a}$$

$$\psi_y = dBx^c y^{d-1} f + bABx^{c+a} y^{d+b-1} f' \tag{3-34b}$$

$$\psi_{xy} = cdBx^{c-1}y^{d-1}f + [(ad + ba + bc)AB] \times$$
$$[x^{c+a-1} y^{d+b-1} f'] + abA^2 B x^{2a+c-1} y^{d+2b-1} f'' \tag{3-34c}$$

$$\psi_{yy} = d(d-1)Bx^c y^{d-2} f + b(2d+b-1)ABx^{a+c} y^{b+d-2} f'$$
$$+ b^2 A^2 B x^{2a+c} y^{2b+d-2} f'' \tag{3-34d}$$

$$\psi_{yyy} = d(d-1)(d-2)Bx^c y^{d-3} f + d(d-1)bABx^{a+c} y^{b+d-3} f'$$
$$+ bd(b+d-2)ABx^{a+c} y^{b+d-3} f'$$
$$+ b^2 dA^2 B x^{2a+c} y^{2b+d-3} f''$$
$$+ b(d+b-1)(b+d-2)ABx^{a+c} y^{d+b-2} f'$$

$$+ b^2(b+d-1)A^2 B x^{2a+c} y^{2b+d-3} f''$$
$$+ b^2 A^2 B(2b+d-2) x^{2a+c} y^{2b+d-3} f''$$
$$+ b^3 A^3 B x^{3a+c} y^{3b+d-3} f''' \qquad (3-34e)$$

where the prime (') indicates differentiation with respect to η.

Substitution of Eqs. (3–34) into Eq. (3–29) gives a very complicated result. Thus, we first consider the boundary conditions, Eqs. (3–30) and (3–31), in order to obtain preliminary values for a, b, c, d. The objective of the analysis is to obtain transformed boundary conditions which are only functions of η. As $y \to \infty$, $\eta \to \infty$ for all $x > 0$. The transformed boundary condition at infinity is then

$$u = \psi_y = \left[B d x^c y^{d-1} f + b A B x^{c+a} y^{d+b-1} f' \right] \to U \qquad (3-35)$$

If b and d are both zero, $u = \psi_y = 0 \to U$, which is true only if there is no flow and hence of no interest. However, it appears convenient to take either b or d equal to zero, since this simplifies the boundary condition by eliminating one of the terms. Referring to Eq. (3–32), and recalling that our objective is to obtain a functional form for η such that the two independent variables, x and y, are combined into one independent variable, implies that neither a nor b can be zero. Thus, we choose $d = 0$. With $d = 0$, Eq. (3–35) becomes

$$b A B x^{c+a} y^{b-1} f' \to U$$

By choosing $c = -a$ the dependence of the boundary condition on x is eliminated. A constant boundary condition at infinity is obtained by eliminating the y dependence by choosing $b = 1$.

Finally, as $\eta \to \infty$ we have that

$$f'(\eta \to \infty) \to \frac{U}{AB} \qquad (3-36)$$

Using these results, the surface boundary conditions at $\eta = 0$ (Eq. 3–30) become

$$f(0) = f'(0) = 0 \qquad (3-37a,b)$$

The governing differential equation (Eq. 3–29) becomes

$$a A^2 B^2 x^{-1} f f'' = \nu A^3 B x^{2a} f'''$$

The choice of $a = -1/2$ eliminates the x dependence, and choosing $1/(2A/B) = \nu$ nondimensionalizes the equation. The governing differential equation is then

$$f''' + f f'' = 0 \qquad (3-38)$$

Notice that this is an ordinary differential equation.

64   Laminar Flow Analysis

Recalling the transformed boundary condition at infinity (Eq. 3–36), we see that a constant boundary condition results if we choose $AB = U$. Solving the two equations in the two unknown constants, A and B, yields

$$A = \sqrt{\frac{U}{2\nu}} \qquad B = \sqrt{2\nu U}$$

The similarity transformations are then

$$\eta = y\sqrt{\frac{U}{2\nu x}} \qquad \psi(x,y) = \sqrt{2\nu U x}\, f(\eta) \qquad (3-39a,b)$$

Collecting our results, we see that the governing nonlinear two point asymptotic boundary value problem is

$$f''' + ff'' = 0 \qquad (3-38)$$

with
$$f(0) = f'(0) = 0 \qquad (3-37a,b)$$

and
$$f'(\eta \to \infty) \to 1 \qquad (3-40)$$

The governing differential equation is called the Blasius equation in honor of H. Blasius, a student of Prandtl's, who obtained the first accurate solution of this equation [Blas08]. Blasius assumed a series expansion valid for small values of η, i.e., near the surface, and an asymptotic expansion for large values of η, i.e., far from the surface. This effectively divides the viscous region into two parts. Sufficient terms were retained in the series expansions to allow for matching the two series in an intermediate region.[†] Subsequently, Töpfer [Topf12] developed another series expansion method of integration which was used by Goldstein [Gold30] and subsequently by Howarth [Howa39] to obtain extremely accurate results.

Recalling the boundary value problems associated with two-dimensional and axisymmetric stagnation point flow (see Sections 2–7 and 2–10) we see that the present boundary value problem is quite similar. Hence, the numerical integration and iteration techniques previously developed in Chapter 2 are applicable. In fact, the three boundary value problems represented by Eqs. (3–37), (3–38), (3–40); (2–65), (2–67), (2–68); and (2–92), (2–94), (2–95) are special cases of the boundary value problem resulting from the analysis of the general two-dimensional incompressible boundary layer discussed in the next section. Thus, we defer a numerical solution of the Blasius equation until that time.

[†]This method is similar to the method of matched asymptotic expansions, or the method of inner and outer expansions. For additional information on these methods the reader is directed to Van Dyke [VanD64].

3–4 General Two-dimensional Incompressible Boundary Layer — The Falkner-Skan Equation

In the previous section we discussed a simplified case of two-dimensional incompressible boundary layer flow, i.e., the semi-infinite flat plate at zero angle of attack. In this simple case the inviscid pressure gradient in the flow is zero. In general this is not the case. Since, as we recall from Euler's equations, the inviscid pressure gradient is related to the inviscid velocity gradient, we ask ourselves the following question: Under what circumstances, i.e., for what types of inviscid flow velocity variations along the body, do similar solutions exist? In order to answer this question we consider the equations governing the general two-dimensional steady incompressible boundary layer, i.e.

$$\frac{\partial u}{\partial x} + \frac{\partial v}{\partial y} = 0 \qquad (3-41)$$

$$u\frac{\partial u}{\partial x} + v\frac{\partial u}{\partial y} = -\frac{1}{\rho}\frac{\partial P}{\partial x} + \nu\frac{\partial^2 u}{\partial y^2} \qquad (3-42)$$

with boundary conditions given by

$$y = 0 \qquad u = v = 0$$
$$y \to \infty \qquad u \to U(x)$$

where $U(x)$ is the inviscid (potential) flow velocity along the body surface. The inviscid flow velocity, U, is assumed to be a function of the coordinate along the surface x and not a function of the coordinate normal to the surface y. Our previous question can be restated as: What functional forms of $U(x)$ yield similarity solutions?

To begin the analysis, we first rewrite the pressure gradient term in Eq. (3–42) using Euler's equation and then introduce the stream function previously defined in Eq. (3–28). The governing boundary value problem is then

$$\psi_y \psi_{xy} - \psi_x \psi_{yy} = U\frac{dU}{dx} + \nu\psi_{yyy} \qquad (3-43)$$

with boundary conditions

$$y = 0 \qquad \psi_x = \psi_y = 0 \qquad (3-44a,b)$$
$$y \to \infty \qquad \psi_y \to U(x) \qquad (3-44c)$$

We now seek to find similar solutions. We choose a somewhat different method than previously used for the Blasius equation. Here we use the 'free parameter method'.[†] We begin by introducing the transformations

$$\xi = x \qquad \eta = \frac{Ay}{g(x)} \qquad \psi(x,y) = BU(x)g(x)f(\xi,\eta) \quad (3-45a,b,c)$$

[†] For a more detailed discussion of the free parameter method and an alternate approach to this problem, see Hansen [Hans64], Section 2.1.

where A and B are as yet unknown constants, and $g(x)$ and $U(x)$ are as yet unknown functions of x only.

There is no a priori reason for introducing transformations of this type except our previous experience with stagnation point flow and the Blasius equation. Recalling the similarity transformations obtained there, we note that the function in x appeared in the denominator of the independent similarity variable; hence, we assume that the generalized independent similarity variable has an unknown function of x in the denominator. In addition, our previous experience indicates that the transformation for the stream function contains the same function of x. Further, recall that the first derivative of the transformed stream function for the Blasius equation gave the velocity profile. This also leads us to the above assumed general form for the transformed stream function. The unknown constants, A and B, are included to allow for nondimensionalization of the resulting equations.

Performing the indicated differentiation and substituting into the governing differential equations (Eq. 3–43) yields

$$f''' + \frac{(AB)^2}{\nu A^3 B} g(Ug)' f f'' + \frac{g^2 U'}{\nu A^3 B}\left[1 - (AB)^2 f'^2\right] = \frac{(AB)^2}{\nu A^3 B} g^2 U(f' f'_\xi - f'' f_\xi) \quad (3-46)$$

where $\quad f' = \dfrac{\partial f}{\partial \eta} \quad f_\xi = \dfrac{\partial f}{\partial \xi} \quad f'_\xi = \dfrac{\partial f'}{\partial \xi} \quad U' = \dfrac{dU}{dx} \quad g' = \dfrac{dg}{dx}$

Note that for all $x > 0$ as $y \to \infty$, $\eta \to \infty$, the boundary condition at infinity becomes

$$\psi_y = AB\, U(x) f'(\xi, \eta \to \infty) \to U(x)$$

or
$$f'(\xi, \eta \to \infty) \to \frac{1}{AB} \quad (3-47)$$

In order to nondimensionalize Eq. (3–46) and to obtain a simple numerical result for the boundary condition at infinity, we choose $AB = 1$ and $\nu A^3 B = U_\infty$, where U_∞ is the potential velocity upstream of $x = 0$. Solving for A and B yields $A = \sqrt{U_\infty/\nu}$ and $B = \sqrt{\nu/U_\infty}$. Equation (3–46) then becomes

$$f''' + \alpha f f'' + \beta(1 - f'^2) = g^2 \frac{U}{U_\infty}(f' f'_\xi - f'' f_\xi) \quad (3-48)$$

where $\quad \alpha = \dfrac{g}{U_\infty}(Ug)' \quad \beta = \dfrac{g^2}{U_\infty} U' \quad (3-49a,b)$

In order for similar solutions to exist, the transformed stream function must be a function of η only, i.e., $f = f(\eta)$ only. Thus, the right hand side of Eq. (3–48) must be zero. Further, α and β must be independent of x. Since g and U were assumed to be functions of x only, this implies that α and β are constants. Hence, similar solutions of the general steady, two-dimensional incompressible boundary layer satisfy the following two-point asymptotic boundary value problem

$$f''' + \alpha f f'' + \beta(1 - f'^2) = 0 \qquad (3-50)$$

with boundary conditions

$$f(0) = f'(0) = 0 \qquad (3-51a,b)$$

$$f(\eta \to \infty) \to 1 \qquad (3-51c)$$

Equation (3–50) is called the Falkner-Skan equation after the two English mathematicians who first studied it [Falk31]. Numerical solutions of Eq. (3–50) subject to Eq. (3–51) were subsequently studied in detail by Hartree [Hart37].

Since α and β are assumed to be constants, Eqs. (3–49a,b) represent two equations in the two unknown functions, $U(x)$ and $g(x)$. $U(x)$ and $g(x)$ can thus be determined. Consider

$$2\alpha - \beta = \frac{2g}{U_\infty}(Ug)' - \frac{g^2}{U_\infty}U'$$

or

$$2\alpha - \beta = \frac{1}{U_\infty}(g^2 U)' \qquad (3-52)$$

Providing $2\alpha - \beta \neq 0$, integration of Eq. (3–52) yields

$$\frac{U}{U_\infty}g^2 = (2\alpha - \beta)x \qquad (3-53)$$

A second algebraic equation for $U(x)$ and $g(x)$ is obtained by considering

$$\alpha - \beta = \frac{g}{U_\infty}(Ug)' - \frac{g^2}{U_\infty}U'$$

Expanding the first term gives

$$\alpha - \beta = \frac{U}{U_\infty}gg' \qquad (3-54)$$

Multiplying both sides of Eq. (3–54) by U' and rewriting yields

$$(\alpha - \beta)\frac{U'}{U} = \frac{g^2}{U_\infty}U'\frac{g'}{g} = \beta\frac{g'}{g}$$

Integration of this equation results in

$$\left(\frac{U}{U_\infty}\right)^{\alpha-\beta} = Kg^\beta \qquad (3-55)$$

where $K = \text{constant } U_\infty^{(\beta-\alpha)}$

Simultaneous solution of Eqs. (3–53) and (3–55) gives $U(x)$ and $g(x)$ directly, i.e.

$$\frac{U(x)}{U_\infty} = K^{\frac{2}{2\alpha-\beta}} \left[(2\alpha - \beta)x\right]^{\frac{\beta}{2\alpha-\beta}} = \text{constant } x^m \qquad (3-56)$$

and

$$g(x) = \left[(2\alpha - \beta)\frac{U_\infty x}{U}\right]^{1/2} \qquad (3-57)$$

Here, we see that similar solutions of the steady two-dimensional incompressible boundary layer exist if the potential velocity, $U(x)$, varies as a power of the distance along the surface. Note that the results are independent of any common factor of α and β, since if a common factor did appear it could be absorbed into $g(x)$. Thus, providing $\alpha \neq 0$, we, without loss of generality, choose $\alpha = 1$. In addition, we introduce

$$m = \frac{\beta}{2 - \beta} \qquad \beta = \frac{2m}{m + 1} \qquad (3-58a, b)$$

$U(x)$ and $g(x)$ then become

$$U(x) = \bar{K} x^m \qquad (3-59)$$

and

$$g(x) = \left(\frac{2}{m+1}\frac{x}{U}U_\infty\right)^{1/2} \qquad (3-60)$$

where

$$\bar{K} = U_\infty K^{m+1}\left(\frac{2}{m+1}\right)^m \qquad (3-61)$$

Using our results for A and $g(x)$ in Eq. (3–45b) yields the appropriate independent similarity variable

$$\eta = y\sqrt{\frac{(m+1)}{2}\frac{U}{\nu x}} \qquad (3-62)$$

In the above analysis we excluded the cases where $\alpha = 0$ and where $2\alpha - \beta = 0$. For $\alpha = 0$ Eq. (3–56) shows that the potential velocity is inversely proportional to the distance along the surface, i.e., $U(x) \sim 1/x$. Depending on the sign of U, this is either a source or sink flow. These flows can be interpreted as the flow in either a convergent or divergent two-dimensional channel. Analytical solutions exist for the boundary layer flow in a convergent channel. Boundary layer solutions do not exist for flow in a divergent channel. However, in both cases these flows have solutions which are exact solutions of the Navier-Stokes equations. When $2\alpha - \beta = 0$, similar solutions exist when the potential velocity has an exponential variation with distance along the surface. These solutions are not discussed further.

3–5 Group Properties of the Blasius Equation

Before considering the numerical solution of the Falkner-Skan equation we discuss a particular property of the Blasius equation first noted by Töepfer [Toep12]. First note that for $\alpha = 1$ the Falkner-Skan equation reduces to the Blasius equation (Eq. 3–38) when $\beta = 0$, i.e.

$$f''' + ff'' = 0 \qquad (3-38)$$

with
$$f(0) = f'(0) = 0 \qquad (3-37)$$

$$f'(\eta \to \infty) \to 1 \qquad (3-40)$$

Thus, for the particular case of $\beta = 0$, if $f(\eta)$ is a solution then $kf(k\eta)$ is also a solution. In order to demonstrate this, let

$$\bar{\eta} = k\eta \qquad f(\eta) = k\bar{f}(\bar{\eta}) \qquad (3-63a,b)$$

After performing the indicated transformation, the governing boundary value problem becomes

$$\bar{f}''' + \bar{f}\bar{f}'' = 0 \qquad (3-64)$$

with
$$\bar{f}(0) = \bar{f}'(0) = 0 \qquad (3-65a,b)$$

and
$$\bar{f}'(\bar{\eta} \to \infty) \to \frac{1}{k^2} \qquad (3-65c)$$

where the prime denotes differentiation with respect to the appropriate argument.

Now looking at $f''(0)$ we have

$$f''(0) = \frac{\bar{f}''(0)}{k^3} \qquad (3-66)$$

Using Eq. (3–65c) to evaluate k, and substituting the results into Eq. (3–66), yields

$$f''(0) = \frac{\bar{f}''(0)}{[\bar{f}'(\eta \to \infty)]^{3/2}} \qquad (3-67)$$

This result shows, for the particular case of $\beta = 0$, that the two-point asymptotic boundary value problem can be treated as an initial value problem. The technique is to choose any reasonable value of $\bar{f}''(0)$, perform the numerical integration out to the 'edge' of the boundary layer, evaluate $\bar{f}'$ at the 'edge' of the boundary layer, and use Eq. (3–67) to calculate the value of $f''(0)$, which yields convergence to $f'(\eta \to \infty) \to 1$. This technique is illustrated in the following discussion of the numerical integration of the Falkner-Skan equation.

3–6 Numerical Integration of the Falkner-Skan Equation

The details of the program for numerically integrating the Falkner-Skan equation are discussed in Appendix D, Section D–13. Here, in order to illustrate the group properties of the Blasius equation we use a slightly modified version of `gblpdriv`. The modification eliminates the `while` loop and the call to `ns` at the end of the program. Only the call to `rk` remains. Thus, only a single integration is performed each time the modified `faskan` program is run.

Comparing Eq. (2–65) for two-dimensional stagnation point flow with the Falkner-Skan equation, Eq. (3–50), we see that for $\beta = 1$ the two equations are identical. Similarly, comparing Eq. (2–92) for axisymmetric stagnation point flow and the Falkner-Skan equation shows that Eq. (2–92) is a special case of the Falkner-Skan equation for $\beta = 1/2$. Previously we found solutions for $\beta = 1/2, 1$ of $f''(0) = 0.92768004$ and $f''(0) = 1.232588$, respectively. For the Blasius equation, $\beta = 0$. Recalling our discussion of the group properties of the Blasius equation (Section 3–5) and the above results for $\beta = 1/2, 1$, we arbitrarily take $f''(0) = 1$ as our first estimate.

```
Beta = 0
f''(0) = 1
 η        f''            f'          f
 0        1              0           0
 6.       4.63499e-10    1.65519     8.365703
```

The integration yields $f'(\eta)$ at η_{max} of 1.65519. Using Eq. (3–67) yields (to six significant figures) a value of $f''(0) = 0.469600$. Thus, we take this as our second estimate.

```
Beta = 0
f''(0) = 0.469600
 η        f''            f'          f
 0        0.469600       0           0
 6.       3.556907e-6    0.9999993   4.783220
```

The second integration yields $f'(\eta_{max}) = 0.9999993$. To six significant figures this is the correct outer boundary condition which verifies our previous analysis. However, the `faskan` program requires convergence on $f'(\eta_{max})$ to within $\pm 5 \times 10^{-7}$. Our third estimate is 0.469601.

```
Beta = 0
f''(0) = 0.469601
 η        f''            f'          f
 0        0.469601       0           0
 6.       3.556843e-6    1.000001    4.783227
```

Finally, the fourth estimate is 0.4696005.

Numerical Integration of the Falkner-Skan Equation

```
Beta = 0
f''(0) = 0.4696005
 η         f''              f'          f
 0        0.4696005         0           0
 6.       3.556875e-6       1.          4.783223
```

This value yields convergence to within the specified degree of accuracy. Here we see that the group property of the Blasius equation allowed us to calculate the required value of $f''(0)$ without resort to the formal Nachtsheim-Swigert iteration scheme. The full results for the Blasius profile are shown in Table 3–1.

Table 3–1. Results for the numerical integration of the Blasius equation.

η	f''	f'	f
0	0.4696005	0	0
0.2	0.46930657	9.3905401e−2	9.391422e−3
0.4	0.4672547	0.18760534	0.03754924
0.6	0.46173493	0.28057576	8.4385663e−2
0.8	0.45119049	0.37196365	0.14967468
1.	0.43437958	0.46063307	0.23299035
1.2	0.41056575	0.54524709	0.33365774
1.4	0.37969252	0.62438697	0.4507241
1.6	0.34248737	0.69670022	0.58295692
1.8	0.3004455	0.76105808	0.728873
2.	0.25566929	0.8166954	0.88679774
2.2	0.21057998	0.86330501	1.0549482
2.4	0.16756036	0.90106625	1.2315289
2.6	0.12861283	0.93060206	1.4148256
2.8	9.5113386e−2	0.95287624	1.6032851
3.	6.7710286e−2	0.96905538	1.7955696
3.2	4.6370361e−2	0.98036575	1.9905828
3.4	3.0535214e−2	0.98797122	2.1874693
3.6	1.9328694e−2	0.99288865	2.3855926
3.8	1.1758678e−2	0.99594502	2.5845011
4.	6.8740853e−3	0.99777083	2.7838889
4.2	3.861352e−3	0.99881904	2.9835579
4.4	2.0840747e−3	0.99939734	3.1833855
4.6	1.0807525e−3	0.99970394	3.3832989
4.8	5.3848399e−4	0.99986013	3.5832571
5.	2.5778052e−4	0.99993659	3.7832377
5.2	1.1856508e−4	0.99997256	3.9832291
5.4	5.2395285e−5	0.99998882	4.1832254
5.6	2.2246211e−5	0.99999588	4.383224
5.8	9.0750329e−6	0.99999883	4.5832235
6.	3.556875e−6	1.	4.7832234

72   Laminar Flow Analysis

Using the unmodified **faskan** program with automatic iteration and an initial estimate for $f''(0) = 1$ requires twelve iterations to converge to a solution within the required accuracy.

The last significant figure in the value of $f''(0)$ is influenced by both the required error values and the value of $\eta_{\max}$. For example, if the error criteria is $f''(\eta_{\max}) = \pm 5 \times 10^{-6}$ and $f'(\eta_{\max}) - 1 = \pm 5 \times 10^{-7}$, then any value $0.4696002^- \leq f''(0) \leq 0.4696008^+$ yields a converged solution at $\eta_{\max} = 6$. Relaxing the error criteria on $f'(\eta_{\max})$ to $f'(\eta_{\max}) - 1 = \pm 5 \times 10^{-6}$ yields converged solutions for $0.469597^- \leq f''(0) \leq 0.04696049^+$ at $\eta_{\max} = 6$.

With an error criteria on $f''(\eta_{\max})$ of $\pm 5 \times 10^{-7}$ the solution fails to converge at $\eta_{\max} = 6$ to within the required accuracy. However, increasing $\eta_{\max}$ to 8 immediately yields convergence for $f''(0) = 0.4696000$. It also narrows the region of convergence (see Appendix B).

These theoretical results for the velocity profile are compared with the experimental results of Nikuradse [Niku42] in Figure 3–2. The comparison is very good. This confirms the validity of the mathematical model of incompressible viscous boundary layer flow.

To more fully illustrate the **faskan** program we seek an additional solution for $\beta = 0.1$. Recalling the procedure used in determining the initial estimate for $f''(0)$ for axisymmetric stagnation point flow (see Section 2–10), we take the initial estimate of $f''(0)$ for $\beta = 0.1$ to be the previously determined value for $\beta = 0$, i.e., $f''(0) = 0.4696005$.

```
Beta = 0.1
f"(0) = 0.4696005
  η          f"              f'              f
  0       .4696005           0               0
  6.    -1.2993635e-2    0.73091529      3.7572104
```

Integration yields $f'(\eta_{\max}) = 0.73091529$, which is too small. From this result, taken in conjunction with the results for $\beta = 0$, we expect that the correct value of $f''(0)$ is larger than 0.4696005. But how much larger? In order to obtain a tentative value for the next estimate of $f''(0)$, recall that the Newton-Raphson iteration scheme (see Appendix B) shows that the estimated correction to $f''(0)$ is

$$\Delta f''(0) = \frac{1 - f'(\eta_{\max})}{\Delta f'(\eta_{\max})/\Delta f''(0)} \qquad (3-68)$$

However, at this point we do not have enough information to evaluate the derivative $\Delta f'(\eta_{\max})/\Delta f''(0)$ for the case of $\beta = 0.1$. Tentatively assuming that the derivative is not significantly affected by the change in β, we evaluate the derivative from our previous results for the Blasius equation. The last two integrations from the Blasius equation yield a value for $\Delta f'(\eta_{\max})/\Delta f''(0) \approx 2$. Using this value of the derivative in Eq. (3–68) shows that $\Delta f''(0) \approx 0.135$.

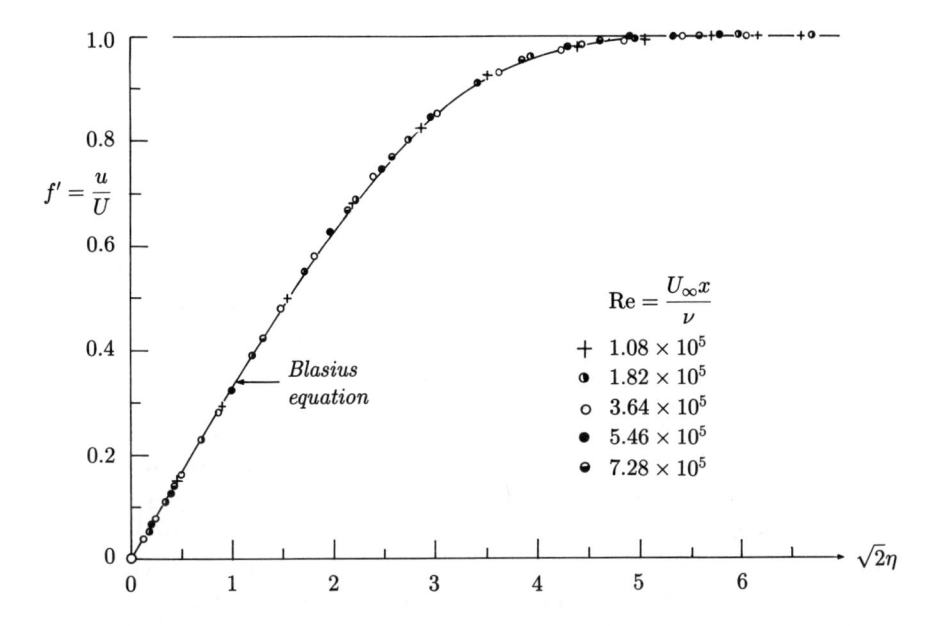

Numerical Integration of the Falkner-Skan Equation

Figure 3-2. Comparison of theoretical and experimental results for the velocity profile on a flat plate. (Adapted from Schlichting [Schl68].)

Thus, our second estimate for $f''(0)$ is 0.605 which, after integration, yields $f'(\eta_{\max}) = 1.0380721$.

```
Beta = 0.1
f"(0) = 0.605
  η         f"            f'           f
  0         0.605         0            0
  6.        1.5810719e-3  1.0380721    5.0886453
```

Continuing this process, the third estimate is taken as 0.585 and the fourth as 0.587. Automatic iteration using the Nachtsheim-Swigert iteration scheme then yields convergence to the correct asymptotic outer boundary condition in one iteration.

```
Beta = 0.1
f"(0) = 0.587
  η         f"             f'           f
  0         0.587          0            0
  6.        -1.8319361e-6  0.99992442   4.9193466

f"(0) = 0.588
  0         0.588          0            0
  6.        8.8218347e-5   1.0020618    4.9288062
```

74 Laminar Flow Analysis

```
Beta = 0.1
f"(0) = 0.587
f"(0) = 0.58703533
0     0.58703533     0              0
6.    1.3541279e-6   0.99999998     4.9196809
Convergence achieved
```

The above discussion illustrates that using information derived from previous solutions provides a powerful technique for generating new solutions. However, it should be noted that the above method is not necessarily successful. If, for example, a solution for $\beta = 0.5$ is attempted using the above technique, $f''(0) = 0.469005$ yields $f'(\eta_{max}) \approx -31.6$! Using Eq. (3–68) and the value $\Delta f'(\eta_{max})/\Delta f''(0) \approx 2$ from the Blasius solution leads to a second estimate of $f''(0) \approx 15$. The result of this second integration is $f'(\eta_{max}) \approx 88.3$. At this point it is obvious that further estimates will yield increasingly divergent results. These results indicate that the above technique is only successful if the desired new solution is essentially a perturbation of the known solution. The desired solution for $\beta = 0.5$ can be found by determining intermediate solutions at $\beta = 0.2, 0.3, 0.4$ and, finally, at 0.5 by successively using the above technique. Figure 3–3 shows a curve of β versus $f''(0)$, and Table C–1 of Appendix C gives

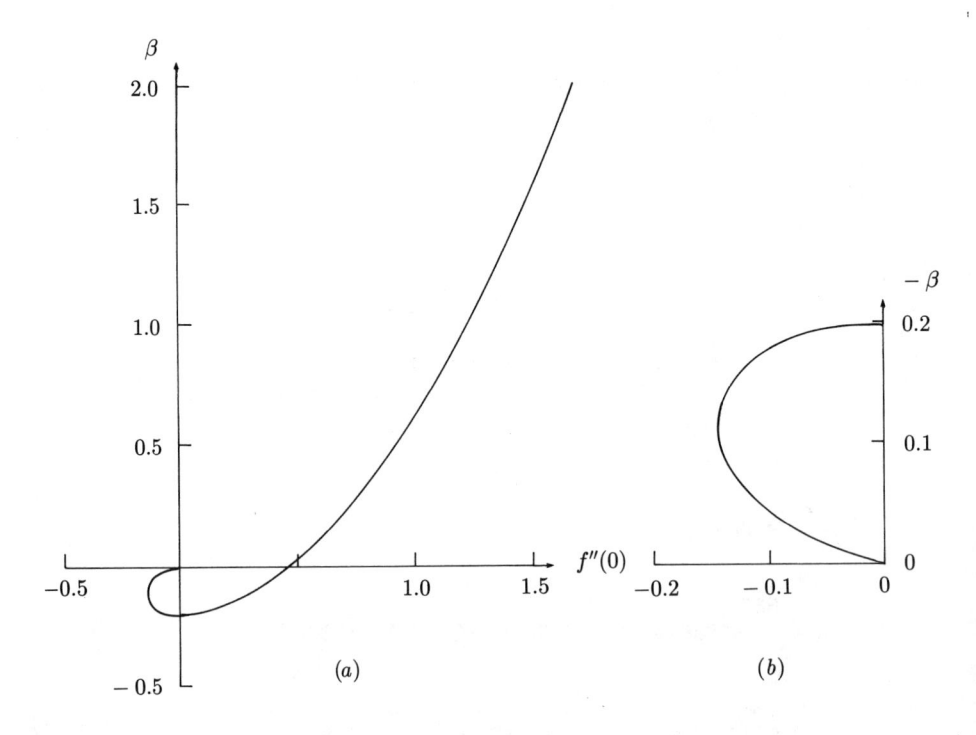

Figure 3–3. The shearing stress parameter $f''(0)$ as a function of β.

the corresponding numerical results obtained using the `faskan` program. These numerical results are in excellent agreement with previously published numerical results (see, for example, Moore [Moor64] and Stewartson [Stew54]). Figure 3–4 shows the velocity profiles $f'(\eta)$ for various values of β.

From the results shown in Figures 3–3 and 3–4 and Table C–1 of Appendix C we note that solutions exist for which $\beta < 0$, and in which the direction of the velocity near the wall is reversed.

Solutions for which $\beta < 0$ are generally referred to as the lower branch solutions. Solutions for which $\beta < 0$ and $f''(0) < 0$ are referred to as the reverse flow solutions. Hartree [Hart37] numerically studied the solutions to the Falkner-Skan equation when $f''(0) > 0$. For $\beta > 0$, $f''(0) > 0$ he found that the solution is unique. Defining β_0 as the value when $f''(0) = 0$, he found that for $f''(0) > 0$ and

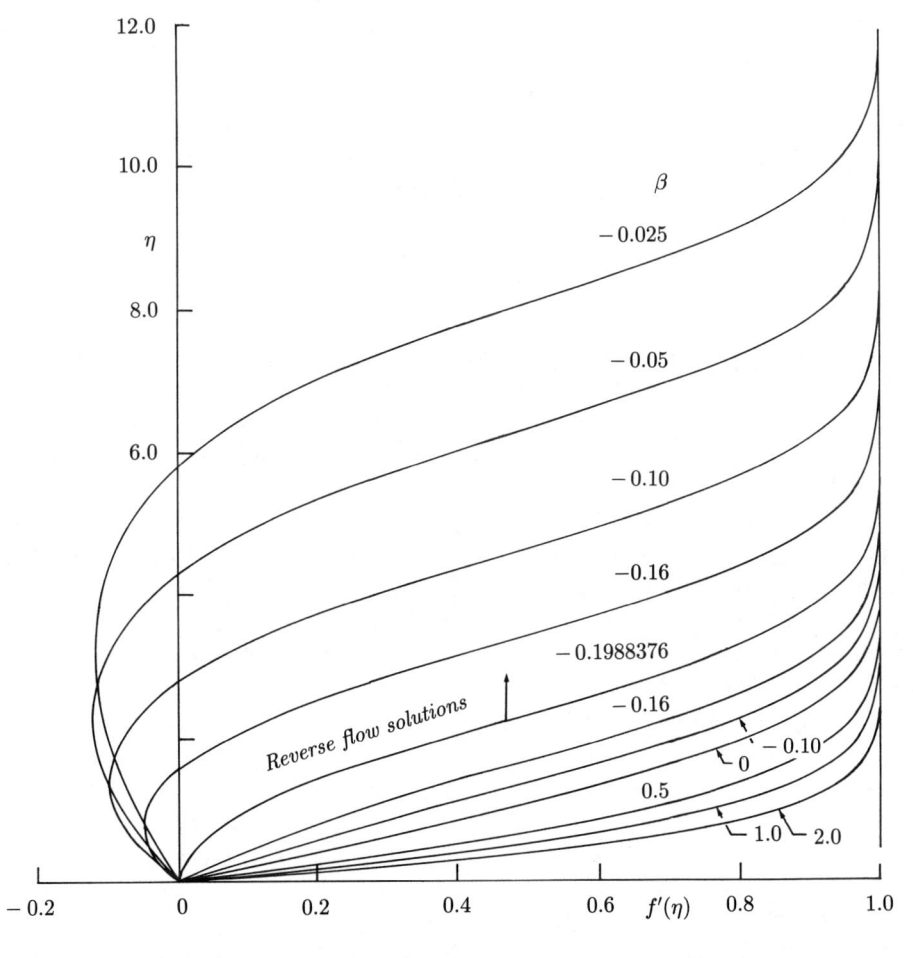

Figure 3–4. Falkner-Skan velocity profiles.

$\beta_0 < \beta < 0$ $f'(\eta \to \infty) \to 1$ for an infinite number of values of $f''(0)$. Thus, the solutions are not unique. However, Hartree found that one of these solutions exhibited exponential decay, i.e., $f' \to 1$ exponentially from below, while all others decayed algebraically. Hartree, basing his argument on continuity requirements, chose the exponential decay solution as the correct one. Stewartson [Stew54] subsequently placed this choice on more solid mathematical foundations. Hartree [Hart37] numerically found, and Stewartson [Stew54] subsequently proved, that for $\beta < \beta_0$ all solutions exhibit the property that $f'(\eta) > 1$. Since there is no physical mechanism in incompressible flow which generates a velocity within the boundary layer which exceeds the external stream velocity, these solutions are rejected on physical grounds.[†]

It is interesting to note that the Nachtsheim-Swigert iteration technique (see Nachtsheim and Swigert [Nach65] and Appendix B) used here yields a unique solution even for the case $\beta_0 \leq \beta < 0$ without the imposition of additional conditions, i.e., the Nachtsheim-Swigert iteration technique automatically yields the exponential decay solution.

The reverse flow solutions ($\beta_0 \leq \beta < 0, f''(0) < 0$) were originally investigated by Stewartson [Stew54]. From Figure 3–4 we see that these solutions contain a region of retarded flow, the extent of which approaches infinity as $\beta \to 0^-$. In these solutions $f' \to 1$ exponentially from below. The Nachtsheim-Swigert iteration technique, as originally presented in [Nach65], does not yield the reverse flow solutions. It always yields solutions with $f''(0) > 0$. In order to illustrate this, recall that the Nachtsheim-Swigert iteration technique, as originally presented, uses a small value of η_{max} to perform the initial iterations (see Section 2–7 and [Nach65]). Subsequently, η_{max} is increased and the final accurate solution obtained. Now consider the reverse flow solution for $\beta = 0.1$, $f''(0) = -0.1405462$ (see Table C–1). Initially choose a small value of η_{max}, say 4, in the **faskan** program. Using this value as the initial estimate for $f''(0)$ yields $f'(\eta_{max} = 4) = 0.2784268$. The first automatic iteration results in $f''(0) = 0.0730903$ and $f'(\eta_{max} = 4) = 0.7155146$. The second, third and fourth iterations yield $f''(0) = 0.1565081, 0.2092102$ and 0.2445931, with $f'(\eta_{max} = 4) = 0.819244, 0.8788328$ and 0.91710764, respectively. If η_{max} is now increased to 9 and automatic iteration continued, convergence to the solution without reverse flow (see Table C–1) takes place automatically, i.e., $f''(0) = 0.3192698$.

However, a slight conceptual modification of the Nachtsheim-Swigert iteration technique allows it to be used to obtain reverse flow solutions. The required modification is suggested by the velocity profiles shown in Figure 3–3 and the behavior of η_{conv} as $\beta \to \beta_0$ for $f''(0) > 0$ (see Table 3–1). Here, η_{conv} is the value of η at which $|1 - f'| < 5 \times 10^{-7}$ and $|f''| < 5 \times 10^{-6}$. From Table C–1[‡] we

---
[†] However, see papers by Libby [Libb68] and Rogers [Roge69a].
[‡] Actually, Table C–1 shows that this modification is also necessary to generate the solutions for $\beta_0 < \beta < 0$, $f''(0) > 0$.

see that η_{conv} increases as $\beta \to \beta_0$ for $f''(0) > 0$. This suggests that η_{max} for the reverse flow solutions must be more than or equal to that for $\beta = \beta_0$.

The investigation then proceeds, using the perturbation technique outlined above, i.e., we first seek a solution for $\beta = -0.19$, $f''(0) < 0$, then for $\beta = -0.18$, $f''(0) < 0$, etc. When a solution is attempted for $\beta = -0.1$, $f''(0) < 0$ with $\eta_{\text{max}} = 9$, it is found that convergence to the specified accuracy is not possible. This suggests that η_{max} is not large enough. When η_{max} is increased to 12, convergence to the specified accuracy occurs. Table C–1 shows that a further increase to $\eta_{\text{max}} = 16$ is required for the solution at $\beta = -0.025$, $f''(0) < 0$. Experience shows that $\eta_{\text{max}} \to \infty$ as $\beta \to 0^-$, $f''(0) \to 0^-$.

3–7 Skin Friction, Displacement and Momentum Thickness

One of the important boundary layer characteristics is the shearing stress at the surface. The shearing stress at the surface is generally called the skin friction. For a two-dimensional steady incompressible flow, the skin friction is given by (see Chapter 1)

$$\tau(x,0) = \mu \left. \frac{\partial u}{\partial y} \right|_{y=0} \tag{3-69}$$

Zero skin friction occurs when $\partial u/\partial y = 0$. This implies that $\partial u/\partial y$ changes sign at this point. Upstream of the point of zero skin friction the velocity is positive, whereas downstream it is negative. Under these conditions the flow ahead of the point of zero skin friction 'lifts' from the surface in order to pass over the region of backflow. The flow is said to separate from the surface. The point of zero shear is called the separation point. The streamlines at a separation point behave as shown in Figure 3–5. Further, if we consider a surface along which the pressure gradient is increasing in the direction of flow, a so-called adverse pressure gradient, then the development of the boundary layer velocity profiles as the separation point is approached is shown in Figure 3–6.

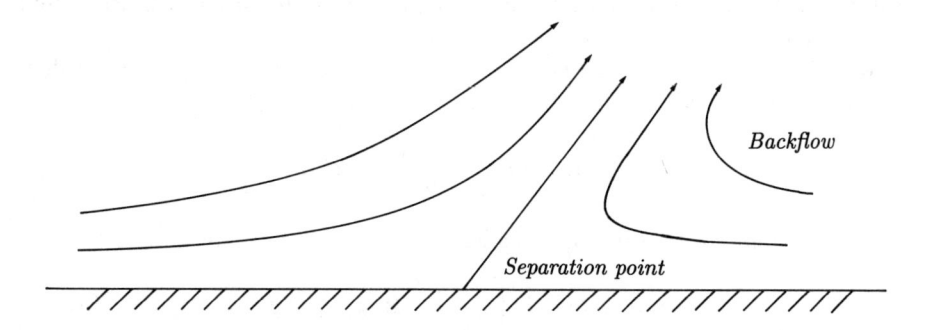

Figure 3–5. Streamlines in the vicinity of a point of zero skin friction, the separation point.

78   Laminar Flow Analysis

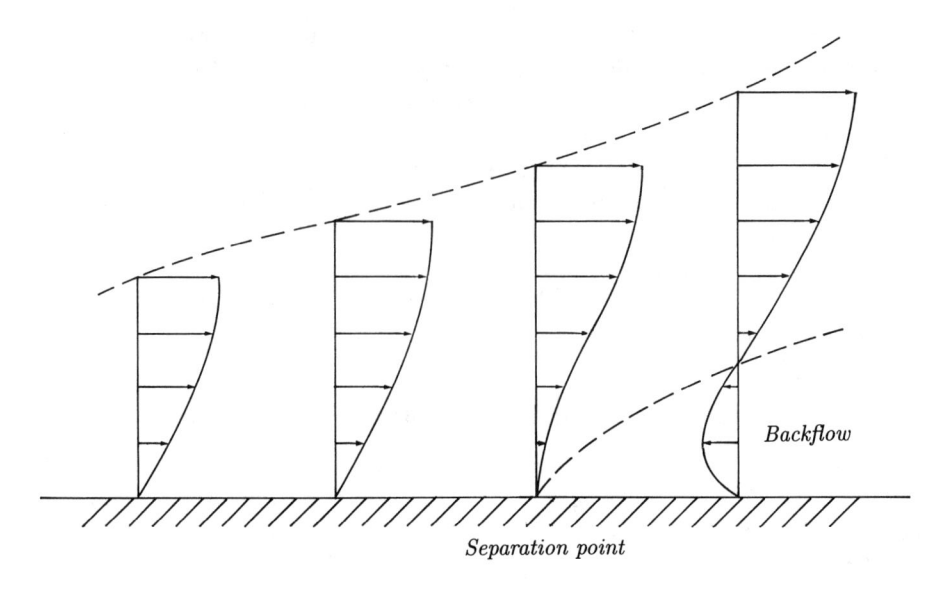

Figure 3–6. Velocity profiles in a region of adverse pressure gradient.

Returning to our similarity solutions, i.e., the Falkner-Skan equations, we see that Eq. (3–49b) shows that β negative corresponds to an adverse pressure gradient. The solutions to the Falkner-Skan equation (see Table C-1) show that the boundary layer can only sustain a small adverse pressure gradient without separation, i.e., $f''(0) = 0$ for $\beta = \beta_0 = -0.1988376$. Table C-1 also shows that the extent of the disturbed region, i.e., η_{conv}, increases rapidly for the reverse flow solutions to the Falkner-Skan equations. This indicates that downstream of a separation point the boundary layer thickness significantly increases. In addition, the vertical velocity downstream of a separation point is no longer of $O(\delta')$ but rather of $O(1)$. Under these conditions, the *thin* boundary layer assumption fails, and the mathematical model no longer corresponds to the physical problem. Thus, the solutions in this region are of more mathematical than physical interest.

Returning to the Falkner-Skan solutions, the shearing stress at the wall in terms of transformed variables becomes

$$\tau(x,0) = \mu U \sqrt{\frac{m+1}{2} \frac{U}{\nu x}} \, f''(0) \qquad (3-70)$$

Introduction of a skin friction coefficient $C_f = \tau(x,0)/(1/2 \rho U_\infty^2)$ yields

$$C_f = \sqrt{2(m+1)} \left(\frac{U}{U_\infty}\right)^{3/2} (\text{Re}_{x_\infty})^{-1/2} f''(0) \qquad (3-71)$$

where

$$\text{Re}_{x_\infty} = \frac{U_\infty x}{\nu} \qquad (3-72)$$

From Table C-1 we see that the skin friction increases with increasing values of β. In addition, Eq. (3–71) shows that for the Falkner-Skan similarity solutions $C_f \sim x^{(3m-1)/2}$. For the Blasius solution $m = \beta = 0$, and this reduces to $C_f \sim x^{-1/2}$. Equation (3–71) yields

$$C_f = \frac{\sqrt{2}f''(0)}{\sqrt{\text{Re}_{x_\infty}}} = \frac{0.6641154}{\sqrt{\text{Re}_{x_\infty}}} \qquad (3-73)$$

In addition to the skin friction, the displacement and momentum thicknesses are important characteristics of the boundary layer. The displacement thickness, δ^*, is interpreted as the decrease, with respect to the equivalent potential flow, in the mass rate of flow between the surface and a streamline at a large distance from the surface due to the effect of viscosity. A graphical interpretation of the displacement thickness is given in Figure 3–7. Here ρ_e is the density of the fluid at the 'edge' of the boundary layer. The mass flow rate in the equivalent potential flow is $\rho_e U y$. The mass flow rate in the boundary layer flow is $\int_0^\infty \rho u\, dy$. The decrease in the mass flow rate due to viscosity is $\int_0^\infty \rho(U - u)\, dy$. From our original definition of the displacement thickness we have

$$\delta^* = \int_0^\infty \left(1 - \frac{\rho u}{\rho_e U}\right) dy \qquad (3-74)$$

Here we see that the displacement thickness can also be interpreted as the distance a streamline at the outer edge of the boundary layer is displaced with respect to its position in the corresponding potential flow.

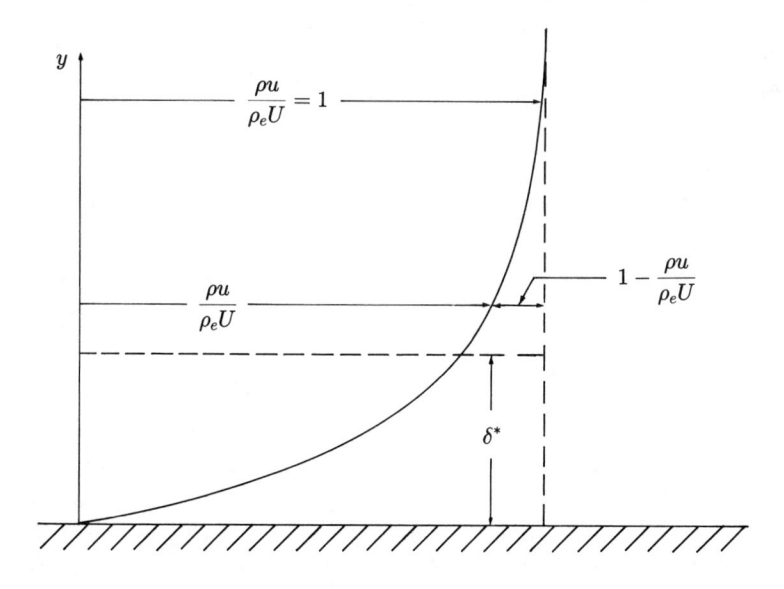

Figure 3–7. Graphical illustration of the displacement thickness.

80 Laminar Flow Analysis

Similarly, defining the momentum thickness, θ, as the decrease in momentum of the flow in the boundary layer with respect to that in the equivalent potential flow yields

$$\theta = \int_0^\infty \frac{\rho u}{\rho_e U}\left(1 - \frac{u}{U}\right)dy \qquad (3-75)$$

For the Falkner-Skan similarity flow, the displacement and moment thicknesses become

$$\delta^* = \sqrt{\frac{2}{m+1}\frac{\nu x}{U}} \int_0^\infty (1 - f')dy = \sqrt{\frac{2}{m+1}\frac{\nu x}{U}}\, I_1 \qquad (3-76)$$

and

$$\theta = \sqrt{\frac{2}{m+1}\frac{\nu x}{U}} \int_0^\infty f'(1 - f')dy = \sqrt{\frac{2}{m+1}\frac{\nu x}{U}}\, I_2 \qquad (3-77)$$

where I_1 and I_2 are definite integrals of the boundary layer solutions. Two particular cases of interest are the flat plate and two-dimensional stagnation point flow, i.e., $\beta = m = 0$ and $\beta = m = 1$. For a flat plate $U = $ constant and the increase in both the displacement and momentum thickness along the surface is proportional to $x^{1/2}$. However, for two-dimensional stagnation point flow $U \sim x$ and Eqs. (3–76) and (3–77) show that the displacement and momentum thicknesses are constant. I_1 and I_2 can be calculated by making the changes to the **faskan** program. Specifically, **eqmotfs** is changed to

equations of motion routine for the Falkner-Skan equations including the displacement and momentum integrals

subroutine eqmot(eta, x(), param(), f())
f(1) = −x(3)∗x(1)− param(1)∗(1−x(2)∗x(2))
f(2) = x(1)
f(3) = x(2)
f(4) = 1−x(2)
f(5) = x(2)∗f(4)
return

In the **faskan** program m = 5, headnam(4) = ''I1'', headnam(5) = ''I2'' are the only additions required.

The results of this calculation are shown in Table C-1 and Figure 3–8. Here we see that the displacement thickness increases with decreasing values of β and approaches infinity as the singularity at $f''(0) = 0^-$ and $\beta = 0^-$ is approached. The momentum thickness increases with decreasing values of β, until it reaches a maximum at the separation point. Subsequently the momentum thickness decreases for the reverse flow solutions as the singular point is approached. Physically this is a result of the reverse flow near the surface.

Physical Interpretation of the Falkner-Skan Equation   81

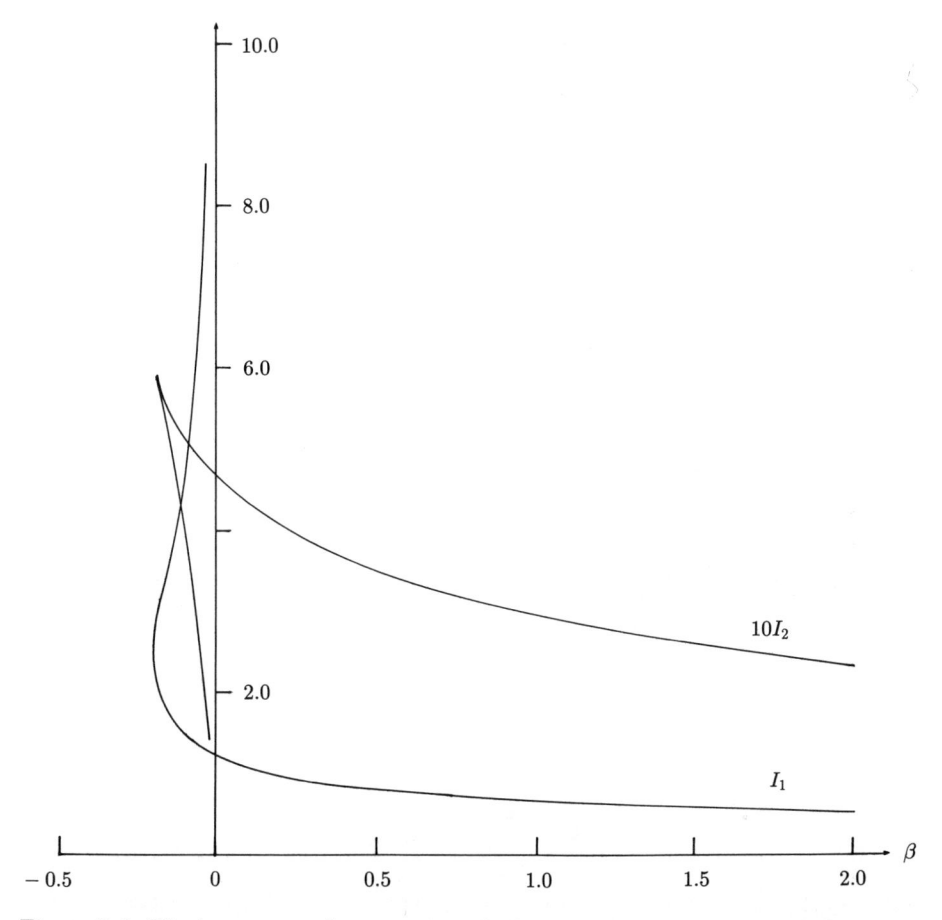

Figure 3–8. Displacement and momentum thickness as functions of the Falkner-Skan parameter, β.

3–8 Physical Interpretation of the Falkner-Skan Equation

Now that we have extensively studied the Falkner-Skan equation, it is appropriate to discuss the physical significance of the solutions. We have previously shown (see Section 3–4) that similarity solutions exist if the potential velocity has a power law variation with the surface distance. The potential flow past a wedge exhibits this characteristic. The complex potential function for this flow can be written as (see Figure 3–9 and [Rose63])

$$F(z) = \phi_i + i\psi_i = \frac{1}{m+1} U_\infty e^{-m\pi i} z^{m+1} \qquad (3-78)$$

where the complex variable $z = xe^{i\theta}$ and $m = \beta/(\pi - \beta) > 0$. The velocity is given by

$$\frac{dF(z)}{dz} = u - iv = U_\infty e^{-m\pi i} z^m = U_\infty x^m e^{m(\pi-\theta)i}$$

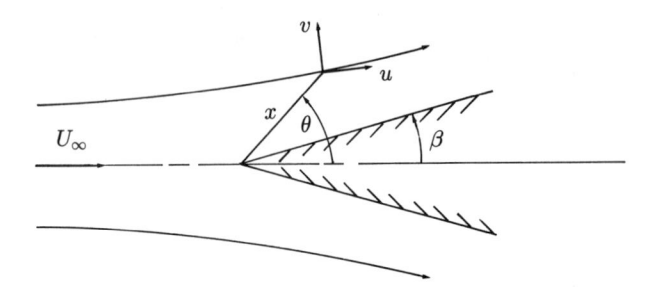

Figure 3–9. Wedge flow.

The velocity components are

$$u = U_\infty x^m \cos m(\pi - \theta) \quad (3-79a)$$

$$v = U_\infty x^m \sin m(\pi - \theta) \quad (3-79b)$$

At the wedge surface $\theta = \beta$, and Eqs. (3–79a,b) reduce to

$$u = U_\infty x^m \cos \beta \quad (3-80a)$$

$$v = U_\infty x^m \sin \beta \quad (3-80b)$$

The magnitude of the velocity is then

$$\frac{U}{U_\infty} = \left(\frac{u^2 + v^2}{U_\infty^2}\right)^{1/2} = x^m \quad (3-81)$$

where in this case x is the distance along the surface. The Falkner-Skan solutions for $m > 0$, i.e., $\beta > 0$, can thus be interpreted as the flow past a wedge of semiangle $\beta\pi/2$. Let us illustrate several of these flows. First we divide the discussion into those regions where $-0.198876 \leq \beta \leq 0$, $0 \leq \beta \leq 1$, and $1 \leq \beta \leq 2$. When β is in the range $0 < \beta < 1$ the flow is accelerated. Three typical potential flows are shown in Figures 3–10a,b,c. When $\beta = 0$ we have an infinitesimally thin flat plate with $x = 0$ at the leading edge. When $\beta = 0.5$ we have a 45° half angle wedge, with $x = 0$ beginning at the apex of the wedge. For $\beta = 1$ we have two dimensional stagnation point flow, with $x = 0$ at the stagnation point of the flow. When $\beta = 1.5$, Figure 3–10d shows that the potential flow corresponds to flow into an acute corner. Here the apex of the corner is a singular point of the flow, and we take $x = 0$ at this point. However, when $\beta = 2$ we have

$$\frac{U}{U_\infty} = x^m = x^{\frac{\beta}{(2-\beta)}}$$

which implies that m increases without bound. From Figure 3–10d the flow folds back on itself. This implies that the physical space through which the fluid flows

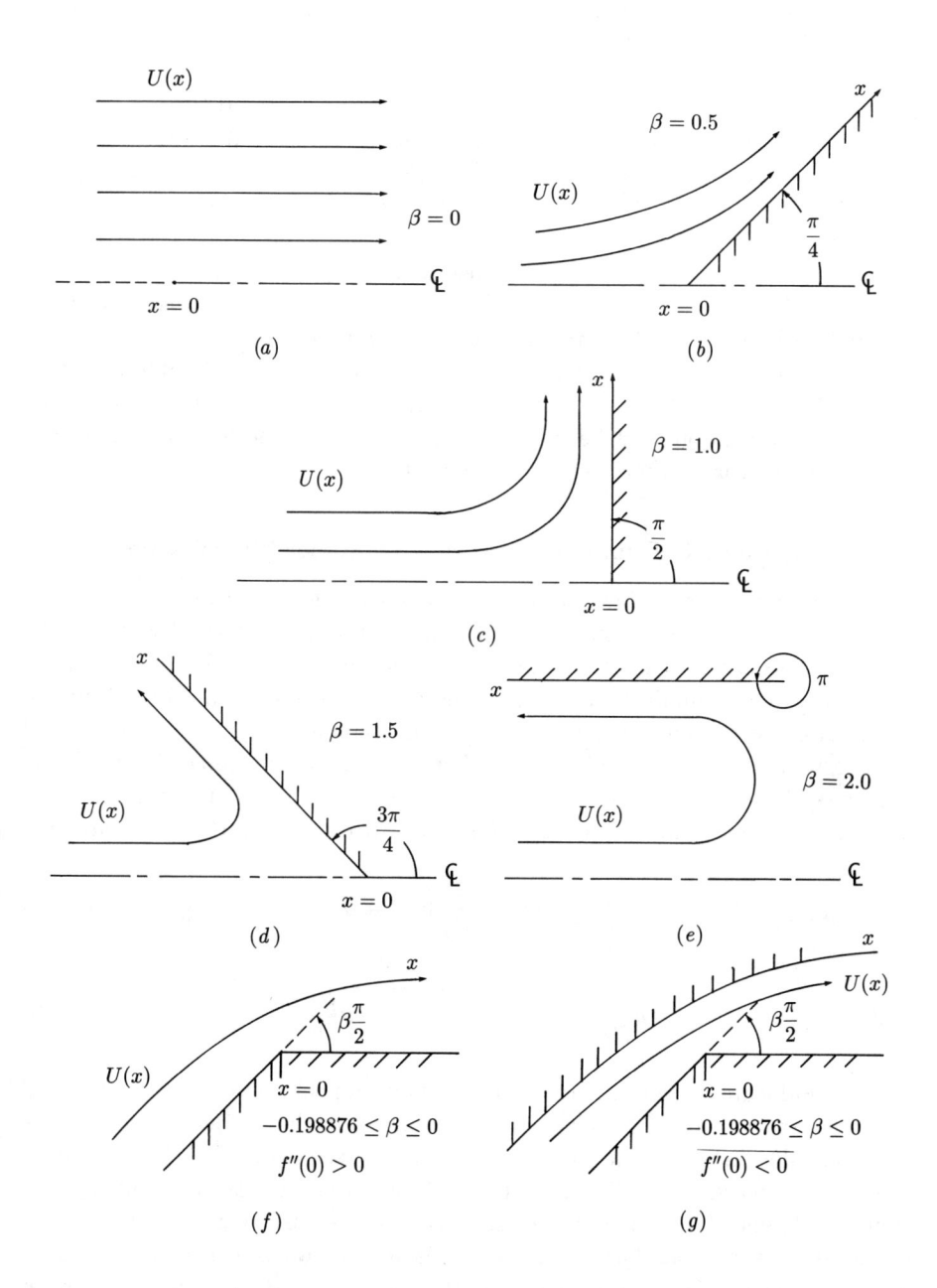

Physical Interpretation of the Falkner-Skan Equation 83

Figure 3–10. Potential flows associated with the Falkner-Skan solutions.

vanishes. However, it can be shown (see [Evan68]) that the flow illustrated in Figure 3–10e is physically possible and represents a solution for $\beta = 2$. In this flow the velocity close to the channel walls has an exponential variation. No singularity occurs in the flow, and $x = 0$ is taken as any arbitrary point in the flow. For further details the reader is referred to Evans [Evan68].

Looking now at possible flows for $-0.198876 \leq \beta \leq 0$, we see that these flows can be either accelerated or decelerated. Flows for which $-0.198876 \leq \beta \leq 0$ and $f''(0) > 0$ are accelerated, whereas those for which $-0.198876 \leq \beta \leq 0$ and $f''(0) < 0$ are decelerated. Evans ([Evan68]) suggests that these flows occur as shown in Figures 3–10f and 3–10g. In Figure 3–10f an accelerated flow with $-0.198876 \leq \beta \leq 0$ occurs downstream of the corner, providing the boundary layer starts at the corner. This flow exists only for $f''(0) > 0$, since separation occurs if $-0.198876 \leq \beta \leq 0$ and $f''(0) < 0$ for the external stream shown in Figure 3–10f. However, if a streamline of the flow is replaced by a solid surface, as shown in Figure 3–10g, the flow does not separate, and a decelerated flow with $-0.198876 < \beta < 0$, $f''(0) < 0$ exists. In addition to these interpretations, the Falkner-Skan solutions can be physically interpreted as the flow past a surface with some externally impressed pressure gradient.

3–9 Axisymmetric Boundary Layers — Mangler Transformation

Many boundary layer problems of interest are concerned with axisymmetric flow. The boundary layer equations for axisymmetric flow are directly derived by the previously discussed order-of-magnitude analysis, or they are obtained from the two-dimensional boundary layer equations by a coordinate transformation. In either case it is customary to assume that the boundary layer thickness is small with respect to the local body radius. For the case of axisymmetric steady incompressible constant property flow, the axisymmetric continuity and boundary layer momentum equations are (see Rosenhead [Rose63] and Hansen [Hans64])

$$\frac{\partial}{\partial x}(ru) + \frac{\partial}{\partial y}(rv) = 0 \qquad (3-82)$$

and

$$u\frac{\partial u}{\partial x} + v\frac{\partial u}{\partial y} = -\frac{1}{\rho}\frac{\partial P}{\partial x} + \nu\frac{\partial^2 u}{\partial y^2} \qquad (3-83)$$

where x and y are the coordinate distances along and normal to the surface, u and v are the components of velocity along and normal to the surface and $r(x)$ is the perpendicular distance from the axis of symmetry (see Figure 3–11). Comparing these equations with Eqs. (3–41) and (3–42), note that only the continuity equation is different. Thus, it is reasonable to seek a transformation that reduces the axisymmetric boundary layer equations to the two–dimensional boundary layer equations. The appropriate transformation was discovered by Mangler

([Mang48a]). The Mangler transformation is also valid for compressible boundary layers; however, here we consider only the incompressible case. Since in deriving the axisymmetric boundary layer equations the assumption is made that the boundary layer thickness is small with respect to the local body radius, it is reasonable to approximate the radial distance to the boundary layer edge by the distance to the surface, r_0 (see Figure 3–11). The Mangler transformations are then

$$\bar{x} = \int_0^x r_0^2(x)dx \qquad \bar{y} = r_0 y \qquad (3-84a,b)$$

$$\bar{u} = u \qquad \bar{U}(\bar{x}) = U(x) \qquad \bar{v} = \frac{1}{r_0}\left(v + \frac{y}{r_0} u \frac{dr_0}{dx}\right) \qquad (3-85a,b,c)$$

From Eqs. (3–84a,b) it is easily shown that the independent variable transformations are

$$\frac{\partial}{\partial x} = r_0^2 \frac{\partial}{\partial \bar{x}} + \frac{\bar{y}}{r_0} \frac{dr_0}{dx} \frac{\partial}{\partial \bar{y}}$$

and

$$\frac{\partial}{\partial y} = r_0 \frac{\partial}{\partial \bar{y}}$$

Equations (3–82) and (3–83) then become

$$\frac{\partial \bar{u}}{\partial \bar{x}} + \frac{\partial \bar{v}}{\partial \bar{y}} = 0 \qquad (3-86)$$

$$\bar{u}\frac{\partial \bar{u}}{\partial \bar{x}} + \bar{v}\frac{\partial \bar{u}}{\partial \bar{y}} = \bar{U}\frac{d\bar{U}}{d\bar{x}} + \nu\frac{\partial^2 \bar{u}}{\partial \bar{y}^2} \qquad (3-87)$$

which are identical with Eq. (3–41) and (3–42) after Euler's equation is used to rewrite Eq. (3–42). Thus, the determination of the laminar axisymmetric

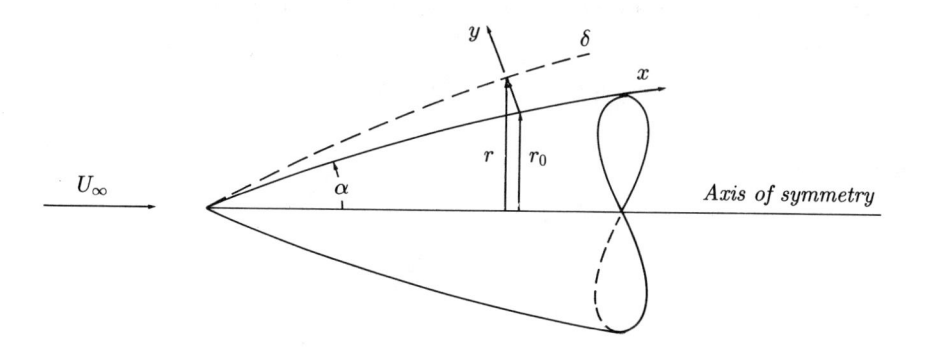

Figure 3–11. Axisymmetric boundary layer coordinate system.

86   Laminar Flow Analysis

boundary layer is reduced to the determination of the equivalent two-dimensional boundary layer. In particular, the Falkner-Skan similarity solutions are applicable. Here we see that $\bar{U}(\bar{x}) = U(x)$ gives the equivalent two-dimensional inviscid flow and allows identification of the surface.

To better illustrate the above concept, consider the axisymmetric analog of the flow past a sharp flat plate, i.e., the axisymmetric flow past a pointed cone. First consider the inviscid flow past the cone. The flow is governed by Laplace's equation. It is convenient to formulate the problem in spherical polar coordinates. The solution of Laplace's equation in spherical polar coordinates (see Goldstein [Gold30] or Evans [Evan68]) shows that

$$U_i = U_\infty x^n \qquad (3-88)$$

The exponent n depends on the cone half angle, α. Determination of the relationship between n and α requires the solution of Laplace's equation. Evans ([Evan68]) shows that this is accomplished by separation of variables in terms of a function $R(r)$ only and a function of $G(\alpha)$ only. The resulting differential equation for $R(r)$ is easily integrated. The resulting differential equation for $G(\alpha)$ is reduced to Legendre's differential equation by a simple trigonometric transformation. Evans further shows that for integer values of $n \geq 0$[†] the solution to Legendre's equation is given in terms of Legendre polynomials. Evans presents the first six of these values. For nonintegral values of n, the solutions are obtained by numerical integration. Evans presents the results of numerical solutions for α as a function of n for $0 \leq 0 \leq 9$. These results are shown in Figure 3–12. Although Evans presents accurate solutions for α as a function of n, the first determination of α as a function of n is due to Mangler [Mang48b].

The governing equations for the corresponding axisymmetric boundary layer problem are Eqs. (3–82) and (3–83). Here we have (see Figure 3–11)

$$r_0 = x \sin \alpha \qquad (3-89)$$

Using the Mangler transformation the corresponding two-dimensional boundary layer problem is

$$f''' + ff'' + \left(\frac{2n}{3+n}\right)(1 - f'^2) = 0 \qquad (3-90)$$

with

$$u = Uf'(\bar{\eta}) \qquad (3-91)$$

$$\bar{\eta} = \left[\frac{(3+n)U_\infty x^{n-1}}{2\nu}\right]^{1/2} y \qquad (3-92)$$

and

$$v = -\frac{1}{2}U\left[\frac{2\nu}{(3+n)U_\infty x^{n+1}}\right]^{1/2}[(3+n)f + (n-1)\bar{\eta}f'] \qquad (3-93)$$

[†]Here our notation for n corresponds to $(n-1)$ in Evans' notation.

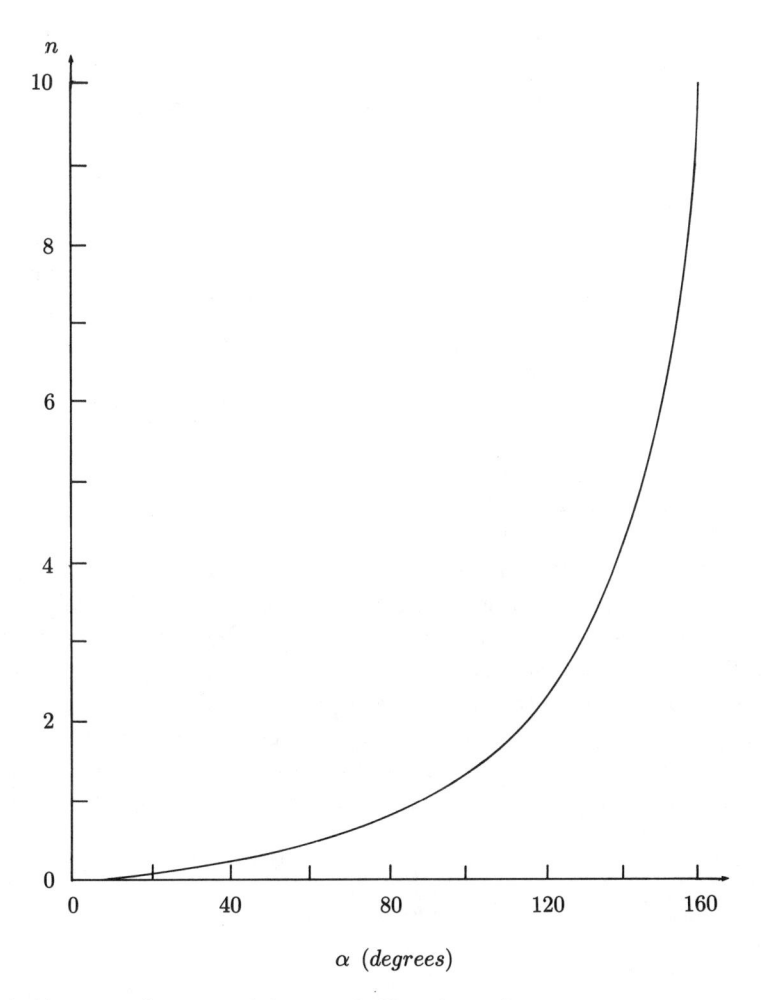

Figure 3-12. n as a function of the cone half angle, α, for axisymmetric potential flow.

Here the prime denotes differentiation with respect to $\bar{\eta}$. The boundary conditions for an impermeable wall are

$$f(0) = f'(0) = 0 \qquad (3-94a)$$
$$f'(\bar{\eta} \to \infty) \to 1 \qquad (3-94b)$$

Comparing Eq. (3-90) with the equation governing two-dimensional boundary layer flow, Eq. (3-50), we see that the axisymmetric boundary layer flow past a cone is related to two-dimensional boundary layer flow with $\beta = [2n/(3+n)]$. This may be interpreted as flow past a wedge with half angle $[2n/(3+n)]\pi/2$. For example, for a 30° half angle cone Figure 3-12 yields $n = 0.125$. β is then $2n/(3+n) = 0.08$, and the flow past a 30° half angle cone is equivalent to the flow past a wedge of half angle 0.04π (see Section 3-8).

88   Laminar Flow Analysis

We now turn our attention to the discussion of several of the flows with integral values of n. We first consider $n = 0$. From Figure 3–12 we see that the cone angle, α, is zero. Under these circumstances the cone is reduced to an infinitely thin needle. For $n = 0$, $\beta = 0$ and the boundary layer solution corresponds to the Blasius solution with one notable exception. The Blasius solution is for the viscous boundary layer on an infinitely thin flat plate in a uniform parallel flow. From Figure 3–1 we note that the streamlines are uniformly spaced. Here the corresponding axisymmetric solution is for the viscous boundary layer on an infinitely thin needle in a uniform flow (see Figure 3–13). For axisymmetric flow the streamlines are not uniformly spaced; they are more closely spaced as the distance from the axis of symmetry increases.

For $n = 1$, $\beta = 1/2$ and from Figure 3–12 $\alpha = 90°$. This corresponds to axisymmetric stagnation point flow previously discussed in Section 3–10. In this case the cone is an infinite circular disk. Figure 3–14 shows that the viscous boundary layer starts at the 'center' of the disk, i.e., at the stagnation point, and spreads radially outward.

For $n > 1$, Figure 3–12 shows that the cone half angle is more than $90°$. This represents flow into a converging conical surface as shown in Figure 3–15. Note that here the viscous boundary layer starts from the apex of the cone. For further discussion of the character of these conical surfaces the reader is referred to Evans [Evan68], where they are extensively discussed.

Before leaving our discussion of axisymmetric flows, it is of interest to obtain the relationship between the nondimensional stretched coordinates, η and $\bar{\eta}$, for two-dimensional and axisymmetric flow. From Eqs. (3–58a) and (3–62) we have

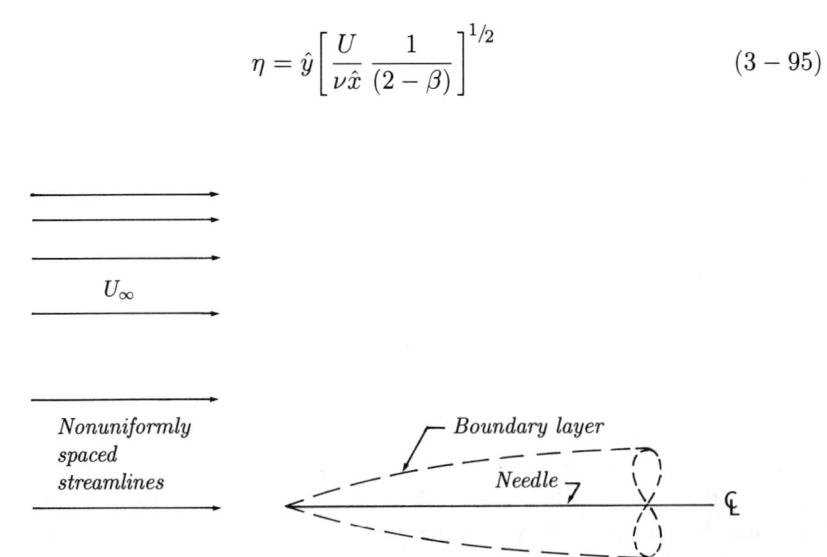

$$\eta = \hat{y} \left[\frac{U}{\nu \hat{x}} \frac{1}{(2 - \beta)} \right]^{1/2} \qquad (3-95)$$

Figure 3–13. Viscous boundary layer on an infinitely thin axisymmetric needle $n = 0$.

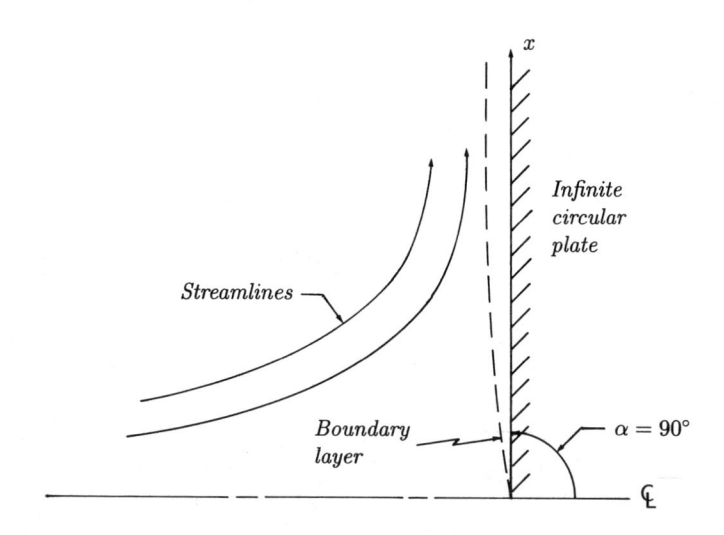

Figure 3-14. Axisymmetric stagnation point flow, $n = 1$.

where here $\hat{x}$ and $\hat{y}$ are two-dimensional coordinates. Similarly, the axisymmetric stretched normal coordinate, $\bar{\eta}$, is

$$\bar{\eta} = y \left[\frac{3}{(2-\beta)} \frac{U}{\nu x} \right]^{1/2} \qquad (3-96)$$

Comparing Eqs. (3-95) and (3-96) and using Eqs. (3-84a,b) yields

$$\eta = S(x)\bar{\eta} \qquad (3-97)$$

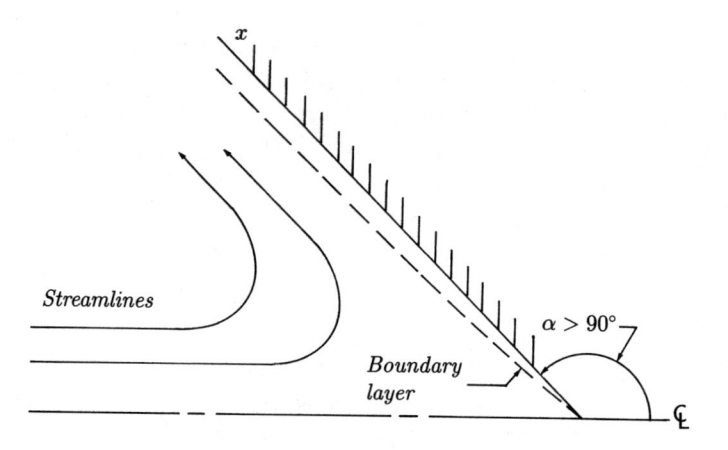

Figure 3-15. Axisymmetric flow for $n > 1$.

Laminar Flow Analysis

where $S(x)$ is given by

$$S(x) = \left(\frac{r_0^2 x}{\int_0^x r_0^2 \, dx} \right)^{1/2} \tag{3-98}$$

For conical flow $S(x) = \sqrt{3}$.

3-10 Boundary Layers with Mass Transfer

In all of our previous discussion we assumed the surface impermeable, i.e., no mass was transferred through the surface. We now consider solutions to the steady two-dimensional incompressible viscous boundary layer equations when there is mass transfer through the surface. Mass can be transferred either into or out of the surface, that is, we consider both suction and injection or blowing. Our previous discussion for zero mass transfer is, of course, included as a special case. Here we limit our discussion to similar solutions to the boundary layer equations. Further, we assume that mass transfer occurs normal to the surface, that in the case of injection the injected fluid and the fluid in the main stream are the same, and finally that the injected fluid has the same temperature as the surface. The governing equations are again Eqs. (3-41) and (3-42), i.e.

$$\frac{\partial u}{\partial x} + \frac{\partial v}{\partial y} = 0 \tag{3-41}$$

$$u\frac{\partial u}{\partial x} + v\frac{\partial u}{\partial y} = -\frac{1}{\rho}\frac{\partial P}{\partial x} + \nu\frac{\partial^2 u}{\partial y^2} \tag{3-42}$$

The boundary conditions with mass transfer through the surface are

$$y = 0 \quad u = 0 \quad v = v_0 \tag{3-98a,b}$$

$$y \to \infty \quad u \to U(x) \tag{3-98c}$$

where, again, $U(x)$ is the inviscid flow velocity along the surface and v_0 is the velocity normal to the surface with which fluid passes through the surface. We now seek similar solutions for these equations. Recalling our previous discussion (see Section 3-4) we note that similar solutions of the general steady two-dimensional incompressible boundary layer with mass transfer at the surface satisfy the two-point asymptotic boundary value problem given below. Here we again restrict ourselves to $\alpha \neq 0$ and $(2\alpha - \beta) \neq 0$ and without loss of generality take $\alpha = 1$. Thus

$$f''' + ff'' + \beta(1 - f'^2) = 0 \tag{3-50}$$

with boundary conditions

$$f(0) = f_0 = \text{constant} \tag{3-99a}$$

$$f'(0) = 0 \tag{3-99b}$$

and

$$f'(\eta \to \infty) \to 1 \tag{3-99c}$$

The nondimensional stream function at the surface, $f(0)$, is set equal to a constant to satisfy similarity conditions. This requires that the normal velocity with which fluid crosses the surface varies as the reciprocal square root of the distance along the surface i.e., as $x^{-1/2}$. In particular, differentiating Eq. (3-45c) yields

$$v_0 = -f_0 \sqrt{\frac{m+1}{2}} \sqrt{\frac{\nu U}{x}} = \frac{-f_0}{\sqrt{2-\beta}} \sqrt{\frac{\nu U}{x}} \qquad (3-100)$$

Negative values of f_0 correspond to injection of fluid through the surface into the boundary layer, whereas positive values correspond to suction. Examination of the governing boundary value problem, Eqs. (3-50) and (3-99), shows that only the boundary condition $f(0) = f_0 = $ constant differs from that governing the 'impermeable' surface case. Solutions to the present two-point asymptotic boundary value problem are obtained by simply adding init(1)= f_0, where f_0 is the numeric value for the mass transfer parameter, e.g., init(1) = 0.1, to the faskan program.

The normal velocity, v_0 and hence f_0, can theoretically have any value between $\pm\infty$. However, in the case of injection large injection rates supply fluid with zero velocity along the surface faster than the available viscous mechanism can accelerate it. Under these circumstances the boundary layer is 'blown away from the surface'. Physically this increases the boundary layer thickness so that the original 'thin' boundary layer approximation is violated. This phenomenon was first discovered by Lock [Lock51] in an investigation of the laminar boundary layer between two two-dimensional jets with different properties. It was subsequently interpreted by Emmons and Leigh [Emmo53] as boundary layer 'blow off'. When $\beta = 0$, Eq. (3-50) reduces to the Blasius equation. The associated boundary value problem is that for uniform parallel flow past an infinitesimally thin flat plate with surface mass transfer. It is this boundary value problem that Emmons and Leigh [Emmo53] investigated. Emmons and Leigh's recomputed limiting value of $f_0 = -0.875745$[†] for which boundary layer 'blow off' occurs is in excellent agreement with Lock's original value. For the limiting injection value of $f_0 = -0.875745$ the nondimensional shearing stress parameter $f''(0) = 0$. This is physically reasonable, since it is at this point that the viscous acceleration mechanisms can no longer accelerate the fluid injected at the surface.

In addition to the limiting value for f_0 Emmons and Leigh obtained more accurate results and extended the original hand calculations of Schlichting and Bussmann [Schl43]. Their results for $\beta = 0$ cover the range from the limiting injection value of $f_0 = -0.875745$ to $f_0 = 7.07$.

Before continuing our discussion of mass transfer, we note that the group property of the Blasius equation previously discussed (see Section 3-5) is still

---

[†]Emmons and Leigh's values of f_0 are divided by $\sqrt{2}$ and their values of $f''(0)$ divided by $2\sqrt{2}$ to correspond to the present notation.

92   Laminar Flow Analysis

valid. With surface mass transfer we have the additional relation

$$f(0) = \frac{\bar{f}(0)}{[\bar{f}'(\bar{\eta} \to \infty)]^{1/2}} \qquad (3-101)$$

The Falkner-Skan equation with mass transfer through the surface has also been extensively investigated (see Evan66; Evan63; Evan62; Spal61a; Spal61b, for example). In particular, Evans [Evan66] conducted an extensive investigation and compilation of the similar solutions. Evans' results were used as a guide in obtaining the solutions presented in the tables in Appendix C. Figure 3-16 shows curves of β vs $f''(0)$ parameterized with the mass transfer parameter f_0. Comparison with Figure 3-3 shows that suction increases the *magnitude* of the shearing stress parameter $f''(0)$ at the surface. Further, Figure 3-17 shows that suction increases the magnitude of the adverse pressure gradient that a similar boundary layer sustains without separation. Injection, on the other hand, decreases the magnitude of the adverse pressure gradient that a similar boundary layer sustains without separation. Further examination of Figure 3-16 shows that for values of the suction parameter less than $f_0 = -0.875745$ solutions do

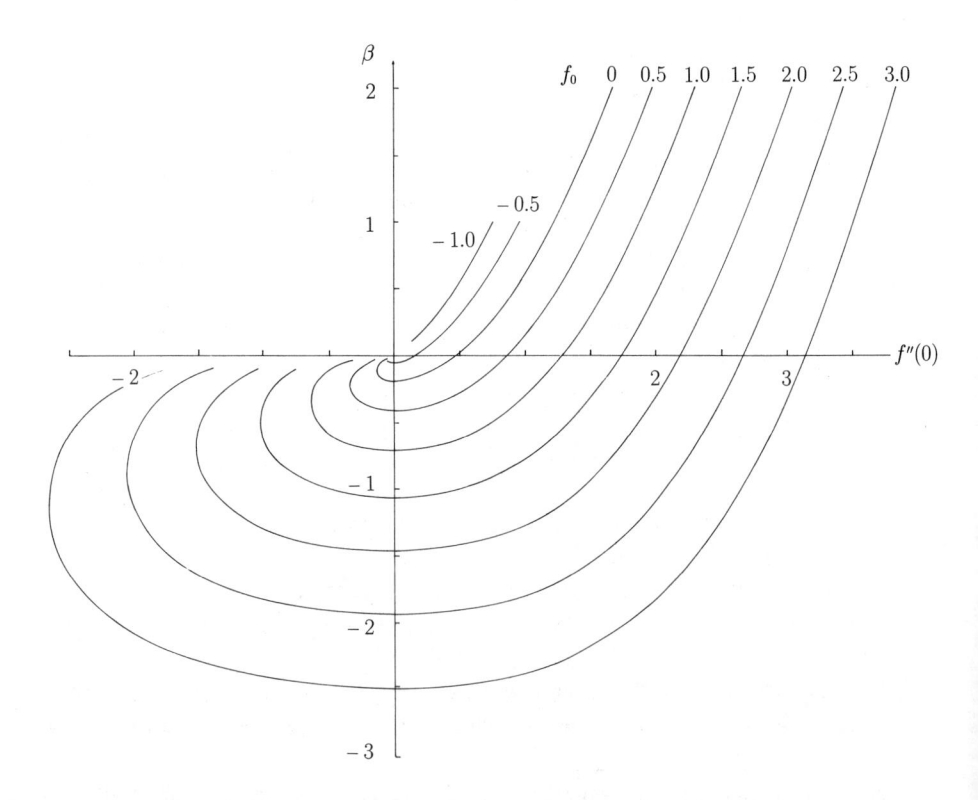

Figure 3-16. Boundary layer solutions with mass transfer.

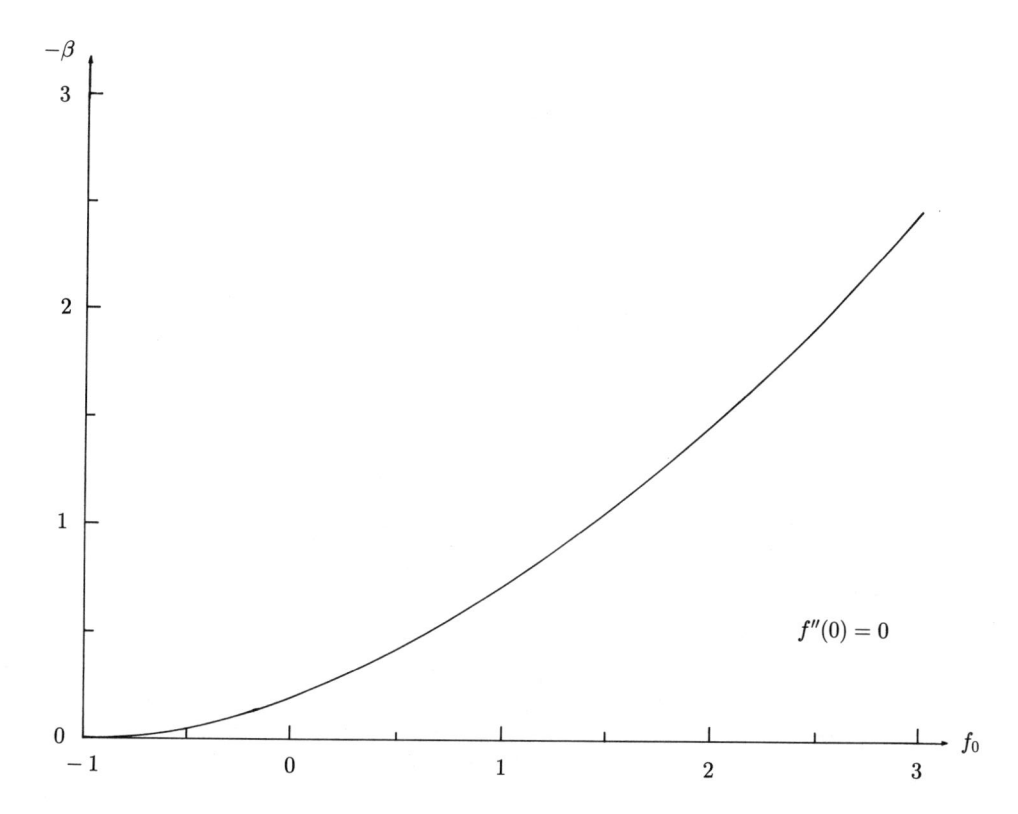

[Graph showing curve of $-\beta$ vs f_0 with label $f''(0) = 0$]

Figure 3–17. Effect of mass transfer on boundary layer separation.

not exist near $\beta = 0$. Figure 3–18 shows the effect of mass transfer on the boundary layer velocity profiles. Comparison with Figure 3–4 shows that the effect of suction on the velocity profile is similar to the effect of pressure gradient. The boundary layer thickness decreases, and the shearing stress at the surface increases, with either increasing pressure gradient or suction. The effect of blowing on the velocity profile is similar to the effect of an adverse pressure gradient. For β or f_0 less than zero and the shearing stress parameter positive, the velocity profile exhibits an inflection point, the boundary layer thickness increases and the shearing stress parameter at the surface decreases with decreasing values of β or f_0. There is no apparent analogy between the reverse flow solutions, i.e., $\beta < 0$ and $f''(0)$ without mass transfer and the effects of blowing on the velocity profiles.

We now turn our attention to nonsimilar boundary layers.

3–11 Nonsimilar Boundary Layers

Previously we were concerned with similar solutions of the boundary layer equations. Since for similar solutions the governing partial differential equations were

94   *Laminar Flow Analysis*

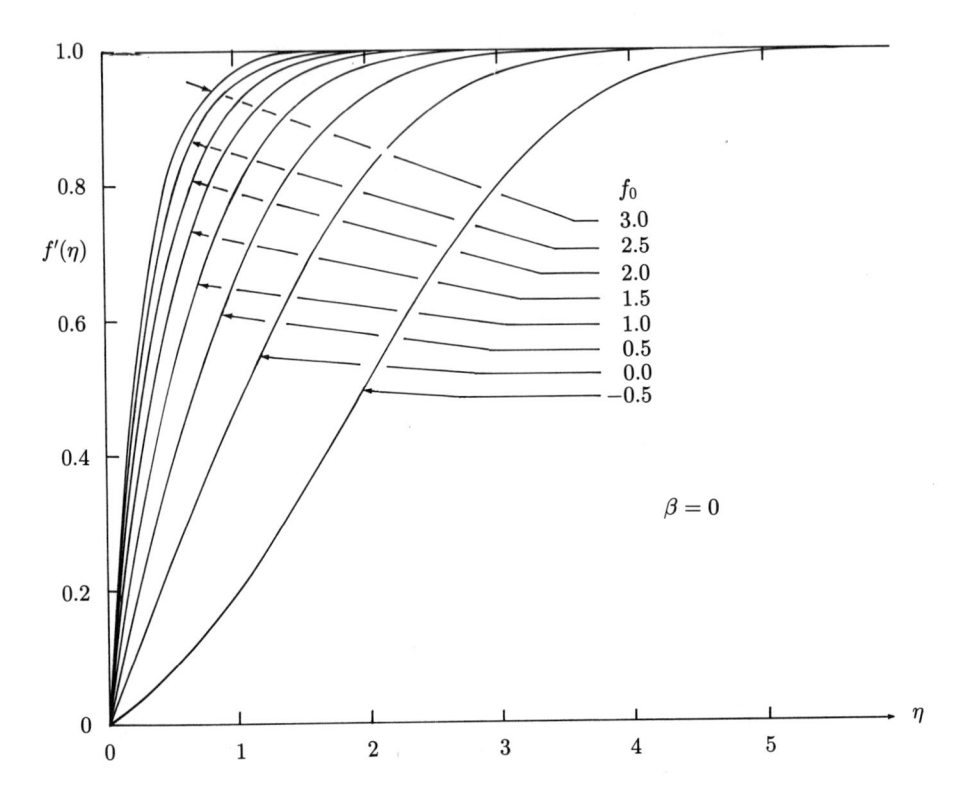

Figure 3–18. The effect of mass transfer on the velocity profiles for $\beta = 0$.

reduced to ordinary differential equations, considerable mathematical simplification resulted. This allowed parametric investigation of the solutions, e.g., the parameterization of the Falkner-Skan solutions with the pressure gradient parameter β. However, many boundary layer flows are not similar. Nonsimilarities are encountered because

> The velocity variation in the potential flow at the edge of the boundary layer is not of the required form (e.g., see Eq. 3–56).
>
> The boundary conditions are not of the required form (e.g., see Eq. 3–100).
>
> Additional terms which are nonsimilar occur in the governing equations.

A number of methods for solving nonsimilar boundary layer problems have been developed. These methods generally fall into the following categories:

> Approximate techniques such as series expansions or integral solutions: Of these the most familiar are the von Karman integral technique and the Blasius or Görtler series expansions (see [Schl68] and [Gort52]).
>
> Finite difference-differential techniques: Here, the governing equations are put into almost similar form, i.e., they are almost ordinary differential equations. The remaining nonsimilar terms or partial derivatives are then treated

in a finite difference form. The resulting finite difference-differential equations are solved at successive stations along the body. This technique is illustrated in Smith and Clutter [Smit63] and in Flügge-Lotz and Baxter [Flug62].

Some ad hoc approximation regarding the contribution of the nonsimilar terms to the complete boundary layer solutions is made. A good summary of these ideas is given in Moore [Moor64]. In particular, Lees and Reeves [Lees64] used this technique to good effect in discussing shock-boundary layer interactions.

The Meksyn-Merk method: The method is fundamentally due to Meksyn [Meks61] but has been clarified and extensively discussed by Merk [Merk59], Evans [Evan68] and Dewey and Gross [Dewe67]. Fundamentally, the method assumes that if the boundary layer is almost similar the direct effect of the nonsimilar terms can be determined by asymptotically expanding the full boundary layer equations in terms of small parameters which measure the departure from similarity. The result is a hierarchy of equations or systems of equations, the first of which is the similar solution. This technique is mathematically sound. Further, it can, in principle, be extended to any arbitrary degree of accuracy by successively solving the higher-order equations or systems of equations. However, it suffers from the practical necessity of very accurately calculating several similar solutions in the neighborhood of the desired nonsimilar solution. For this reason very few solutions are known.

Local similarity. Another method for analyzing nonsimilar boundary layer flows is to assume that the flow is 'locally similar'. This implies that the previous (upstream) history of the flow is ignored except as it influences the calculation of the similarity variables.

The accuracy of the locally similar solutions depends upon the rate of change of the external inviscid flow and the body geometry in the streamwise direction. When these change slowly, the derivatives of the boundary layer dependent variables with respect to the streamwise coordinate are small with respect to the derivatives in the normal direction and can be neglected. For example, this implies that the derivatives of f with respect to ξ are small, and hence the term $(f'f_\xi - f''f_\xi)$ on the right hand side of Eq. (3–48) is negligible. Alternately, it is assumed that the coefficient of the terms on the right hand side of Eq. (3–48) is sufficiently small that the right hand side is negligible. The terms on the left hand side of Eq. (3–48), which are functions of ξ, are assumed to take their local values. This allows the boundary layer equations to be considered as ordinary differential equations in η, with ξ as a parameter. Locally similar solutions are then determined for various positions along the surface, i.e., for various values of β.

Marvin and Sinclair [Marv67] showed that for the compressible boundary layer the concept of local similarity yields acceptable results for favorable and mildly adverse pressure gradients. A subsequent study by Gallo, Marvin, and Gnos [Gall69] confirmed that local similarity does not yield acceptable results for strong adverse pressure gradients.

Each of these techniques is useful for certain types of problems. However, each has significant limitations. We now proceed to develop a general nonsimilar boundary layer method which is simple, internally self-consistent, provides the ability to internally determine the accuracy of the results, and which can be extended to any arbitrary degree of accuracy. This method, originally proposed by Sparrow, Quack and Boerner [Spar70] for the velocity field (see Eqs. 3–48 and 3–51), is called the locally nonsimilar boundary layer method.

3-12 Locally Nonsimilar Boundary Layer Method

We begin the development of the locally nonsimilar boundary layer method by recalling the governing equations and associated boundary conditions for the general two-dimensional steady constant property boundary layer (see Section 3–4). In terms of the stream function these are

$$\psi_y \psi_{xy} - \psi_x \psi_{yy} = U \frac{dU}{dx} + \nu \psi_{yyy} \tag{3-43}$$

and
$$\psi_x(x, 0) = \psi_y(x, 0) = 0 \tag{3-44a,b}$$

$$\psi_y(x, y \to \infty) \to U(x) \tag{3-44c}$$

In order to make the governing equations almost similar, i.e., almost ordinary differential equations, and thus decrease the dependence of the solutions on the streamwise coordinate, a similarity transformation is introduced. The transformations are similar in form to those previously derived for the similar boundary layer equations (Eqs. 3–45). However, the streamwise coordinate x is now explicitly transformed. In particular, following Görtler [Gort52], Meksyn [Meks61] and Sparrow, Quack and Boerner [Spar70], we introduce transformations of the form

$$\xi = \int_0^x \left(\frac{U(x)}{U_\infty}\right) d\left(\frac{x}{\ell}\right) \tag{3-102}$$

$$\eta = \frac{yU(x)}{(2\nu \ell U_\infty \xi)^{1/2}} \tag{3-103}$$

$$\psi = (2\nu \ell U_\infty \xi)^{1/2} f(\xi, \eta) \tag{3-104}$$

where U_∞ is some constant reference velocity and ℓ is some characteristic length. Equations (3–43) and (3–44) then become

$$f''' + ff'' + \beta(1 - f'^2) = 2\xi(f' f'_\xi - f'' f_\xi) \tag{3-105}$$

with
$$f(\xi, 0) = f'(\xi, 0) = 0 \tag{3-106a,b}$$

and
$$f'(\xi, \eta \to \infty) \to 1 \tag{3-106c}$$

where the primes denote differentiation with respect to η and

$$\beta = \frac{2}{U^2} \frac{dU}{dx} \int_0^x U \, dx \tag{3-107}$$

Equations (3–105) and (3–106) are of the same functional form as Eqs. (3–48) and (3–51). However, here f, ξ, η, and β have somewhat different definitions. The terms on the right hand side of the momentum equation, Eq. (3–105), represent the nonsimilar boundary layer effect. In the locally similar approximation these terms are neglected. The essence of the locally nonsimilar method is to introduce an auxiliary function

$$S(\xi, \eta) = f_\xi \qquad (3-108)$$

which allows writing the momentum equation as

$$f''' + ff'' + \beta(1 - f'^2) = 2\xi(f'S' - f''S) \qquad (3-109)$$

where $S' = f'_\xi = \partial f'/\partial \xi$. Physically, $S(\xi, \eta)$ is considered the variation of the stream function in the streamwise direction. In order to determine $S(\xi, \eta)$ the momentum equation, Eq. (3–109), and the boundary conditions, Eq. (3–105), are differentiated with respect to ξ to yield

$$S''' + fS'' - 2\beta f'S' + f''S + \frac{d\beta}{d\xi}(1 - f'^2) =$$
$$2(f'S' - f''S) + 2\xi(S'^2 - S''S) + 2\xi(f'S'_\xi - f''S_\xi) \qquad (3-110)$$

with boundary conditions

$$S(\xi, 0) = S'(\xi, 0) = 0 \qquad (3-111a, b)$$

and
$$S'(\xi, \eta \to \infty) \to 0 \qquad (3-111c)$$

Equations (3–109) and (3–110), subject to the boundary conditions given in Eqs. (3–106) and (3–111), could be considered as a pair of coupled ordinary differential equations for the dependent functions f and S, except for the appearance of the term $(f'S'_\xi - f''S_\xi)$ in Eq. (3–110). Thus, we assume that this term is negligible.[†] This assumption is justifiable provided ξ itself, or alternately $(f'S'_\xi - f''S_\xi)$, is sufficiently small. Notice that in the locally similar method part of the momentum equation itself is lost when the term on the right hand side is neglected, whereas in the locally nonsimilar method part of an auxiliary equation is neglected. Thus, the locally similar method is inherently more accurate.

Gathering together the governing equations for the locally nonsimilar model, we have

$$f''' + ff'' + \beta(1 - f'^2) = 2\xi(f'S' - f''S) \qquad (3-109)$$

$$S''' + fS'' - 2\beta f'S' + f''S + \frac{d\beta}{d\xi}(1 - f'^2) =$$
$$2(f'S' - f''S) + 2\xi(S'^2 - S''S) \qquad (3-112)$$

with boundary conditions

$$f(\xi, 0) = f'(\xi, 0) = S(\xi, 0) = S'(\xi, 0) = 0 \qquad (3-113a, b, c, d)$$
$$f'(\xi, \eta \to \infty) \to 1 \qquad S'(\xi, \eta \to \infty) \to 0 \qquad (3-113e, f)$$

[†]Sparrow, Quack and Boerner [Spar70] also neglect the term $2\xi(S'^2 - S''S)$ in their original formulation.

98   Laminar Flow Analysis

Equations (3–109), (3–112) and (3–113) constitute the two-equation locally nonsimilar model. For a known inviscid flow at the edge of the boundary layer $U(x)$ and hence $\xi, \beta(\xi)$ and $d\beta/d\xi$ are known constants at any position $\xi(x)$ on the surface. Equations (3–109), (3–112) subject to (3–113) are then a pair of coupled ordinary differential equations with known constant coefficients subject to two-point asymptotic boundary conditions. Notice that it is now necessary to satisfy two asymptotic boundary conditions (Eqs. 3–113e and f). These equations can be integrated using a simple extension of the techniques previously developed. The extension of the Nachtsheim-Swigert iteration technique to a pair of asymptotic boundary conditions is discussed in Appendix B.

The two-equation locally nonsimilar boundary layer model is the first in a series of systems of simultaneous equations. Each successive system of equations increases the accuracy of the results. The three-equation locally nonsimilar boundary layer model is developed by introducing a second auxiliary function

$$h = S_\xi = f_{\xi\xi} \qquad (3-114)$$

into Eqs. (3–110) and (3–111). In order to determine $h(\xi, \eta)$ the first auxiliary equation and the associated boundary conditions, i.e., Eqs. (3–110) and (3–111), are differentiated with respect to ξ to yield

$$h''' + fh'' - 2\beta(S'^2 + f'h') + f''h + 2SS''$$
$$- 4\frac{d\beta}{d\xi}f'S' + \frac{d^2\beta}{d\xi^2}(1 - f'^2) = 4(S'^2 - SS'' + f'h' - f''h)$$
$$+ 2\xi(3h'S' - 2S''h - h''S) + 2\xi(f'h'_\xi - f''h_\xi) \qquad (3-115)$$

Consistent with the two-equation model, the last term on the right hand side is neglected.[†] Gathering the governing equations for the three-equation model together we have

$$f''' + ff'' + \beta(1 - f'^2) = 2\xi(f'S' - f''S) \qquad (3-109)$$

$$S''' + fS' - 2\beta f'S' + f''S + \frac{d\beta}{d\xi}(1 - f'^2) =$$
$$2(f'S' - f''S) + 2\xi(S'^2 - SS'' + f'h' - f''h) \qquad (3-116)$$

$$h''' + fh'' - 2\beta(S'^2 + f'h') + f''h - 2SS'' - 4\frac{d\beta}{d\xi}f'S' + \frac{d^2\beta}{d\xi^2}(1 - f'^2) =$$
$$4(S'^2 - SS'' + f'h' - f''h) + 2\xi(3h'S' - 2S''h - h''S) \qquad (3-117)$$

[†]Sparrow, Quack and Boerner [Spar70] also neglect the term $2\xi(3h'S' - 2S''h - h''S)$ in their original formulation.

Locally Nonsimilar Boundary Layer Method 99

with boundary conditions

$$f(\xi, 0) = f'(\xi, 0) = 0 \qquad (3-118a, b)$$

$$S(\xi, 0) = S'(\xi, 0) = 0 \qquad (3-118c, d)$$

$$h(\xi, 0) = h'(\xi, 0) = 0 \qquad (3-118e, f)$$

and

$$f'(\xi, \eta \to \infty) \to 1 \qquad (3-118g)$$

$$S'(\xi, \eta \to \infty) \to 0 \qquad (3-118h)$$

$$h'(\xi, \eta \to \infty) \to 0 \qquad (3-118i)$$

Here, we note that the terms previously neglected in the two-equation model (see Eq. 3–110) are now included in Eq. (3–117). The boundary value problem for the three-equation locally nonsimilar model now involves three simultaneous equations subject to two-point asymptotic boundary conditions, three of which are specified at infinity. The solution of this boundary value problem is obtained by means of a straightforward extension of the previously discussed methods.

We now illustrate the locally nonsimilar boundary layer method, using the nonsimilar boundary layer on a circular cylinder whose axis is perpendicular to the direction of a uniform parallel flow. The analysis is developed for the two-equation model (see Eqs. 3–109, 3–112 and 3–113). The inviscid flow velocity distribution for two-dimensional irrotational steady flow about a circular cylinder (see, for example, Kuethe and Schetzer [Kuet59]) is

$$U(x) = 2U_\infty \sin\left(\frac{x}{R}\right) \qquad (3-119)$$

where U_∞ is the velocity in the uniform parallel flow ahead of the cylinder, and R is the radius of the cylinder. Recalling Eqs. (3–102), (3–103) and (3–107), and using Eq. (3–119), yields

$$\xi = \int_0^x \frac{U(x)}{U_\infty} d\left(\frac{x}{R}\right) = 2\int_0^x \sin\left(\frac{x}{R}\right) d\left(\frac{x}{R}\right) = 2\left[1 - \cos\left(\frac{x}{R}\right)\right] \qquad (3-120)$$

$$\eta = \frac{yU(x)}{(2\nu R U_\infty \xi)^{1/2}} = \left(\frac{y}{R}\right)\left(\frac{U_\infty R}{\nu}\right)^{1/2} \frac{\sin\left(\frac{x}{R}\right)}{\left[1 - \cos\left(\frac{x}{R}\right)\right]^{1/2}} \qquad (3-121)$$

and

$$\beta = \frac{2}{U^2}\frac{dU}{dx}\int_0^x U dx = \frac{2\cos\left(\frac{x}{R}\right)}{\left[1 + \cos\left(\frac{x}{R}\right)\right]}$$

where we have taken the characteristic length $\ell = R$.

Before considering the solution in detail, we note that the nondimensional velocity profile is again given by

$$u = \psi_y = \frac{\partial \eta}{\partial y}\psi_\eta = U f'(\xi, \eta) \qquad (3-122)$$

Further, the shearing stress at the surface is

$$\tau(\xi, 0) = \mu \frac{\partial u}{\partial y} = \frac{\mu U^2}{(2\nu \ell U_\infty \xi)^{1/2}} f''(\xi, 0) \qquad (3-123)$$

In the particular case of the circular cylinder we have

$$\tau(\xi, 0) = 2\rho U_\infty^2 \left(\frac{\nu}{RU_\infty}\right)^{1/2} \left[1 - \cos\left(\frac{x}{R}\right)\right]^{3/2} f''(\xi, 0) \qquad (3-124)$$

Introducing a local skin friction coefficient defined as

$$C_{f_x} = \frac{2\tau(\xi, 0)}{\rho U_\infty^2}$$

yields
$$C_{f_x} = 4\left(\frac{\nu}{RU_\infty}\right)^{1/2} \left[1 - \cos\left(\frac{x}{R}\right)\right]^{3/2} f''(\xi, 0) \qquad (3-125)$$

These results show that the shearing stress and the skin friction coefficient are functions of the square root of the Reynolds number based on the radius of the cylinder, i.e., $U_\infty R/\nu$ as well as the position on the cylinder surface. Finally, notice that neither the velocity profile nor the shearing stress directly involve the auxiliary function S.

The solution of the two-equation model for the nonsimilar boundary layer on a circular cylinder is obtained by developing a suitable main program and eqmot routine for the gblpdriv routine. The routines for the free convection and compressible boundary layer equations (see Chapters 5 and 6 and Appendix D–15 and D–16) can be used as a guide.

Comparison of the results for the nondimensional shearing stress at the surface obtained using a local similarity model, the two- and three-equation models for the locally nonsimilar boundary layer method, a finite difference–differential method [Terr60], and a finite difference method [Scho64] taken from Sparrow, Quack and Boerner [Spar70] is shown in Figure 3–19. Numerical results for the nondimensional shearing stress parameter $f''(0)$ are given in Table C-10 of Appendix C. These numerical results were obtained using the modifications to the faskan program given above. If we take the finite difference results as potentially the most accurate, Figure 3–19 shows that the local similarity model is considerably in error for x/R values more than about 0.8. The two-equation model gives good results for values of x/R up to approximately 1.2 and the three-equation model up to approximately 1.4, which corresponds to a position about 80° from the forward stagnation point. For larger values of x/R the effects of separation become more important, with a resulting decrease in accuracy of the results.

We now turn our attention to nonsimilarities in the boundary conditions. In particular, consider the nonsimilar boundary layer with mass transfer at the surface. In our previous discussion (see Section 3–10) we assumed that at the surface the distribution of the normal velocity component $v(x)$ varied such that $f(0) = f_w = $ constant. This was the similarity requirement. Relaxing this requirement and re-examining the normal velocity component at the surface $v(x)$ with the help of Eqs. (3–102) to (3–104) yields

$$-v(x) = \psi_x = \frac{\partial \eta}{\partial x}\psi_\eta + \frac{\partial \xi}{\partial x}\psi_\xi$$

or

$$-v(x) = \psi_x = \frac{\partial \eta}{\partial x}(2\nu \ell U_\infty \xi)^{1/2} f'(\xi, 0)$$
$$+ \frac{\partial \xi}{\partial x}\left[(\nu \ell U_\infty)(2\nu \ell U_\infty \xi)^{-1/2} f(\xi, 0)\right.$$
$$\left. + (2\nu \ell U_\infty \xi)^{1/2} f_\xi(\xi, 0)\right] \qquad (3-126)$$

Recalling that $f'(\xi, 0) = 0$ and $(\partial \xi/\partial x) = (U/U_\infty)(1/\ell)$ allows writing the surface mass transfer boundary condition as

$$f + 2\xi f_\xi = -v(x)\frac{U_\infty}{U(x)}\left(\frac{2\ell\xi}{\nu U_\infty}\right)^{1/2} \qquad (3-127)$$

In order to obtain the necessary mass transfer boundary condition for the first

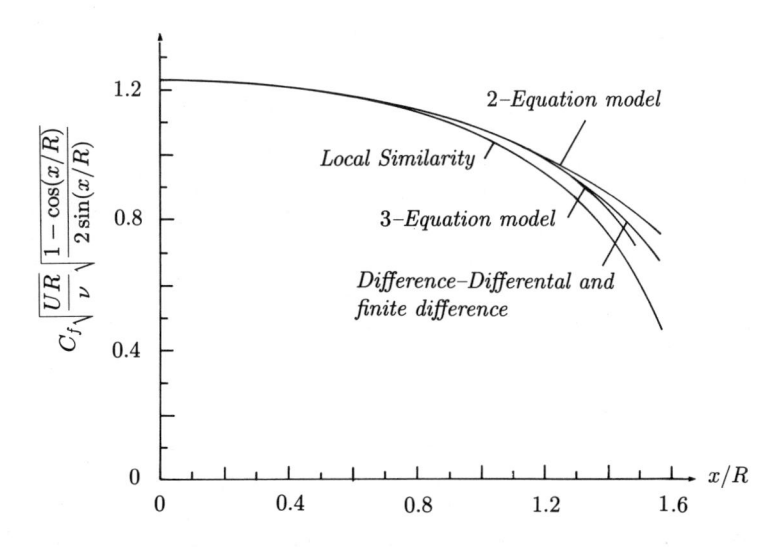

Figure 3–19. Shearing stress on a circular cylinder in cross flow. Adapted from [Spar70].

102   Laminar Flow Analysis

auxiliary momentum equation, Eq. (3–127) is differentiated with respect to ξ. This yields

$$S + 2S + 2\xi S_\xi = -\left(\frac{2\ell}{\nu U_\infty}\right)^{1/2} U_\infty \frac{\partial}{\partial \xi}\left[(v^2 U^{-2}\xi)^{1/2}\right] \quad (3-128)$$

where again $S(\xi, \eta) = f_\xi(\xi, \eta)$. Consistent with the two-equation model, the term $2\xi S_\xi$ is neglected. The mass transfer boundary condition for the first auxiliary momentum equation is then

$$S(\xi, 0) = -\frac{1}{3}\left(\frac{2\ell U_\infty}{\nu}\right)^{1/2} \frac{\partial}{\partial \xi}\left[(v^2 U^{-2}\xi)^{1/2}\right] \quad (3-129)$$

Now, considering the mass transfer boundary condition for the second auxiliary momentum equation (Eq. 3–117), we again introduce the auxiliary function $h(\xi, \eta) = S_\xi(\xi, \eta) = f_{\xi\xi}(\xi, \eta)$ into Eq. (3–128) to obtain

$$3S + 2\xi h = -\left(\frac{2\ell}{\nu U_\infty}\right)^{1/2} U_\infty \frac{\partial}{\partial \xi}\left[(v^2 U^{-2}\xi)^{1/2}\right] \quad (3-130)$$

Differentiating Eq. (3–130) with respect to ξ and, consistent with the three-equation model, neglecting the term $2\xi h_\xi$ yields

$$h(\xi, 0) = -\frac{1}{5}\left(\frac{2\ell U_\infty}{\nu}\right)^{1/2} \frac{\partial^2}{\partial \xi^2}\left[(v^2 U^{-2}\xi)\right] \quad (3-131)$$

which is the mass transfer boundary condition for the three-equation model.

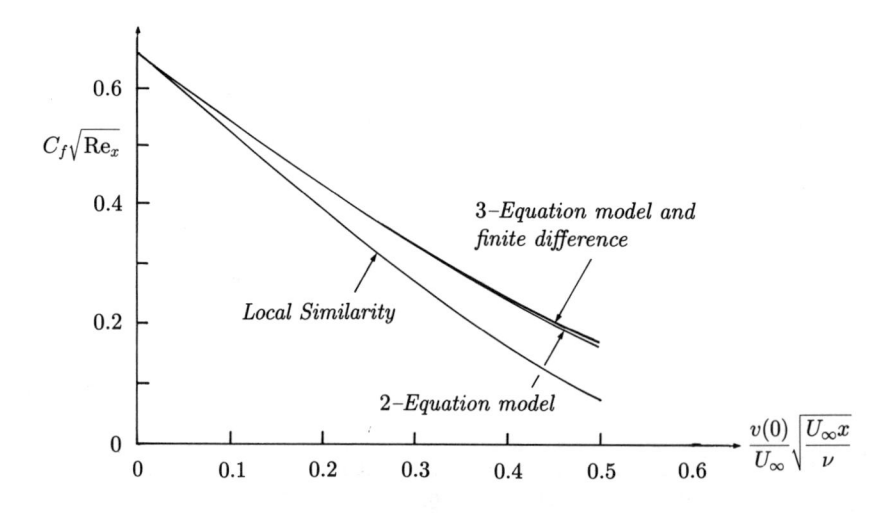

Figure 3–20. Shearing stress on a flat plate with uniform surface mass transfer. Adapted from [Spar70].

A particularly simple result of considerable practical interest occurs for a flat plate with uniform surface mass transfer. Here $U(x) = U_\infty =$ constant and $v(x) = v_0 =$ constant. For the two-equation model the mass transfer boundary conditions, i.e., Eqs. (3–127) and (3–129), become

$$S(\xi,0) = -\frac{1}{3}\left(\frac{\ell U_\infty}{\nu}\right)^{1/2} \frac{v_0}{U_\infty} \left(\frac{\ell}{2x}\right)^{1/2} \qquad (3-132)$$

and
$$f(\xi,0) = -\frac{2}{3}\frac{v_0}{U_\infty}\left(\frac{2U_\infty x}{\nu}\right)^{1/2} = -\frac{2}{3}(\mathrm{Re}_x)^{1/2}\frac{v_0}{U_\infty} \qquad (3-133)$$

Note that Eq. (3–133) shows that the functional form of $f(\xi,0) \sim x^{1/2}$ is similar to that for $f(0)$ in the similar boundary layer analysis (see Eq. 3–100). However, the nonsimilar behavior appears in Eq. (3–132), where we see that $S(\xi,0) \sim x^{-1/2}$.

Sparrow, Quack and Boerner [Spar70] investigated the flat plate with uniform surface mass transfer using both two- and three-equation models for the locally nonsimilar boundary layer method.[†]

A comparison of these results with the local similarity method and with a finite difference solution [Cath65] is shown in Figure 3–20. The results of the two- and three-equation models for the locally nonsimilar boundary layer method are in excellent agreement with each other and with the finite difference solution. Figure 3–20 shows that the local similarity method is considerably in error.

The additional modifications required to allow for surface mass transfer, along with those for the two-equation locally nonsimilar boundary layer method, can be incorporated into the **faskan** program.

[†]In Sparrow, Quack and Boerner [Spar70] an additional transformation $\xi = (2\xi)^{1/2}$ is made prior to deriving the mass transfer boundary condition for the two-equation model. Subsequently the term $\xi f_{\xi\xi}$ is neglected. This in effect neglects part of the contribution of $S(\xi,0)$ to Eq. (3–129). However, their results are still of considerable qualitative interest.

FOUR

THERMAL LAYERS AND FORCED CONVECTION BOUNDARY LAYERS

4–1 Introduction

In Chapter 1 the boundary value problems associated with the Navier-Stokes equations were broadly classified into three types:

1. Uncoupled—dynamics of the flow;
2. Uncoupled—dynamics and thermodynamics of the flow;
3. Coupled—dynamics and thermodynamics of the flow.

Chapters 2 and 3 were concerned with Type 1 problems. There we obtained solutions for the velocity field, with the uncoupled energy or temperature field reserved for later solution. Having investigated the velocity field, we are now concerned with investigating the temperature field which exists about a body.

We first investigate selected analytical solutions of the temperature field associated with the Navier-Stokes equations. These solutions serve to illustrate the solution technique as well as several basic concepts associated with convective heat transfer. Forced and free convection boundary layers are then investigated.

4–2 General Characteristics of Thermal Layers

To illustrate some of the characteristics of thermal layers we consider two-dimensional steady compressible viscous flow with constant properties, i.e., the viscosity, thermal conductivity and specific heat at constant pressure are not functions of temperature. Generally a fluid is considered incompressible when constant properties are assumed; however, this is not inherently necessary. Although the temperature changes in low speed compressible flow are generally small enough that property variations are considered negligible, small temperature changes resulting from heating of the fluid cause related density changes. These density changes result in volume changes which generate a buoyancy force. This buoyancy force is considered as a body force due to the gravitational field, and is accounted for by assuming variable density in a fluid with otherwise constant properties.

The volume coefficient of thermal expansion at constant pressure, $\bar{\beta}$, is a thermodynamic property of the fluid relating density changes to temperature changes and is given as

$$\bar{\beta} = -\frac{1}{\rho}\left(\frac{\partial \rho}{\partial T}\right)_P \qquad (4-1)$$

The temperature difference between a given fluid element and the surrounding fluid is

$$\Delta T = T - T_e \qquad (4-2)$$

Here, the subscript e is used for a reference condition in the surrounding fluid. Introducing the volume coefficient of thermal expansion, the density difference of the fluid element is given by $\rho\bar{\beta}\Delta T$. Here ρ is the density before heating. This assumes that the partial derivative in Eq. (4–1) is adequately approximated by its finite difference representation. Further, this approximation assumes that the density and temperature changes are small enough that the fluid properties can be considered constant. The body force per unit volume is then $\rho\vec{g}\bar{\beta}\Delta T$, where $\vec{g}$ is the vector acceleration of gravity given by

$$\vec{g} = g_x\vec{i} + g_y\vec{j} + g_z\vec{k} \qquad (4-3)$$

Introducing these body forces into the Navier-Stokes equations (Eqs. 1–1, 1–33, 1–41) for constant property steady two-dimensional compressible heat conducting viscous flow without bulk heat addition, the continuity equation becomes

$$\frac{\partial}{\partial x}(\rho u) + \frac{\partial}{\partial y}(\rho v) = 0 \qquad (4-4)$$

The momentum equations become

x momentum

$$\rho\left(u\frac{\partial u}{\partial x} + v\frac{\partial u}{\partial y}\right) = -\frac{\partial P}{\partial x} + \rho g_x\bar{\beta}\Delta T + \mu\left(\frac{4}{3}\frac{\partial^2 u}{\partial x^2} + \frac{\partial^2 u}{\partial y^2} + \frac{1}{3}\frac{\partial^2 v}{\partial x\partial y}\right) \qquad (4-5a)$$

y momentum

$$\rho\left(u\frac{\partial v}{\partial x} + v\frac{\partial v}{\partial y}\right) = -\frac{\partial P}{\partial y} + \rho g_y\bar{\beta}\Delta T + \mu\left(\frac{4}{3}\frac{\partial^2 v}{\partial y^2} + \frac{\partial^2 v}{\partial x^2} + \frac{1}{3}\frac{\partial^2 u}{\partial x\partial y}\right) \qquad (4-5b)$$

The energy equation becomes

$$\rho c_p\left(u\frac{\partial T}{\partial x} + v\frac{\partial T}{\partial y}\right) = k\left(\frac{\partial^2 T}{\partial x^2} + \frac{\partial^2 T}{\partial y^2}\right) + u\frac{\partial P}{\partial x} + v\frac{\partial P}{\partial y} + \mu\Phi \qquad (4-6)$$

where
$$\Phi = 2\left[\left(\frac{\partial u}{\partial x}\right)^2 + \left(\frac{\partial v}{\partial y}\right)^2\right] - \frac{2}{3}\left(\frac{\partial u}{\partial x} + \frac{\partial v}{\partial y}\right)^2 + \left(\frac{\partial v}{\partial x} + \frac{\partial u}{\partial y}\right)^2 \quad (4-7)$$

Introducing the equation of state

$$P = P(\rho, T) \quad (4-8)$$

yields five equations in the five unknown functions u, v, T, P, ρ. The system is thus determinant.

Before considering specific solutions of these equations subject to appropriate boundary conditions, a knowledge of the important nondimensional parameters is desired. Previously (see Section 3–2) we nondimensionalized the general compressible Navier-Stokes equations and found that the Reynolds number and Prandtl number are significant nondimensional parameters. Because of the particular form of the body force term in the present case, it is more convenient to nondimensionalize the equations in the following manner

$$\bar{x} = \frac{x}{\ell} \quad \bar{y} = \frac{y}{\ell} \quad \bar{u} = \frac{u}{U_e} \quad \bar{v} = \frac{v}{U_e} \quad \bar{P} = \frac{P}{\rho_e U_e^2}$$

$$\bar{\rho} = \frac{\rho}{\rho_e} \quad \theta = \frac{T - T_e}{T_w - T_e} \quad (4-9)$$

Here, this particular form of the numerator of the nondimensional temperature is based on the form of the buoyancy force, and the form of the denominator is based on the boundary conditions, e.g., at the surface the fluid temperature is equal to the surface temperature, T_w, and $\theta = 1$.

Using these nondimensional parameters, the continuity equation becomes

$$\frac{\partial}{\partial \bar{x}}(\bar{\rho}\bar{u}) + \frac{\partial}{\partial \bar{y}}(\bar{\rho}\bar{v}) = 0 \quad (4-10)$$

which is identical with Eq. (3–2) for steady flow. The x and y momentum equations become

x momentum
$$\bar{u}\frac{\partial \bar{u}}{\partial \bar{x}} + \bar{v}\frac{\partial \bar{u}}{\partial \bar{y}} = -\frac{1}{\bar{\rho}}\frac{\partial \bar{P}}{\partial \bar{x}} + \frac{\mu}{\rho_e U_e \ell}\frac{1}{\bar{\rho}}\left(\frac{4}{3}\frac{\partial^2 \bar{u}}{\partial \bar{x}^2} + \frac{\partial^2 \bar{u}}{\partial \bar{y}^2} + \frac{1}{3}\frac{\partial^2 \bar{v}}{\partial \bar{x}\partial \bar{y}}\right)$$
$$+ \frac{T_w - T_e}{U_e^2}\ell g_x \bar{\beta}\theta \quad (4-11a)$$

y momentum
$$\bar{u}\frac{\partial \bar{v}}{\partial \bar{x}} + \bar{v}\frac{\partial \bar{v}}{\partial \bar{y}} = -\frac{1}{\bar{\rho}}\frac{\partial \bar{P}}{\partial \bar{y}} + \frac{\mu}{\rho_e U_e \ell}\frac{1}{\bar{\rho}}\left(\frac{4}{3}\frac{\partial^2 \bar{v}}{\partial \bar{y}^2} + \frac{\partial^2 \bar{v}}{\partial \bar{x}^2} + \frac{1}{3}\frac{\partial^2 \bar{u}}{\partial \bar{x}\partial \bar{y}}\right)$$
$$+ \frac{T_w - T_e}{U_e^2}\ell g_y \bar{\beta}\theta \quad (4-11b)$$

108   Laminar Flow Analysis

The energy equation becomes

$$\bar{u}\frac{\partial \theta}{\partial \bar{x}} + \bar{v}\frac{\partial \theta}{\partial \bar{y}} = \frac{k}{\ell\rho_e c_p U_e}\left(\frac{\partial^2 \theta}{\partial \bar{x}^2} + \frac{\partial^2 \theta}{\partial \bar{y}^2}\right) + \frac{U_e^2}{c_p(T_w - T_e)}\left(\bar{u}\frac{\partial \bar{P}}{\partial \bar{x}} + \bar{v}\frac{\partial \bar{P}}{\partial \bar{y}}\right)$$
$$+ \frac{\mu U_e}{\rho_e \ell c_p(T_w - T_e)}\bar{\Phi} \qquad (4-12)$$

where $\bar{\Phi}$ is the nondimensional viscous dissipation function.

Here we see that the coefficient of the second term on the right side of the x and y momentum equations, $\mu/\rho_e V_e \ell$, is the reciprocal of the Reynolds number, i.e., $1/\text{Re}_\ell$. The Reynolds number is interpreted as the ratio of the effects of the inertia of the fluid represented by $\rho_e V_e \ell$ to the effects of the viscosity of the fluid represented by μ.

The coefficient of the buoyancy term in the momentum equations, i.e., the coefficient of θ, is also a nondimensional parameter. It is usually written in a modified form obtained by multiplication of the numerator and denominator by $(\rho_e \mu \ell)^2$. For the x momentum equation this yields

$$\left(\frac{\mu}{\rho_e U_e \ell}\right)^2 \frac{\rho_e^2 \ell^3 (T_w - T_e)g_x \bar{\beta}}{\mu^2}$$

which can be written as

$$\left(\frac{1}{\text{Re}_\ell}\right)^2 \text{Gr}_x$$

where Gr_x is a nondimensional parameter called the Grashof number, i.e.

$$\text{Gr}_x = \frac{\rho_e^2 \ell^3 (T_w - T_e)g_x \bar{\beta}}{\mu^2} \qquad (4-13)$$

The results for the y momentum equation are identical except for the substitution of g_y for g_x. The Grashof number is interpreted as the ratio of the buoyancy effects to the viscous effects in the fluid.

Considering the energy equation, we see that multiplying the coefficient of the first term on the right hand side by μ/μ allows it to be written as

$$\frac{k}{\mu c_p}\frac{\mu}{\rho_e U_e \ell} = \frac{1}{\text{Pr}}\frac{1}{\text{Re}_\ell}$$

The Prandtl number, $\text{Pr} = \mu c_p/k$, is interpreted as the ratio of the viscous effects to the thermal conductivity effects. If we divide the numerator and denominator by the density, the Prandtl number can be written as $\text{Pr} = \nu/\alpha$, where ν is the kinematic viscosity and α is the thermal diffusivity.

The coefficient of the second term on the right hand side of the energy equation is in the form of the Eckert number, i.e.

$$\text{E} = \frac{U_e^2}{c_p(T_w - T_e)} \qquad (4-14)$$

The Eckert number is interpreted as a ratio of the energy generated by viscous dissipation to the energy transfer due to heat conduction through the fluid. Physically viscous dissipation is a transformation of kinetic energy into internal energy.

The coefficient of the last term on the right hand side of the energy equation is written as

$$\frac{\mu}{\rho_e U_e \ell} \frac{U_e^2}{c_p(T_w - T_e)} = \frac{\text{E}}{\text{Re}_\ell} \qquad (4-15)$$

Rewriting the momentum and energy equations in terms of nondimensional parameters yields

x momentum

$$\bar{u}\frac{\partial \bar{u}}{\partial \bar{x}} + \bar{v}\frac{\partial \bar{u}}{\partial \bar{y}} = -\frac{1}{\bar{\rho}}\frac{\partial \bar{P}}{\partial \bar{x}} + \frac{1}{\text{Re}_\ell}\frac{1}{\bar{\rho}}\left(\frac{4}{3}\frac{\partial^2 \bar{u}}{\partial \bar{x}^2} + \frac{\partial^2 \bar{u}}{\partial \bar{y}^2} + \frac{1}{3}\frac{\partial^2 \bar{v}}{\partial \bar{x}\partial \bar{y}}\right) + \frac{1}{\text{Re}_\ell^2}\text{Gr}_x \theta$$

$$(4-16a)$$

y momentum

$$\bar{u}\frac{\partial \bar{v}}{\partial \bar{x}} + \bar{v}\frac{\partial \bar{v}}{\partial \bar{y}} = -\frac{1}{\bar{\rho}}\frac{\partial \bar{P}}{\partial \bar{y}} + \frac{1}{\text{Re}_\ell}\frac{1}{\bar{\rho}}\left(\frac{4}{3}\frac{\partial^2 \bar{v}}{\partial \bar{y}^2} + \frac{\partial^2 \bar{v}}{\partial \bar{x}^2} + \frac{1}{3}\frac{\partial^2 \bar{u}}{\partial \bar{x}\partial \bar{y}}\right) + \frac{1}{\text{Re}_\ell^2}\text{Gr}_y \theta$$

$$(4-16b)$$

energy equation

$$\bar{u}\frac{\partial \theta}{\partial \bar{x}} + \bar{v}\frac{\partial \theta}{\partial \bar{y}} = \frac{1}{\Pr \text{Re}_\ell}\left(\frac{\partial^2 \theta}{\partial \bar{x}^2} + \frac{\partial^2 \theta}{\partial \bar{y}^2}\right) + \text{E}\left(\bar{u}\frac{\partial \bar{P}}{\partial \bar{x}} + \bar{v}\frac{\partial \bar{P}}{\partial \bar{y}}\right) + \frac{\text{E}}{\text{Re}_\ell}\bar{\Phi}$$

$$(4-17)$$

Here, we see that any solution of the governing equation is a function of the Reynolds, Prandtl, Grashof and Eckert numbers, as well as the nondimensional spacial coordinates $\bar{x}$, $\bar{y}$, $\bar{z}$ and the boundary conditions. Hence, the solutions are complex and difficult to obtain. Under these circumstances we look for simplifications which make the problem more tractable. Studying the equations more thoroughly reveals that the momentum and energy equations are coupled through both the nondimensional temperature θ and the nondimensional pressure gradients $\partial \bar{P}/\partial \bar{x}$ and $\partial \bar{P}/\partial \bar{y}$. If the Grashof number is small with respect to the square of the Reynolds number, then the buoyancy effects are negligible with respect to the viscous effects. However, the equations are still coupled through the pressure gradient terms. If the nondimensional pressure gradient terms $\partial \bar{P}/\partial \bar{x}$ and $\partial \bar{P}/\partial \bar{y}$ themselves are negligible with respect to the other terms in the equations, then the momentum equations are uncoupled from the energy equation. Alternately, the momentum and energy equations are uncoupled if the Eckert number approaches zero. This assumes the physically reasonable condition that the nondimensional pressure gradient parameters are always finite. Physically this occurs for very low speed flow. Note that the alternate interpretation of a small Eckert number due to a large temperature difference

110   Laminar Flow Analysis

$(T_w - T_e)$ is excluded by the assumptions used in deriving the buoyancy force terms in the equations.

If the Eckert number is small with respect to the Reynolds number, then the viscous dissipation terms in the energy equation are negligible. This implies that viscous heating of the fluid is negligible and is a common assumption in low speed convective heat transfer. However, this assumption does not necessarily allow the pressure gradient terms in the energy equation to be neglected, since the nondimensional pressure gradient terms can themselves be sufficiently large that their product with the Eckert number is not negligible.

From the above discussion we see that analytical solutions of these equations, subject to appropriate boundary conditions, are, in general, only possible for rather restrictive geometries and/or physical assumptions.

4–3 Simple Couette Flow

We begin our discussion of the analytical solutions by considering simple Couette flow, when the fluid is incompressible (see Section 2–4). Simple Couette flow assumes that the pressure gradient is zero, i.e., $dP/dx = 0$ (see Figure 4–1). A discussion of simple Couette flow serves to illustrate many of the important aspects of the solution of the energy equation.

In our previous discussion of the velocity field and the solution of the momentum equation with $dP/dx = 0$, we found that the momentum equations reduced to

$$\frac{d^2 u}{dy^2} = 0$$

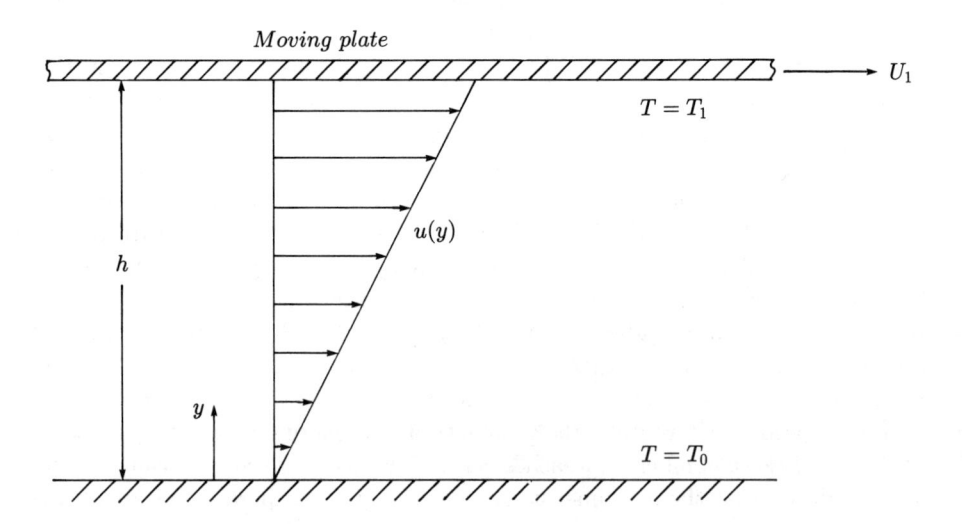

Figure 4–1. Simple Couette flow.

with boundary conditions

$$y = 0 \quad u = 0$$
$$y = h \quad u = U_1$$

where U_1 is the velocity of the moving plate. The solution of this boundary value problem is given as

$$u(y) = U_1 \frac{y}{h} \qquad (2-10)$$

Recall the previously discussed assumptions for simple Couette flow, i.e., $v = \partial u/\partial x = \partial v/\partial y = \partial P/\partial y = \partial P/\partial x = 0$. Further, assuming that the temperature gradient in the flow direction is negligible compared to the temperature gradient normal to the flow yields $\partial T/\partial x = 0$. The limitations of this particular assumption are discussed in Sections 4–8 and 4–11. Thus, the energy equation reduces to

$$\frac{d^2T}{dy^2} = -\frac{\mu}{k}\left(\frac{du}{dy}\right)^2 \qquad (4-18)$$

First, note that this is an ordinary differential equation. Since $u \neq u(x)$ and $T \neq T(x)$, then the velocity and temperature fields must only be functions of y. Further, this is a linear ordinary differential equation. Again (see Section 2–3) we observe that because of the geometry all of the nonlinear convective terms vanish.

The boundary conditions for constant wall temperature are given by

$$y = 0 \quad T = T_0 \qquad (4-19a)$$
$$y = h \quad T = T_1 \qquad (4-19b)$$

where here the subscripts 0 and 1 refer to the stationary wall and the moving wall, respectively.

Since the momentum and energy equations are uncoupled, we differentiate our previous results for the velocity distribution (Eq. 2–10) and substitute into the energy equation. Thus, we have

$$\frac{d^2T}{dy^2} = -\frac{\mu}{k}\frac{U_1^2}{h^2} \qquad (4-20)$$

Notice that the right hand side is a constant. The solution is

$$T(y) = Ay^2 + By + C$$

Imposing the boundary conditions (Eq. 4–19) and using the differential equation to evaluate A yields

$$\frac{T - T_0}{T_1 - T_0} = \frac{y}{h} + \frac{\mu U_1^2}{2k(T_1 - T_0)}\frac{y}{h}\left(1 - \frac{y}{h}\right) \qquad (4-21)$$

The coefficient of the second term on the right hand side is rewritten as

$$\frac{\mu U_1^2}{2k(T_1 - T_0)} = \frac{\mu c_p}{2k}\frac{U_1^2}{c_p(T_1 - T_0)} = \frac{\Pr E}{2}$$

Introducing nondimensional variables $\theta = (T - T_0)/(T_1 - T_0)$ and $\bar{y} = y/h$, we write the nondimensional temperature distribution as

$$\theta = \bar{y} + \frac{\Pr E}{2}\bar{y}(1-\bar{y}) \qquad (4-22)$$

Here we see that the temperature distribution consists of a linear term which is the same as for a fluid at rest with no viscous heating plus a superimposed parabolic distribution due to viscous heating. Figure 4–2 shows the nondimensional temperature distributions for several values of $\Pr E$. Note that the nondimensional temperature distribution is dependent on the value of the $\Pr E$ product. Since the transport properties μ, k, c_p are assumed constant, the Prandtl number $\Pr = \mu c_p/k$ is also a constant. Further, the Prandtl number for real fluids is not equal to zero. Hence, a $\Pr E$ product equal to zero corresponds to a fluid at rest, and Eq. (4–20) reduces to the heat conduction equation for one-dimensional steady state heat conduction without internal generation. The linear temperature distribution that exists in a fluid at rest between two walls at different temperatures is shown in Figure 4–2 for $\Pr E = 0$. For values of $\Pr E$ greater than zero, the internal friction within the fluid modifies the temperature distribution. Of particular interest is the fact that the slope of the temperature distribution at the moving wall changes sign. Recall that the heat transfer per unit area from the fluid to the wall is given by (see Eq. 1–35)

$$\left(\frac{q}{A}\right)_y = -k\frac{\partial T}{\partial y} \qquad (1-35)$$

Hence, for a given temperature difference $(T_1 - T_0) > 0$, heat transfer from the upper wall to the fluid occurs only as long as the velocity of the moving wall does not exceed a certain critical value. The direction of heat transfer changes when the sign of the temperature gradient at the wall changes, i.e., when the temperature gradient at the wall is zero. From Eq. (1–35) we see that this corresponds to zero heat transfer from the wall. Such a surface is referred to as an adiabatic wall. The condition for zero heat transfer at the wall is obtained by differentiating either the dimensional equation (Eq. 4–21) or the nondimensional equation (Eq. 4–22) and evaluating the result at the upper wall. Performing the differentiation, we have

$$\left.\frac{\partial T}{\partial y}\right|_{y=h} = 0 = (T_1 - T_0) - \frac{\mu U_1^2}{2k}$$

and the condition for zero heat transfer is

$$T_1 - T_0 = \frac{\mu U_1^2}{2k} \qquad (4-23)$$

or in nondimensional form

$$\Pr E = 2 \qquad (4-24)$$

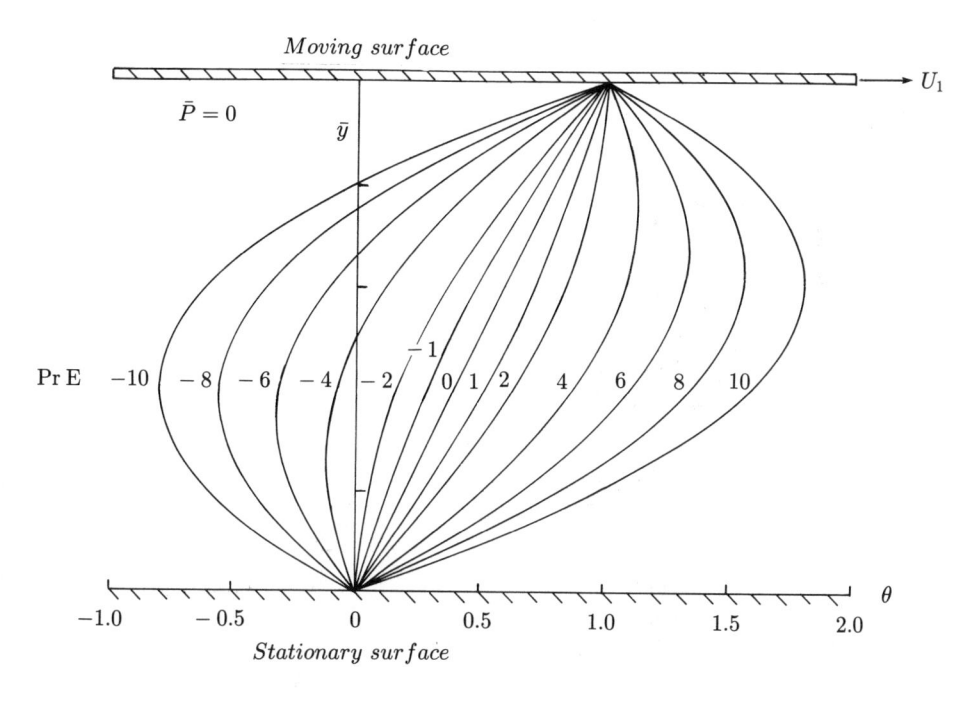

Figure 4–2. Nondimensional temperature distribution for simple Couette flow.

Thus, heat transfer takes place from the upper wall to the fluid with consequent cooling of the upper wall if

$$\frac{\mu U_1^2}{2k} < (T_1 - T_0) \qquad \text{or} \qquad \Pr E < 2 \qquad (4-25a)$$

Heat transfer takes place from the fluid to the upper wall with consequent heating of the upper wall if

$$\frac{\mu U_1^2}{2k} > (T_1 - T_0) \qquad \text{or} \qquad \Pr E > 2 \qquad (4-25b)$$

These results show the importance of viscous heating on the heat transfer at the upper wall. With the upper wall at a higher temperature than the lower wall and the fluid at rest, heat transfer always occurs from the upper surface to the fluid, i.e., the fluid is heated by the upper wall. However, as the fluid velocity increases the heat transfer from the upper wall to the fluid decreases, until at a Prandtl-Eckert product of two it is zero. Further increases in fluid velocity result in a reversal in the direction of heat transfer, i.e., the upper wall is now further heated by the fluid. This is a result of the internal heat generated due to the viscosity of the fluid. Since the Eckert number is proportional to the square of the velocity, viscous heating is of particular importance for high velocity flow. Similar results are obtained for the heat transfer at the stationary wall. Here, the adiabatic wall condition occurs for a Prandtl-Eckert product $\Pr E = -2$ (see Figure 4–2).

A special case of simple Couette flow occurs when the wall temperatures are equal, i.e., $T_1 = T_0$. The temperature distribution is then given by a symmetrical parabola

$$T - T_0 = \frac{\mu U_1^2}{2k} \frac{y}{h} \left(1 - \frac{y}{h}\right) \qquad (4-26)$$

The maximum temperature in the fluid occurs at the center of the flow and is given by

$$T_{\max} = T_0 + \mu \frac{U_1^2}{8k} \qquad (4-27)$$

Using Eq. (4–26) and (1–35) the heat transfer rate per unit area is given by

$$\left(\frac{q}{A}\right)_{\bar{y}} = \frac{\mu U_1^2}{2h}(2\bar{y} - 1) \qquad (4-28)$$

This shows that at the lower ($\bar{y} = 0$) and upper ($\bar{y} = 1$) surfaces the heat transfer is from the fluid to the surface. Further, for a given fluid the heat transfer increases as the square of the increase in velocity and directly as the decrease in gap width, h. These results have obvious application in hydrodynamic lubrication of bearings.

An additional special case of simple Couette flow occurs when we assume that the lower wall ($y = 0$) is adiabatic, or insulated. Thus, the boundary conditions for the governing equation (Eq. 4–20) become

$$y = 0 \qquad \frac{dT}{dy} = 0 \qquad (4-29a)$$

$$y = h \qquad T = T_1 \qquad (4-29b)$$

The solution of the governing boundary value problem is then

$$T(y) = T_1 + \mu \frac{U_1^2}{2k}\left[1 - \left(\frac{y}{h}\right)^2\right] \qquad (4-30)$$

The temperature at the lower wall ($y = 0$) is the adiabatic wall temperature

$$T(0) = T_{aw} = T_1 + \frac{\mu U_1^2}{2k} \qquad (4-31)$$

This is the steady state equilibrium temperature at the lower wall due to the viscous heating. Further, for a given temperature of the upper wall, T_1, the adiabatic wall temperature increases as the square of the velocity of the upper wall.

Consistent with our previous definition, the nondimensional temperature distribution is given by

$$\bar{\theta} = \frac{T - T_{aw}}{T_1 - T_{aw}} = \bar{y}^2 \qquad (4-32)$$

and shown in Figure 4–3. Using Eq. (4–31) we generalize the heat transfer criteria given in Eq. (4–25). Hence, if

$$(T_{aw} - T_1) < (T_1 - T_0) \tag{4-33a}$$

then heat transfer takes place from the upper wall to the fluid, i.e., the upper wall is cooled. However, if

$$(T_{aw} - T_1) > (T_1 - T_0) \tag{4-33b}$$

then heat transfer takes place from the fluid to the upper wall, i.e., the upper wall is heated.

4–4 Nonsimple Couette Flow

In the previous section we considered simple Couette flow. There solutions to the energy equations were obtained with zero pressure gradient. We now consider nonsimple Couette flow. Here we obtain solutions to the energy equation for nonzero pressure gradient. Our purpose is to investigate and analyze the effects of adverse and favorable pressure gradients on the temperature field and on the heat transfer at the surface.

In Section (2–4) we investigated the velocity field for nonsimple Couette flow. There we found that the nondimensional velocity distribution is given by

$$\bar{u} = \bar{y}\left[1 + \bar{P}(1 - \bar{y})\right] \tag{2-13}$$

where
$$\bar{u} = \frac{u}{U_1} \quad \text{and} \quad \bar{y} = \frac{y}{h} \tag{2-12}$$

and $\bar{P}$ is a nondimensional pressure gradient parameter.

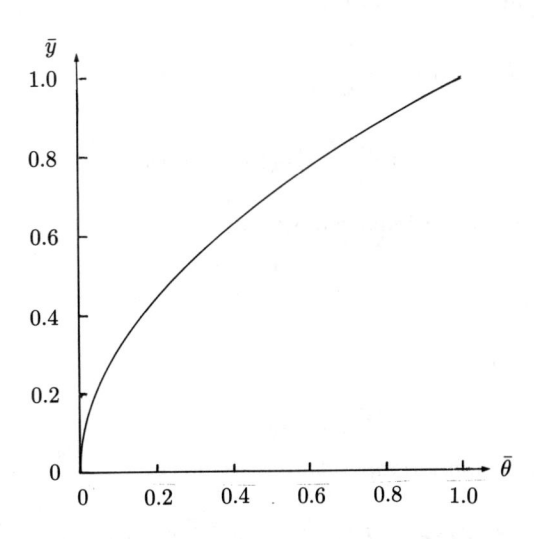

Figure 4–3. Adiabatic wall ($\bar{y} = 0$) nondimensional temperature distribution.

116   Laminar Flow Analysis

Again, the energy equation for incompressible steady constant property flow reduces to Eq. (4–18), with boundary conditions given by Eq. (4–19 a, b). Rewriting Eq. (4–18) in terms of the nondimensional variables, $\bar{u}$ and $\bar{y}$, yields

$$\frac{d^2 T}{d\bar{y}^2} = -\frac{\mu U_1^2}{k}\left(\frac{d\bar{u}}{d\bar{y}}\right)^2 \qquad (4-34)$$

Differentiating the nondimensional velocity distribution, Eq. (2–13), and substituting it into Eq. (4–34) yields, after some rearrangement, the ordinary differential equation

$$\frac{d^2 T}{d\bar{y}^2} = -\frac{\mu U_1^2}{k}\left[(1+\bar{P})^2 - 4\bar{P}(1+\bar{P})\bar{y} + 4\bar{P}^2\bar{y}^2\right] \qquad (4-35)$$

Integrating twice gives the temperature

$$T = -\frac{\mu U_1^2}{k}\left[\frac{(1+\bar{P})^2}{2}\bar{y}^2 - \frac{2}{3}\bar{P}(1+\bar{P})\bar{y}^3 + \frac{\bar{P}^2\bar{y}^4}{3}\right] + A\bar{y} + B$$

where A and B are constants of integration.

Using the boundary condition given in Eq. (4–19 a) shows that

$$B = T_0$$

The second boundary condition, Eq. (4–19 b), gives

$$A = (T_1 - T_0) + \frac{\mu U_1^2}{6k}\left(3 + 2\bar{P} + \bar{P}^2\right)$$

Thus, the temperature distribution becomes

$$(T - T_0) = (T_1 - T_0)\bar{y} - \frac{\mu U_1^2}{k}\left[\frac{(1+\bar{P})^2}{2}\bar{y}^2 - \frac{2}{3}\bar{P}(1+\bar{P})\bar{y}^3 + \frac{\bar{P}^2\bar{y}^4}{3}\right]$$

$$+ \frac{\mu U_1^2}{6k}\left[3 + 2\bar{P} + \bar{P}^2\right]\bar{y} \qquad (4-36)$$

Introducing the nondimensional temperature

$$\theta = \frac{T - T_0}{T_1 - T_0}$$

and recognizing that

$$\frac{\mu U_1^2}{k(T_1 - T_0)} = \Pr\mathrm{E}$$

yields the nondimensional temperature distribution

$$\theta = \bar{y} - \frac{\Pr\mathrm{E}}{6}\left[2\bar{P}^2\bar{y}^4 - 4\bar{P}(1+\bar{P})\bar{y}^3 + 3(1+\bar{P})^2\bar{y}^2 - (3+2\bar{P}+\bar{P}^2)\bar{y}\right] \qquad (4-37)$$

Figure 4–4 shows the nondimensional temperature distribution for various values of the nondimensional pressure gradient parameter and the Prandtl-Eckert product. In analyzing these curves it is essential to keep in mind that values of $\bar{P} < 0$ correspond to adverse pressure gradients, whereas $\bar{P} > 0$ indicates favorable pressure gradients. Further, for a finite Prandtl number and a fixed value of $(T_1 - T_0)$, the Prandtl-Eckert product increases with increasing velocity of the moving plate. Negative values of the Prandtl-Eckert product are physically interpreted as the case when the stationary wall is at a higher temperature than the moving wall, i.e., $T_0 > T_1$. In this case, negative values of the nondimensional temperature θ correspond to positive temperature differences $(T - T_0)$.

Comparing the nondimensional temperature profiles shown in Figures 4–4, we note that for any particular value of the nondimensional pressure gradient $\bar{P}$ the maximum value of the nondimensional temperature, θ_{max}, increases for increasingly positive Prandtl-Eckert product. Conversely, the minimum value of the nondimensional temperature, θ_{min}, increases for increasingly negative Prandtl-Eckert products. Note that the location of the maximum and minimum values of θ move toward the moving surface as the nondimensional pressure gradient decreases. Further, from Figure 4–4 we see that the temperature gradient at both the moving and stationary surfaces change sign. This implies that the direction of heat transfer also changes and hence is zero at some point. Since this point

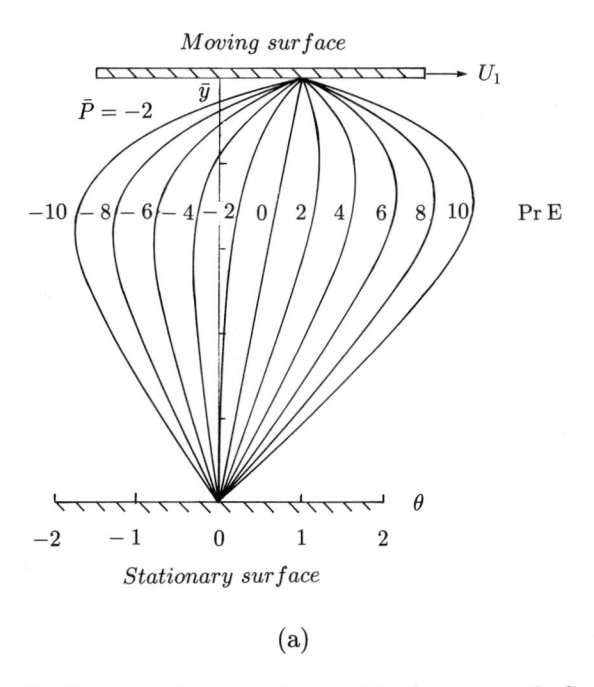

(a)

Figure 4–4. Nondimensional temperature profiles for nonsimple Couette flow. (a) $\bar{P} = -2$; (b) $\bar{P} = 2$.

118 Laminar Flow Analysis

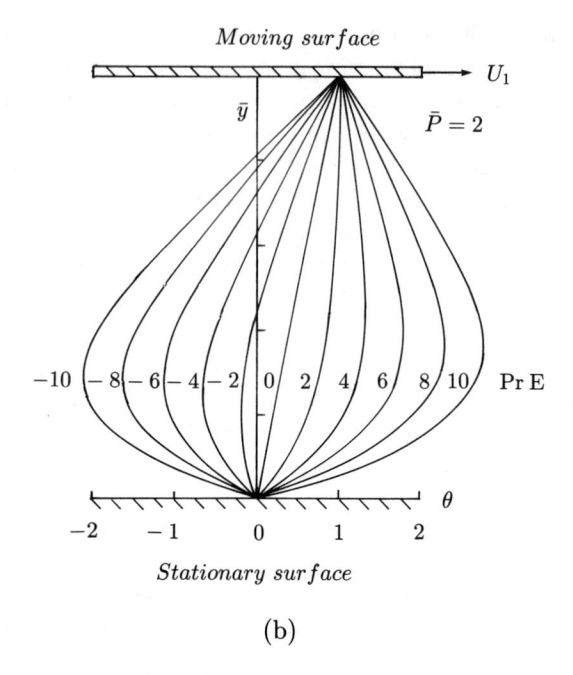

Figure 4–4. (cont.)

is a function of the nondimensional pressure gradient $\bar{P}$ and the Prandtl-Eckert product, the adiabatic wall temperature is also a function of these two variables. Differentiating Eq. (4–37) with respect to $\bar{y}$ we have

$$\frac{d\theta}{d\bar{y}} = 1 - \frac{\Pr E}{6} \Big[8\bar{P}^2 \bar{y}^3 - 12\bar{P}(1+\bar{P})\bar{y}^2$$
$$+ 6(1+\bar{P})^2 \bar{y} - (3 + 2\bar{P} + \bar{P}^2) \Big] \quad (4-38)$$

The position $\bar{y}$ of the relative maximum or minimum values of θ shown in Figure 4–4 is determined by setting $d\theta/d\bar{y} = 0$. As previously mentioned, this position is, of course, a function of $\bar{P}$ and $\Pr E$. However, the adiabatic wall condition is of more immediate interest. For zero heat transfer at the moving surface, i.e., at $\bar{y} = 1$, the nondimensional pressure gradient parameter $\bar{P}$ as a function of the Prandtl-Eckert product $\Pr E$ is

$$\bar{P} = 3\left(\frac{\Pr E}{2} - 1\right) \quad \begin{pmatrix} \text{For zero heat transfer} \\ \text{at the moving surface} \end{pmatrix} \quad (4-39a)$$

Conversely, the $\Pr E$ as a function of $\bar{P}$ is

$$\Pr E = 2\left(1 + \frac{\bar{P}}{3}\right) \quad \begin{pmatrix} \text{For zero heat transfer} \\ \text{at the moving surface} \end{pmatrix} \quad (4-39b)$$

Here, we see that the nondimensional pressure gradient parameter and the Prandtl-Eckert product *for zero heat transfer at the moving surface* are linearly related.

Similarly, the relation between $\bar{P}$ and $\Pr E$ *for zero heat transfer at the stationary surface* is shown to be parabolic. Thus

$$\bar{P} = -1 \pm \sqrt{\frac{-2(\Pr E + 3)}{\Pr E}} \quad \begin{pmatrix} \text{For zero heat transfer} \\ \text{at the lower surface} \end{pmatrix} \quad (4-40a)$$

or

$$\Pr E = \frac{-6}{\bar{P}^2 + 2\bar{P} + 3} \quad \begin{pmatrix} \text{For zero heat transfer} \\ \text{at the lower surface} \end{pmatrix} \quad (4-40b)$$

These results show that if imaginary values of $\bar{P}$ are excluded on physical grounds the range of values of $\Pr E$ which yields zero heat transfer is $-3 \leq \Pr E < 0$. Further, as $\Pr E$ approaches zero $\bar{P}$ approaches plus and minus infinity, i.e., $\Pr E \to 0$, $\bar{P} \to \pm\infty$.

4–5 Flow in a Channel with Straight Parallel Walls — Poiseuille Flow

Continuing our discussion of the analytical solutions for the energy equation, we consider the temperature distribution for the steady incompressible constant property flow of a viscous flow through a channel with flat parallel walls (see Figure 4–5). This flow is generally called Poiseuille flow. Recall that in Section 2–3 we showed that the momentum and energy equations for this flow are uncoupled. Further recall that the velocity distribution (see Section 2–3) is given by

$$u(y) = u_m \left[1 - \left(\frac{y}{h}\right)^2\right] \quad (2-4)$$

where

$$u_m = -\frac{h^2}{2} \frac{1}{\mu} \frac{dP}{dx} \quad (2-5)$$

is the maximum velocity in the channel.

Using our previous assumptions with respect to the velocity and pressure fields, i.e., $v = \partial u/\partial x = \partial v/\partial y = \partial P/\partial y = 0$ and $d\bar{P}/dx = $ constant, and assuming that the temperature gradient in the flow direction is negligible compared to the temperature gradient normal to the flow, i.e., $\partial T/\partial x = 0$, the incompressible energy equation (Eq. 1–51) becomes

$$\frac{d^2 T}{dy^2} = -\frac{\mu}{k}\left(\frac{du}{dy}\right)^2 \quad (4-41)$$

Differentiating the velocity profile and substituting the result into the energy equation yields

$$\frac{d^2 T}{dy^2} = -4\frac{\mu}{k}\frac{u_m^2}{h^2}\left(\frac{y}{h}\right)^2 \quad (4-42)$$

The solution of this ordinary differential equation is

$$T = -\frac{\mu u_m^2}{3k}\left(\frac{y}{h}\right)^4 + Ay + B \qquad (4-43)$$

where appropriate boundary conditions are required to evaluate the integration constants, A and B.

For isothermal wall boundary conditions with the upper wall at a temperature T_1 and the lower wall at a temperature T_0 we have

$$y = -h \qquad T = T_0 \qquad y = +h \qquad T = T_1 \qquad (4-44)$$

The solution of this boundary value problem is then

$$T - T_0 = -\mu\frac{u_m^2}{3k}\left(\frac{y}{h}\right)^4 + \frac{\mu u_m^2}{3k} + \frac{T_1 - T_0}{2}\left(\frac{y}{h}\right) + \frac{T_1 - T_0}{2} \qquad (4-45)$$

When the wall temperatures are equal, i.e., $T_1 = T_0$

$$T - T_0 = \frac{\mu u_m^2}{3k}\left[1 - \left(\frac{y}{h}\right)^4\right] \qquad (4-46)$$

Here we see that the temperature distribution is a fourth-degree polynomial. The maximum temperature occurs at the center of the channel and is given by

$$T_{\max} - T_0 = \frac{\mu u_m^2}{3k} \qquad (4-47)$$

Hence, we write the nondimensional temperature distribution as

$$\frac{T - T_0}{T_{\max} - T_0} = \left[1 - \left(\frac{y}{h}\right)^4\right] \qquad (4-48)$$

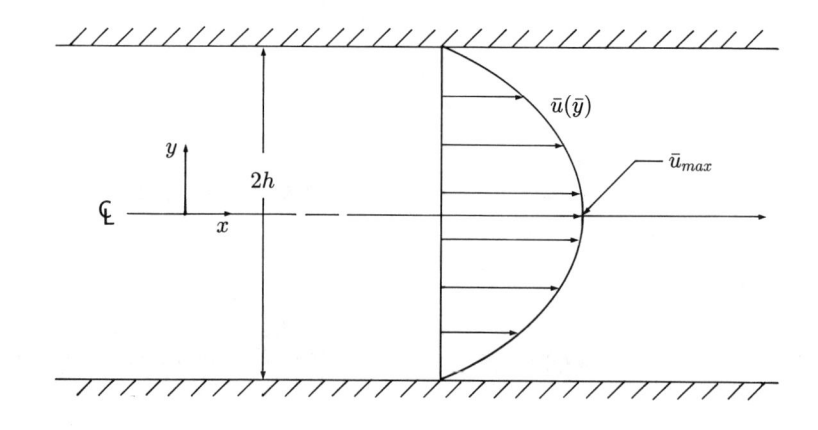

Figure 4–5. Poiseuille flow.

This result is shown in Figure 4–6.

Now, returning to the more general case when the wall temperatures are not equal, we introduce more appropriate nondimensional variables

$$\theta = \frac{T - T_0}{T_1 - T_0} \quad \text{and} \quad \bar{y} = \frac{y}{h}$$

Further, defining an appropriate Eckert number as

$$\mathrm{E} = \frac{u_m^2}{c_p(T_1 - T_0)}$$

allows writing Eq. (4–45) as

$$\theta = \frac{\Pr \mathrm{E}}{3}\left(1 - \bar{y}^4\right) + \frac{1}{2}(\bar{y} + 1) \tag{4-49}$$

The nondimensional maximum or minimum temperature ratio now occurs at

$$\bar{y}\bigg|_{\theta_{\substack{\max \\ \min}}} = \pm 3\left(\frac{3}{8}\frac{1}{\Pr \mathrm{E}}\right)^{1/3} \tag{4-50}$$

and is given by

$$\theta_{\substack{\max \\ \min}} = \frac{\Pr \mathrm{E}}{3} + \frac{3}{16}\left(\frac{3}{\Pr \mathrm{E}}\right)^{1/3} + \frac{1}{2} \tag{4-51}$$

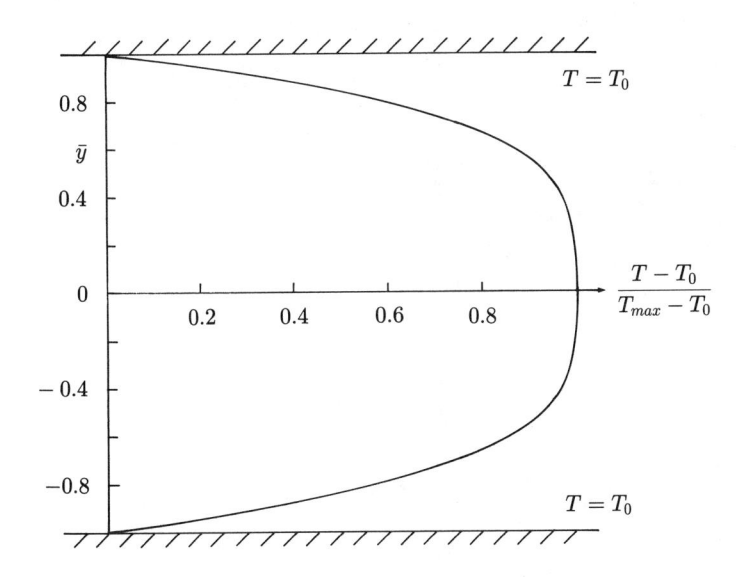

Figure 4–6. Nondimensional temperature distribution in Poiseuille flow for equal wall temperature.

Figure 4–7 shows that the nondimensional temperature gradient at both surfaces changes sign as a function of the Prandtl-Eckert product. Hence, the direction of the heat transfer to the surface also changes. Differentiating the nondimensional temperature and equating it to zero yields the condition for zero heat transfer to the wall, i.e., the adiabatic wall temperature.

At the surface where $\bar{y} = 1$ the Prandtl-Eckert product for zero heat transfer is

$$\Pr E = \frac{3}{8} \qquad (4-52a)$$

At the surface where $\bar{y} = -1$ we have

$$\Pr E = -\frac{3}{8} \qquad (4-52b)$$

The surface where $\bar{y} = 1$ is heated if $\Pr E > 3/8$ and cooled if $\Pr E < 3/8$. Similarly, the surface where $\bar{y} = -1$ is heated if $\Pr E > -3/8$ and cooled if $\Pr E < -3/8$.

4–6 The Energy Equation for the Boundary Layer

In the previous sections we discussed one form of the body force term, i.e., the buoyancy term which occurs in a fluid, as well as some general aspects of obtaining solutions of the energy equation. Subsequently, we obtained and

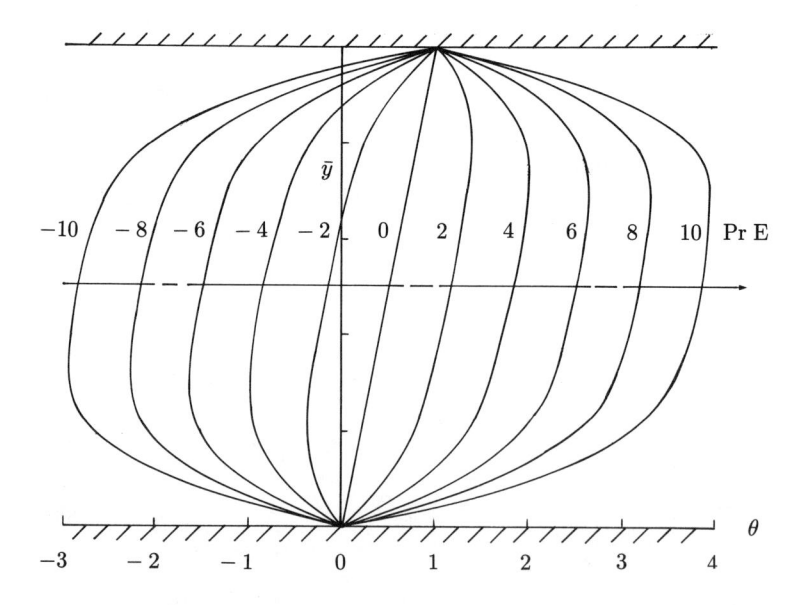

Figure 4–7. Nondimensional temperature distribution in Poissuille flow for unequal wall temperatures.

The Energy Equation for the Boundary Layer

discussed several analytical solutions of the energy equation. Here we turn our attention to the energy equation associated with the thin viscous boundary layer. From Chapter 3 (Eqs. 3–16, 3–19, 3–22, 3–23) and the discussion of Section 4–2, the steady two-dimensional compressible boundary layer equations with constant properties are

continuity
$$\frac{\partial}{\partial x}(\rho u) + \frac{\partial}{\partial y}(\rho v) = 0 \tag{3–23}$$

x momentum
$$u\frac{\partial u}{\partial x} + v\frac{\partial v}{\partial y} = g_x \bar{\beta} \Delta T - \frac{1}{\rho}\frac{\partial P}{\partial x} + \nu\left(\frac{\partial^2 u}{\partial y^2}\right) \tag{4–53}$$

energy
$$u\frac{\partial h}{\partial x} + v\frac{\partial h}{\partial y} = \frac{u}{\rho}\frac{\partial P}{\partial x} + \frac{\nu}{\Pr}\left(\frac{\partial^2 h}{\partial y^2}\right) + \nu\left(\frac{\partial u}{\partial y}\right)^2 \tag{4–54}$$

where we used the form of the buoyancy term given in Eq. (4–5a) in the x momentum equation. In addition, we assumed that the bulk rate of heat addition, Q, is zero in arriving at Eq. (4–54).

When the Grashof number is small with respect to the square of the Reynolds number (see Section 4–2) the buoyancy term in the momentum equation is negligible. When this approximation is used it is generally reasonable to make the further assumption that the fluid is incompressible. When the fluid is incompressible the work due to compression is zero. Thus, the $u(\partial P/\partial x)$ term in the energy equation is zero. The governing equations then become

continuity
$$\frac{\partial u}{\partial x} + \frac{\partial v}{\partial y} = 0 \tag{4–55}$$

momentum
$$u\frac{\partial u}{\partial x} + v\frac{\partial v}{\partial y} = -\frac{1}{\rho}\frac{dP}{dx} + \nu\left(\frac{\partial^2 u}{\partial y^2}\right) \tag{4–56}$$

energy
$$u\frac{\partial h}{\partial x} + v\frac{\partial h}{\partial y} = \frac{\nu}{\Pr}\left(\frac{\partial^2 h}{\partial y^2}\right) + \nu\left(\frac{\partial u}{\partial y}\right)^2 \tag{4–57}$$

Flows for which these equations represent an adequate mathematical model generally are associated with large Reynolds numbers and small temperature differences. Such flows are generally called *forced*; the associated heat transfer process is called *forced convection*. Forced convection flows are discussed in Sections 4–8 to 4–11.

For flows in which the Grashof number is not small with respect to the square of the Reynolds number the buoyancy term cannot be neglected. However, in such flows the effects due to the pressure field are generally not significant, and hence the pressure gradient term in the momentum equation can be neglected. Many such flows occur in nature. They are generally called natural flows, and the heat transfer process is called *natural* or *free convection*. This boundary layer analysis is applied to flows which occur for small (but not sufficiently small that the boundary layer concept is not valid) Reynolds numbers, and relatively large (but not so large that the variation in properties must be accounted for) temperature differences. Free convection flows are discussed in Chapter 5.

Flows in which both buoyancy and pressure field effects are important are generally termed *combined* and the associated heat transfer *combined forced and free convection*. These flows are discussed in Chapter 5.

Returning to the governing equations for forced convection, we immediately note that the energy equation is uncoupled from the continuity and momentum equations, i.e., we have a Type 2 flow (see Section 1–10). Consequently, the velocity field is independent of the temperature field. However, note that the temperature field still depends on the velocity field. Further, examination of the energy equation reveals that it is a linear differential equation. Thus, the solution of the energy equation is considerably easier than that of the momentum equation. For example, superposition of solutions is possible. This characteristic is effectively exploited to obtain solutions of the energy equation.

Recall that in the derivation of the energy equation for the boundary layer we assumed that the Prandtl number was of order unity (see Section 3–2). This implied that the thickness of the thermal boundary layer is of the same order-of-magnitude as the thickness of the velocity boundary layer. However, for many fluids the Prandtl number is not of order unity, and the thickness of the thermal boundary layer is not of the same order-of-magnitude as the thickness of the velocity boundary layer. Using subscripts V and T to represent the velocity and thermal boundary layers, and returning to Eqs. (3–15) and (3–20), we again argue that for consistency the nondimensional thickness of the velocity boundary layer is

$$\bar{\delta}'_V \sim \frac{1}{\sqrt{\text{Re}_\ell}} \qquad (3-15)$$

In a similar manner we argue, from Eq. (3–20), that the thickness of the thermal boundary layer is

$$\bar{\delta}'_T \sim \frac{1}{\sqrt{\text{Re}_\ell \, \text{Pr}}}$$

These results show that

$$\bar{\delta}'_T \sim \frac{\bar{\delta}'_V}{\sqrt{\text{Pr}}}$$

Here we see that when the Prandtl number is of order unity the thicknesses of the velocity and thermal boundary layers are of the same order. However, when the Prandtl number is much less than one the thermal boundary layer is significantly thicker than the velocity boundary layer. This implies a further restriction on the minimum value of the Reynolds number in order to satisfy the concept of a thin boundary layer and hence adequately represent the flow by the boundary layer equations. Conversely, when the Prandtl number is very large the velocity boundary layer is much thicker than the thermal boundary layer.

Returning again to Eqs. (4–56) and (4–57), i.e., the momentum and energy equations, we see that the terms on the left hand side represent convection (see Section 1–1) of momentum and thermal energy in the boundary layer. Further, the terms $\nu(\partial^2 u/\partial y^2)$ and $(\nu/\text{Pr})(\partial^2 h/\partial y^2)$ represent the diffusion of momentum and thermal energy, respectively, in the boundary layer. Here the momentum diffusion coefficient is ν, the kinematic viscosity. The diffusion coefficient for thermal energy is $\nu/\text{Pr} = \alpha$. Recalling from our previous discussion of the velocity boundary layer (see Chapter 3) that near the body surface the velocity is low and the change in the velocity gradient $\partial^2 u/\partial y^2$ large, we see that the transport of momentum near the surface is dominated by diffusion, whereas near the outer edge of the boundary layer the velocity is nearly equal to the external stream velocity with a correspondingly small change in velocity gradient $\partial^2 u/\partial y^2$. Here, momentum and energy transport by convection predominates.

Examining the momentum and energy equations further, and recalling from our previous nondimensional discussion in Section 4–2 that the viscous dissipation term $\nu(\partial u/\partial y)^2$ is negligible when the Eckert number is small with respect to the Reynolds number, we see that for zero pressure gradient the momentum and energy equations have similar forms, i.e.

momentum
$$u\frac{\partial u}{\partial x} + v\frac{\partial u}{\partial y} = \nu\left(\frac{\partial^2 u}{\partial y^2}\right) \qquad (4-58)$$

energy
$$u\frac{\partial T}{\partial x} + v\frac{\partial T}{\partial y} = \frac{\nu}{\text{Pr}}\left(\frac{\partial^2 T}{\partial y^2}\right) \qquad (4-59)$$

Here, we used the fact that for a thermally perfect fluid $dh = c_p dT$, which for an incompressible fluid becomes $dh = c\, dT$.

Introducing the nondimensional variables defined in Eq. (4–9), we see that for unit Prandtl number the momentum and energy equations have exactly the same form, i.e.

momentum
$$\bar{u}\frac{\partial \bar{u}}{\partial \bar{x}} + \bar{v}\frac{\partial \bar{u}}{\partial \bar{y}} = \frac{1}{\text{Re}_\ell}\left(\frac{\partial^2 \bar{u}}{\partial \bar{y}^2}\right) \qquad (4-60)$$

energy
$$\bar{u}\frac{\partial \theta}{\partial \bar{x}} + \bar{v}\frac{\partial \theta}{\partial \bar{y}} = \frac{1}{\text{Re}_\ell}\left(\frac{\partial^2 \theta}{\partial \bar{y}^2}\right) \qquad (4-61)$$

Thus, it follows that solutions for $\bar{u}$ and θ have identical nondimensional forms when the nondimensional boundary conditions are also identical. That is, a

126 Laminar Flow Analysis

nondimensional solution of the momentum equation is also a nondimensional solution of the energy equation. This further implies that $\partial \bar{u}/\partial \bar{y} = \partial \theta/\partial \bar{y}$. When this result is applied at the surface, an important relationship between the shearing stress (skin friction) and the heat transfer at the surface is obtained. This relationship was first discovered by O. Reynolds and is called Reynolds analogy.

Recall that the shearing stress is given by

$$\tau = \mu \frac{\partial u}{\partial y}$$

which in terms of the nondimensional variables given in Eq. (4–9) becomes

$$\tau = \mu \frac{U_e}{\ell} \frac{\partial \bar{u}}{\partial \bar{y}} \qquad (4-62)$$

Similarly, in nondimensional variables (see Eq. 1–35) the heat transfer rate per unit area becomes

$$\left(\frac{q}{A}\right)_y = -\frac{k(T_w - T_e)}{\ell} \frac{\partial \theta}{\partial \bar{y}} \qquad (4-63)$$

Using the results of our previous discussion, i.e., $\partial \bar{u}/\partial \bar{y} = \partial \theta/\partial \bar{y}$, we have

$$\left(\frac{q}{A}\right)_y = -\frac{k(T_w - T_e)}{\mu U_e} \tau \qquad (4-64)$$

This result shows that for the particular conditions assumed here there is a definite relationship between the heat transfer rate and the shearing stress.

Introducing the nondimensional *local* Nusselt number and the *local* skin friction coefficient, we arrive at a more common statement of Reynolds analogy. The Nusselt number is considered as a nondimensional heat transfer rate. Physically it is interpreted as the ratio of the convective heat transfer rate at a surface in a moving fluid, to the conductive heat transfer rate between two surfaces when the fluid is at rest. The conductive heat transfer rate between two surfaces at temperatures, T_w and T_e, separated by a distance, ℓ, is

$$\left(\frac{q}{A}\right) = \frac{k(T_w - T_e)}{\ell}$$

Thus, we define the local Nusselt number as

$$\mathrm{Nu}_x = \frac{\ell \dfrac{\partial T}{\partial y}}{T_w - T_e} \qquad (4-65)$$

In nondimensional variables we have

$$\mathrm{Nu}_x = -\frac{\partial \theta}{\partial \bar{y}} \qquad (4-66)$$

The nondimensional local skin friction coefficient is defined as the ratio of the local shearing stress to the local dynamic pressure. Thus

$$C_{f_x} = \frac{\tau}{\frac{1}{2}\rho U_e^2} \qquad (4-67)$$

Combining our results, we write Reynolds analogy as

$$\text{Nu}_x = \frac{1}{2}\text{Re}_x C_{f_x} \qquad (4-68)$$

Although this result is derived for the special case of incompressible steady flow in which the pressure gradient is zero, the Prandtl number unity and the viscous dissipation negligible, Reynolds analogy in a modified form is valid for other than zero pressure gradient flows.

An additional nondimensional parameter used as a heat transfer parameter is the Stanton number. The Stanton number is physically interpreted as the ratio of the convective heat transfer rate between the moving fluid and the surface, and the heat transfer rate which results if the flow is subjected to the temperature $T_w - T_e$. This heat transfer rate is $-\rho c_p U_e (T_w - T_e)$. Hence, the Stanton number is

$$\text{St}_x = \frac{k\dfrac{\partial T}{\partial y}}{\rho c_p U_e (T_w - T_e)} \qquad (4-69)$$

Introducing the Nusselt number we have

$$\text{St}_x = \frac{\text{Nu}_x}{\Pr \text{Re}} \qquad (4-70)$$

Note that for the particular case of unit Prandtl number Reynolds analogy can be stated as

$$\text{St}_x = \frac{C_{f_x}}{2} \qquad (4-71)$$

4–7 Forced Convection Boundary Layer — Parallel Flow Past a Flat Plate

We first consider the steady parallel flow of an incompressible thermally perfect fluid past a semi-infinite flat plate (see Figure 4–8). We assume that the fluid

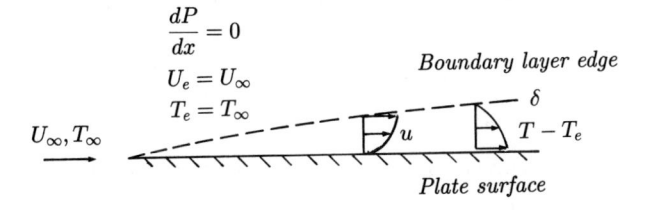

Figure 4–8. Forced convection boundary layer flow past a flat plate.

properties are constant and neglect buoyancy forces, i.e., we consider the forced convecton boundary layer. For a flat plate the pressure gradient is zero. Hence, the governing equations (Eqs. 4–55 to 4–57) are

continuity
$$\frac{\partial u}{\partial x} + \frac{\partial v}{\partial y} = 0 \qquad (4-55)$$

momentum
$$u\frac{\partial u}{\partial x} + v\frac{\partial u}{\partial y} = \nu\left(\frac{\partial^2 u}{\partial y^2}\right) \qquad (4-56)$$

energy
$$\rho c_p\left(u\frac{\partial T}{\partial x} + v\frac{\partial T}{\partial y}\right) = k\frac{\partial^2 T}{\partial y^2} + \mu\left(\frac{\partial u}{\partial y}\right)^2 \qquad (4-74)$$

The surface boundary conditions for the velocity field are given by the no-slip and no-mass-transfer conditions. The surface boundary conditions for the temperature field require that the surface and the fluid have the same temperature, or that there be no heat transfer from the surface to the fluid or, conversely, from the fluid to the surface. At the edge of the boundary layer the viscous flow merges smoothly into the inviscid flow. Thus, we write

$$y = 0 \qquad u = v = 0 \qquad T = T_w \quad \text{or} \quad \frac{\partial T}{\partial y} = 0 \qquad (4-75a,b,c)$$

$$y \to \infty \qquad u \to U_\infty \qquad T \to T_\infty \qquad (4-75d,e)$$

where here we use the subscript ∞ to indicate conditions in the inviscid flow at the edge of the boundary layer. This implies that the existence of the boundary layer on the surface does not affect the initial free stream conditions.

Again, we recognize that the continuity and momentum equations are uncoupled from the energy equation. Hence, the associated boundary value problem for the continuity and momentum equations is solved independently of that associated with the energy equation. This solution is then used in the subsequent solution of the energy equation. We recognize the boundary value problem associated with the continuity and momentum equations as one we previously studied in Section 3–3. Consequently, the solution of the velocity field is known. In Section 3–3 we found, after introducing a stream function, that the partial differential equations are reduced to a single ordinary differential equation using the similarity transformations

$$\eta = y\left(\frac{U}{2\nu x}\right)^{1/2} \qquad \psi(x,y) = (2\nu U x)^{1/2} f(\eta) \qquad (3-39a,b)$$

The governing nonlinear two-point asymptotic boundary value problem for the velocity field is then the Blasius equation

$$f''' + ff'' = 0 \qquad (3-38)$$

with boundary conditions

$$\eta = 0 \qquad f(0) = f'(0) = 0 \qquad (3-37a,b)$$

and

$$\eta \to \infty \qquad f'(\eta \to \infty) \to 1 \qquad (3-40)$$

Using these same similarity transformations for the energy equation (Eq. 4–74) yields

$$\frac{d^2T}{d\eta^2} + \Pr f \frac{dT}{d\eta} = -\Pr \frac{U_\infty^2}{c_p} f''^2 \qquad (4-76)$$

where here the temperature is still assumed to be a function of x and y. Recalling the boundary conditions for the energy equation, we introduce the nondimensional temperature

$$\theta = \frac{T - T_\infty}{T_w - T_\infty} \qquad (4-77)$$

Equation (4–76) then becomes

$$\theta'' + \Pr f \theta' = -\Pr \mathrm{E}\, f''^2 \qquad (4-78)$$

where we defined the appropriate Eckert number as

$$\mathrm{E} = \frac{U_\infty^2}{c_p(T_w - T_\infty)} \qquad (4-79)$$

and the $'$ indicates differentiation with respect to η. If we assume that $\theta = \theta(\eta)$ only, then Eq. (4–78) is an ordinary differential equation, and the nondimensional temperature profiles are similar provided the Eckert number is constant. Since the inviscid flow velocity $U(x) = U_\infty$ and temperature $T(x) = T_\infty$ are constant, the Eckert number is constant if the wall temperature is constant. Thus, we assume an isothermal wall.

The nondimensional boundary conditions for Eq. (4–78) are

$$\eta = 0 \qquad \theta = 1 \quad \text{or} \quad \theta' = 0 \qquad (4-80a,b)$$

and

$$\eta \to \infty \qquad \theta(\eta \to \infty) \to 0 \qquad (4-80c)$$

Equation (4–78) is a second-order linear nonhomogeneous differential equation subject to two-point asymptotic boundary conditions. Since it is linear, the principle of superposition applies. Hence, we first seek a solution to the homogeneous equation and then seek a particular solution to the nonhomogeneous equation. By virtue of the principle of superposition the sum of these solutions is also a solution. Therefore we assume that

$$\theta(\eta) = K\theta_1(\eta) + \mathrm{E}\,\theta_2(\eta) \qquad (4-81)$$

where $\theta_1(\eta)$ denotes the solution of the homogeneous equation, $\theta_2(\eta)$ denotes the solution of the nonhomogeneous equation and K is a constant. We further choose the boundary conditions such that $\theta_1(\eta)$ is the solution when there is heat transfer at the surface, and $\theta_2(\eta)$ is the solution when there is no heat transfer at the surface, i.e., for an adiabatic wall. $\theta_1(\eta)$ and $\theta_2(\eta)$ then satisfy the following homogeneous and nonhomogeneous two-point asymptotic boundary value problems:

homogeneous

$$\theta_1'' + \Pr f \theta_1' = 0 \qquad (4-82)$$

with
$$\theta_1(0) = 1 \qquad (4-83a)$$
$$\theta_1(\eta \to \infty) \to 0 \qquad (4-83b)$$

nonhomogeneous

$$\theta_2'' + \Pr f \theta_2' = -\Pr f''^2 \qquad (4-84)$$

with
$$\theta_2'(0) = 0 \qquad (4-85a)$$
$$\theta_2(\eta \to \infty) \to 0 \qquad (4-85b)$$

Notice that the boundary value problem governing $\theta_1(\eta)$ is the same as is obtained by neglecting the viscous dissipation term, $-\Pr f''^2$. This is a frequent assumption in low speed flow.

Both the homogeneous and the nonhomogeneous boundary value problems are amenable to analytical solution. The solution of the homogeneous problem is

$$\theta_1(\eta) = \frac{\int_{\xi=\eta}^{\infty} (f''(\xi))^{\Pr} d\xi}{\int_{\xi=0}^{\infty} (f''(\xi))^{\Pr} d\xi} \qquad (4-86)$$

This solution was originally given in 1921 by the German investigator E. Pohlhausen [Pohl21]. Realizing that the definite integral is a constant, determined from our previous solutions of the Blasius equation (see Section 3–3), it is convenient to define α_0 as

$$\alpha_0(\Pr) = \frac{1}{\int_0^\infty (f''(\xi))^{\Pr} d\xi} \qquad (4-87)$$

$\alpha_0(\Pr)$ is evaluated by modifying the **faskan** program given in Section 3–6.

Since for the special case of unit Prandtl number $\alpha_0(1) = 1$, Eq. (4–86) reduces to

$$\theta_1(\eta) = 1 - f'(\eta) \qquad (4-88)$$

Thus, when there is heat transfer at the surface the nondimensional temperature distribution has the same functional form as the nondimensional velocity distribution. Figure 4–9 shows the nondimensional temperature distribution $\theta_1(\eta)$ for the homogeneous boundary value problem for several values of the Prandtl number. Note that, as expected from our order-of-magnitude analysis, the thickness of the boundary layer decreases as the Prandtl number increases. These nondimensional temperature distributions are generated by suitable modifications of the **faskan** program of Chapter 3. The gradient of the nondimensional temperature distribution at the wall for the homogeneous case is determined by differentiating Eq. (4–86). Hence

$$\theta_1'(0) = -\alpha_0(\text{Pr}) \left(f''(0)\right)^{\text{Pr}} \qquad (4-89)$$

Recalling our previous results for the Blasius equation (see Table C–1, $\beta = 0$ of Appendix C), we have $f''(0) = 0.4696005$, or

$$\theta_1'(0) = -[0.4696005]^{\text{Pr}} \alpha_0(\text{Pr}) \qquad (4-90)$$

Figure 4–10 gives $\theta_1'(0)$ as a function of Prandtl number. The data in this figure

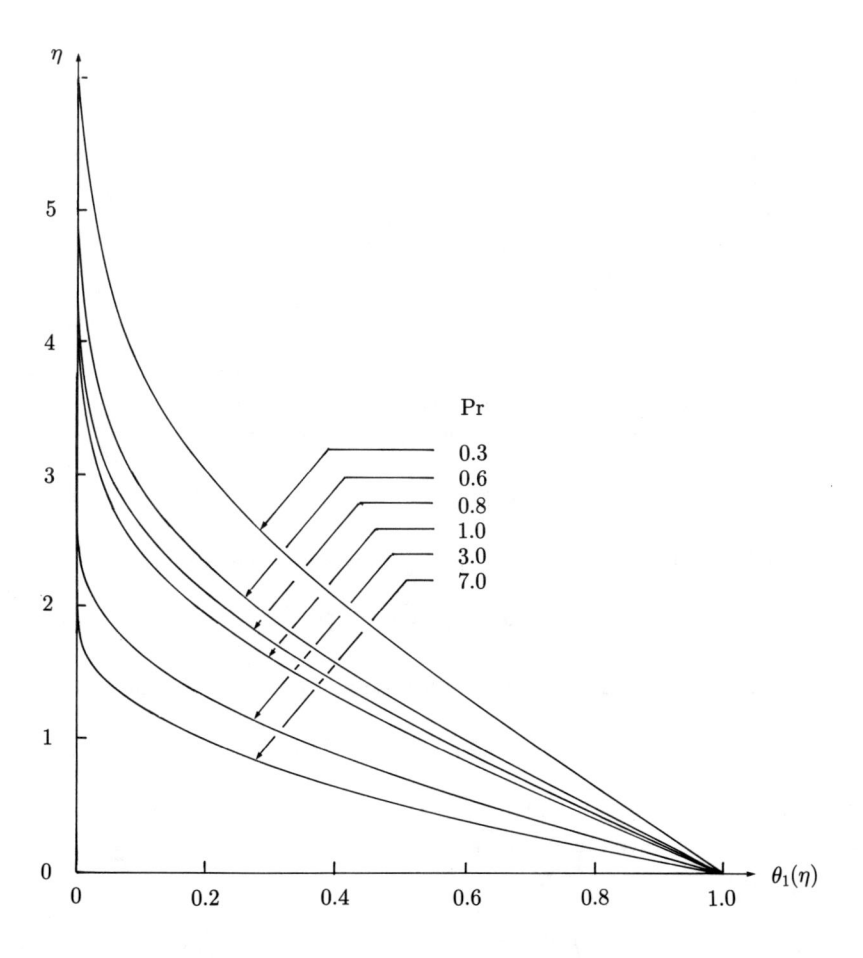

Figure 4–9. Nondimensional temperature distribution for the homogeneous case.

132 **Laminar Flow Analysis**

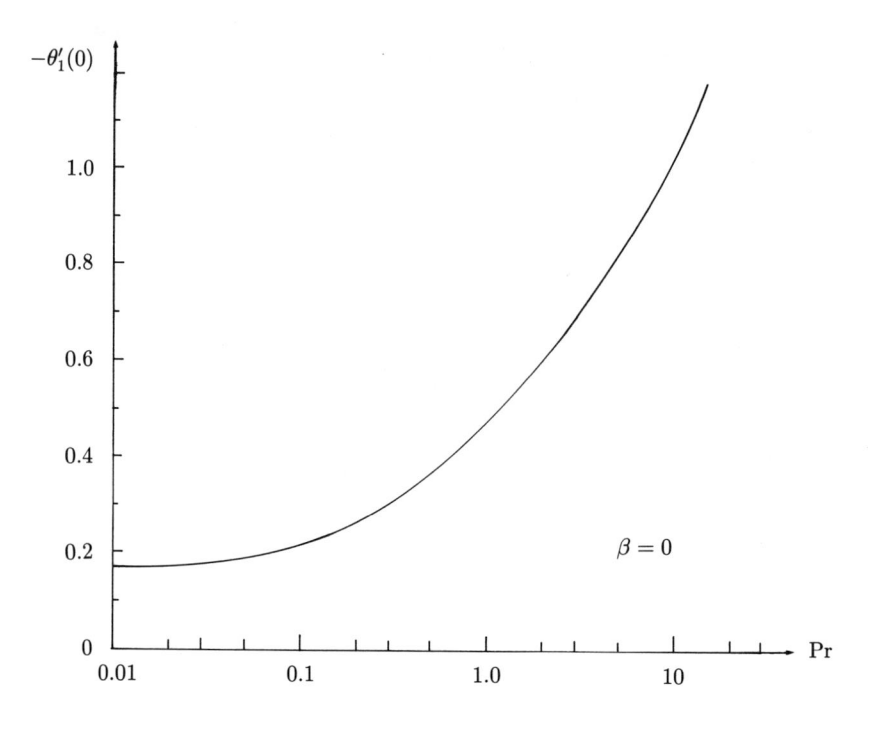

Figure 4–10. Nondimensional heat transfer parameter $\beta = 0$.

is obtained by modifying the **faskan** program discussed in Chapter 3. Note that $\theta'_1(0)$ decreases with decreasing Prandtl number. For Prandtl numbers from approximately 0.6 to 15, $\theta'_1(0)$ is adequately represented by

$$\theta'_1(0) \approx -0.4696005 \sqrt[3]{\Pr} \qquad 0.6 \leq \Pr \leq 15 \qquad (4-91)$$

As the Prandtl number approaches zero an adequate approximation is

$$\theta'_1(0) \approx -0.798 \sqrt[3]{\Pr} \qquad \Pr \to 0 \qquad (4-92)$$

As the Prandtl number approaches infinity an adequate approximation is

$$\theta'_1(0) \approx -0.479 \sqrt[3]{\Pr} \qquad \Pr \to \infty \qquad (4-93)$$

The nonhomogeneous boundary value problem for the adiabatic wall (see Eqs. 4–84, 4–85) is solved using the method of variation of a parameter [Spie58]. The solution is

$$\theta_2(\eta) = \Pr \int_{\xi=\eta}^{\infty} \left(f''(\xi)\right)^{\Pr} \left(\int_0^{\xi} \left(f''(\tau)\right)^{2-\Pr} d\tau \right) d\xi \qquad (4-94)$$

For the special case of unit Prandtl number, this result reduces to

$$\theta_2(\eta) = \frac{1}{2}\left(1 - f'^2(\eta)\right)$$

The value of $\theta_2(0)$ is determined from the integral

$$\theta_2(0) = \Pr \int_0^\infty (f''(\xi))^{\Pr} \left(\int_0^\xi (f''(\tau))^{2-\Pr} d\tau \right) d\xi \qquad (4-95)$$

Selected numerical results are given in Table C–11 of Appendix C. For unit Prandtl number $\theta_2(0) = 1/2$. For Prandtl numbers near unity, an adequate approximation is

$$2\theta_2(0, \Pr) \approx (\Pr)^{1/2}$$

Figure 4–11 shows the nondimensional temperature distribution for the nonhomogeneous or adiabatic wall boundary value problem for various values of

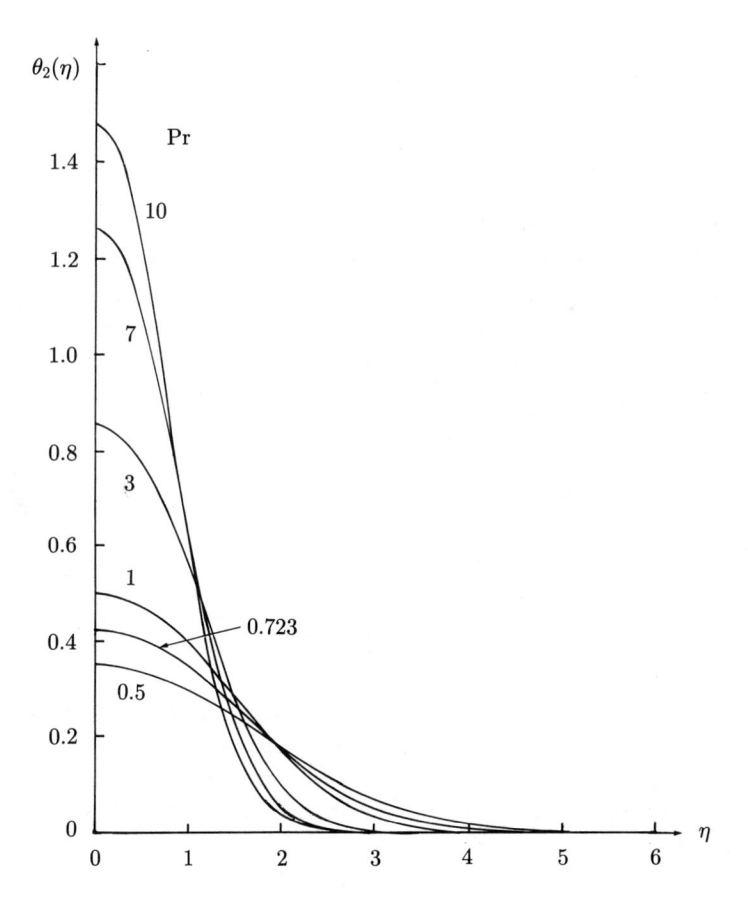

Figure 4–11. Nondimensional temperature distribution for the adiabatic wall.

the Prandtl number. These curves essentially represent the contribution of viscous dissipation to the temperature distribution in the thermal boundary layer. From these results, we see that for small Prandtl numbers the viscous dissipation makes a significant contribution to the temperature distribution. Further, from Eq. (4–81) we see that this contribution becomes particularly important as the Eckert number increases. These particular conditions generally occur for the flow of gases such as air.

The constant, K, in Eq. (4–81) is evaluated by considering the requirement that the general solution for the temperature distribution satisfies the condition $T = T_w$ at $y = 0$, which corresponds to $\theta = 1$ at $\eta = 0$. Thus

$$K = 1 - \text{E}\,\theta_2(0)$$

and Eq. (4–81) becomes

$$\theta(\eta) = [1 - \text{E}\,\theta_2(0)]\theta_1(\eta) + \text{E}\,\theta_2(\eta) \qquad (4-96)$$

For the special case of unit Prandtl number

$$\theta(\eta) = \theta_1(\eta) + \text{E}\left(\theta_2(\eta) - \frac{\theta_1(\eta)}{2}\right) \qquad (4-97)$$

Figure 4–12 shows the combined nondimensional temperature profiles for several values of the Prandtl number and Prandtl-Eckert product.

The heat transfer rate at the surface is proportional to the gradient of the nondimensional temperature distribution at the surface (see Eq. 1–35). In particular, the local Nusselt number (Eq. 4–65) based on the distance from the leading edge as the characteristic length is

$$\text{Nu}_x = \left(\frac{U_\infty x}{2\nu}\right)^{1/2} \theta_1'(0) = (\text{Re}_x)^{1/2} \frac{\theta_1'(0)}{\sqrt{2}} \qquad (4-98)$$

Similarly, the local Stanton number (Eq. 4–69) is

$$\text{St}_x = \frac{1}{\text{Pr}} \frac{1}{\sqrt{\text{Re}_x}} \frac{\theta_1'(0)}{\sqrt{2}} \qquad (4-99)$$

Since $\theta_1'(0)$ is a function of the Prandtl number, this result gives the Nusselt number as a function of the Reynolds and Prandtl numbers. For forced convection over a flat surface with zero pressure gradient, isothermal boundary conditions, constant fluid properties and without viscous dissipation, the Nusselt number is approximated by

$$\frac{h_x x}{k} = \text{Nu}_x \approx 0.332\,(\text{Re}_x)^{1/2}\,(\text{Pr})^{1/3}$$

for $0.6 \leq \text{Pr} \leq 15$.

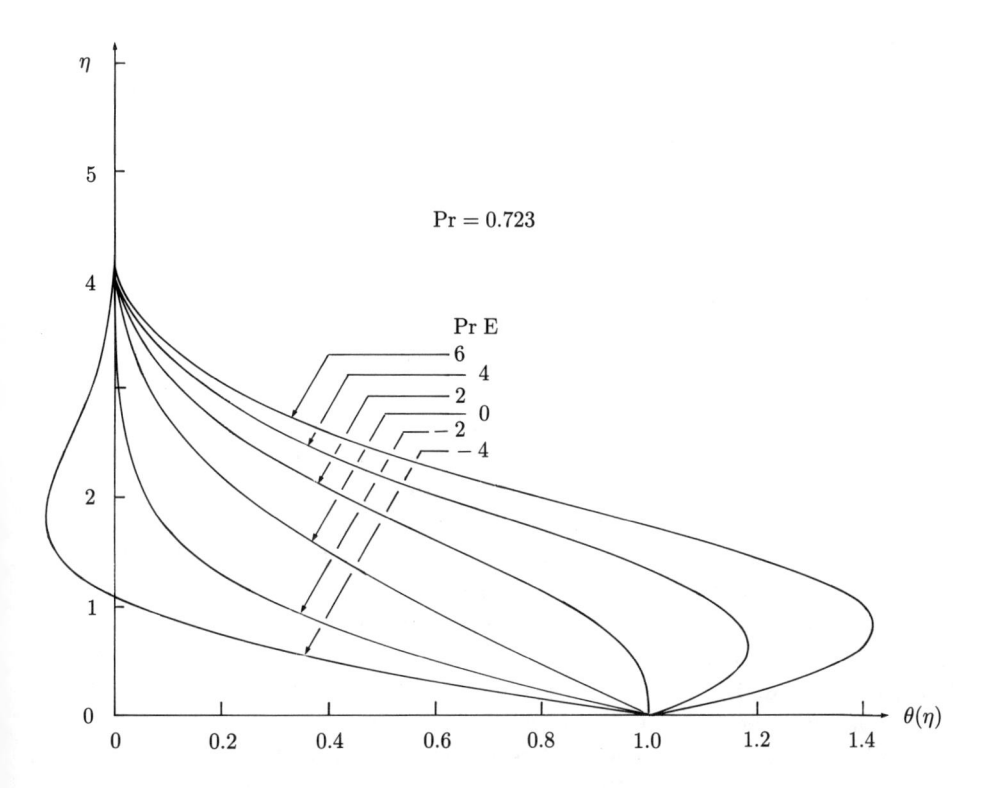

Figure 4–12. The nondimensional temperature distribution on a heated or cooled flat plate with frictional heating.

The total heat transfer by convection from a surface of area A_s is given by $q = \bar{h} A_s (T_w - T_\infty)$, where $\bar{h}$ is the average convective heat transfer coefficient defined as

$$\bar{h} = \frac{\int_0^L h_x \, dx}{L}$$

and L is the total length of the surface measured along the flow direction. Using our previous expression for h_x and carrying out the integration yields an average Nusselt number

$$\frac{\bar{h}L}{k} = \overline{\mathrm{Nu}}_L = 0.664 \mathrm{Nu}_x|_{x=L} = 2\mathrm{Nu}_L$$

Thus, for this special situation, the average Nusselt number is simply twice the value of the local Nusselt number evaluated at the rear edge of the surface.

When there is a significant difference between T_w and T_∞, what values of the fluid properties such as k, ν and α should be used to calculate the Nusselt, Reynolds and Prandtl numbers? For constant fluid properties it is common practice to evaluate the fluid properties at a film temperature defined as the

average of T_w and T_∞, i.e.
$$T_f = \frac{T_w + T_\infty}{2}$$

Differentiating Eq. (4–96) and evaluating the result at $\eta = 0$ yields the nondimensional temperature gradient at the surface. Thus

$$\theta'(0) = \left(1 - \mathrm{E}\,\theta_2(0)\right) \theta'_1(0)$$

Using Eq. (4–89) yields

$$\theta'(0) = -\left(1 - \mathrm{E}\,\theta_2(0)\right) \alpha_0(\mathrm{Pr}) \left(f''(0)\right)^{\mathrm{Pr}} \qquad (4-100)$$

When the nondimensional temperature gradient at the surface is negative, heat transfer occurs from the surface to the fluid, i.e., the surface is cooled. When the gradient is positive, heat transfer occurs from the fluid to the surface, i.e., the surface is heated. Examination of Eq. (4–100) shows that if $\mathrm{E}\,\theta_2(0) < 1$ the surface is cooled. If $\mathrm{E}\,\theta_2(0) > 1$ the surface is heated. Further, when $\mathrm{E}\,\theta_2(0) = 1$ no heat transfer takes place at the surface. Since $\theta_2(0)$ is a function of the Prandtl number, the direction of heat transfer at a surface is also a function of the Prandtl number as well as the Eckert number. Figure 4–12 clearly reveals this phenomenon. Further, Figure 4–12 shows that when $\mathrm{E}\,\theta_2(0) > 1$ there is a layer of hot fluid near the surface which results in heat transfer to the surface.

Finally, recalling that when the Eckert number is zero frictional heating is neglected, $\theta'(0)$ is always negative and hence the surface is always cooled. Using Eq. (4–100), the Nusselt and Stanton numbers, Eqs. (4–98) and (4–99), become

$$\mathrm{Nu}_x = \sqrt{\mathrm{Re}_x}\, \frac{\alpha_0(\mathrm{Pr})}{\sqrt{2}} \left(\mathrm{E}\,\theta_2(0,\mathrm{Pr}) - 1\right) \left(f''(0)\right)^{\mathrm{Pr}} \qquad (4-101)$$

and
$$\mathrm{St}_x = \frac{1}{\sqrt{\mathrm{Re}_x}}\, \frac{1}{\mathrm{Pr}}\, \frac{\alpha_0(\mathrm{Pr})}{\sqrt{2}} \left(\mathrm{E}\,\theta_2(0,\mathrm{Pr}) - 1\right) \left(f''(0)\right)^{\mathrm{Pr}} \qquad (4-102)$$

where the notation $\theta_2(0,\mathrm{Pr})$ is used to indicate that the adiabatic wall temperature is a function of the Prandtl number.

We now turn our attention to forced convection boundary layer flows which are subject to a pressure gradient. For these flows the velocity field is governed by the Falkner-Skan equation.

4–8 Forced Convection Boundary Layer Flows with Pressure Gradient and Nonisothermal Surface Conditions

Having discussed the zero pressure gradient isothermal wall solution, we now turn our attention to more general boundary layer flows subjected to an external pressure gradient and to nonisothermal variations in surface temperature. For these flows, the governing boundary layer equations again are Eqs. (4–55) to (4–57), i.e.

Forced Convection Boundary Layer — Pressure & Temperature Gradients

continuity
$$\frac{\partial u}{\partial x} + \frac{\partial v}{\partial y} = 0 \qquad (4-55)$$

momentum
$$u\frac{\partial u}{\partial x} + v\frac{\partial u}{\partial y} = -\frac{1}{\rho}\frac{dP}{dx} + \nu\left(\frac{\partial^2 u}{\partial y^2}\right) \qquad (4-56)$$

energy
$$u\frac{\partial h}{\partial x} + v\frac{\partial h}{\partial y} = \frac{\nu}{\text{Pr}}\left(\frac{\partial^2 h}{\partial y^2}\right) + \nu\left(\frac{\partial u}{\partial y}\right)^2 \qquad (4-57)$$

The surface boundary conditions for the velocity field are no-velocity-slip and no-mass-transfer at the surface. Here the surface boundary condition for the energy equation is that the temperature of the surface is a function of the distance from the leading edge. The boundary conditions at the outer edge of the boundary layer are given by the usual requirement that the viscous flow smoothly transition into the inviscid flow at the edge of the boundary layer. Mathematically, these boundary conditions are represented by

$$y = 0 \qquad u = v = 0 \qquad T = T(x) \qquad (4-103a,b,c)$$
$$y \to \infty \qquad u \to U(x) \qquad T \to T_e \qquad (4-103d,e)$$

In Section 3–4 we showed that similar solutions of the equations for the velocity field are governed by the Falkner-Skan equation with suitable boundary conditions. There, the similarity condition for the external velocity field is shown to be $U(x) = U_\infty x^m$. We therefore restrict our present discussion to flows which satisfy this condition. Our immediate purpose is to determine the corresponding conditions for which similar solutions of the energy equation exist. Following our discussion of the Falkner-Skan equation, we first introduce the similarity transformations

$$\eta = y\left[\frac{(m+1)}{2}\frac{U}{\nu x}\right]^{1/2} \qquad (3-62)$$

and
$$\psi(x,y) = \left[\frac{2}{m+1}\nu x U\right]^{1/2} f(\eta) \qquad (4-104)$$

We previously showed that introducing these similarity transformations into the continuity and momentum equations, Eqs. (4–55) and (4–56), yields

$$f''' + ff'' + \beta(1 - f'^2) = 0 \qquad (3-50)$$

with boundary conditions
$$f(0) = f'(0) = 0 \qquad (3-51a,b)$$

and
$$f'(\eta \to \infty) \to 1 \qquad (3-51c)$$

Introduction of these same similarity transformations into the energy equation (Eq. 4–57) yields

$$\frac{\partial^2 T}{\partial \eta^2} + \Pr f \frac{\partial T}{\partial \eta} + 2\Pr \frac{m-1}{m+1} \eta f' \frac{\partial T}{\partial \eta} - \frac{2\Pr}{m+1} f' x \frac{\partial T}{\partial x} = -\Pr \frac{U^2}{c_p} f''^2$$

where the relation $h = c_p T$ is used to arrive at this result. Recalling that $m = \beta/(2 - \beta)$ allows this equation to be rewritten as

$$\frac{\partial^2 T}{\partial \eta^2} + \Pr f \frac{\partial T}{\partial \eta} + 2\Pr (\beta - 1) \eta f' \frac{\partial T}{\partial \eta} - (2 - \beta) \Pr x f' \frac{\partial T}{\partial x} = -\Pr \frac{U^2}{c_p} f''^2 \quad (4-105)$$

We now introduce a nondimensional temperature

$$\theta = \frac{T - T_w}{T_e - T_w} \quad (4-106)$$

which is a function of η only.

The particular form of θ is chosen from an analysis of the boundary conditions (see Eq. 4–75). At the surface, i.e., at $y = 0$, the temperature of the fluid is equal to the temperature of the surface. Hence, $T = T_w$ and $\theta = 0$. At a large distance from the surface, i.e., as $y \to \infty$, the temperature of the fluid in the boundary layer approaches the inviscid flow temperature, hence $T = T_e$ and $\theta = 1$. We further assume that the surface heat transfer is sufficiently small that it does not change the temperature in the inviscid flow at the edge of the boundary layer. Thus, T_e is a constant equal to T_∞. Using these assumptions, Eq. (4–105) becomes

$$\theta'' + \Pr f \theta' - \Pr (2 - \beta)(1 - \theta) f' \frac{x}{T_e - T_w} \frac{dT_w}{dx} = -\Pr \frac{U_\infty^2}{c_p (T_e - T_w)} x^{2m} f''^2 \quad (4-107)$$

This is an ordinary differential equation. Similar solutions with θ only a function of η exist if

$$\frac{x}{T_e - T_w} \frac{dT_w}{dx} = \text{constant}$$

and

$$\frac{x^{2m}}{T_e - T_w} = \text{constant}$$

Examination of these requirements leads one to investigate a power law surface temperature distribution. In particular, we assume that

$$T_w - T_e = T_1 x^n \quad (4-108)$$

where T_1 is a constant associated with the initial temperature distribution. Note that Eq. (4–108) reduces to the isothermal wall condition when $n = 0$. Using this assumption, Eq. (4–107) becomes

$$\theta'' + \Pr f\theta' + n\Pr(2-\beta)(1-\theta)f' = \Pr E_1 x^{2m-n} f''^2 \qquad (4-109)$$

where the Eckert number is chosen as $E = U_\infty^2/c_p T_1$. The associated boundary conditions are

$$\theta(0) = 0 \qquad \theta(\eta \to \infty) \to 1 \qquad (4-110a,b)$$

Equation (4–109) shows that there are two classes of similar solutions of the energy equation for forced convection: those with viscous dissipation, and those without viscous dissipation. For low speed incompressible flow the Eckert number is small, since U_∞ is small. Under these conditions, the viscous dissipation term on the right hand side of Eq. (4–109) can be neglected. Equation (4–109) thus reduces to

$$\theta'' + \Pr f\theta' + n\Pr(2-\beta)(1-\theta)f' = 0 \qquad (4-111)$$

with boundary conditions given by Eq. (4–110).

When $n = 0$, i.e., the isothermal wall case, this equation reduces to the same form as the homogeneous solution for the flat plate isothermal wall case (see Eq. 4–82). Although Eq. (4–111) is of the same form as Eq. (4–82), its solution $\theta = \theta(\eta)$ is not the same. Here, the nondimensional stream function $f(\eta)$, given by the solution of the Falkner-Skan equation, depends on the value of β; consequently, the solution of Eq. (4–111) also depends on the value of β.

Examination of Eqs. (4–111) and (4–109) shows that in the special case of $\beta = 2$ (see Figure 3–10) similar solutions of the energy equation exist for arbitrary wall temperature variations when viscous dissipation is neglected. If $\beta \neq 2$ and $n \neq 0$, then similar solutions of the energy equation depend on both the pressure gradient parameter β and the surface temperature parameter n.

When viscous dissipation is not neglected, i.e., when the Eckert number is not negligible, similar solutions of the energy equation exist only if $2m - n = 0$. Since m is related to the pressure gradient parameter β, similar solutions of the energy equation exist for only one wall temperature variation, in particular for $n = 2\beta/(2-\beta)$, $\beta \neq 2$. When $0 < \beta < 2$, the surface temperature increases in the direction of the flow. For $\beta < 0$, the surface temperature decreases in the direction of the flow. When $\beta = 0$, $n = 0$ and the surface temperature is constant.

4-9 Numerical Integration of the Energy Equation for a Forced Convection Boundary Layer

Since for a forced convection boundary layer the velocity field is uncoupled from the temperature field, the governing boundary value problem for the velocity field, i.e., the Falkner-Skan equation, is solved first. The numerical solution of the Falkner-Skan equation and the program **faskan** were previously discussed

Laminar Flow Analysis

Table 4-1. Nondimensional velocity and temperature profile for a forced convection boundary layer without viscous dissipation for an isothermal wall.

E = 0, Pr = 1

η	f″	f′	f	θ'	θ
0	0.4696005	0	0	0.4696005	0
0.2	0.46930657	0.09390540	0.00939142	0.46930657	0.09390540
0.4	0.4672547	0.18760534	0.03754924	0.4672547	0.18760534
0.6	0.46173493	0.28057576	0.08438566	0.46173493	0.28057576
0.8	0.45119049	0.37196365	0.14967468	0.45119049	0.37196365
1.	0.43437958	0.46063307	0.23299035	0.43437958	0.46063307
1.2	0.41056575	0.54524709	0.33365774	0.41056575	0.54524709
1.4	0.37969252	0.62438697	0.4507241	0.37969252	0.62438697
1.6	0.34248737	0.69670022	0.58295692	0.34248737	0.69670022
1.8	0.3004455	0.76105808	0.728873	0.3004455	0.76105808
2.	0.25566929	0.8166954	0.88679774	0.25566929	0.8166954
2.2	0.21057998	0.86330501	1.0549482	0.21057998	0.86330501
2.4	0.16756036	0.90106625	1.2315289	0.16756036	0.90106625
2.6	0.12861283	0.93060206	1.4148256	0.12861283	0.93060206
2.8	9.5113386e−2	0.95287624	1.6032851	9.5113386e−2	0.95287624
3.	6.7710286e−2	0.96905538	1.7955696	6.7710286e−2	0.96905538
3.2	4.6370361e−2	0.98036575	1.9905828	4.6370361e−2	0.98036575
3.4	3.0535214e−2	0.98797122	2.1874693	3.0535214e−2	0.98797122
3.6	1.9328694e−2	0.99288865	2.3855926	1.9328694e−2	0.99288865
3.8	1.1758678e−2	0.99594502	2.5845011	1.1758678e−2	0.99594502
4.	6.8740853e−3	0.99777083	2.7838889	6.8740853e−3	0.99777083
4.2	3.861352e−3	0.99881904	2.9835579	3.861352e−3	0.99881904
4.4	2.0840747e−3	0.99939734	3.1833855	2.0840747e−3	0.99939734
4.6	1.0807525e−3	0.99970394	3.3832989	1.0807525e−3	0.99970394
4.8	5.3848399e−4	0.99986013	3.5832571	5.3848399e−4	0.99986013
5.	2.5778052e−4	0.99993659	3.7832377	2.5778052e−4	0.99993659
5.2	1.1856508e−4	0.99997256	3.9832291	1.1856508e−4	0.99997256
5.4	5.2395285e−5	0.99998882	4.1832254	5.2395285e−5	0.99998882
5.6	2.2246211e−5	0.99999588	4.383224	2.2246211e−5	0.99999588
5.8	9.0750329e−6	0.99999883	4.5832235	9.0750329e−6	0.99999883
6.	3.556875e−6	1.	4.7832234	3.556875e−6	1.

in Section 3-6. The **forcb** program uses the same logic as the **faskan** program to obtain the required value of $f''(0, \beta)$ and thus a solution of the velocity field. This value of $f''(0, \beta)$ is then used to generate the required values of $f(\eta)$, $f'(\eta)$ and $f''(\eta)$ for the solution of the energy equation.[†]

[†]See Appendix D-14 for an alternate technique for solving the forced convection boundary value problem.

Once the value of $f''(0, \beta)$ is determined, solutions of the Falkner-Skan equation are again obtained simultaneously with the solution of the energy equation. This procedure is more efficient and more flexible than storing $f(\eta)$, $f'(\eta)$ and $f''(\eta)$ for specific values of η and using these stored values to obtain the subsequent solution of the energy equation.

A typical run of the `forcb` program is illustrated below. The Falkner-Skan pressure gradient parameter β is taken as zero. This corresponds to the solution for the forced convection boundary layer discussed in Section 4–8. Recalling that the solution to the Falkner-Skan equation with zero pressure gradient (see Section 3–4 and Table C–1) is 0.4696005, we use this value. A Prandtl number of 1, an Eckert number of zero and an isothermal wall correspond to the analytical solution obtained in Section 4–7. Recalling Eqs. (4–87), (4–97) and (4–100) shows that $\bar{\theta}'(0) = -f''(0)$, where here we use $\bar{\theta}$ to designate the nondimensional temperature used in Section 4–7, i.e., $\bar{\theta} = (T - T_\infty)/(T_w - T_\infty)$. Comparing the definition of $\bar{\theta}$ with the present definition of the nondimensional temperature (see Eq. 4–106) yields $\bar{\theta} = 1 - \theta$. Thus, $\bar{\theta}' = -\theta' = f''(0)$. Hence we choose 0.4696005 for $\theta'(0)$. As expected and shown below, convergence is achieved immediately. The velocity and temperature profiles are shown in Table 4–1.

```
Beta = 0
Prandtl Number = 1
f"(0) = .4696005
theta'(0) = .4696005
 η      f"           f'    f         θ'           θ
 0     .4696005      0     0         .4696005     0
 6.    3.55687e-6    1.    4.78322   3.55687e-6   1.
```

In order to more fully illustrate the `forcb` program we seek an additional solution for $\beta = 0.1$, an Eckert number $E = 0.1$ and a Prandtl number $Pr = 0.7$. The actual computer run is shown below. Recalling the results of our study of the Falkner-Skan equation, we use $f''(0) = 0.5870352$. Assuming an Eckert number not equal zero includes viscous dissipation in the analysis. Thus, there is a coupling between the pressure gradient parameter β and the surface temperature variation. Here, for $\beta = 0.1$ the temperature variation exponent is 0.105263. Assuming that the effects of Prandtl number, viscous dissipation and surface temperature variation are small, we initially take $\theta'(0)$ as 0.4696005.

```
Beta = 0.1
Prandtl Number = 0.7
n = 0.105263
Eckert Number = 0.1
f"(0) = 0.5870352
theta'(0) = 0.4696005
 η      f"           f'    f         θ'           θ
 0      0.5870352    0     0         0.4696005    0
 6.     1.34222e-6   1.    4.91968   2.16714e-3   1.04907
```

142 Laminar Flow Analysis

The integration shows that $\theta'(0)$ is less than our estimated value, i.e., $\theta(\eta_{max} = 6) > 1$. Assuming the derivative $\Delta\theta'(\eta_{max})/\Delta\theta'(0) \approx 2$, our second estimate is 0.4446. As shown below, integration yields $\theta(\eta_{max} = 6) = 0.970217$.

```
Beta = 0.1
Prandtl Number = 0.7
n = 0.105263
Eckert Number = 0.1
f''(0) = 0.5870352
theta'(0) = 0.4446
```

η	f''	f'	f	θ'	θ
0	0.5870352	0	0	0.4446	0
6.	1.34222e−6	1.	4.91968	−1.2267e−3	0.970217

We can now obtain a more accurate value of $\Delta\theta'(\eta_{max})/\Delta\theta'(0)$. Using the results of the previous runs, we estimate that $\Delta\theta(\eta_{max})/\Delta\theta(0) \approx 3.15$. Using this value we estimate $\theta'(0) = 0.4540$, which upon integration yields $\theta(\eta_{max} = 6) = 0.999867$, i.e.

```
Beta = 0.1
Prandtl Number = 0.7
n = 0.105263
Eckert Number = 0.1
f''(0) = 0.5870352
theta'(0) = 0.4540
```

η	f''	f'	f	θ'	θ
0	.5870352	0	0	0.4540	0
6.	1.34222e−6	1.	4.91968	4.93586e−5	0.999867

Automatic iteration using the Nachtsheim-Swigert technique shows that convergence within a required error of $\pm 5 \times 10^{-6}$ on all asymptotic boundary conditions does not occur for $\eta_{max} = 6$. Increasing η_{max} to 7 yields convergence to the correct outer boundary condition ($\theta'(0) = 0.454038$) in one iteration, as illustrated below.

```
Beta = 0.1
Prandtl Number = 0.7
n = 0.105263
Eckert Number = 0.1
f''(0) = 0.5870352
theta'(0) = 0.454038
```

η	f''	f'	f	θ'	θ
0	.5870352	0	0	0.454038	0
7.	4.26628e−9	1.	5.91968	1.24473e−6	1.

Again (see Section 3–6), we see that using information derived from previous solutions provides a powerful technique for obtaining additional solutions. A note of caution, however; the above method is not necessarily successful (see Section 3–6).

Similar solutions of the energy equation for an isothermal wall without viscous dissipation have been extensively investigated. The most complete compilations are those of Evans [Evan68] and of Elzy and Sisson [Elzy67]. Selected values for $\theta'(0)$ taken from Elzy and Sisson are shown in Figure 4–13 and in Table C–12. The results obtained using `forcb` agree with the results of Elzy and Sisson to within six significant figures. Similar solutions of the energy equation when there is reverse flow (i.e., $\beta < 0$, $f''(0) < 0$) apparently have not been investigated. Figure 4–13 clearly shows that the heat transfer parameter $\theta'(0)$ is a strong function of the Prandtl number. However, Figure 4–13 also shows that it is only a weak function of the pressure gradient parameter β for small Prandtl numbers (Pr < 1). For larger Prandtl numbers the pressure gradient effect becomes more important, especially for retarded flows ($\beta < 0$).

Similar solutions of the energy equation for a nonisothermal wall and without viscous dissipation have also been investigated (see [Chap49; Levy52; Schu53; Punn53; Fett54; Ligh50; Koh61]). Figure 4–14 and Table C–13 show the heat

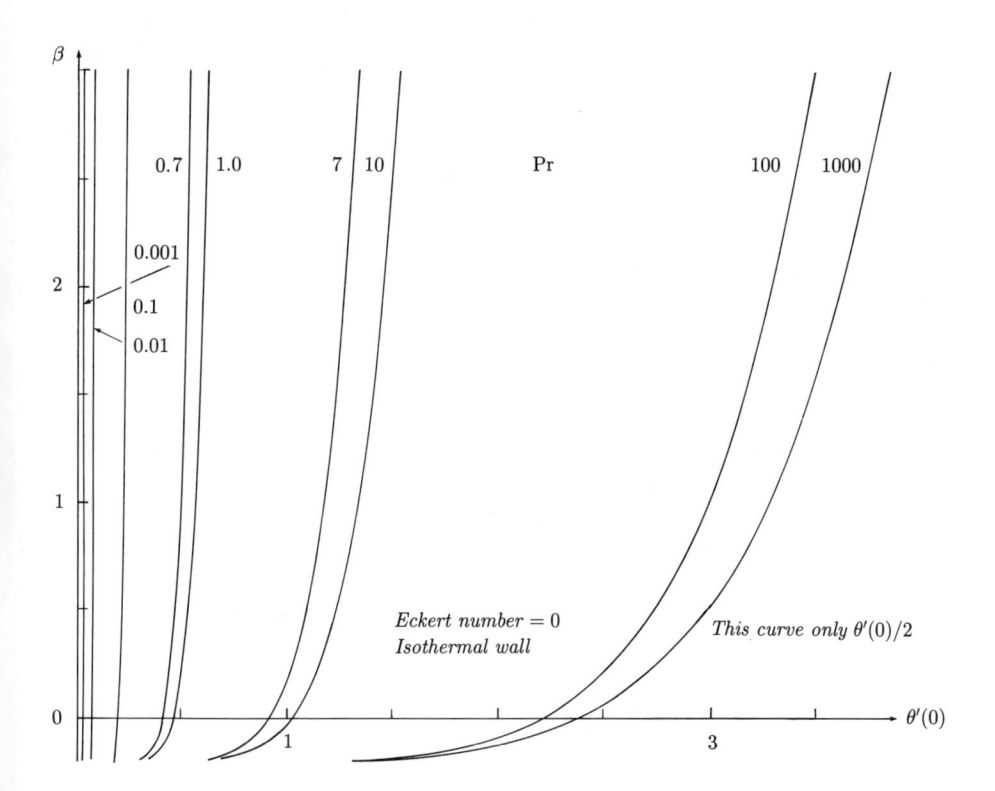

Figure 4–13. The heat transfer parameter $\theta'(0)$ for an isothermal wall without viscous dissipation.

transfer parameter $\theta'(0)$ as a function of the wall temperature variation exponent n for several Prandtl numbers.

For the special case of $n(2 - \beta) = -1$, Eq. (4–111) has an analytical solution given by

$$\theta = 1 - e^{-\int_0^\eta \mathrm{Pr} f d\eta} \qquad (4-112)$$

In this case $\theta'(0) = 0$, which corresponds to an adiabatic wall. This result shows that the adiabatic wall condition for nonisothermal surfaces is independent of the Prandtl number. Figure 4–14 shows the variation of $\theta'(0)$ as a function of n for unit Prandtl number. For an isothermal wall ($n = 0$) the value of $\theta'(0)$ increases with increasing β, as shown. For $n < 0$ the temperature potential $(T_w - T_\infty)$ increases with x, and vice versa. The curve for $\beta = 1$, which physically corresponds to two-dimensional stagnation point flow, shows an interesting behavior for $n > 0$. This behavior has not yet been satisfactorily explained.

Figure 4–14, in particular the results for $\beta = -0.1$ along with Eq. (4–112), also shows that for $n > -1/(2 - \beta)$ heat transfer occurs from the surface to the fluid, and for $n < -1/(2 - \beta)$ heat transfer occurs from the fluid to the surface, even though the surface temperature exceeds the temperature at the edge of the boundary layer. Due to the required pressure gradient and wall temperature variation, the behavior for $n = -0.5$ and $\beta = -0.1$ is difficult to observe experimentally.

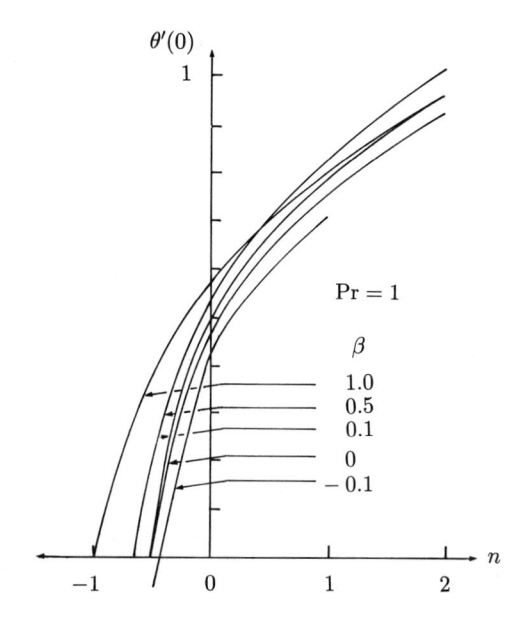

Figure 4–14. The heat transfer parameter $\theta'(0)$ as a function of the wall temperature variation exponent n when there is no viscous dissipation.

Figure 4–15 illustrates the effect of the wall temperature variation on the nondimensional temperature profiles for two-dimensional stagnation point flow, i.e., for $\beta = 1$. Here we see that the temperature near the surface is increased when the surface temperature increases in the flow direction. When the surface temperature decreases in the flow direction, the temperature near the surface is decreased.

When the effects of viscous dissipation are considered, the surface temperature variation is coupled to the potential flow at the edge of the boundary layer by

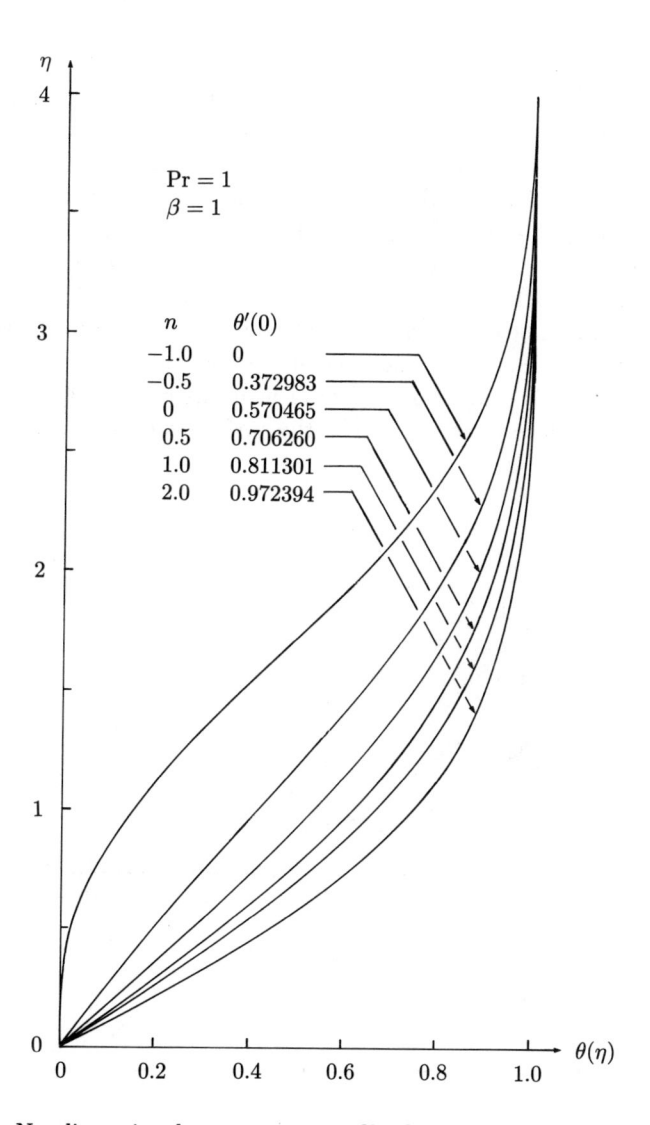

Pr = 1
$\beta = 1$

n	$\theta'(0)$
−1.0	0
−0.5	0.372983
0	0.570465
0.5	0.706260
1.0	0.811301
2.0	0.972394

Figure 4–15. Nondimensional temperature profiles for two-dimensional stagnation point flow, with surface temperature variation but without viscous dissipation.

the relationship $n = 2\beta/(2 - \beta)$. Table C–14 and Figure 4–16 show the effects of viscous dissipation and nonisothermal surface temperature variation for unit Prandtl number, and $\beta = 1$, 0.5 and -0.1 on the heat transfer parameter $\theta'(0)$. Figure 4–16 reveals that $\theta'(0)$ is a linear function of the Eckert number. For

$\beta = 1$ $\qquad\qquad \theta'(0, E_1) = 0.972394 - 0.486197 \, E_1$

$\beta = 0.5$ $\qquad\quad \theta'(0, E_1) = 0.759906 - 0.379953 \, E_1$

$\beta = -0.1$ $\qquad \theta'(0, E_1) = 0.389712 - 0.194856 \, E_1$

When the Eckert number is 2 the heat transfer parameter $\theta(0) = 0$, i.e., the wall is adiabatic. Further, the Eckert number for zero heat transfer is independent of the pressure gradient parameter. These results show that we need to calculate $\theta'(0)$ for only one value of the Eckert number other than 2 in order to determine $\theta'(0, E_1)$. Figure 4–17 shows the nondimensional temperature profiles for two-dimensional stagnation point flow, $\beta = 1$, for various Eckert numbers. For large negative Eckert numbers, the temperature in the thermal boundary layer exceeds the free stream temperature, i.e., $\theta(\eta) > 1$. Thus, there is a layer of hot fluid near the surface, and heat transfer takes place from the fluid to the surface, i.e., the surface is heated. For Eckert numbers more than two $\theta(\eta) < 1$ and $\theta'(0) < 0$. Consequently, there is a layer of cool fluid near the surface, and heat transfer occurs from the surface to the fluid, i.e., the surface is cooled.

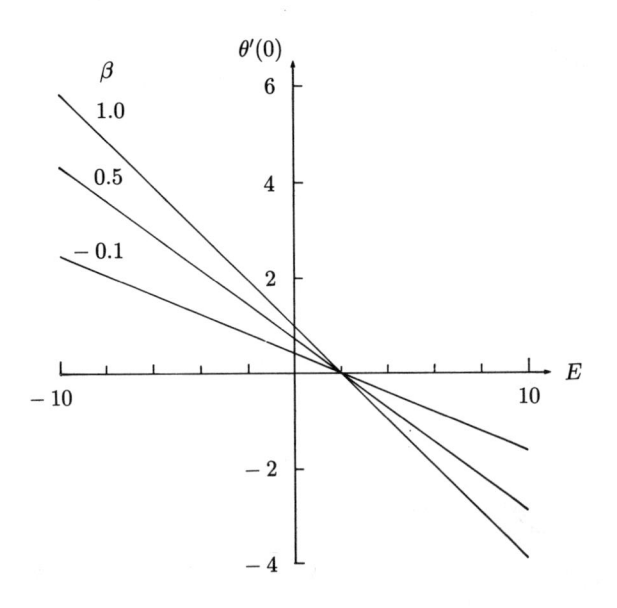

Figure 4–16. The effect of viscous dissipation on the nondimensional heat transfer parameter $\theta'(0)$ for forced convection from a nonisothermal surface $\Pr = 1$.

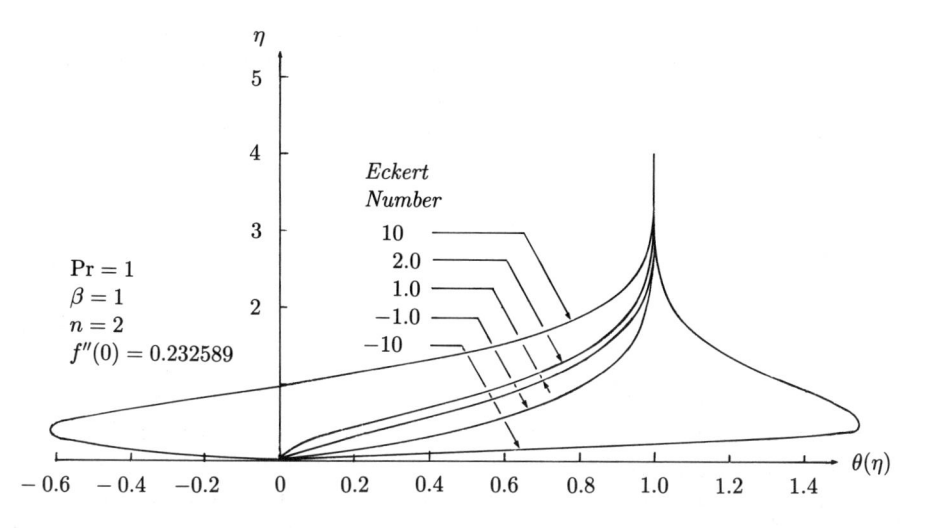

Figure 4-17. The effect of viscous dissipation on the nondimensional temperature profiles for forced convection for a two-dimensional stagnation point flow with a nonisothermal surface, i.e., $\beta = 1$, Pr $= 1$, $n = 2$.

4-10 Forced Convection Boundary Layer Flows With Mass Transfer

We now turn our attention to forced convection boundary layer flows with mass transfer. Recalling our previous discussion of the momentum equation with mass transfer (see Section 3–10), we confine our discussion to similar solutions. We assume that mass is transferred either into or out of the surface, that the mass transfer occurs normal to the surface, that in the case of injection the fluid is injected normal to the surface, that the injected fluid and the main stream fluid are the same and, finally, that the injected fluid is at the same temperature as the surface. With these assumptions the continuity and momentum equations in similarity variables are (see Section 3–10)

$$f''' + ff'' + \beta(1 - f'^2) = 0 \qquad (3-50)$$

with boundary conditions

$$f(0) = f_0 = \text{constant} \qquad f'(0) = 0 \qquad (3-99a,b)$$

$$f'(\eta \to \infty) \to 1 \qquad (3-99c)$$

The nondimensional stream function at the surface $f(0)$ is set equal to a constant to satisfy similarity conditions. This implies that the normal velocity component varies as

$$v_0 = -f_0 \sqrt{\frac{m+1}{2}} \sqrt{\frac{\nu U}{x}} \sim x^{\frac{m-1}{2}}$$

Here we see that negative values of f_0 correspond to blowing or injection and positive values to suction. Further, application of these transformations, including the effects of mass transfer at the surface, do not change the form of the energy equation. Thus, neglecting viscous dissipation, we have

$$\theta'' + \Pr f\theta' + n\Pr(2-\beta)(1-\theta)f' = 0 \qquad (4-111)$$

with boundary conditions

$$\theta(0) = 0 \qquad (4-113a)$$

and

$$\theta(\eta \to \infty) \to 1 \qquad (4-113b)$$

Solutions of the forced convection boundary layer equations with mass transfer are obtained by modifying the `forcb` program. In particular, the only change to the `forcb` program is to set `init(3)` = f_0, where f_0 is the numerical value of the mass transfer parameter.

Solutions of these equations for an isothermal wall with mass transfer were extensively investigated by Elzy and Sisson [Elzy67] for $f''(0) > 0$. Figures 4–18 and Table C–15 through C–20 give $\theta'(0)$ as a function of β for various values of the Prandtl number and mass transfer parameter. Examination of these results reveals that for moderate suction, i.e, for $\beta > 0$, $f_0 > 0$ $\theta'(0)$ is not a strong function of β for Prandtl numbers more than one. For strong suction $f_0 \gg 0$ the heat transfer parameter $\theta'(0)$ increases with increasing Prandtl number for constant f_0. $\theta'(0)$ also increases for any given Prandtl number with increasing f_0. Further examination of the results reveals that for blowing, $f_0 < 0$, $\theta'(0)$ increases with decreasing Prandtl number.

Solutions of Eqs. (3–50) and (4–111) subject to the boundary conditions of Eqs. (3–99) and (4–113) were investigated by Koh and Hartnett [Koh61] for a nonisothermal wall with suction. Their investigations were carried out for a Prandtl number of 0.73, representative of air. They considered values of $\beta = 0$, $1/2$ and 1, of n from -0.5 to 10, and of f_0 from 0 to $12/\sqrt{2-\beta}$. For $\beta = 1$ the results for $n = -1$ are also given. Figures 4–19 show $\theta'(0)/\sqrt{2-\beta}$ as a function of the surface temperature variation exponent n. Examination of these results shows that the nondimensional heat transfer parameter increases with increasing suction and with the magnitude of n. Further, it increases with increasing values of β for any given set of n and f_0. We also observe that for large values of f_0, $\theta'(0)$ is almost a linear function of n.

Figures 4–20 show the nondimensional temperature profiles $\theta(\eta)$ vs η for two-dimensional stagnation point flow, $\beta = 1$. Recalling that, for $\beta = 1$ without mass tranfer, $\theta'(0) = 0$ when $n = -1$, we see that this feature is illustrated in Figure 4–20a. Further examination shows that the thickness of the thermal boundary layer increases with decreasing values of the surface temperature variation exponent n. Examination of Figures 4–20b and 4–20c reveals that suction removes the adiabatic wall condition for $n = -1$, with resulting heat transfer at the surface. Suction also decreases the effect of the nonisothermal wall on the nondimensional temperature profiles.

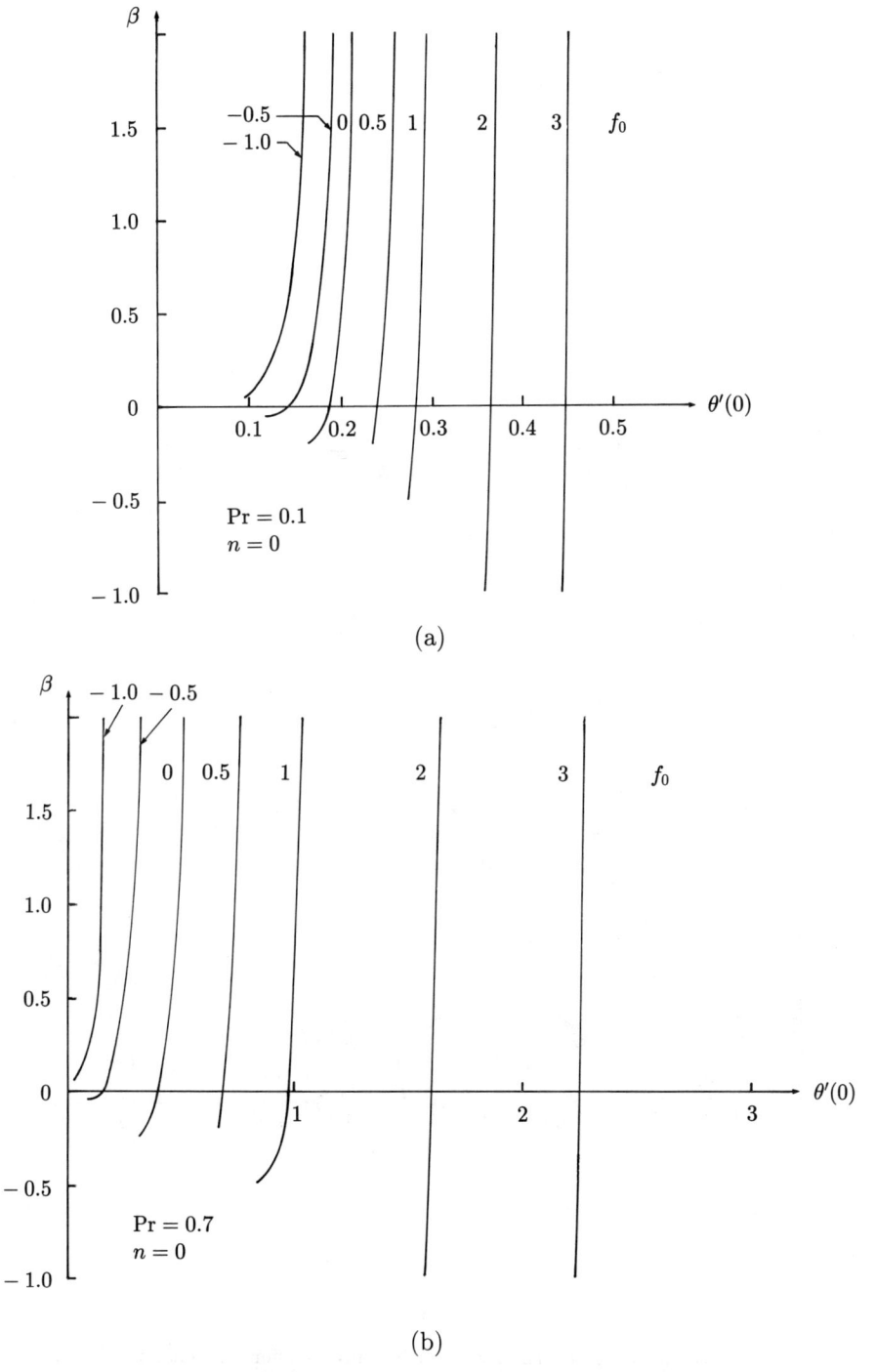

Figure 4–18. $\theta'(0)$ for an isothermal wall with mass transfer but without viscous dissipation. (a) $\text{Pr} = 0.1$; (b) $\text{Pr} = 0.7$.

150 *Laminar Flow Analysis*

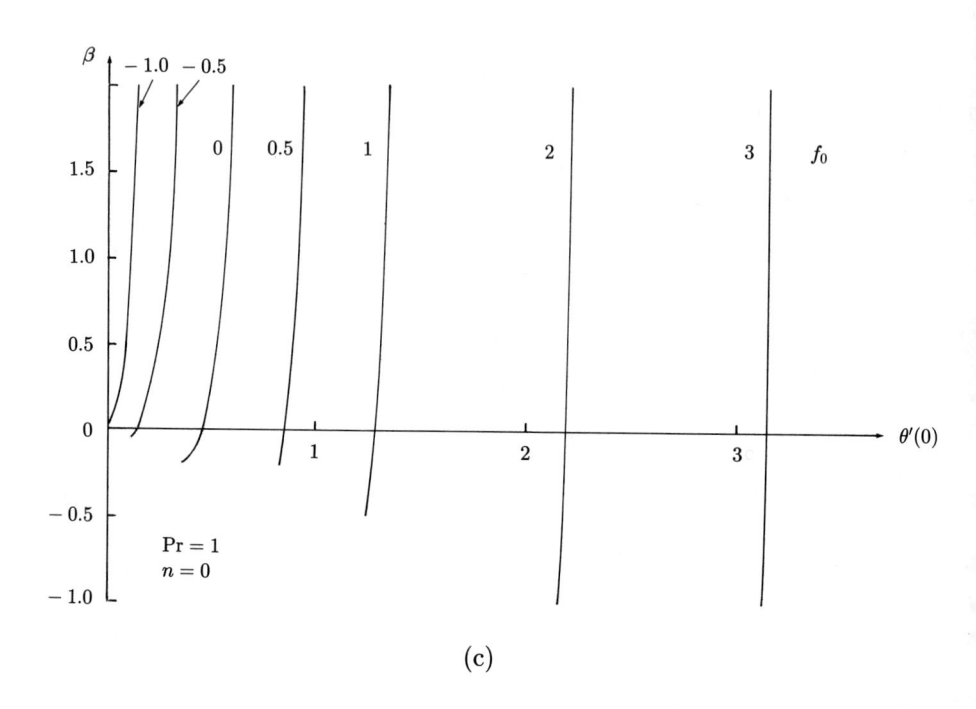

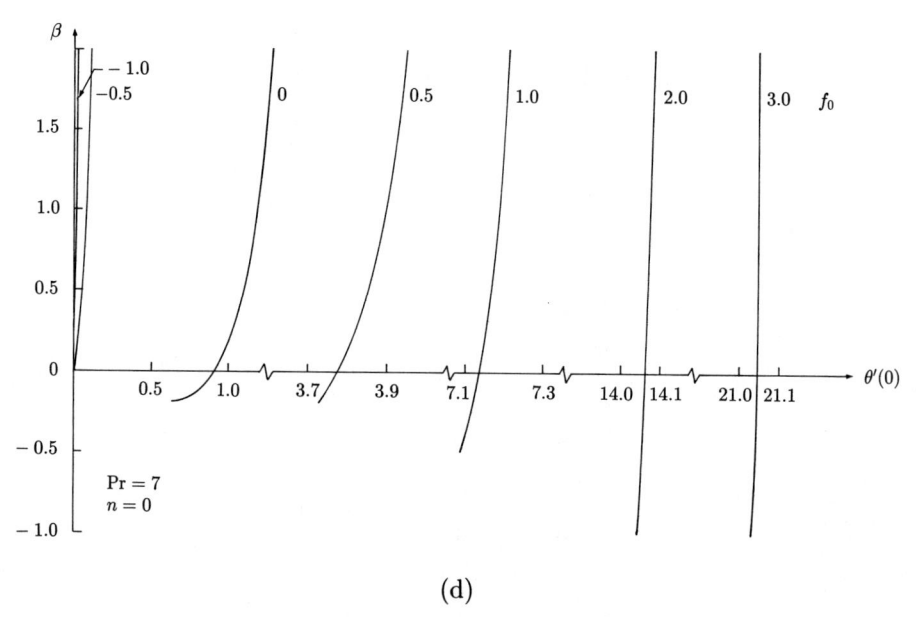

(c)

(d)

Figure 4–18. $\theta'(0)$ for an isothermal wall with mass transfer but without viscous dissipation. (c) Pr = 1; (d) Pr = 7.

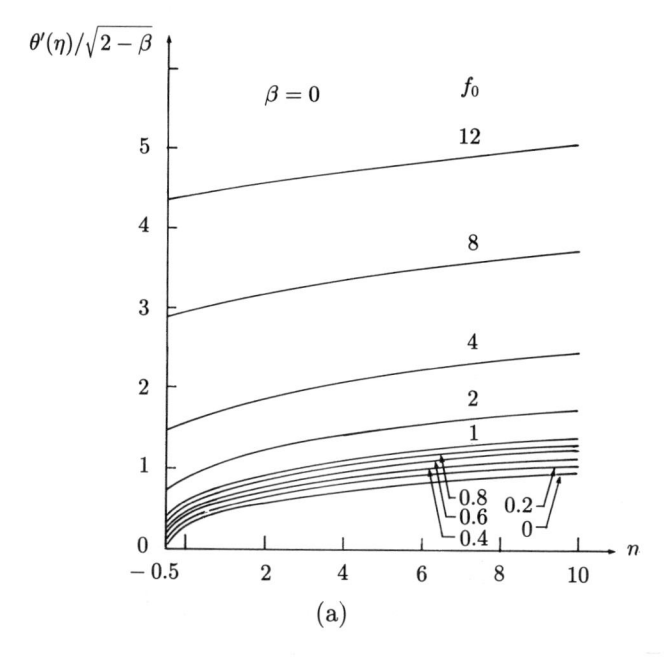

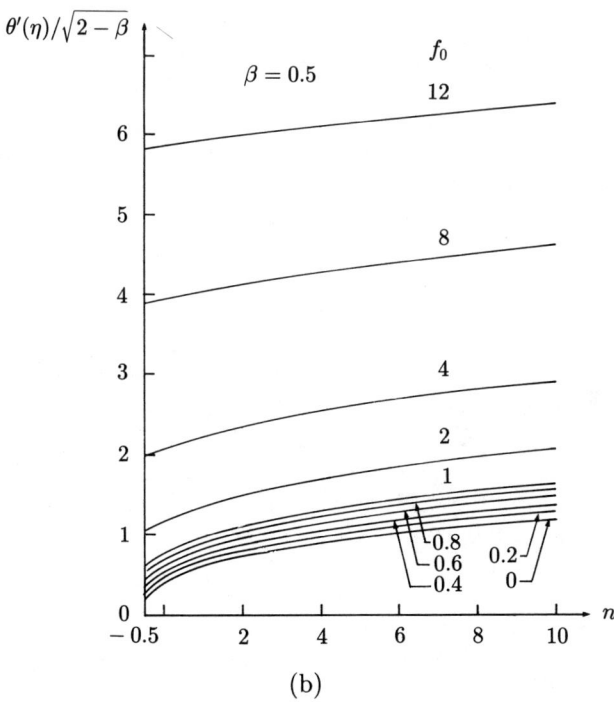

Figure 4-19. Nondimensional heat transfer parameter for forced convection with suction and variable wall temperature but without viscous dissipation, Prandtl number of 0.73. (a) $\beta = 0$; (b) $\beta = 1/2$; (c) $\beta = 1$. From [Koh61].

152   *Laminar Flow Analysis*

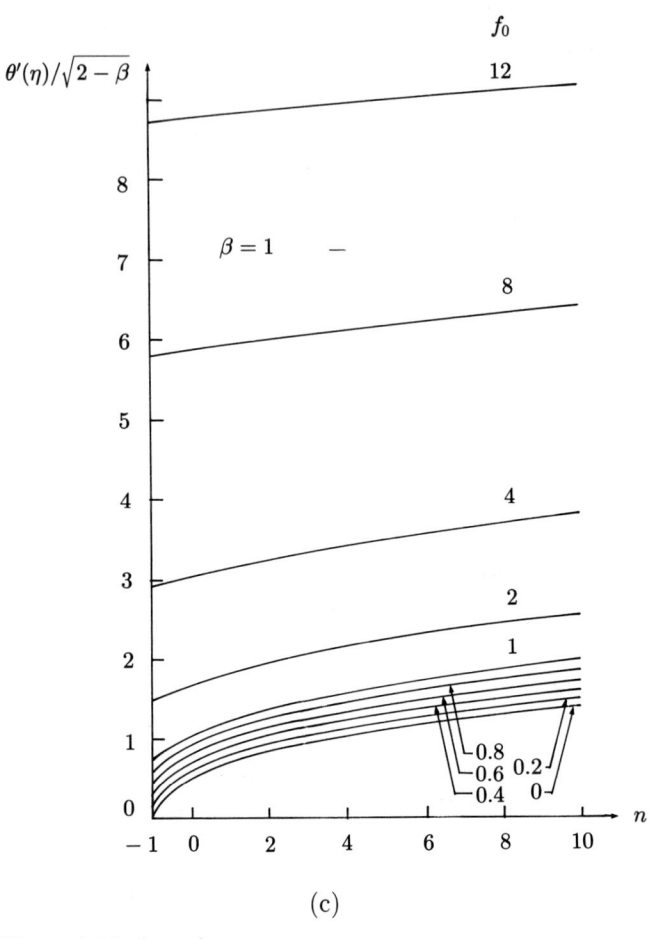

(c)

Figure 4–19. (*cont.*)

Introducing the Nusselt and Stanton numbers (see Eqs. 4–65 and 4–69) yields in similarity variables

$$\mathrm{Nu}_x = -\left[\frac{1}{2-\beta}\frac{Ux}{\nu}\right]^{1/2}\theta'(0) \qquad (4-114a)$$

or introducing the local Reynolds number based on local conditions at the edge of the boundary layer, i.e., $(\mathrm{Re}_x)_e = U_e x/\nu$ yields

$$\frac{\mathrm{Nu}_x}{\sqrt{(\mathrm{Re}_x)_e}} = -\frac{\theta'(0)}{\sqrt{2-\beta}} \qquad (4-114b)$$

Rewriting this result in terms of a Reynolds number based on free stream conditions, i.e.

$$\frac{\mathrm{Nu}_x}{\sqrt{(\mathrm{Re}_x)_e}} = -\frac{x^{\frac{\beta}{2-\beta}}}{\sqrt{2-\beta}}\theta'(0) \qquad (4-114c)$$

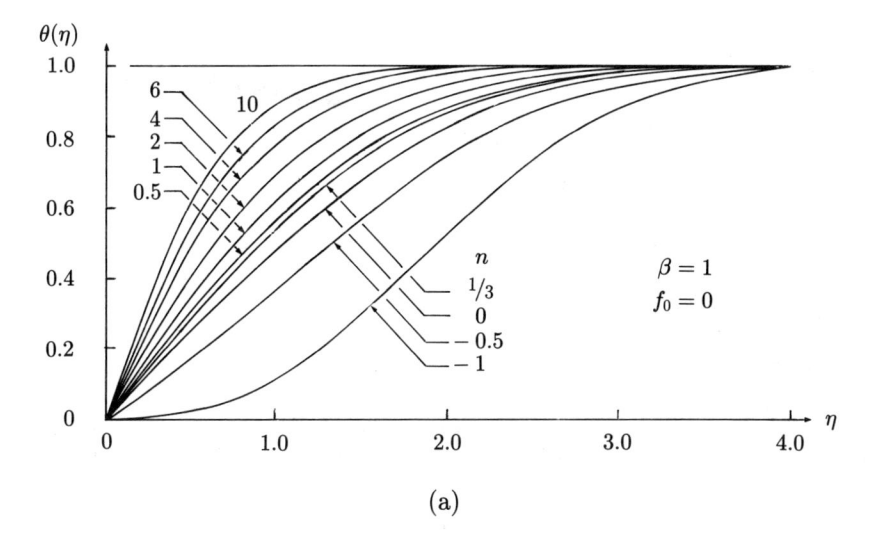

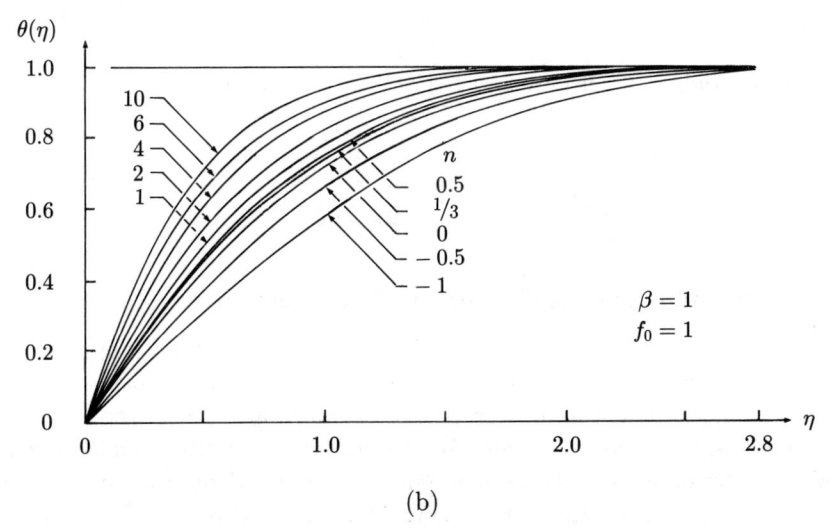

Figure 4–20. Nondimensional temperature profiles for two-dimensional stagnation point flow, $\beta = 1$, for forced convection with suction and variable wall temperature but without viscous dissipation, Prandtl number of 0.73. (a) $f_0 = 0$; (b) $f_0 = 1$; (c) $f_0 = 8$. From [Koh61].

reveals that the common ratio of Nusselt number to the square root of the free stream Reynolds number is not constant.

Similarly, we have that the Stanton number is

$$\sqrt{(\mathrm{Re}_x)_e}\, \mathrm{St}_x = -\frac{\theta'(0)}{\mathrm{Pr}\sqrt{2-\beta}} \qquad (4-115a)$$

154   Laminar Flow Analysis

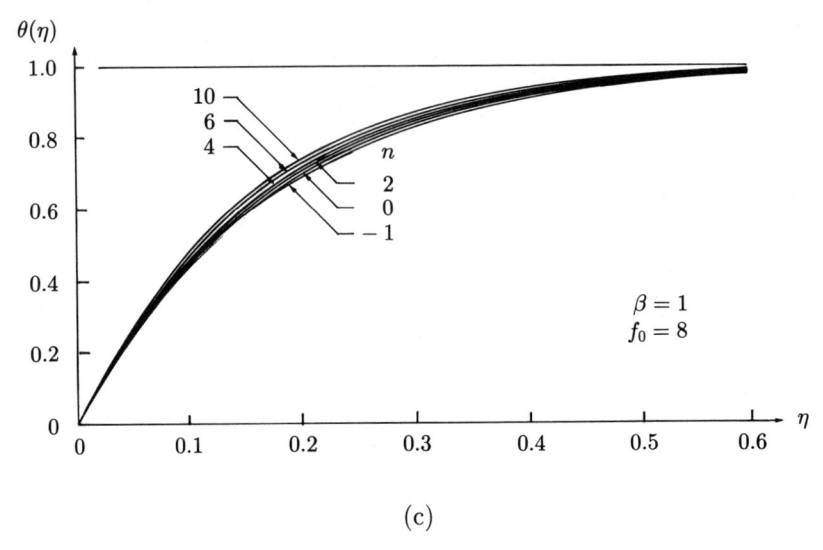

(c)

Figure 4–20. (cont.)

or for a Reynolds number based on free stream conditions

$$\sqrt{(\text{Re}_x)_\infty}\, \text{St}_x = -\frac{x^{-\frac{\beta}{2(2-\beta)}}}{\text{Pr}\sqrt{2-\beta}}\, \theta'(0) \qquad (4-115b)$$

4–11 Nonsimilar Forced Convection Boundary Layer

The nonsimilar steady two-dimensional incompressible constant property forced convection boundary layer is of considerable interest. A number of methods for solving this problem have been proposed. Generally the methods are classified in the same manner as those for the nonsimilar hydrodynamic boundary layer (see Section 3–11). These methods are compared by Spalding and Pun [Spal62]. They indicate that the Meksyn-Merk method is one of the more accurate. Evans [Evan68] described and provided extensive tabulations of the results for the Meksyn-Merk method. However, as previously pointed out in the discussion of the nonsimilar hydrodynamic boundary layer (see Section 3–11), the Meksyn-Merk method suffers from computational difficulties. We therefore consider the locally nonsimilar boundary layer method for forced convection.

We begin the analysis by recalling the governing equations for the steady two-dimensional incompressible constant property boundary layer

continuity

$$\frac{\partial u}{\partial x} + \frac{\partial v}{\partial y} = 0 \qquad (4-55)$$

momentum

$$u\frac{\partial u}{\partial x} + v\frac{\partial v}{\partial y} = -\frac{1}{\rho}\frac{dP}{dx} + \nu\left(\frac{\partial^2 u}{\partial y^2}\right) \qquad (4-56)$$

energy

$$u\frac{\partial h}{\partial x} + v\frac{\partial h}{\partial y} = \frac{\nu}{\Pr}\left(\frac{\partial^2 h}{\partial y^2}\right) + \nu\left(\frac{\partial u}{\partial y}\right)^2 \qquad (4-57)$$

with boundary conditions

$$u(x,0) = 0 \qquad v(x,0) = v(x) \qquad h(x,0) = h_w(x) \qquad (4-116a,b,c)$$

and

$$u(x, y \to \infty) \to U(x) \qquad h(x, y \to \infty) \to h_e(x) \qquad (4-116d,e)$$

Introducing the independent and dependent variable transformations used in our previous analysis of the nonsimilar hydrodynamic boundary layer, i.e.,

$$\xi = \int_0^x \frac{U(x)}{U_\infty} d\left(\frac{x}{\ell}\right) \qquad (3-102)$$

$$\eta = \frac{yU(x)}{(2\nu\ell U_\infty \xi)^{1/2}} \qquad (3-103)$$

$$\psi(x,y) = (2\nu\ell U_\infty \xi)^{1/2} f(\xi, \eta) \qquad (3-104)$$

allows writing the momentum equation as (see Section 3–11)

$$f''' + ff'' + \beta(1 - f'^2) = 2\xi(f' f'_\xi - f'' f_\xi) \qquad (3-105)$$

with boundary conditions

$$f(\xi,0) + 2\xi f_\xi(\xi,0) = -v(x)\frac{U_\infty}{U(x)}\left(\frac{2\ell\xi}{\nu U_\infty}\right)^{1/2} \qquad (3-127)$$

$$f'(\xi,0) = 0 \qquad (3-106b)$$

and

$$f'(\xi, \eta \to \infty) \to 1 \qquad (3-106c)$$

Introducing these transformations, the energy equation becomes

$$h_{\eta\eta} + \Pr f h_\eta = \Pr U^2 (f_{\eta\eta})^2 + 2\xi \Pr (f_\xi h_\eta - f_\eta h_\xi) \qquad (4-117)$$

Further, introducing the nondimensional temperature ratio

$$\theta = \frac{T - T_w}{T_e - T_w}$$

156   Laminar Flow Analysis

and recalling the previous definition of β, i.e.

$$\beta = \frac{2}{U^2}\frac{dU}{dx}\int_0^x U dx \qquad (3-107)$$

the energy equation finally becomes

$$\theta'' + \Pr f\theta' = \Pr \mathrm{E}_1 f''^2$$
$$+ 2\xi\Pr\left[(f_\xi\theta' - f'\theta_\xi) + \alpha_1(1-\theta)f' + \alpha_2 f'\theta\right] \qquad (4-118)$$

where the primes denote differentiation with respect to η and

$$\alpha_1(\xi) = \frac{1}{T_w(x) - T_e(x)}\frac{\partial T_w}{\partial \xi} \qquad (4-119)$$

$$\alpha_2(\xi) = \frac{1}{T_w(x) - T_e(x)}\frac{\partial T_e}{\partial \xi} \qquad (4-120)$$

$$\mathrm{E}_1 = \frac{(U(x))^2}{c_p(T_w(x) - T_e(x))} \qquad (4-121)$$

Notice that this result is similar to Eq. (4–109) and that here the Eckert number $\mathrm{E}_1(x)$ has a different definition from that previously used in the discussion of the similar boundary layer. In particular, the definition of the Eckert number is based on the variable velocity $U_e(x)$ and the variable temperatures $T_w(x)$ and $T_e(x)$. The boundary conditions are

$$\theta(\xi, 0) = 0 \qquad \theta(\xi, \eta \to \infty) \to 1 \qquad (4-122a,b)$$

Alternately, zero heat transfer at the surface, i.e., an adiabatic wall, can be specified. Under these circumstances Eq. (4–122a) is replaced by $\theta'(\xi, 0) = 0$.

In our previous similarity analysis, which is equivalent to a local similarity approximation, the term $2\xi\Pr(f_\xi\theta' + f'\theta_\xi)$ was neglected because f and θ were assumed to be functions of η only. The term $\alpha_2 f'\theta$ was zero because T_e was assumed to be constant. Further, α_1 and E_1 were of a particular form based on the assumed form of the temperature variation for the nonisothermal wall (see Eq. 4–108). We now develop the locally nonsimilar boundary layer method for the forced convection boundary layer.

Following development of the two-equation locally nonsimilar model for the hydrodynamic boundary layer (see Section 3–11), we introduce auxiliary functions

$$S(\xi, \eta) = f_\xi \qquad (3-108)$$

and

$$t(\xi, \eta) = \theta_\xi \qquad (4-123)$$

The momentum equation for the two-equation locally nonsimilar model is then given by Eq. (3–109) and the first auxiliary momentum equation by Eq. (3–112) with no-mass-transfer boundary conditions given by Eq. (3–113). Introducing Eqs. (3–108) and (4–123) into the energy equation yields

$$\theta'' + \Pr f\theta' = \Pr E_1 f''^2$$
$$+ 2\xi\Pr\left[(S\theta' - f't) + \alpha_1(1 - \theta)f' + \alpha_2 f'\theta\right] \qquad (4-124)$$

In order to evaluate $t(\xi, \eta)$, the energy equation is differentiated with respect to ξ. Thus

$$t'' + \Pr(S\theta' + ft') = 2\Pr\left[(S\theta' - f't) + \alpha_1(1 - \theta)f' + \alpha_2 f'\theta\right]$$
$$+ \Pr\left(\frac{dE_1}{d\xi} f''^2 + 2E_1 f'' S''\right) + 2\xi\Pr\left\{(St' - S't)\right.$$
$$+ \alpha_1\left[(1 - \theta)S' - tf'\right] + \alpha_2(f't + S'\theta) + \frac{d\alpha_1}{d\xi}(1 - \theta)f'$$
$$\left. + \frac{d\alpha_2}{d\xi} f'\theta\right\} + 2\xi\Pr(S_\xi\theta' + f't_\xi) \qquad (4-125)$$

Recalling the system of momentum equations for the locally nonsimilar method, i.e., Eqs. (3–109) and (3–112), we see that the system of momentum equations and the system of energy equations, i.e., Eqs. (4–124) and (4–125), are uncoupled. Thus, as was the case for the similar solutions, the system of momentum equations is solved independent of the system of energy equations. The results are then used in the subsequent solution of the system of energy equations.[†]

Assuming that the solution for the system of momentum equations is known, Eqs. (4–124) and (4–125) could be considered as a pair of coupled ordinary differential equations for the dependent functions θ and t, except for the appearance of the term $2\xi\Pr(S_\xi\theta' + f't_\xi)$ in Eq. (4–125). Consistent with the previous development of the locally nonsimilar method for the hydrodynamic boundary layer, we neglect this term. The resulting system of energy equations for the two-equation locally nonsimilar model is

$$\theta'' + \Pr f\theta' = \Pr E_1 f''^2$$
$$+ 2\xi\Pr\left[(S\theta' - f't) + \alpha_1(1 - \theta)f' + \alpha_2 f'\theta\right] \qquad (4-124)$$

and

$$t'' + \Pr(S\theta' + ft') = 2\Pr\left[(S\theta' - f't) + \alpha_1(1 - \theta)f' + \alpha_2 f'\theta\right]$$

[†]Although only the two-equation model is discussed here, this is also true for all higher-order models for the locally nonsimilar method for forced convection.

$$+ \Pr\left(\frac{dE_1}{d\xi}f''^2 + 2E_1 f'' S''\right) + 2\xi\Pr\Big\{(St' - S't)$$
$$+ \alpha_1[(1-\theta)S' - tf'] + \alpha_2(f't + S'\theta)$$
$$+ \frac{d\alpha_1}{d\xi}(1-\theta)f' + \frac{d\alpha_2}{d\xi}f'\theta\Big\} \tag{4-126}$$

The appropriate boundary conditions for no-mass transfer at the surface are

$$\theta(\xi, 0) = 0 \qquad t(\xi, 0) = 0 \qquad (4-122a, 4-127a)$$

and
$$\theta(\xi, \eta \to \infty) \to 0 \qquad t(\xi, \eta \to \infty) \to 0 \quad (4-122b, 4-127b)$$

where Eqs. (4–127 a, b) are obtained by differentiating Eqs. (4–122 a, b) with respect to ξ. Since for a known inviscid flow and a known surface temperature distribution ξ, β, α_1, α_2, E_1, $d\beta/d\xi$, $d\alpha_1/d\xi$ and $d\alpha_2/d\xi$ are known functions, Eqs. (4–124) and (4–126), subject to Eqs. (4–122) and (4–127), are considered as a pair of coupled ordinary differential equations subject to two-point asymptotic boundary conditions. Note that the locally nonsimilar boundary layer method results in a Class 3 problem (see Section 1–10), i.e., the momentum and the first auxiliary momentum equations and the energy and the first auxiliary energy equations are coupled. Numerical integration is straightforward. In each case the pair of coupled equations is reduced to a system of simultaneous first-order ordinary differential equations (see Appendix A).

For mass transfer through the surface, it is convenient to use Eq. (3–128). In this case, only the boundary conditions for the momentum and first auxiliary momentum equations are affected. The boundary conditions for the energy equations remain the same. The nondimensional heat transfer at the surface, i.e., the Nusselt number, is given by Eq. (4–65).

The three-equation locally nonsimilar boundary layer model for forced convection is obtained by introducing additional auxiliary functions

$$h(\xi, \eta) = S_\xi = f_{\xi\xi} \qquad \text{and} \qquad j(\xi, \eta) = t_\xi = \theta_{\xi\xi}$$

into Eqs. (4–124) and (4–126), and subsequently differentiating with respect to ξ. The equations necessary to evaluate h and j are obtained by neglecting those terms involving h_ξ and j_ξ. The boundary conditions for h and j are obtained in a similar manner. The results for $h(\xi, \eta)$ are given in Eqs. (3–117) and (3–118 e, f, i). The result for $j(\xi, \eta)$ is left for the problems.

In the preceding sections we have investigated the incompressible steady constant property forced convection boundary layer in considerable detail. We now turn our attention to the free convection or natural boundary layer.

FIVE

FREE CONVECTION BOUNDARY LAYERS

5-1 Introduction

Recalling our previous discussion of the general characteristics of thermal boundary layers (see Section 4-2), we now turn our attention to solution of the governing equations for flows in which the body force terms are not negligible, i.e., to flows in which the Grashof number is not small with respect to the square of the Reynolds number. Such flows are generally called free or natural convection flows. In such flows the effects due to the pressure field are generally insignificant. Hence, the pressure gradient terms in the momentum equations are neglected. Physically such flows arise in the following manner: Consider an object, such as a plate, in a fluid at rest but subjected to a body force such as gravity. If the plate and the surrounding fluid are at the same temperature, the body forces acting on the fluid are in equilibrium with the hydrostatic pressure, and no flow results. However, if the plate is heated or cooled an unbalancing body force results from the decrease or increase of density due to heat transfer from or to the fluid near the plate surface. The fluid near the plate surface is thus accelerated. The result is that a boundary layer flow develops. Although free convection boundary layers are generally associated with body forces generated by the gravitational field, they also result from the centrifugal forces found in rotating flows.

5-2 Free Convection Boundary Layers

To simplify the discussion of free convection boundary layers, we limit ourselves to the special case of a semi-infinite flat plate parallel to the direction of the generating body force. Further, we consider that the flow is incompressible, except for the density changes in the body forces term, two-dimensional, steady and

160 Laminar Flow Analysis

with constant properties.† With these assumptions, the governing Eqs. (3–23), (4–53) and (4–54) become

$$\frac{\partial u}{\partial x} + \frac{\partial v}{\partial y} = 0 \tag{3-23}$$

$$u\frac{\partial u}{\partial x} + v\frac{\partial u}{\partial y} = g\bar{\beta}(T - T_\infty) + \nu\frac{\partial^2 u}{\partial y^2} \tag{5-1}$$

and

$$u\frac{\partial T}{\partial x} + v\frac{\partial T}{\partial y} = \alpha\frac{\partial^2 T}{\partial y^2} \tag{5-2}$$

where $\alpha = k/\rho c_p$ is the thermal diffusivity. Comparison of Eqs. (5–2) and (4–54) reveals that viscous dissipation, as well as the work due to the body force term, are neglected in the energy equation. Consideration of the boundary conditions reveals that there are two physical situations that are of interest. These are shown in Figure 5–1.

When the surface temperature, T_w, is greater than the temperature of the surrounding fluid, T_∞, the resulting heat transfer from the surface decreases the density of the fluid near the plate surface with a resulting upward motion of the fluid.‡ Conversely, when the surface temperature is less than that of the surrounding fluid, heat transfer to the surface increases the density of the fluid near the surface with a resulting downward motion of the fluid. When the coordinate systems are taken as shown in Figure 5–1, the governing equations given above apply to both situations.

When there is no-mass transfer through the plate, the boundary conditions at the surface are given by

$$y = 0 \qquad v = u = 0 \qquad T_w = T_w(x) \tag{5-3a,b,c}$$

Here we allowed for a nonisothermal wall by specifying that the surface temperature is a function of the distance along the plate surface. Since motion of the fluid near the surface is a result of heat transfer between the fluid and the surface, it is reasonable to assume that the fluid far from the surface is unaffected by this motion. Thus, the boundary conditions at infinity are

$$y \to \infty \qquad u \to 0 \qquad T \to T_\infty \tag{5-3d,e}$$

---

†Ostrach [Ostr53] considers the more general case of compressible, steady, two-dimensional viscous free convection flow with variable properties about a semi-infinite vertical flat plate. When the relative temperature difference between the surface and the surrounding fluid, $(T_w - T_\infty)/T_\infty$, is small, the resulting equations reduce to those studied here.

‡This assumes that the fluid density decreases with increasing temperature.

Free Convection Boundary Layers 161

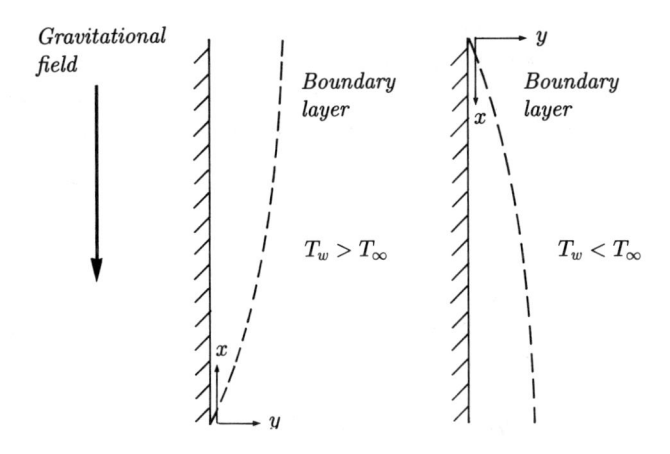

Figure 5-1. Free convection flow on a vertical semi-infinite flat plate.

We now seek solutions to this boundary value problem. Recalling our previous analyses, we satisfy the continuity equation by introducing a stream function such that
$$u = \frac{\partial \psi}{\partial y} \quad \text{and} \quad v = -\frac{\partial \psi}{\partial x}$$

Further, we introduce a nondimensional temperature
$$\theta(x,y) = \frac{T - T_\infty}{T_w - T_\infty} \qquad (5-3)$$

which simplifies the thermal boundary conditions. The governing equations and boundary conditions then become
$$\psi_y \psi_{xy} - \psi_x \psi_{yy} = g\bar{\beta}(T_w - T_\infty)\theta + \nu \psi_{yyy} \qquad (5-4)$$

and
$$\psi_y \theta_x - \psi_x \theta_y + \frac{\theta \psi_y}{T_w - T_\infty} \frac{\partial T_w}{\partial x} = \alpha \theta_{yy} \qquad (5-5)$$

with boundary conditions
$$y = 0 \quad \psi_y = \psi_x = 0 \quad \theta(x,0) = 1 \qquad (5-6a,b,c)$$

and
$$y \to \infty \quad \psi_y \to 0 \quad \theta \to 0 \qquad (5-6d,e)$$

where we assume T_∞ is a constant.

We now seek similar solutions of these equations. Comparing the form of Eq. (5-4) with that of Eq. (3-29), and recalling our discussion of the Blasius equation (see Section 3-3), we consider the general affine transformations given by
$$\eta = Ax^a y^b \qquad (3-32)$$
$$\psi(x,y) = Bx^c y^d f(\eta) \qquad (3-33)$$

where the transformed stream function $f(\eta)$ is assumed to be a function of η only. A, B, a, b, c, d are as yet undetermined constants. Again, the constants A and B are included to nondimensionalize the resulting equations. Recalling our discussion of the forced convection boundary layer for a nonisothermal wall (see Section 4-8), we assume a power law variation of surface temperature, i.e.

$$T_w - T_\infty = T_1 x^n \tag{5-7}$$

where T_1 and n are constants. The isothermal wall case is included for $n = 0$. Sparrow and Gregg [Spar58] have shown that similarity solutions also exist when $T_w - T_\infty = T_1 e^{mx}$. Further, Yang [Yang60] has found a number of other possible similarity solutions for laminar free convection on vertical plates and cylinders. These solutions are not discussed in detail.

Substitution of Eqs. (3-32), (3-33) and (5-7) into the governing equations yields a very complicated result. Recalling our discussion of the Blasius equation, we first consider the velocity boundary condition at infinity. From Eq. (3-34b) and (5-6d) we have

$$\psi_y = dBx^c y^{d-1} f + bAB x^{c+a} y^{d+b-1} f' \to 0$$

where the prime denotes differentiation with respect to η. Since taking $b = d = 0$ yields the trivial result $u = \psi_y = 0$ everywhere, b and d cannot both be zero. However, it is convenient to take either d or b equal to zero, since this simplifies the boundary condition. Since the functional form of the similarity variable η must contain both of the independent variables x and y, neither a nor b can be zero. Thus we take $d = 0$, which yields

$$bAB x^{c+a} y^{b-1} f' \to 0 \quad \text{as} \quad \eta \to \infty$$

It is necessary to eliminate the y dependence of the boundary condition by choosing $b = 1$, since for $b < 1$ the boundary condition at infinity is satisfied for any arbitrary value of f', while for $b > 1$ the boundary condition as $y \to \infty$ can be indeterminant. Here, unlike for the Blasius solution, we have a zero boundary condition at the edge of the boundary layer. Hence, it is *not* necessary to eliminate the x dependence from the boundary condition by choosing $a = -c$. Thus, a and c are still arbitrary at this point.

Turning now to the momentum and energy equations, with $d = 0$ and $b = 1$ we have

$$(AB)^2(a+c)x^{2(a+c)-1}f'^2 - c(AB)^2 x^{2(a+c)-1} ff''$$
$$= g\bar{\beta} T_1 x^n \theta + \nu A^3 B x^{3a+c} f''' \tag{5-8}$$

and
$$nAB x^{a+c-1} f'\theta - cAB x^{a+c-1} f\theta' = \alpha A^2 x^{2a} \theta'' \tag{5-9}$$

Free Convection Boundary Layers

In order to reduce these equations to ordinary differential equations the x dependence must be eliminated. This is accomplished if

$$2(a+c) - 1 = n$$
$$3a + c = n$$

Solving for a and c in terms of n yields $a = (n-1)/4$ and $c = (n+3)/4$. Using these results, Eqs. (5–8) and (5–9) become

$$f''' + \frac{(n+3)}{4}\frac{B}{\nu A}ff'' - \frac{(n+1)}{2}\frac{B}{\nu A}f'^2 + \frac{g\bar{\beta}T_1}{\nu A^3 B}\theta = 0 \qquad (5-10)$$

and

$$\theta'' + \frac{n+3}{4}\frac{B}{\alpha A}f\theta' - \frac{n}{\alpha}\frac{B}{A}f'\theta = 0 \qquad (5-11)$$

These equations are nondimensionalized by choosing appropriate values for A and B. For convenience, we choose $B/(4\nu A) = 1$ and $(g\bar{\beta}T_1)/(\nu A^3 B) = 1$.[†] Solution of these two equations for the constants A and B yields

$$A = \left(\frac{g\bar{\beta}T_1}{4\nu^2}\right)^{1/4}$$

and

$$B = 4\nu\left(\frac{g\bar{\beta}T_1}{4\nu^2}\right)^{1/4}$$

Thus, Eqs. (5–10) and (5–11) finally become

$$f''' + (n+3)ff'' - 2(n+1)f'^2 + \theta = 0 \qquad (5-12)$$

and

$$\theta'' + \Pr\left[(n+3)f\theta' - 4nf'\theta\right] = 0 \qquad (5-13)$$

with boundary conditions at $\eta = 0$

$$f(0) = f'(0) = 0 \qquad \theta(0) = 1 \qquad (5-14a,b,c)$$

and as $\eta \to \infty$

$$f'(\eta \to \infty) \to 0 \qquad \theta(\eta \to \infty) \to 0 \qquad (5-14d,e)$$

where

$$\eta = \left(\frac{g\bar{\beta}T_1}{4\nu^2}\right)^{1/4} yx^{\frac{n-1}{4}} \qquad (5-15)$$

and

$$\psi = 4\nu\left(\frac{g\bar{\beta}T_1}{4\nu^2}\right)^{1/4} x^{\frac{n+3}{4}} f(\eta) \qquad (5-16)$$

[†]The choice $B/(4\nu A) = 1$ and $B/(4\alpha A) = 1$ yields no useful information.

164 Laminar Flow Analysis

Before proceeding, note that the relationship between the physical velocity, u, and the nondimensional velocity, $f'(\eta)$, is given by

$$u = 4\nu \left(\frac{g\bar{\beta}T_1}{4\nu^2}\right)^{1/2} x^{\frac{n+1}{2}} f'(\eta)$$

Examination of these equations shows that the free convection boundary layer flow on a vertical flat plate is governed by a pair of coupled nonlinear ordinary differential equations subject to two-point asymptotic boundary conditions. Specifically, the momentum and energy equations are coupled through the nondimensional temperature function, θ. Here we encounter our first example of a Class 3 problem (see Section 1–10).

5–3 Numerical Integration of the Free Convection Boundary Layer Equations

Since for a free convection boundary layer the velocity and temperature fields are coupled, the momentum and energy equations must be solved simultaneously. With two asymptotic boundary conditions, i.e., $f'(\eta \to \infty) \to 0$ and $\theta(\eta \to \infty) \to 0$, to be satisfied, it is necessary to estimate two unknown initial conditions at $y = \eta = 0$. These are $f''(0)$ and $\theta'(0)$. Recalling that the momentum and energy equations are coupled, we see that a change in $f''(0)$ affects the value of both $f'(\eta_{\max})$ and $\theta(\eta_{\max})$. Similarly, a change in $\theta'(0)$ also affects the value of both $f'(\eta_{\max})$ and $\theta(\eta_{\max})$. Here we again are using the notation $\eta_{\max}$ for the value of η chosen to represent infinity. Thus, the iteration scheme used to insure asymptotic convergence to the required outer boundary conditions must account for this coupling between the value of $f''(0)$ and $\theta'(0)$. The Nachtsheim-Swigert iteration scheme used in the faskan and forcb programs was extended to include this coupling effect. This form of the Nachtsheim-Swigert iteration scheme is discussed in Appendix B. The details of the freeb program are given in Appendix D–15.

5–4 Results for a Free Convection Boundary Layer on an Isothermal Vertical Flat Plate

The boundary value problem governing the free convection boundary layer flow on an isothermal vertical plate is included in the more general analysis for a nonisothermal surface for $n = 0$. A typical run of the freeb program for unit Prandtl number, $\Pr = 1$, and an isothermal wall, $n = 0$, is shown below. Although the magnitude of the unknown surface gradients, $f''(0)$ and $\theta'(0)$, are not known, we expect, on the basis of our previous analysis of the forced convection boundary layer, that they are of order one. The negative sign for $\theta'(0)$ is predicated on the fact that for a heated plate we expect heat transfer to occur from the plate to the fluid.

Free Convection Boundary Layer on an Isothermal Vertical Flat Plate

Choosing $\eta_{max} = 4$ for the initial integration yields $f''(0) = 0.63775$ and $\theta'(0) = -0.56902$ in five iterations. Increasing η_{max} to 10 and using these values for $f''(0)$ and $\theta'(0)$ yields convergence within $\pm 5 \times 10^{-6}$ on f'', f', θ' and θ in five iterations, as shown below.

```
Prandtl Number = 1
n = 0
f''(0) = 0.63775
theta'(0) = -0.56902
```

η	f''	f'	θ'	θ
0	0.63775	0	−0.56902	1
10.	9.407228e−3	3.986160e−2	−1.429782e−6	−1.319635e−2

```
f''(0) = 0.63875
theta'(0) = -0.56902
```

0	0.63875	0	−0.56902	1
10.	7.299972e−3	3.360635e−2	−1.093337e−6	−1.055700e−2

```
f''(0) = 0.63775
theta'(0) = -0.56802
```

0	0.63775	0	−0.56802	1
10.	8.934567e−3	3.167793e−2	−2.242599e−6	−1.228930e−2

```
f''(0) = 0.6420631
theta'(0) = -0.56744371
```

0	0.6420631	0	−0.56744371	1
10.	3.857655e−4	3.264830e−3	−8.639731e−7	−6.066140e−4

```
f''(0) = 0.64218313
theta'(0) = -0.56713645
```

0	0.64218313	0	−0.56713645	1
10.	−3.829511e−6	−7.033459e−5	−9.610531e−7	−1.747123e−6

```
f''(0) = 0.6421859
theta'(0) = -0.56714713
```

0	0.6421859	0	−0.56714713	1
10.	−1.332160e−6	1.827635e−5	−9.554661e−7	−5.62148e−6

```
f''(0) = 0.64218631
theta'(0) = -0.5671452
```

0	0.64218631	0	−0.5671452	1
10.	−3.217811e−6	−1.33437e−6	−9.562096e−7	−2.670774e−6

Convergence achieved

The values of $f''(0)$ and $\theta'(0)$ which yield asymptotic convergence to the correct outer boundary conditions are: $f''(0) = 0.642186$ and $\theta'(0) = -0.567145$.

166   Laminar Flow Analysis

The first four figures agree with those of Ostrach [Ostr53]. The complete velocity and thermal boundary layer profiles are given in Table 5–1.

Ostrach [Ostr53] investigated free convection boundary layer flow on an isothermal vertical flat plate. The results for the wall shearing stress and heat transfer parameters, $f''(0)$ and $\theta'(0)$, are shown as a function of Prandtl number in Figure 5–2. Numerical results based on those of Ostrach [Ostr53] are given in Table C–21 of Appendix C. Here we see that the nondimensional heat transfer parameter, $\theta'(0)$, increases with increasing Prandtl number, whereas the nondimensional shearing stress at the surface decreases. Introducing the local Nusselt number

$$\frac{h_x x}{k} = \mathrm{Nu}_x = -\frac{x\dfrac{\partial T}{\partial y}}{T_w - T_\infty}$$

which, after transforming to similarity variables, yields

$$\frac{\mathrm{Nu}_x}{\left(\dfrac{\mathrm{Gr}_x}{4}\right)^{1/4}} = -\theta'(0) \tag{5-17}$$

where the local Grashof number is

$$\mathrm{Gr}_x = \frac{x^3(T_w - T_\infty)g\bar{\beta}}{\nu^2}$$

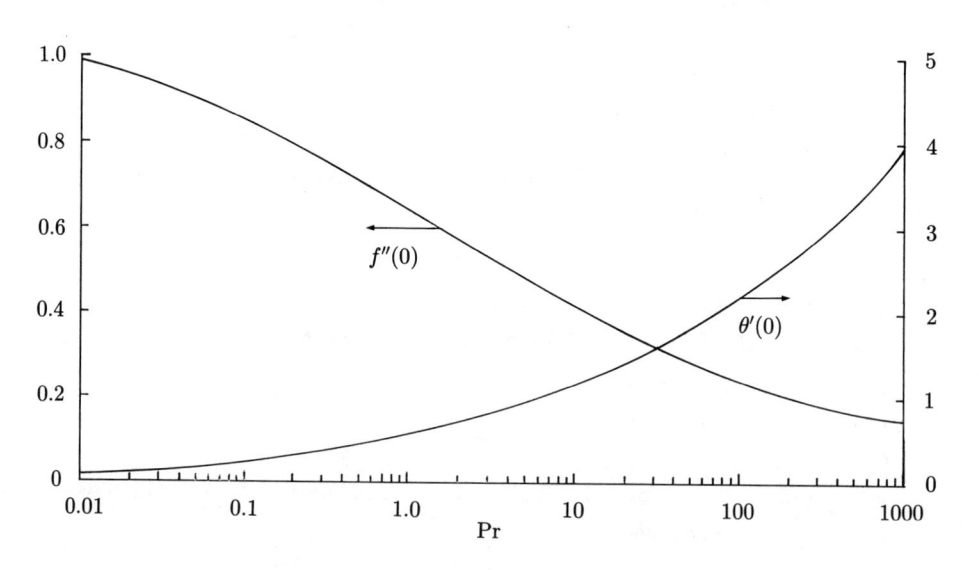

Figure 5–2. Nondimensional wall shearing stress and heat transfer parameter for free convection flow on an isothermal vertical flat plate.

Free Convection Boundary Layer on an Isothermal Vertical Flat Plate

Table 5–1. Nondimensional profiles for a free convection boundary layer on a vertical isothermal plate.

Pr = 1, $f''(0) = 0.642161$, $\theta'(0) = -0.567139$

η	$f''(\eta)$	$f'(\eta)$	$f(\eta)$	$\theta'(\eta)$	θ
0	0.64218631	0	0	−0.5671452	1
0.25	0.41096031	0.13083967	0.01755979	−0.56457447	0.85837781
0.5	0.22094988	0.20892194	0.06101993	−0.54884831	0.71882894
0.75	7.5464404e−2	0.24503667	0.11852243	−0.5133999	0.58561802
1.	−2.6193982e−2	0.25031827	0.18097084	−0.45889659	0.4637327
1.25	−8.8827517e−2	0.23519996	0.24198615	−0.39149685	0.3572507
1.5	−0.12011945	0.20851915	0.29761324	−0.31964438	0.26835263
1.75	−0.12880491	0.17702225	0.34585044	−0.25099094	0.19715435
2.	−0.12303508	0.1453167	0.38611218	−0.19064914	0.1421564
2.25	−0.10938888	0.1161557	0.41872475	−0.14091614	0.10093958
2.5	−9.2566449e−2	9.0882113e−2	0.4445166	−0.10190637	7.0801214e−2
2.75	−7.5553719e−2	6.9884625e−2	0.4645237	−7.2445627e−2	4.9189262e−2
3.	−6.0004424e−2	5.2980774e−2	0.47980083	−5.0828164e−2	3.3924760e−2
3.25	−4.6653986e−2	3.9697628e−2	0.49131608	−3.5306758e−2	2.3267806e−2
3.5	−3.5669266e−2	2.9455799e−2	0.49990306	−2.4342099e−2	1.5892499e−2
3.75	−2.6904681e−2	2.1677607e−2	0.50624911	−1.6689485e−2	1.0821575e−2
4.	−2.0071100e−2	1.5842440e−2	0.51090356	−1.1395922e−2	7.3518651e−3
4.25	−1.4837296e−2	1.1508788e−2	0.51429523	−7.7581072e−3	4.9861813e−3
4.5	−1.0884898e−2	8.3171204e−3	0.51675291	−5.2700800e−3	3.3774053e−3
4.75	−7.9339113e−3	5.9830202e−3	0.51852507	−3.5743376e−3	2.2854147e−3
5.	−5.7510344e−3	4.2862842e−3	0.51979738	−2.4214886e−3	1.5452066e−3
5.25	−4.1487921e−3	3.0592302e−3	0.52070724	−1.6391439e−3	1.0439429e−3
5.5	−2.9803866e−3	2.1758297e−3	0.52135555	−1.1089202e−3	7.0472699e−4
5.75	−2.1330761e−3	1.5423550e−3	0.52181592	−7.4990227e−4	4.7528614e−4
6.	−1.5215590e−3	1.0897017e−3	0.52214174	−5.0696994e−4	3.2015029e−4
6.25	−1.0820570e−3	7.6728652e−4	0.52237158	−3.4266531e−4	2.1528175e−4
6.5	−7.6734000e−4	5.3830893e−4	0.52253315	−2.3157675e−4	1.4440533e−4
6.75	−5.4270950e−4	3.7613511e−4	0.52264628	−1.5648600e−4	9.6508737e−5
7.	−3.8284240e−4	2.6157634e−4	0.52272517	−1.0573653e−4	6.4144175e−5
7.25	−2.6936225e−4	1.8086237e−4	0.52277988	−7.1441919e−5	4.2276199e−5
7.5	−1.8899840e−4	1.2414576e−4	0.52281759	−4.8268775e−5	2.7501139e−5
7.75	−1.3220791e−4	8.4406150e−5	0.52284336	−3.2611383e−5	1.7518686e−5
8.	−9.2153903e−5	5.6652464e−5	0.52286079	−2.2032572e−5	1.0774388e−5
8.25	−6.3954322e−5	3.7345187e−5	0.52287239	−1.4885263e−5	6.2178985e−6
8.5	−4.4133216e−5	2.3979860e−5	0.52287995	−1.0056451e−5	3.1395332e−6
8.75	−3.0222274e−5	1.4788038e−5	0.52288473	−6.7940853e−6	1.0598011e−6
9.	−2.0472888e−5	8.5233310e−6	0.52288759	−4.5900353e−6	−3.4525287e−7
9.25	−1.3648978e−5	4.3088016e−6	0.52288916	−3.1009898e−6	−1.2944965e−6
9.5	−8.8784804e−6	1.5284309e−6	0.52288986	−2.0950013e−6	−1.9357973e−6
9.75	−5.5472616e−6	−2.4987912e−7	0.52289001	−1.4153639e−6	−2.3690544e−6
10.	−3.2235441e−6	−1.3288060e−6	0.5228898	−9.5620712e−7	−2.6617590e−6

168   Laminar Flow Analysis

From Eq. (5–17) we see that h_x varies as $x^{-1/4}$ for an isothermal surface. The average convective heat transfer coefficient, $\bar{h}$, for free convection is then $\bar{h} = (4/3)h_x|_{x=L}$, and the total heat transfer by free convection is calculated from $q = \bar{h}A_s(T_w - T_\infty)$. An empirical result that is sometimes used to approximate the numerical solution given by Eq. (5–17) is

$$\frac{\mathrm{Nu}_x}{\left(\dfrac{\mathrm{Gr}_x}{4}\right)^{1/4}} = \frac{0.676(\mathrm{Pr})^{1/2}}{(0.861 + \mathrm{Pr})^{1/4}}$$

Figure 5–3 shows the nondimensional velocity and temperature profiles on the vertical isothermal flat plate for various Prandtl numbers. Here we see that the maximum value of the nondimensional velocity and the value of η at which it occurs increase with decreasing Prandtl number. Further, comparing the nondimensional velocity and temperature profiles shows that for large Prandtl numbers (Pr $\gg$ 1) the velocity boundary layer is much thicker than the thermal boundary layer.

Figure 5–4 shows a comparison of experimental data from Schmidt and Beckmann [Schm30] for free convection boundary layers and the theory discussed

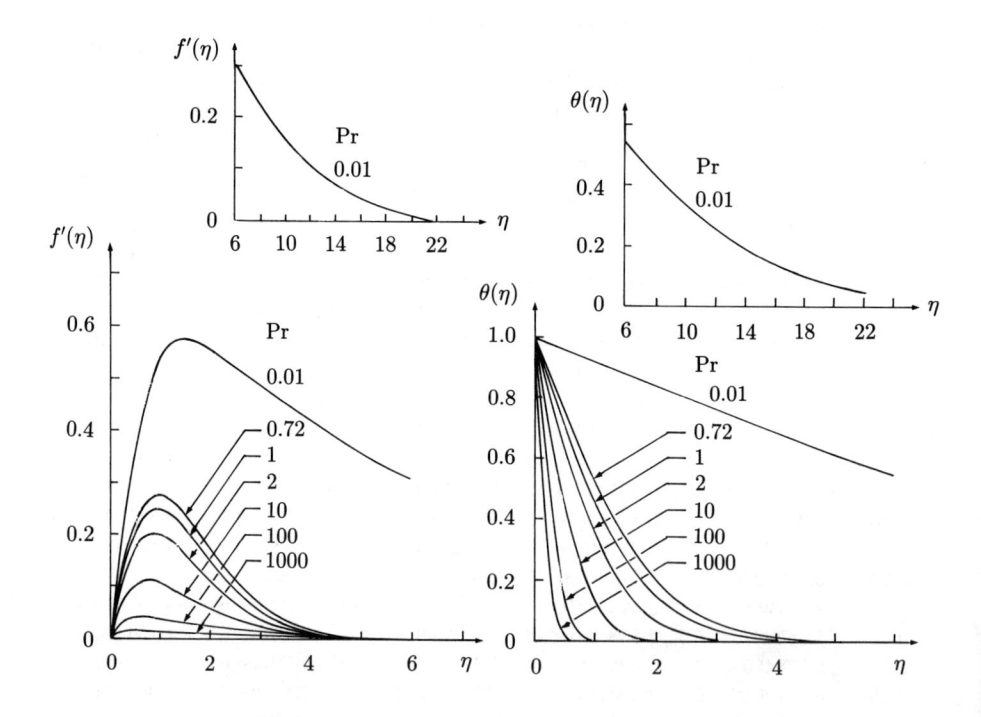

Figure 5–3. Nondimensional velocity and temperature profiles for various Prandtl numbers on a vertical isothermal flat plate. (Adapted from Ostrach [Ostr53].)

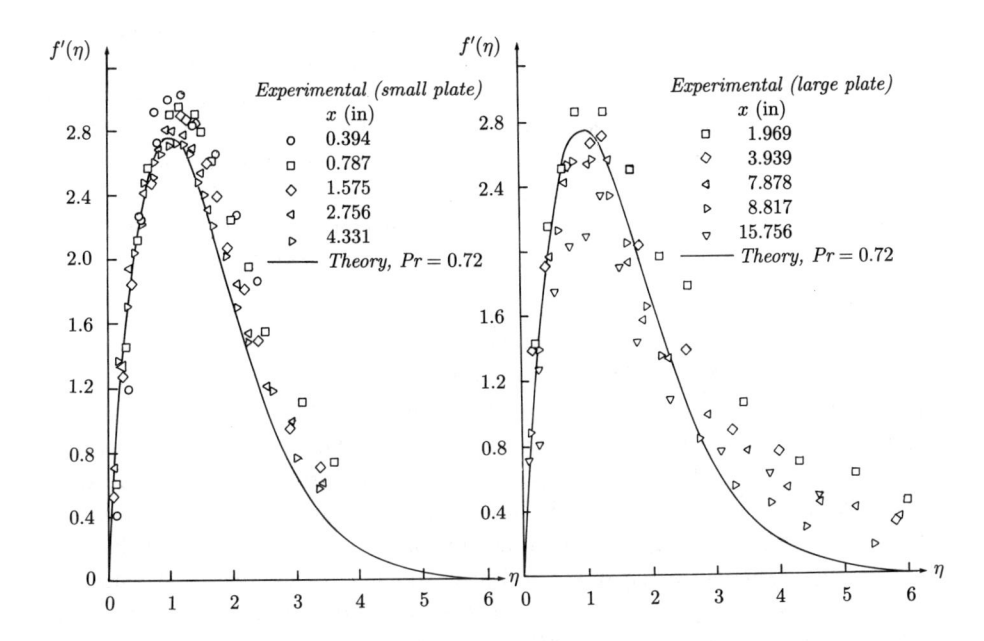

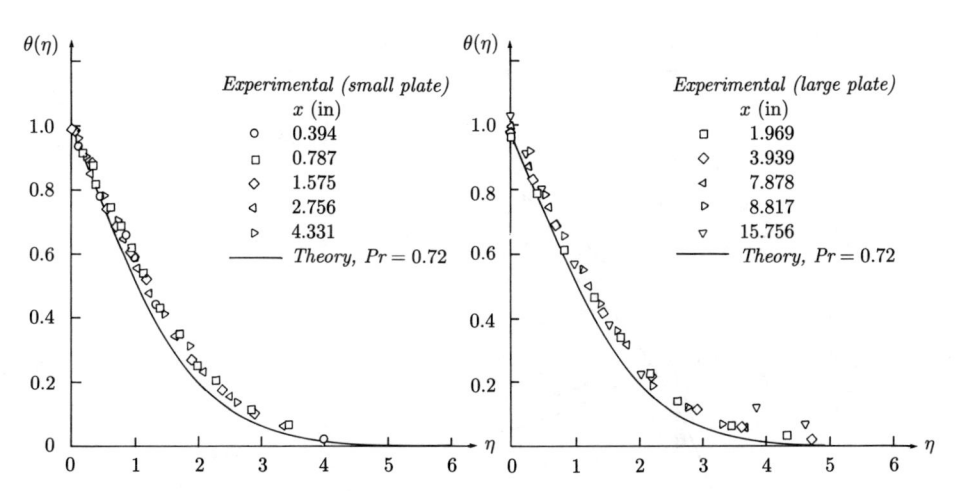

Figure 5-4. Comparison of experimental and theoretical temperature distributions on an isothermal vertical flat plate for Prandtl number 0.72. Adapted from [Spar58].

above for a Prandtl number of 0.72. Here we see that the agreement between the experimental data and the theory is quite good, especially for the nondimensional temperature profile.

The analyses in this section are limited to laminar boundary layer flow in free convection on an isothermal surface. Since the Grashof number represents a ratio of buoyancy forces to viscous forces, its value can be used to specify flow regimes in a manner similar to the Reynolds number in forced convection. For air in free convection on an isothermal vertical plate, laminar flow occurs for $10^4 < \mathrm{Gr}_x < 10^9$. Below the lower limit the buoyancy forces are too small for a velocity boundary layer to form, and the full Navier-Stokes equations must be used to determine the flow field. At the upper limit, transition from laminar to turbulent flow occurs. In turbulent flow, integral methods, empirical results or numerical methods are normally used to estimate heat transfer rates.

5-5 Results for a Free Convection Boundary Layer on a Nonisothermal Vertical Flat Plate

The laminar free convection boundary layer on a nonisothermal vertical flat plate was investigated by Sparrow and Gregg [Spar58]. These studies correspond to the analysis developed above for $n \neq 0$. Sparrow and Gregg confined their analysis to Prandtl numbers of 0.7 and 1 and wall temperature variation exponents $-0.8 < n < 3$. Since Yang [Yang60] does not present exact numerical values for the nondimensional shearing stress and heat transfer parameters at the surface, these solutions are regenerated using the `freeb` program described in Appendix D-15. The results for $f''(0)$ and $\theta'(0)$ are shown in Figure 5-5. The numerical

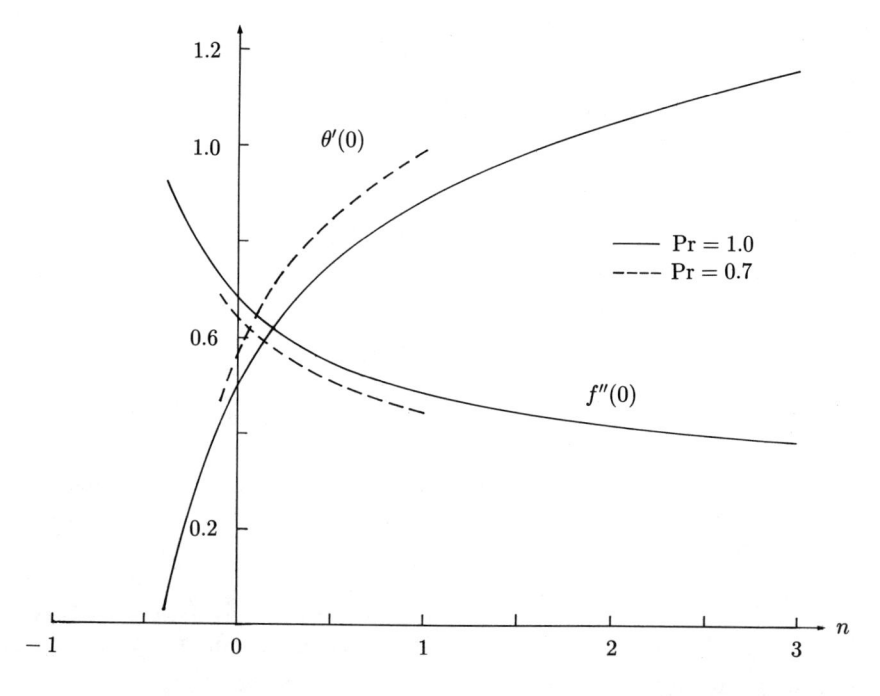

Figure 5-5. Nondimensional shearing stress and heat transfer parameters for free convection on a nonisothermal wall. (Adapted from [Spar58]).

Free Convection Boundary Layer on a Nonisothermal Vertical Flat Plate 171

values are given in Table C–22 in Appendix C. Note the nondimensional heat transfer parameter increases significantly with increasing values of n, whereas the nondimensional shearing stress parameter decreases. Examination of Figure 5–5 and the numerical results shows that locally applying the isothermal wall assumption to calculate results for a nonisothermal wall yields unacceptable values, except for slowly varying wall temperatures, say $-0.1 < n < 0.1$. Further examination of the results reveals that for $n < -0.6$, $\theta'(0)$ is positive. Physically this corresponds to heat transfer from the fluid to the wall and to an infinite source of heat at the leading edge. Since such sources of heat are not readily available, there is some question about the applicability of these solutions.

Figure 5–6 shows the nondimensional temperature and velocity profiles for

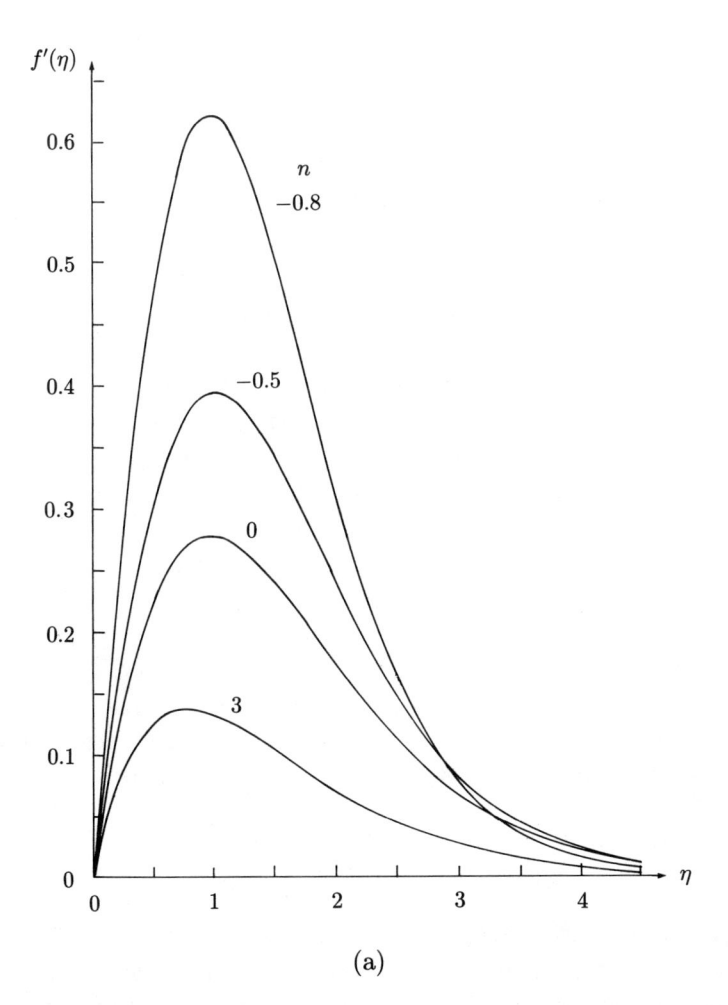

Figure 5–6. Nondimensional velocity and temperature profiles for free convection on a nonisothermal wall, Pr = 0.7.

172 *Laminar Flow Analysis*

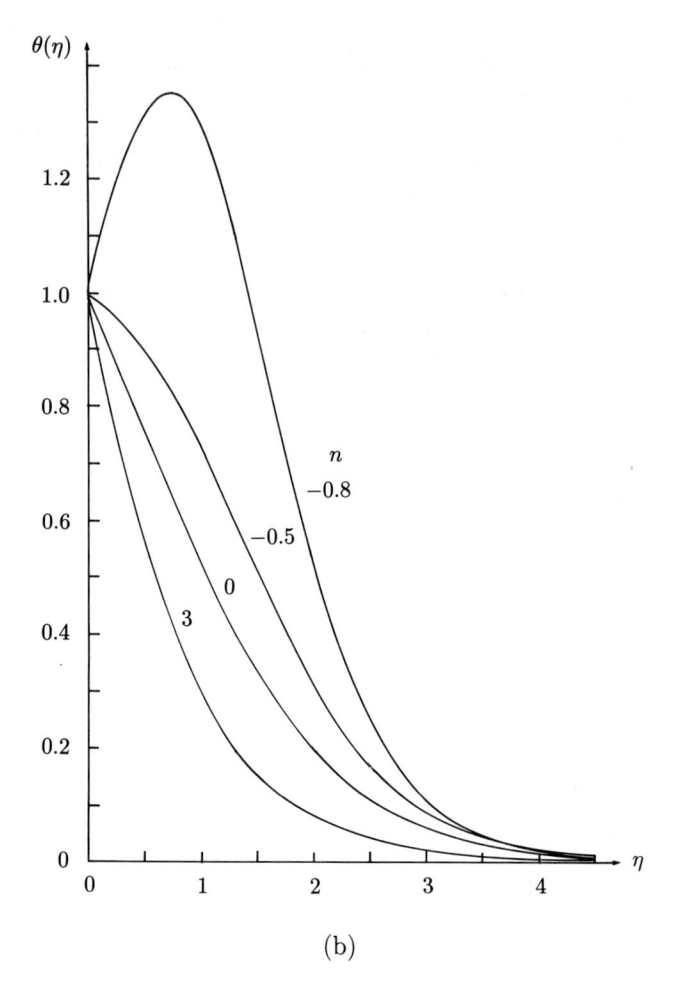

(b)

Figure 5–6. (*cont.*)

a Prandtl number of 0.7 and various values of the wall temperature variation exponent n. For $n > 0$, we see that the nondimensional profiles are similar to those for an isothermal plate. However, note that for $n < 0$ the profiles exhibit a definite inflection point, while for $n < -0.6$ the temperature near the surface is greater than the surface temperature. The nondimensional velocity profiles are similar to those for the isothermal wall. However, also note that the maximum observed velocity increases with decreasing wall temperature variation exponent n. Figure 5–7 shows the relationship between the nondimensional temperature and velocity functions for Prandtl numbers of 0.7 and 1. Significant differences are observed except for $n = -0.8$, where a crossover in the temperature profiles is observed.

$f'(\eta)$ axis with curves labeled Pr = 0.7 and Pr = 1.

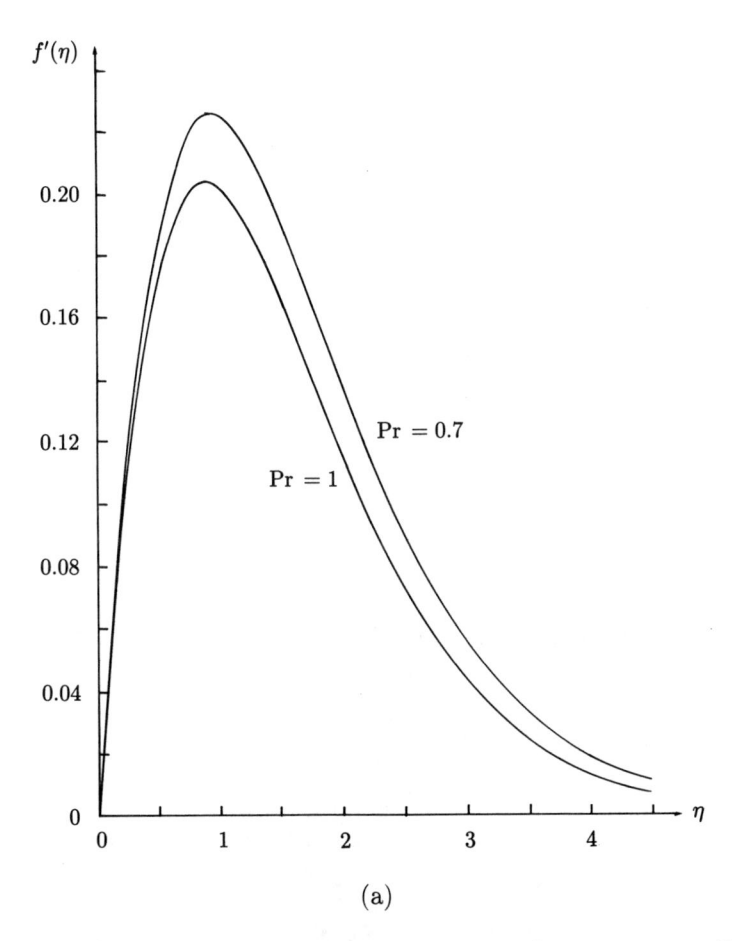

(a)

Figure 5-7. Comparison of the nondimensional velocity and temperature profiles for free convection on a nonisothermal wall for Pr = 0.7 and 1.

5-6 The Free Convection Boundary Layer on a Nonisothermal Vertical Flat Plate with Mass Transfer

Here we consider the free convection boundary layer on a nonisothermal vertical flat plate with mass transfer through the surface (see Figure 5-8). Considering our previous discussion, we limit ourselves to similar solutions. When mass is transferred into the fluid, the injected fluid is considered to have the same physical properties as the surrounding fluid. The injected fluid is assumed to have the same temperature as the surface and to be injected normal to the surface. Under these conditions the mass-diffusion equation need not be considered.[†]

[†]Free convection solutions which require the mass-diffusion equations are discussed in Adams and Lowell [Adam68].

$\theta(\eta)$

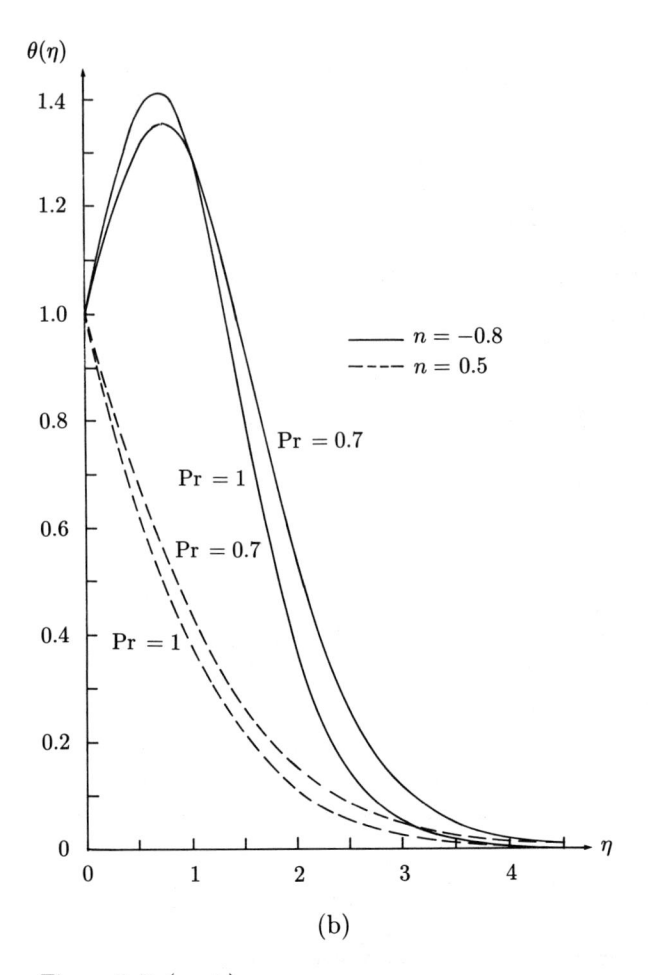

n = −0.8
n = 0.5

Pr = 0.7
Pr = 1
Pr = 0.7
Pr = 1

(b)

Figure 5–7. (*cont.*)

Hence the only difference between the mathematical model for the problem under consideration and that previously discussed is the boundary condition on the normal velocity at the surface of the plate. Thus, the governing equations are

$$f''' + (n+3)ff'' - 2(n+1)f'^2 + \theta = 0 \qquad (5-12)$$

and
$$\theta'' + \Pr\left[(n+3)f\theta' - 4nf'\theta\right] = 0 \qquad (5-13)$$

where $f(\eta)$, $\theta(\eta)$ and η are again defined as in Eqs. (5–3), (5–15) and (5–16). The appropriate boundary conditions at the surface, i.e., at $\eta = 0$, are

$$f'(0) = 0 \qquad \theta(0) = 1 \qquad (5-14b,c)$$

and
$$f(0) = \text{constant} \qquad (5-18)$$

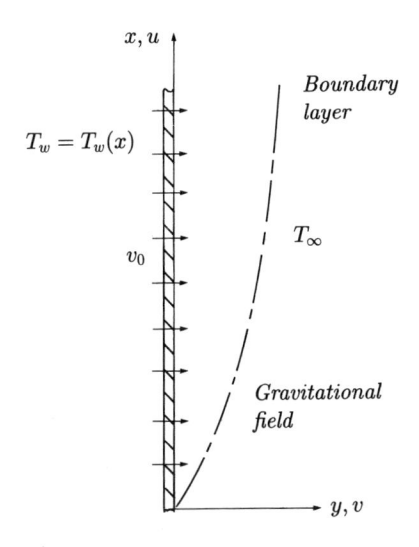

Figure 5-8. Free convection boundary layer on a nonisothermal vertical flat plate with mass transfer.

where the nondimensional stream function at the surface is constant to satisfy similarity conditions. The boundary conditions at infinite distance from the plate are again given by

$$f'(\eta \to \infty) \to 0 \qquad \theta(\eta \to \infty) \to 0 \qquad (5-14d, e)$$

Using Eqs. (5–15) and (5–16) allows writing the normal component of the velocity as

$$v = -\frac{\nu}{x}\left[\frac{g\bar{\beta}x^3(T_w - T_\infty)}{\nu^2}\right]^{1/4}[(n+3)f + (n-1)f'\eta] \qquad (5-19)$$

At the surface this becomes

$$v_0 = -\frac{\nu}{x}\left[\frac{g\bar{\beta}x^3(T_w - T_\infty)}{\nu^2}\right]^{1/4}(n+3)f_0 \qquad (5-20)$$

In order for f_0 to be constant and hence have similar solutions, the mass-transfer distribution, i.e., the distribution of normal velocity at the surface, must have a power law variation with distance along the plate. In particular

$$v_0 \sim x^{\frac{n-1}{4}}$$

Here we see that the distribution of normal velocity depends on the surface temperature variation exponent, n, whereas for the forced convection boundary layer with mass transfer it depends on the pressure gradient parameter, β, and not on the surface temperature variation exponent.

176   *Laminar Flow Analysis*

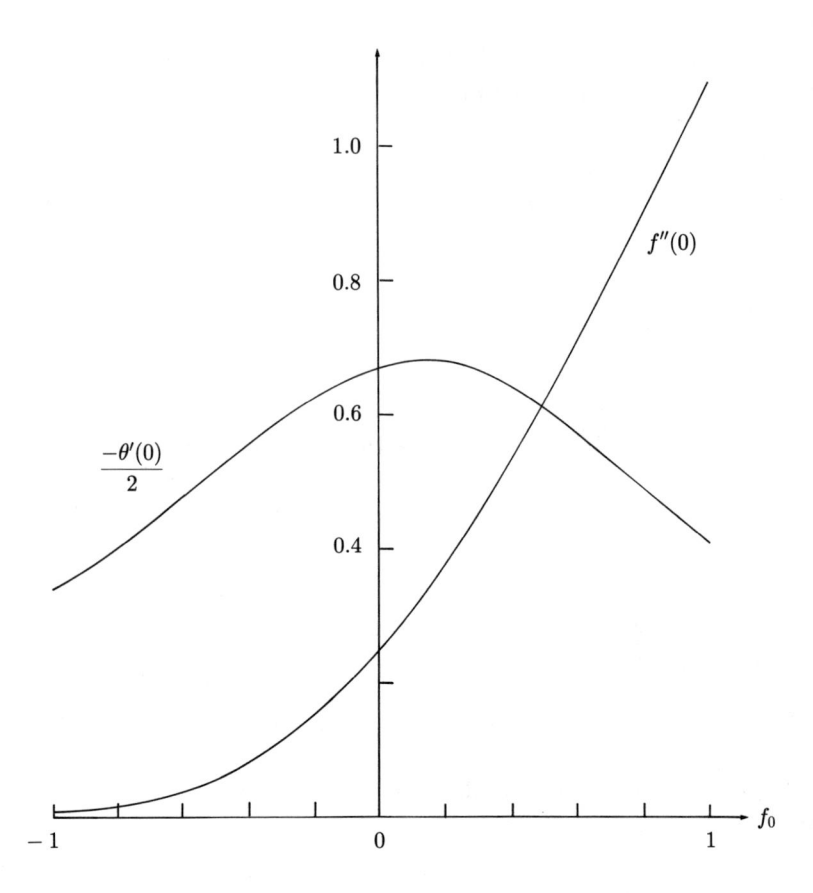

[Figure: plot with vertical axis marked 0.4, 0.6, 0.8, 1.0 and horizontal axis f_0 from -1 to 1. Curve labeled $f''(0)$ increases; curve labeled $\dfrac{-\theta'(0)}{2}$ peaks near $f_0 = 0$.]

Figure 5–9. Nondimensional shearing stress and heat transfer parameters for free convection from an isothermal wall with mass transfer, $\mathrm{Pr} = 0.73$.

Solutions of the free convection boundary layer equations and the associated boundary conditions given above are obtained by adding to the **freeb** program `init(3) = `f_0 where f_0 is the numerical value of the nondimensional mass-transfer parameter.

Solutions of these equations for a Prandtl number of 0.73 for an isothermal wall were investigated by Eichhorn [Eich60]. The nondimensional shearing stress and heat transfer parameters $f''(0)$ and $\theta'(0)$ as determined by Eichhorn are given in Figure 5–9 as functions of the mass-transfer parameter f_0. Numerical results are given in Table C–23 of Appendix C. Here $f_0 < 0$ corresponds to blowing or injection and $f_0 > 0$ to suction. Figure 5–9 reveals that the nondimensional shearing stress at the surface, i.e., the skin friction, is not significantly affected by mass transfer at the surface. However, note that $f''(0)$ has a maximum at a small positive value of the mass-transfer parameter, f_0. Figure 5–9 also shows that for an isothermal wall the nondimensional heat transfer parameter, $\theta'(0)$, significantly increases for increasing suction and approaches zero for significant blowing, e.g., $-\theta'(0) = 0.00748$ for $f_0 = -1$.

Physical explanation of the behavior of the nondimensional shearing stress and heat transfer parameters is aided by the nondimensional velocity and temperature profiles shown in Figure 5–10. As the blowing rate increases ($f_0 < 0$) the heated fluid near the surface is forced outward. Here the buoyancy forces can more readily accelerate the flow, since the viscosity effects are smaller. Since the maximum velocity in the layer is increased by this effect, the shearing stress is also increased. However, the point of maximum velocity is further removed from the surface. This effect decreases the shear. These two effects balance one another, with the result that the nondimensional shear is somewhat decreased as blowing increases. For suction, the maximum velocity decreases and the point of maximum velocity moves closer to the surface. The net result of these phenomena is that the point of maximum shear occurs for a small positive value of f_0.

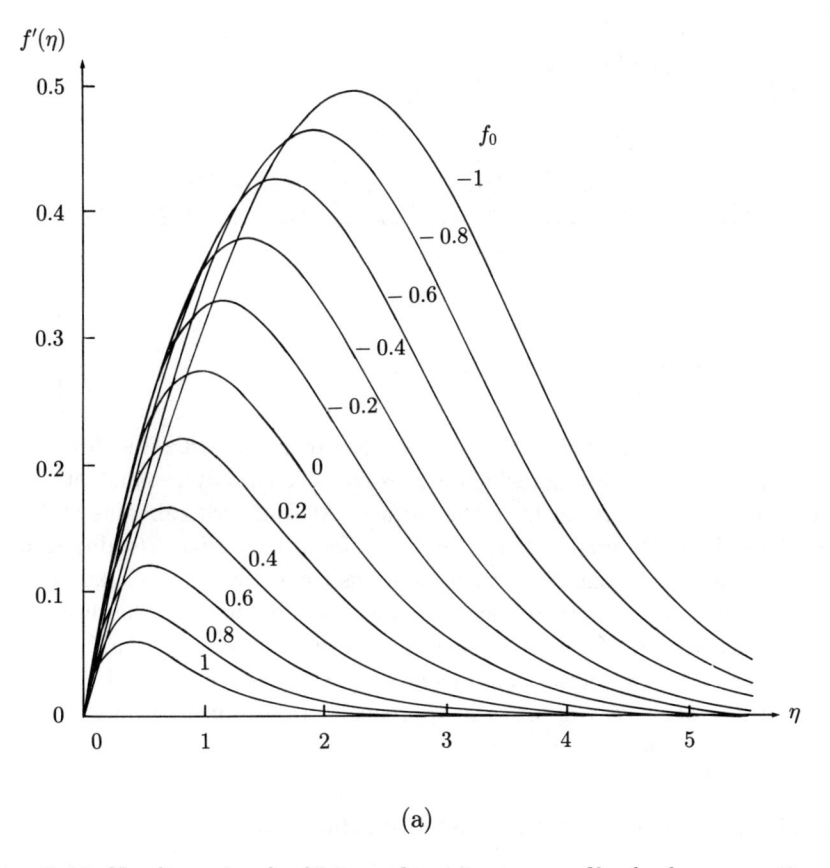

(a)

Figure 5–10. Nondimensional velocity and temperature profiles for free convection on a nonisothermal wall, $\text{Pr} = 0.7$.

$\theta(\eta)$

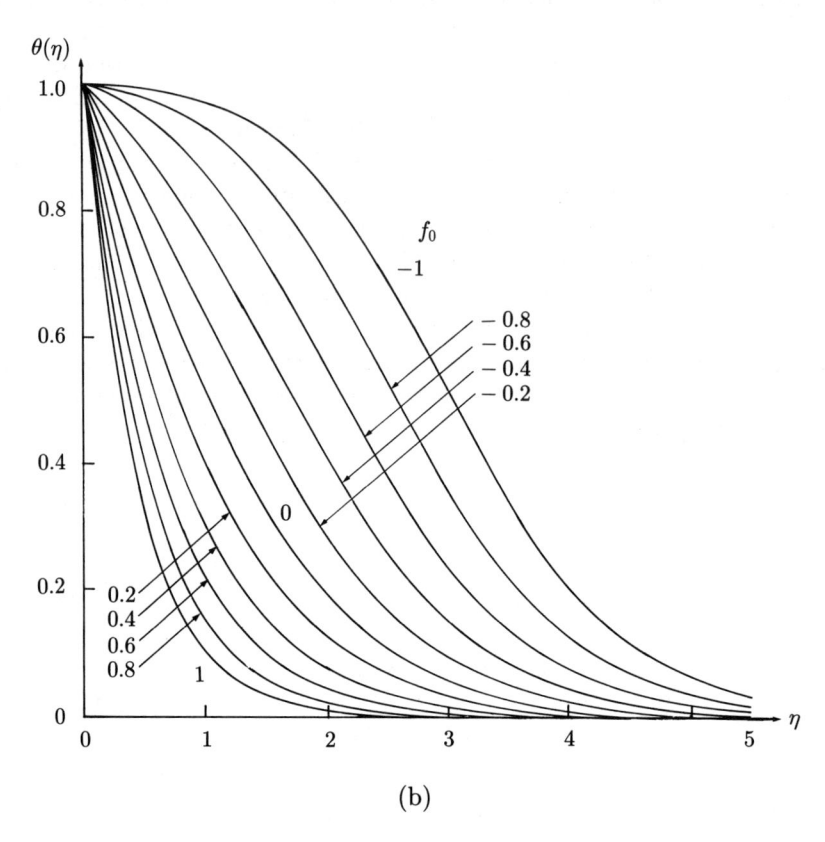

(b)

Figure 5–10. (cont.)

Blowing introduces a layer of fluid at surface temperature. This fluid forms a layer near the surface through which heat is transferred only by conduction, thus decreasing the heat transfer from the surface. At some critical value of f_0, e.g., $f_0 \approx -1$ for the isothermal wall, the heat transfer at the surface is reduced to zero (see Figure 5–10). As one expects, suction has the opposite effect, i.e., the heat transfer at the surface increases, since removal of the fluid near the surface thins the boundary layer and brings the surrounding fluid closer to the surface. As the suction increases the normal component of the velocity becomes less effective, and the heat transfer asymptotically approaches some limiting value. This is the so-called asymptotic suction value.

5–7 Nonsimilar Free Convection Boundary Layer

The previous sections discussed similar solutions for the free convection boundary layer. Here we consider solutions for the nonsimilar free convection boundary layer. In particular, consider steady, two-dimensional, incompressible, except for density changes in the body force term, constant property flow about a semi-infinite flat plate parallel to the direction of the generating body force (see Section 5–1 and Figure 5–1). We again neglect viscous dissipation and assume the temperature in the external inviscid fluid, T_∞, is constant. With these assumptions the governing equations in terms of the stream function, ψ, and the nondimensional temperature $\theta = (T - T_\infty)/(T_w - T_\infty)$ are

$$\psi_y \psi_{xy} - \psi_x \psi_{yy} = g\bar{\beta}(T_w - T_\infty)\theta + \nu\psi_{yyy} \qquad (5-4)$$

and
$$\psi_y \theta_x - \psi_x \theta_y + \frac{\theta\psi_y}{T_w - T_\infty}\frac{\partial T_w}{\partial x} = \alpha\theta_{yy} \qquad (5-5)$$

with no-mass-transfer boundary conditions

$$\psi_y(x,0) = \psi_x(x,0) = 0 \qquad \theta(x,0) = 1 \qquad (5-6a,b,c)$$

and
$$\psi_y(x, y \to \infty) \to 0 \qquad \theta(x, y \to \infty) \to 0 \qquad (5-6d,e)$$

Following our previous developments of the locally nonsimilar method for the hydrodynamic and forced convection boundary layers (see Section 3–11 and Section 4–11) we introduce dependent and independent variable transformations based on the free convection similarity transformations, Eqs. (5–15) and (5–16). Thus, introducing

$$\psi(x, y) = 4\nu \left[\frac{g\bar{\beta}\xi^3}{4\nu^2}(T_w - T_\infty)\right]^{1/4} f(\xi, \eta) \qquad (5-21)$$

$$\eta = y\left[\frac{g\bar{\beta}(T_w - T_\infty)}{4\nu^2\xi}\right]^{1/4} \qquad (5-22)$$

$$\xi = x \qquad (5-23)$$

the momentum and energy equations become

$$f''' + 3ff'' - 2f'^2 + \theta = \alpha_3\, \xi(2f'^2 - ff'') + 4\xi(f'f'_\xi - f''f_\xi) \qquad (5-24)$$

and
$$\theta'' + 3\Pr f\theta' = \alpha_3\, \xi\Pr(4f'\theta - f\theta') + 4\xi\Pr(\theta'f_\xi - f'\theta_\xi) \qquad (5-25)$$

where the prime denotes differentiation with respect to η and

$$\alpha_3(\xi) = \frac{1}{T_w - T_\infty}\frac{\partial T_w}{\partial \xi} \qquad (5-26)$$

The transformed no-mass-transfer boundary conditions are

$$f(\xi, 0) = f'(\xi, 0) = 0 \qquad \theta(\xi, 0) = 1 \qquad (5-27a,b,c)$$

and
$$f'(\xi, \eta \to \infty) \to 0 \qquad \theta(\xi, \eta \to \infty) \to 1 \qquad (5-27d,e)$$

180   Laminar Flow Analysis

The similarity solution discussed in previous sections (see Eqs. 5–12 and 5–13) is obtained by neglecting the last term on the right hand side of both the momentum and energy equations. For those solutions the surface temperature distribution is given by $(T_w - T_\infty) = T_1 x^n$. Hence, $\alpha_3(\xi)$ is $n/x = n/\xi$, and Eqs. (5–24) and (5–25) reduce to Eqs. (5–12) and (5–13).

We begin the development of the two-equation model for the locally nonsimilar method by introducing auxiliary functions

$$S(\xi, \eta) = f_\xi \qquad (5-28)$$

and
$$t(\xi, \eta) = \theta_\xi \qquad (5-29)$$

The momentum and energy equations (Eqs. 5–24 and 5–25) are now written in terms of f, θ, S and t. To evaluate $S(\xi,\eta)$ and $t(\xi,\eta)$ we differentiate Eqs. (5–24) and (5–25) with respect to ξ. Differentiating the momentum equation and introducing $S(\xi,\eta)$ and $t(\xi,\eta)$ yields

$$S''' + 3(fS'' + Sf'') - 4f'S' + t + 4(f''S - f'S') =$$
$$4\xi(S'^2 - S''S) - \alpha_3\left(1 + \frac{\xi}{\alpha_3}\frac{d\alpha_3}{d\xi}\right)(ff'' - 2f'^2)$$
$$- \alpha_3\xi(Sf'' + fS'' - 4f'S') + 4\xi(f'S'_\xi - f''S_\xi) \qquad (5-30)$$

Differentiating the energy equation and introducing $S(\xi,\eta)$ and $t(\xi,\eta)$ yields

$$f'' + 3\Pr(S\theta' + St') + 4\Pr(f't - f'\theta') =$$
$$\Pr\left(1 + \frac{\xi}{\alpha_3}\frac{d\alpha_3}{d\xi}\right)(f'\theta' - 4f'\theta) + 4\xi\Pr(t'S - S't)$$
$$+ \alpha_3\xi\Pr(S\theta' + ft' - 4S'\theta - 4f't) + 4\xi\Pr(\theta'S_\xi - f't_\xi) \qquad (5-31)$$

Consistent with our previous development of the two-equation locally nonsimilar model for the hydrodynamic and forced convection boundary layers, we neglect the last term appearing on the right hand side of the first auxiliary momentum and energy equations, i.e., we neglect the term $4\xi(f'S'_\xi - f''S_\xi)$ in the first auxiliary momentum equation and $4\xi\Pr(\theta'S_\xi - f't_\xi)$ in the first auxiliary energy equation. Gathering the system of governing equations for the two-equation locally nonsimilar model together, we have

momentum
$$f''' + 3ff' - 2f'^2 + \theta = \alpha_3\xi(2f'^2 - ff'') + 4\xi(f'S' - f''S) \qquad (5-32)$$

first auxiliary momentum
$$S''' + 3(fS'' + Sf'') - 4f'S' + t + 4(f''S - f'S') = 4\xi(S'^2 - S''S)$$
$$- \alpha_3\left(1 + \frac{\xi}{\alpha_3}\frac{d\alpha_3}{d\xi}\right)(ff'' - 2f'^2) - \alpha_3\xi(Sf'' + fS'' - 4f'S') \qquad (5-33)$$

energy
$$\theta'' + 3\Pr f\theta' = \alpha_3 \xi \Pr (4f'\theta - f\theta') + 4\xi \Pr (\theta'S - f't) \qquad (5-34)$$

first auxiliary energy
$$t'' + 3\Pr(S\theta' + St') + 4\Pr(f't - f'\theta') =$$
$$\Pr\left(1 + \frac{\xi}{\alpha_3}\frac{d\alpha_3}{d\xi}\right)(f'\theta' - 4f'\theta) + 4\xi\Pr(t'S - S't)$$
$$+ \alpha_3 \xi \Pr(S\theta' + ft' - 4S'\theta - 4f't) \qquad (5-35)$$

The no-mass-transfer boundary conditions for the auxiliary equations are obtained by differentiating Eq. (5–27) with respect to ξ. Thus, for no-mass transfer at the surface, the boundary conditions for Eqs. (5–32) to (5–35) are

$$f(\xi, 0) = f'(\xi, 0) = 0 \qquad (5-27a, b)$$

$$S(\xi, 0) = S'(\xi, 0) = 0 \qquad (5-36a, b)$$

$$\theta(\xi, 0) = 1 \qquad t(\xi, 0) = 0 \qquad (5-27c, 5-36c)$$

and
$$f'(\xi, \eta \to \infty) \to 0 \qquad S'(\xi, \eta \to \infty) \to 0 \qquad (5-27d, 5-36d)$$
$$\theta(\xi, \eta \to \infty) \to 0 \qquad t(\xi, \eta \to \infty) \to 0 \qquad (5-27e, 5-36e)$$

Examining Eqs. (5–32) to (5–35) and the boundary conditions, Eqs. (5–27) and (5–36), we see that they represent a tenth-order system of coupled ordinary differential equations for the dependent variables f, θ, S and t subject to two-point asymptotic boundary conditions, four of which are specified at infinity. Numerical integration of this system of equations is accomplished by suitable modifications and additions to the **freeb** program. The necessary modifications to the Nachtsheim-Swigert iteration technique for four asymptotic boundary conditions are given in Appendix B.

The three-equation locally nonsimilar boundary layer model for free convection is obtained by introducing additional auxiliary functions, $h(\xi, \eta) = S_\xi = f_{\xi\xi}$ and $j(\xi, \eta) = t_\xi = \theta_{\xi\xi}$, into Eqs. (5–30) and (5–31) and subsequently differentiating with respect to ξ. The equations necessary to evaluate $h(\xi, \eta)$ and $j(\xi, \eta)$ are obtained by neglecting those terms involving h_ξ and j_ξ. The boundary conditions are obtained in a similar manner. The resulting system of equations is of fifteenth-order, with six asymptotic boundary conditions satisfied at infinity.

Defining a local Nusselt number as
$$\text{Nu}_x = \frac{x\dfrac{\partial T}{\partial y}}{T_w - T_\infty}$$

182   Laminar Flow Analysis

and using Eqs. (5–22) and (5–23) as well as the definition of θ yields

$$\mathrm{Nu}_x = \left(\frac{\mathrm{Gr}_x}{4}\right)^{1/4} \theta'(\xi, 0) \tag{5-37}$$

where the Grashof number is defined as

$$\mathrm{Gr}_x = \frac{x^3 g \bar{\beta}(T_w - T_\infty)}{\nu^2} \tag{5-38}$$

5–8 Nonsimilar Free Convection Boundary Layer with Mass Transfer

We now consider the nonsimilar free convection boundary layer with mass transfer through the surface. Again, when fluid is injected into the boundary layer the injected fluid has the same properties as the surrounding fluid, the same temperature as the surface and is injected normal to the surface. With these assumptions the momentum and energy equations are not changed. Further, neither the boundary conditions for the energy equation (Eqs. 5–27 c, e) nor those for the first auxiliary energy equation (Eqs. 5–36 c, e) are affected. Thus, it is only necessary to consider the effect of mass transfer on the normal boundary conditions for the momentum and first auxiliary momentum equations.

Recalling that $\psi_x = -v(x)$, and using Eqs. (5–21) to (5–23) and (5–27 b), the normal component of the velocity at the surface is written as

$$v(x) = -\psi_x = 4\nu \left(\frac{\mathrm{Gr}_x}{4}\right)^{1/4} \left[f_\xi(\xi, 0) + \frac{1}{4}\left(\frac{3}{\xi} + \alpha_3\right) f(\xi, 0)\right]$$

where the Grashof number is defined in Eq. (5–38). Using $S(\xi, 0) = f_\xi(\xi, 0)$, the mass-transfer boundary condition for the momentum equation is written as

$$f(\xi, 0) + \frac{4}{3}\xi(\alpha_3 f(\xi, 0) + S(\xi, 0)) = -\frac{v(x)}{3\nu}\left(\frac{\mathrm{Gr}_x}{4}\right)^{1/4} \xi \tag{5-39}$$

The mass-transfer boundary condition for the first auxiliary momentum equation is obtained by differentiating Eq. (5–39) with respect to ξ and, consistent with the two-equation locally nonsimilar model, neglecting the terms involving $S_\xi(\xi, 0)$. This yields

$$S(\xi, 0) + \frac{4}{7}\alpha_3 f(\xi, 0) + \frac{4}{7}\xi\left(\alpha_3 S(\xi, 0) + \frac{d\alpha_3}{d\xi} f(\xi, 0)\right) =$$
$$- \frac{v(x)}{7\nu}\left(\frac{\mathrm{Gr}_x}{4}\right)^{-1/4} \xi \left(\frac{1}{v}\frac{dv}{d\xi} - \frac{2}{\xi} - \frac{1}{T_w - T_\infty}\frac{dT_w}{d\xi}\right) \tag{5-40}$$

Since it is assumed that the normal velocity and wall temperature variations are known as functions of x and hence of ξ, the nonsimilar boundary conditions given by Eqs. (5–39) and (5–40) are known at any position on the surface. The governing equations for the nonsimilar free convection boundary layer with mass transfer is then a system of coupled ordinary differential equations subjected to known boundary conditions at any position on the surface. Numerical integration is straightforward and is accomplished by suitable modifications to the `freeb` program.

5–9 Combined Forced and Free Convection Boundary Layer Flows on a Nonisothermal Surface

Having independently discussed forced and free convection boundary layer flows, we now turn our attention to consideration of flows where the combined effects of forced and free convection are important. From our discussion in Section 4–6 we anticipate that these flows are governed by the ratio of the Grashof number to the square of the Reynolds number, i.e., $\mathrm{Gr}_x/\mathrm{Re}_x^2$.

We formulate the governing boundary value problem by considering two-dimensional, steady flow with constant transport properties. The fluid is considered incompressible except for the essential density variations which give rise to buoyancy forces. Pressure gradient and buoyancy force effects are included. Under these conditions the governing boundary layer equations (see Eqs. 3–23, 4–53 and 4–54) become

continuity
$$\frac{\partial u}{\partial x} + \frac{\partial v}{\partial y} = 0 \tag{5-41}$$

momentum
$$u\frac{\partial u}{\partial x} + v\frac{\partial u}{\partial y} = U\frac{dU}{dx} + g_x\bar{\beta}\Delta T + \nu\frac{\partial^2 u}{\partial y^2} \tag{5-42}$$

energy
$$u\frac{\partial T}{\partial x} + v\frac{\partial T}{\partial y} = \frac{k}{\rho c_p}\frac{\partial^2 T}{\partial y^2} + \frac{\nu}{c_p}\left(\frac{\partial u}{\partial y}\right)^2 \tag{5-43}$$

In keeping with our previous analyses of forced and free convection boundary layer flows, we again seek similar solutions of these equations. For forced convection flows similar solutions existed when the velocity in the inviscid flow varied as $U(x) = U_\infty x^m$ and the surface temperature varied as $T_w - T_e = T_1 x^n$. For free convection flows, similar solutions also existed when the surface temperature varied as $T_w - T_e = T_1 x^n$. We therefore make these two assumptions.

We begin the analysis by introducing a stream function to satisfy the continuity equation and a nondimensional temperature function to simplify the thermal boundary conditions, i.e.

$$u = \psi_y \qquad v = -\psi_x \qquad \text{and} \qquad \theta = \frac{T - T_w}{T_e - T_w} \tag{5-41a,b,c}$$

Since the governing equations, except for the buoyancy term, are essentially the same as those for forced convection boundary layer flow, we introduce the similarity transformations

$$\psi(x,y) = \left(\frac{2}{m+1}\nu x U\right)^{1/2} f(\eta) \qquad (4-104)$$

and
$$\eta = y\left(\frac{m+1}{2}\frac{U}{\nu x}\right)^{1/2} \qquad (3-62)$$

used in our analysis of the forced convection boundary layer problem. Under these circumstances Eqs. (5–41) to (5–43) reduce to

$$f''' + ff'' + \beta(1 - f'^2) = (2 - \beta)\frac{g_x \bar{\beta} T_1}{U_\infty^2}(x^{n+1-2m})(1 - \theta) \qquad (5-44)$$

and
$$\theta'' + \Pr f\theta' + n(2 - \beta)\Pr f'(1 - \theta) = \Pr E_1 x^{2m-n} f''^2 \qquad (5-45)$$

where β is the Falkner-Skan pressure gradient parameter and $E_1 = U_\infty^2/c_p T_1$ is the appropriate Eckert number. Here we recognize the left hand side of Eq. (5–44) as the Falkner-Skan equation. The term on the right hand side is the coupling term associated with the buoyancy force. Further, we recognize Eq. (5–45) as the energy equation previously derived for forced convection flow with viscous dissipation over a nonisothermal surface. Equations (5–44) and (5–45) reduce to ordinary differential equations provided the coefficients of the terms on the right hand side of the equations are constant. Since β, Pr, E_1 and $g_x\bar{\beta}T_1/U_\infty^2$ are constant, this requirement reduces to $2m - n = 0$ and $n + 1 - 2m = 0$. These two requirements cannot be satisfied simultaneously. This implies that except for the particular case of $\beta = 2$, similar solutions to combined forced and free convection boundary layer flows with viscous dissipation do not exist. Thus we neglect viscous dissipation in our analysis. This assumption makes the right hand side of Eq. (5–45) zero and allows imposing the restriction that

$$n = 2m - 1 = \frac{3\beta - 2}{2 - \beta}$$

Under these circumstances, similar solutions to combined forced and free convection boundary layer flows are possible. However, note the coupling between the inviscid velocity distribution, $U(x) = U_\infty x^m$, and the surface temperature distribution, $T_w - T_e = T_1 x^n$. In particular, a similar solution of combined forced and free convection flow does not exist for the isothermal flat plate at zero angle of attack, i.e., for $n = 0$, $\beta = 0$. However, a similar solution for an isothermal wall is possible for $m = 1/2$, $\beta = 2/3$. This implies that the inviscid velocity at the edge of the boundary layer varies as $x^{1/2}$.

Turning our attention to the coefficient of the buoyancy term, we have

$$\frac{g_x \bar{\beta} T_1}{U_\infty^2} \frac{x^n}{x^{2m-1}} = \frac{g_x \bar{\beta}(T_w - T_e)x^3}{\nu^2 \left(\dfrac{U^2 x^2}{\nu^2}\right)} = \frac{\text{Gr}_x}{\text{Re}_x^2}$$

where the local Grashof number is defined as

$$\text{Gr}_x = \frac{g_x \bar{\beta}(T_w - T_e)x^3}{\nu^2}$$

and the local Reynolds number as

$$\text{Re}_x = \frac{Ux}{\nu}$$

Using these results, Eqs. (5–44) and (5–45) become

$$f''' + ff'' + \beta(1 - f'^2) = (2 - \beta)\frac{\text{Gr}_x}{\text{Re}_x^2}(1 - \theta) \tag{5 – 46]}$$

and

$$\theta'' + \Pr f\theta' + (3\beta - 2)\Pr f'(1 - \theta) = 0 \tag{5 – 47}$$

The appropriate boundary conditions for no-slip and no-mass transfer at the surface, with the additional stipulation that the fluid temperature at the surface be the same as the surface temperature, are

$$f(0) = f'(0) = \theta(0) = 0 \tag{5 – 48a, b, c}$$

Requiring that the viscous flow smoothly transition into the inviscid flow at the edge of the boundary layer yields

$$f'(\eta \to \infty) \to 1 \qquad \theta(\eta \to \infty) \to 1 \tag{5 – 48d, e}$$

Here we have demonstrated not only that the ratio of the Grashof number to the square of the Reynolds number is the controlling parameter for combined flows, but also that it is independent of the distance along the surface, x, for similar flows. When $\text{Gr}_x/\text{Re}_x^2$ is small forced convection effects predominate, while for large values free convection effects predominate. For $\text{Gr}_x/\text{Re}_x^2 = O(1)$ combined forced and free convection flow occurs.

When $T_w > T_e$ the Grashof number and hence $\text{Gr}_x/\text{Re}_x^2$ is negative. Such flows are called opposing flows, since the buoyancy force acts in a direction opposite to the free stream direction. Flows for which $T_w < T_e$ are designated aiding flows. In this case the Grashof number and hence $\text{Gr}_x/\text{Re}_x^2$ is positive, and the buoyancy force has a component in the same direction as the free stream velocity (see Figure 5–1). Examination of Eq. (5–46) shows that for opposing flows the term $(2 - \beta)(\text{Gr}_x/\text{Re}_x^2)(1 - \theta)$ acts like an adverse pressure gradient in purely forced convection flows. Hence, it is reasonable to expect that for increasingly negative values of $\text{Gr}_x/\text{Re}_x^2$ a situation similar to separation occurs.

186   *Laminar Flow Analysis*

Before considering the numerical integration of Eqs. (5–46) and (5–47) subject to the boundary conditions given in Eqs. (5–48), we remark that the similarity transformations used above are by no means unique. Sparrow, Eichhorn and Gregg [Spar59] point out that the free convection similarity transformations also yield similar solutions. However, in this case the parameter Re_x^2/Gr_x appears not only in the momentum equation but also in the boundary conditions. For this reason these transformations are not considered further.

5–10 Numerical Integration of the Governing Equations for Combined Forced and Free Convection

The governing equations and associated boundary conditions for combined forced and free convection boundary layer flow (Eqs. 5–46 and 5–47) are numerically integrated by making suitable modifications to the **freeb** program and the **eqmot** routine.

For the **freeb** program the modifications are

```
nparams = 3
paramnam (2) = ''Beta''
paramnam (3) = ''Gr/Re Ratio''
paramnam (1) = ''Prandtl Number''
init (5) = 0
asymbc (3) = 1
asymbc (4) = 1
```

The modified **eqmot** routine is now

equations of motion routine for combined forced and free convection boundary layer flows
subroutine eqmot(eta, x(), param(), f())
```
f(1) = x(3)*x(1) + param(1)*(1−x(2)*x(2) +
                            (2 − param(2))*param(3)*(1−x(5))
f(2) = x(1)
f(3) = x(2)
f(4) = −param(1)*x(3)*x(4) − (3*param(2) − 2)*x(2)*(1−x(5))
f(5) = x(4)
```
return

Sparrow, Eichhorn and Gregg [Spar59] obtained numerical solutions of these equations for the isothermal wall with $m = 1/2$, $\beta = 2/3$, with a Prandtl number of 0.7.[†] Their results for the nondimensional shearing stress and heat transfer parameters, i.e., $f''(0)$ and $\theta'(0)$ for an isothermal wall, are shown in Figure 5–11

[†]They also obtained numerical results for uniform heat transfer from the surface with $n = 1/5$, $m = 3/5$, $\beta = 3/4$.

Numerical Integration for Combined Forced and Free Convection 187

as functions of the $|\text{Gr}_x/\text{Re}_x^2|$. The corresponding numerical results are given in Table C–24 of Appendix C. Also shown in Figure 5–11 are the asymptotic values for aiding flows with pure forced and free convection. The corresponding equations are

$$\left.\begin{array}{l}-\theta'(0) = 0.4806 \\ f''(0) = 1.0389\end{array}\right\} \quad \text{for pure forced convection}$$

$$\left.\begin{array}{l}-\theta'(0) = 0.4079\left(\text{Gr}_x^{1/2}/\text{Re}_x\right)^{1/2} \\ f''(0) = 1.1085\left(\text{Gr}_x^{1/2}/\text{Re}_x\right)^{3/2}\end{array}\right\} \quad \text{for pure free convection}$$

The significant feature illustrated in Figure 5–11 is that the nondimensional heat transfer parameter for aiding combined flows is not significantly different from the appropriate asymptotic values for pure flow. This observation has important practical implications. However, the effect on the nondimensional shearing stress parameter is considerably larger. Turning our attention to opposing flows, we find that as the $\text{Gr}_x/\text{Re}_x^2$ parameter becomes increasingly negative the nondimensional shearing stress parameter approaches zero. Sparrow, Eichhorn and Gregg's numerical results show that $f'(0) = 0$ for $\text{Gr}_x/\text{Re}_x^2$ slightly less than -0.95. This behavior is similar to the pure forced convection boundary layer with an adverse pressure gradient. Figure 5–11 also shows that the nondimensional heat transfer parameter decreases as $\text{Gr}_x/\text{Re}_x^2$ becomes more negative.

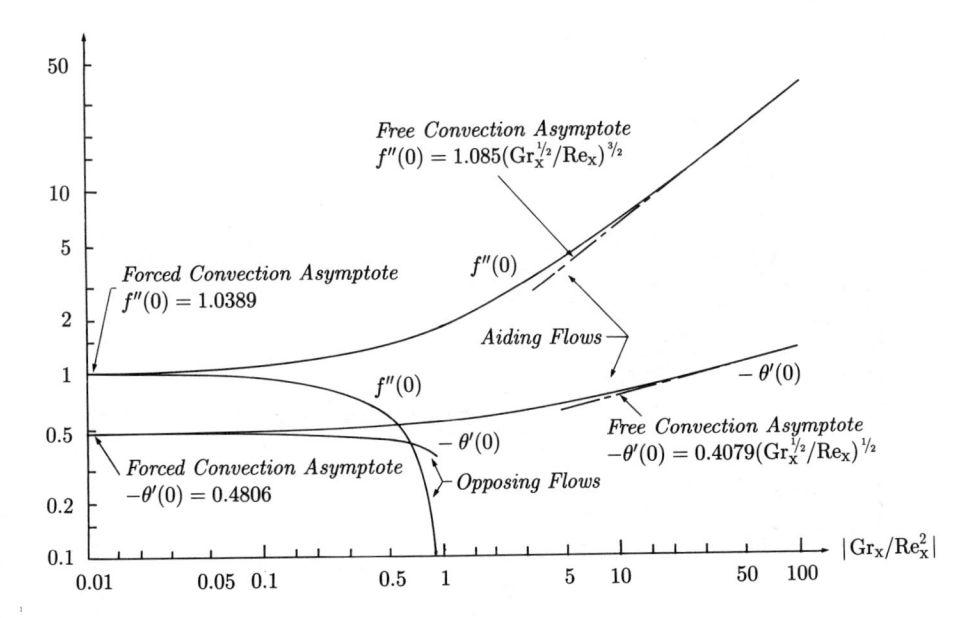

Figure 5–11. Nondimensional shearing stress and heat transfer parameters for combined forced and free convection with an isothermal wall, $\beta = 2/3$, $\text{Pr} = 0.7$.

Figures 5–12 and 5–13 show the nondimensional velocity and temperature profiles for an isothermal wall with a Prandtl number of 0.7 obtained by Sparrow, Eichhorn and Gregg [Spar59]. The nondimensional velocity profiles in Figure 5–12 clearly show how the buoyancy force modifies the forced convection velocity profile. For small positive values of the Gr_x/Re_x^2 parameter the velocity profiles are similar to those for forced convection (see Figure 3–4), while for large values they are similar to those for free convection (see Figure 5–3). For negative values of the Gr_x/Re_x^2 parameter the velocity profiles show the inflection point characteristic of profiles associated with adverse pressure gradients ($\beta < 0$) (see Figure 3–4). The temperature profiles in Figure 5–12 show the characteristic shape found in both forced and free convection boundary layer flows.

Using Figure 5–11 and the numerical results given in Appendix C, criteria for considering flows to be effectively pure are established. If a flow is considered

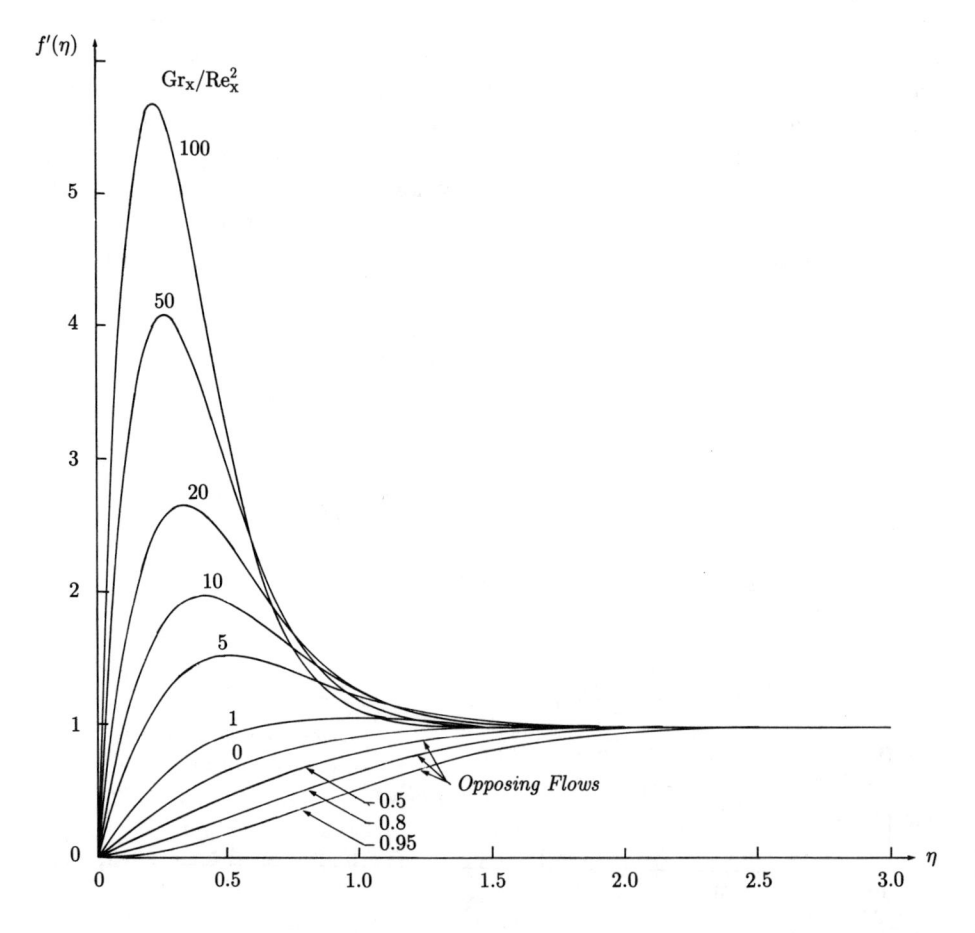

Figure 5–12. Nondimensional velocity profiles for combined forced and free convection flow with an isothermal wall, $\beta = 2/3$, Pr $= 0.7$.

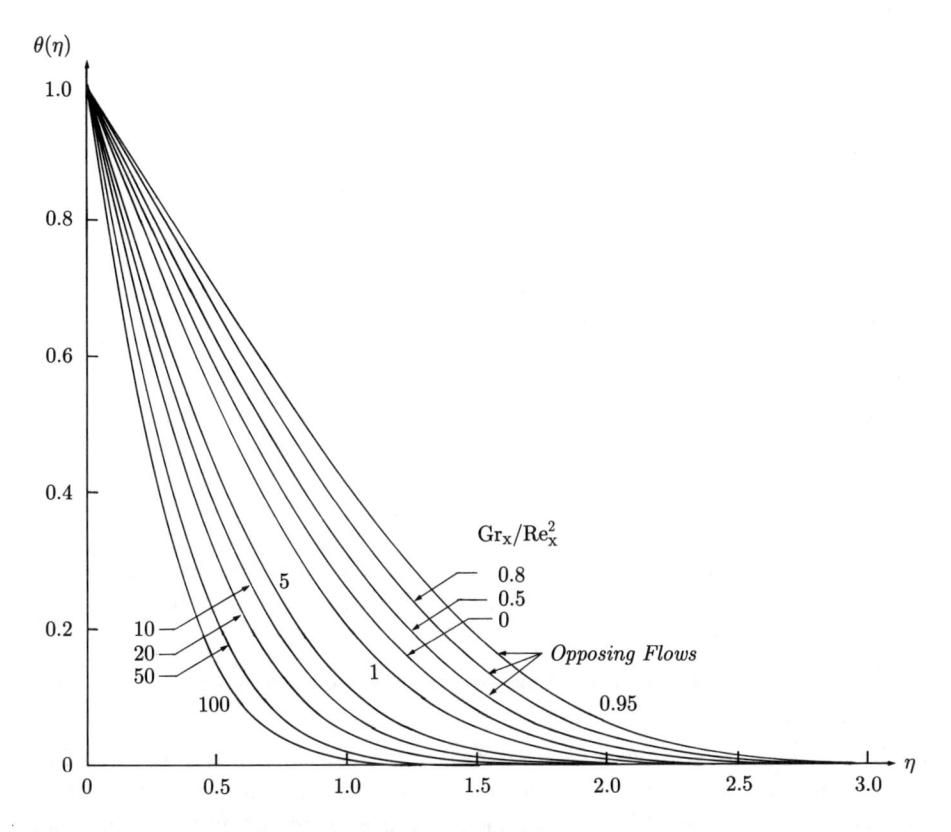

Figure 5–13. Nondimensional temperature profiles for combined forced free convection for an isothermal wall, $\beta = 2/3$, $\Pr = 0.7$.

pure when the nondimensional shearing stress and heat transfer parameters are within five percent of their pure values, then for an isothermal wall with a Prandtl number of 0.7

$0 < \text{Gr}_x/\text{Re}_x^2 < 0.06$ pure forced convection

$0.06 < \text{Gr}_x/\text{Re}_x^2 < 16$ combined forced and free convection

$16 < \text{Gr}_x/\text{Re}_x^2$ pure free convection

For opposing flows under the same conditions

$0 < \text{Gr}_x/\text{Re}_x^2 < 0.06$ pure forced convection

$0.06 < \text{Gr}_x/\text{Re}_x^2$ combined forced and free convection

These criteria are somewhat stringent when only the nondimensional heat transfer parameter is considered. If only $\theta'(0)$ is considered, 0.06 in the above relations is replaced by 0.3. Here a note of caution is in order: Since these criteria are developed for specific conditions, their application for flows which do not meet those conditions is suspect.

Finally, introduction of a local Nusselt number yields

$$\mathrm{Nu}_x = -\left(\frac{1}{2-\beta}\frac{Ux}{\nu}\right)^{1/2}\theta'(0)$$

or after introducing the Reynolds number

$$\frac{\mathrm{Nu}_x}{\sqrt{\mathrm{Re}_x}} = -\frac{\theta'(0)}{\sqrt{2-\beta}} \qquad (5-49)$$

Similarly, the Stanton number is

$$\sqrt{\mathrm{Re}_x}\,\mathrm{St}_x = -\frac{\theta'(0)}{\mathrm{Pr}\sqrt{2-\beta}} \qquad (5-50)$$

Finally, we note that $\theta'(0) = \theta'(0, \mathrm{Pr}, \mathrm{Gr}/\mathrm{Re}_x^2)$, and thus Nusselt and Stanton numbers are functions of the Prandtl, Reynolds and Grashof numbers.

5–11 Nonsimilar Combined Forced and Free Convection Boundary Layers

In the previous two sections we considered the similar combined forced and free convection boundary layer. There we noted that similar solutions with viscous dissipation are not possible for combined forced and free convection. In order to consider viscous dissipation as well as other nonsimilar effects, the locally nonsimilar method for combined forced and free convection is developed.

We begin the analysis by recalling the governing equations for steady two-dimensional incompressible, except for density changes in the body force term, constant property flow. The pressure gradient and buoyancy force terms are included. The governing equations are given by Eqs. (5–41) to (5–43). Recalling our previous analysis of the nonsimilar forced convection boundary layer, we introduce independent and dependent variable transformations given by

$$\xi = \int_0^x \frac{U(x)}{U_\infty}\,d\left(\frac{x}{\ell}\right) \qquad (3-102)$$

$$\eta = \frac{yU(x)}{(2\nu\ell U_\infty\xi)^{1/2}} \qquad (3-103)$$

and
$$\psi(x,y) = (2\nu\ell U_\infty\xi)^{1/2} f(\xi,\eta) \qquad (3-104)$$

$$\theta(\xi,\eta) = \frac{T-T_w}{T_e-T_w} \qquad (4-106)$$

where $T_w = T_w(x)$ and $T_e = T_e(x)$.

With these transformations the continuity equation is satisfied by the stream function, and the momentum and energy equations become

$$f''' + ff'' + \beta(1 - f'^2) = 2\xi\alpha_4(1 - \theta) + 2\xi(f'f'_\xi - f''f_\xi) \quad (5-51)$$

and
$$\theta'' + \Pr f\theta' = E_1\Pr f''^2 + 2\xi\Pr(f_\xi\theta' - \theta_\xi f')$$
$$+ 2\xi\Pr[\alpha_1(1-\theta) + \alpha_2\theta]f' \quad (4-118)$$

where the primes denote differentiation with respect to η and

$$\beta = \frac{2}{U^2}\frac{dU}{dx}\int_0^x U\,dx \quad (3-107)$$

$$E_1(x) = \frac{[U(x)]^2}{c_p[T_w(x) - T_e(x)]} \quad (4-121)$$

$$\alpha_1(\xi) = \frac{1}{T_w - T_e}\frac{\partial T_w}{\partial \xi} \quad (4-119)$$

$$\alpha_2(\xi) = \frac{1}{T_w - T_e}\frac{\partial T_e}{\partial \xi} \quad (4-120)$$

$$\alpha_4(\xi) = \frac{U_\infty}{U^3}\ell\bar{\beta}g_x(T_w - T_e) \quad (5-52)$$

Here note that the energy equation has the same form as that for the nonsimilar forced convection boundary layer. Further, the momentum equation, except for the term $2\xi\alpha_4(1-\theta)$, is also of the same form as that for the forced convection boundary layer. This additional term represents the contribution of the buoyancy force term. Also note that the viscous dissipation term, $E_1\Pr f''^2$, is included. The appropriate boundary conditions, including mass transfer at the surface, are given by Eqs. (3–106 b, c), (3–127) and (4–122 a, b).

In the previous similar analysis of combined forced and free convection, the term $2\xi\alpha_4(1-\theta)$ in the momentum equation, and the terms $2\xi\Pr(f_\xi\theta' - \theta_\xi f')$ and $E_1\Pr f''^2$, as well as the $2\xi f_\xi$ term in the boundary condition, were neglected. Further, β, α_1, α_2 and α_4 were required to have particular forms. The locally nonsimilar boundary layer method allows us to remove these restrictions.

Following the previous analyses (see Section 4–11 and 5–7) we introduce the auxiliary functions

$$S(\xi, \eta) = f_\xi \quad (3-108)$$

and
$$t(\xi, \eta) = \theta_\xi \quad (4-123)$$

The momentum equation, Eq. (5–51), is then written as

$$f''' + ff'' + \beta(1 - f'^2) = 2\xi\alpha_4(\xi)(1 - \theta) + 2\xi(f'S' - f''S) \quad (5-53)$$

Laminar Flow Analysis

Differentiating Eq. (5–53) with respect to ξ and consistent with the previous developments of the two-equation locally nonsimilar boundary layer model, neglecting the term $2\xi(f'S'_\xi - S_\xi f'')$ yields the first auxiliary momentum equation, i.e.

$$S''' + fS'' - 2\beta f'S' + f''S + \frac{d\beta}{d\xi}(1-f'^2) = 2(f'S' - f''S)$$
$$+ 2\xi(S'^2 - S''S) + 2\xi\alpha_4(1-\theta)\left(1 + \frac{1}{\alpha_4}\frac{d\alpha_4}{d\xi}\right) - 2\xi\alpha_4 t \qquad (5-54)$$

The energy equation and the first auxiliary energy equation are identical to those for the two-equation locally nonsimilar forced convection boundary layer model, i.e., Eqs. (4–124) and (4–125). For completeness they are repeated here

$$\theta'' + \Pr f\theta' = E_1 \Pr f''^2 + 2\xi\Pr(S\theta' - tf')$$
$$+ 2\xi\Pr[\alpha_1(1-\theta) + \alpha_2\theta]f' \qquad (4-124)$$

and $\quad t'' + \Pr(S\theta' + ft') = \Pr\left(\dfrac{dE_1}{d\xi}f''^2 + 2E_1 f''S''\right)$

$$+ 2\Pr\left[(S\theta' - tf') + \alpha_1(1-\theta)f' + \alpha_2\theta f'\right]$$
$$+ 2\xi\Pr\Big\{(St' - S't) + \alpha_1[(1-\theta)S' - tf']$$
$$+ \alpha_2(tf' + \theta S') + \frac{d\alpha_1}{d\xi}(1-\theta)f' + \frac{d\alpha_2}{d\xi}\theta f'\Big\}$$
$$(4-125)$$

The appropriate boundary conditions, including the effects of mass transfer at the surface, are given by Eqs. (3–106b, c), (3–127), (3–118), (3–132), (4–122a, b) and (4–127a, b). In particular, at the surface the boundary conditions are

$$f(\xi, 0) + 2\xi S(\xi, 0) = -\frac{v(x)}{U(x)}\left(\frac{2\ell U_\infty \xi}{\nu}\right)^{1/2} \qquad (3-127)$$

$$f'(\xi, 0) = 0 \qquad S'(\xi, 0) = 0 \qquad (3-118b, d)$$

$$S(\xi, 0) = -\frac{1}{3}(\mathrm{Re}_\ell)^{1/2}(2\xi)^{-1/2}\frac{v_0}{U_\infty} \qquad (3-132)$$

$$\theta(\xi, 0) = 0 \qquad t(\xi, 0) = 0 \qquad (4-122a, 4-127a)$$

and as $\eta \to \infty$

$$f'(\xi, \eta \to \infty) \to 1 \qquad S'(\xi, \eta \to \infty) \to 0 \qquad (3-118g, h)$$
$$\theta(\xi, \eta \to \infty) \to 1 \qquad t(\xi, \eta \to \infty) \to 0 \qquad (4-122b, 4-127b)$$

Examination of these equations and the associated boundary conditions shows that in contrast to the two-equation locally nonsimilar model for the forced convection boundary layer the momentum, first auxiliary momentum, the energy and the first auxiliary energy equations are coupled through the buoyancy terms. Further, we see that there are four simultaneous asymptotic boundary conditions to be satisfied. This is similar to the results for the nonsimilar free convection boundary layer investigated in Sections 5–7 and 5–8. Numerical integration of this system of equations is accomplished by suitable modifications and additions to the `freeb` program. The Nachtsheim-Swigert iteration technique for four asymptotic boundary conditions is described in Appendix B.

The three-equation locally nonsimilar boundary layer model for combined forced and free convection is obtained by introducing the additional auxiliary functions, $h(\xi, \eta) = S_\xi = f_{\xi\xi}$ and $j(\xi, \eta) = t_\xi = \theta_{\xi\xi}$. The resulting system of equations is of fifteenth-order, with six asymptotic boundary conditions to be satisfied at infinity.

In this and the previous chapters we extensively investigated similar solutions of the momentum and energy equations when the fluid is essentially incompressible. In the next chapter we discuss solutions of the governing equations when the compressibility of the fluid must be considered.

SIX

THE COMPRESSIBLE BOUNDARY LAYER

6-1 Introduction

In Chapter 5 we first considered the effects of the compressibility of the fluid by investigating the free convection boundary layer. There we considered the effects of compressibility in an approximate manner. In particular, the compressibility of the fluid (i.e., the density change) was considered in deriving the buoyancy force acting on a fluid particle. The effects of compressibility on the rest of the velocity and temperature field were neglected. Further, we considered the fluid properties to be constant. Here we consider the effects of compressibility on the entire velocity and temperature field.

We begin our investigation with a discussion of the effects of viscous heating on the transport properties of the fluid. This is followed by a discussion of an exact solution of the compressible Navier-Stokes equations, i.e., compressible Couette flow. This discussion serves to introduce several physical differences between compressible and incompressible flows. Finally, we consider the compressible boundary layer. Here, again, we concentrate on the similar solutions, since they serve to illustrate the fundamental aspects of the flow.

6-2 Variation of Transport Properties

The study of compressible fluid flows generally concentrates on gas flows. In particular, the study of aerodynamics generates considerable interest in 'high speed' gas flows, principally air. Because we are interested in high speed gas flows, the Eckert number is sufficiently large that viscous heating within the fluid results in significant temperature changes. Thus, the fluid transport properties are variable. The transport properties of importance in a viscous compressible flow are the viscosity, the thermal conductivity, the specific heat at constant pressure and the combination of these variables called the Prandtl number.

Theoretical calculations for monatomic gases show that the viscosity of gases depends only on the temperature and is independent of the pressure. Experimental measurements confirm that this result is essentially correct for all gases (see Figures 6–1 and 6–2). For gases the viscosity increases with increasing temperature. In contrast, the viscosity of liquids depends on both temperature and pressure and decreases with increasing temperature.

Experimental measurements of the viscosity of air are well correlated for a wide range of temperature by a two-parameter equation called the Sutherland equation, i.e.

$$\frac{\mu}{\mu_r} = \frac{T_r + S}{T + S}\left(\frac{T}{T_r}\right)^{3/2} \qquad (6-1)$$

For air between 180 °R and 3400 °R, $S = 198.6$ °R, $T_r = 491.6$ °R and $\mu_r = 3.58 \times 10^{-7}$ lb sec/ft^2. Thus, for air

$$\mu = 2.270\,\frac{T^{3/2}}{T + 198.6} \times 10^{-8} \quad \frac{\text{lb sec}}{\text{ft}^2}$$

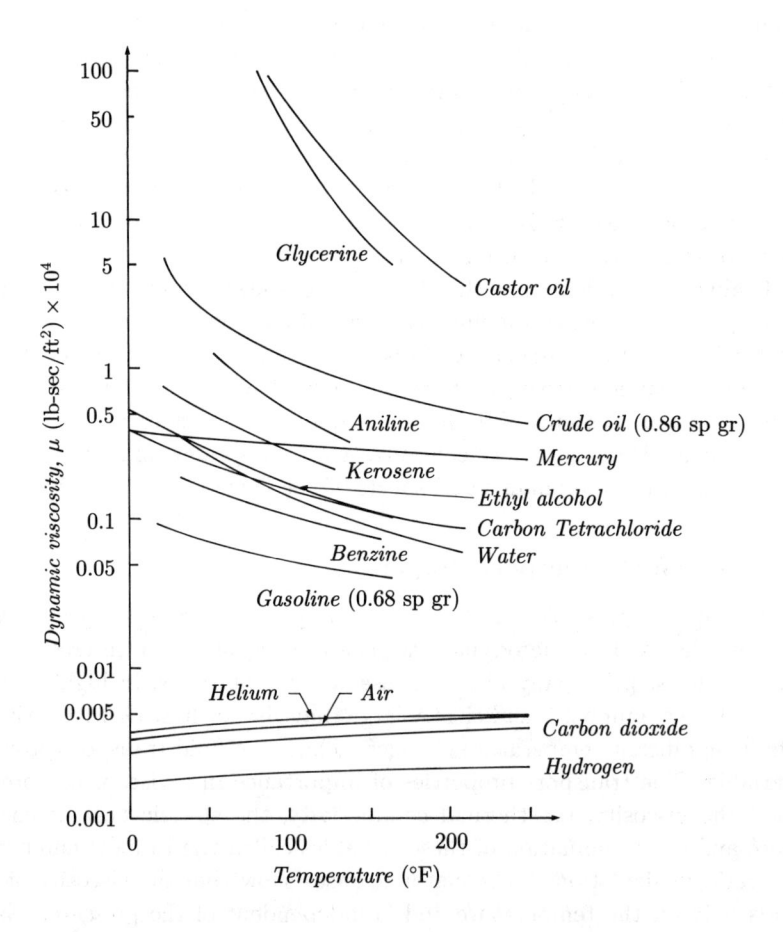

Figure 6–1. Absolute viscosity, μ, of certain gases and liquids.

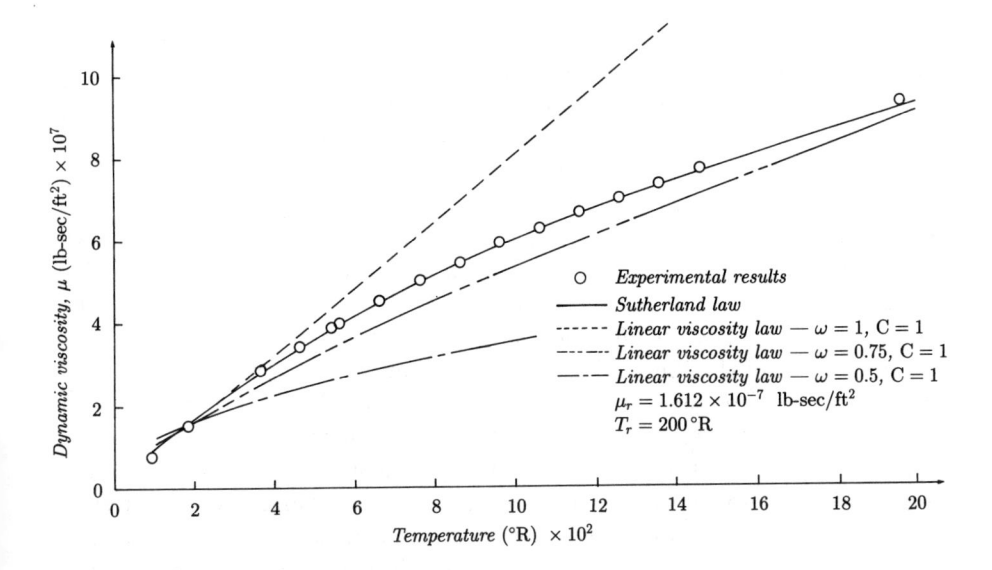

Figure 6-2. Power law viscosity relationship.

Because Sutherland's equation is quite complicated it is not generally used in analytical investigations. For analytical investigations, approximations based on the empirical equation

$$\ln \frac{\mu}{\mu_r} = A \ln \frac{T}{T_r} + B \qquad (6-2)$$

are generally used. When $A = \omega$ and $B = \ln C$, Eq. (6-2) becomes

$$\frac{\mu}{\mu_r} = C \left(\frac{T}{T_r} \right)^\omega \qquad (6-3)$$

Since we have two free parameters, C and ω, it is possible to precisely match the viscosity at two conditions, e.g., the free stream and the surface, or the stagnation condition and the surface. This equation is frequently referred to as the Chapman-Rubesin viscosity law (see [Chap39]). Equation (6-3) is shown in Figure 6-2 for various constant values of C and ω, along with the Sutherland equation and experimental data for air (see [Stre58; Weas87; Chap49]). A simple and useful case of the Chapman-Rubesin viscosity law [Chap49] occurs when $C = 1$ and $\omega = 1$. With these values, and using the surface as the reference condition, we have

$$\mu = \mu_w(x) \frac{T}{T_w(x)}$$

for the variation of viscosity in the fluid. For an isothermal wall this reduces to $\mu =$ (constant) T. This particularly convenient form of the viscosity law generally yields acceptable results for analytical investigations.

The thermal conductivity of gases, k, is also essentially a function of temperature and is independent of pressure. Figure 6–3 shows that the variation of the thermal conductivity of air with temperature is fundamentally the same as that of the dynamic viscosity. Figure 6–3 further shows that the specific heat at constant pressure for air is almost constant for a wide range of temperatures. The behavior of these three transport properties with temperature combine to make the Prandtl number, $\text{Pr} = \mu c_p/R$, essentially invariant with temperature (see Figure 6–3). Thus, it is common practice to assume that the Prandtl number for gases is constant. As we shall subsequently see, this assumption eliminates the need to formally specify the functional variation of the coefficient of thermal conductivity, k, and of the specific heat at constant pressure, c_p, with temperature. Further, we shall subsequently see that considerable mathematical simplification occurs if we choose a unit Prandtl number and a Chapman-Rubesin viscosity law with $C = \omega = 1$. Although the transport properties of air and most gases do not precisely conform to these assumptions, the relative simplicity of the resulting governing boundary value problems more than justifies their use. Further, the results of these simplified analyses serve to illustrate the fundamental ideas and concepts of compressible viscous flow, as well as providing analytical results which, in many cases, are in agreement with experimental observations.

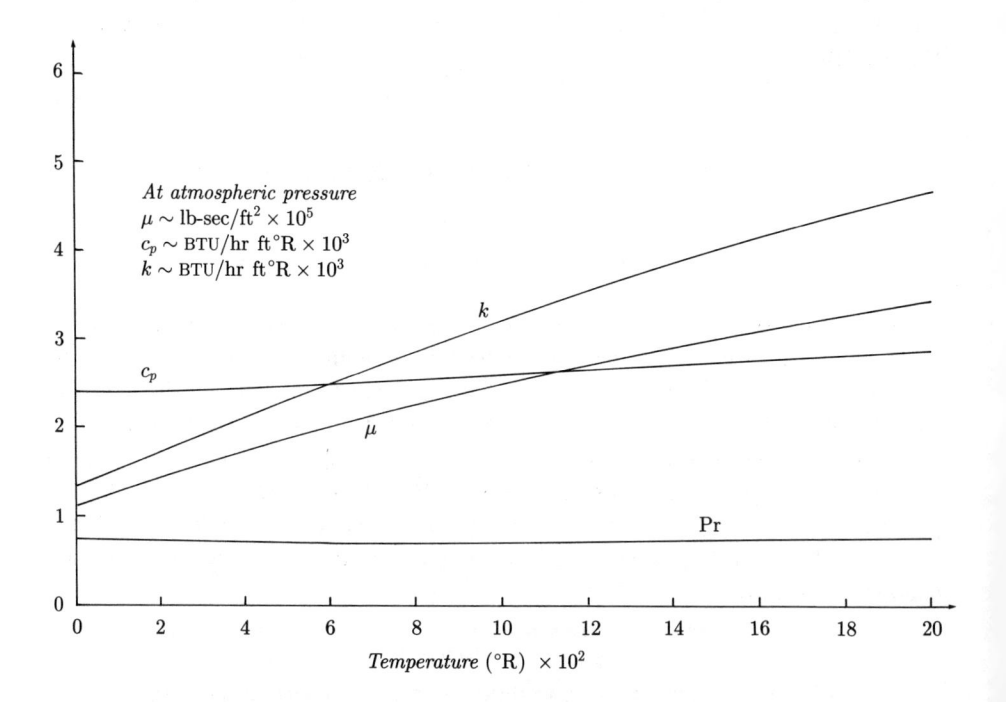

Figure 6–3. Variation of k, c_p and the Prandtl number with temperature.

6–3 Analytical Solutions of the Compressible Navier-Stokes Equations—Couette Flow

Analytical solutions of the compressible Navier-Stokes equations are much more difficult than those for the corresponding incompressible equations. Fundamentally this is because the continuity, momentum and energy equations, along with the equation of state of the fluid, must be solved simultaneously. All of the known analytical solutions of the compressible Navier-Stokes equations are for steady flow, which is a function of only one spatial coordinate. Some of these are: compressible two-dimensional (plane) Couette flow (see Illingworth [Illi50]); compressible axisymmetric (circular) Couette flow ([Illi50]); compressible flow past a semi-infinite porous flat plate parallel to the flow direction, with steady constant suction at the plate surface [Illi50]; circular compressible flow around a circular cylinder with suction at the surface [Illi50]; and steady one-dimensional compressible shock wave structure (see Howarth [Howa53]). Further, Illingworth showed that, for compressible flow, analytic solutions are not possible for the suddenly accelerated plane wall (see Section 2–5) nor for steady flow between parallel plates (see Section 2–3). Here we present the solution for two-dimensional compressible Couette flow in detail.

Consider steady, two-dimensional compressible viscous flow between two horizontal parallel plates induced by the motion of one plate moving with a constant velocity, U, with respect to the other. The distance between the two plates is L (see Figure 6–4). Since, as in the incompressible case, the conditions in the flow depend only on y and not on x, the continuity equation (Eq. 1-1) reduces to

$$\frac{d}{dy}(\rho v) = 0 \qquad (6-4)$$

The x and y momentum equations (Eqs. 1–33a,b) without body forces thus reduce to

$$v\frac{du}{dy} - \frac{d}{dy}\left(\mu \frac{du}{dy}\right) = 0$$

and

$$v\frac{dv}{dy} + \frac{\partial P}{\partial y} = 0$$

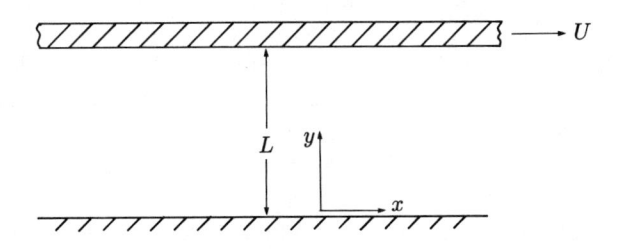

Figure 6–4. Compressible Couette flow.

respectively. Finally, the energy equation (Eq. 1–41) without bulk heat transfer reduces to

$$\frac{1}{\Pr}\frac{d}{dy}\left(\mu\frac{dh}{dy}\right) + \mu\left[\left(\frac{du}{dy}\right)^2 + \frac{4}{3}\left(\frac{dv}{dy}\right)^2\right] = 0$$

The appropriate boundary conditions are given by the no-slip, no-mass-transfer and no-temperature-jump conditions at both the stationary and moving surfaces. The boundary conditions are thus

$$u(0) = 0 \quad v(0) = 0 \quad h(0) = h_w \qquad (6-5a,b,c)$$

$$u(L) = U \quad v(L) = 0 \quad h(L) = h_e \quad P(L) = P_e \quad (6-5d,e,f,g)$$

where the subscripts w and e are used to indicate the stationary and moving surfaces, respectively. Notice that the energy equation is written in terms of the enthalpy rather than the temperature. In addition, we assume that the boundary conditions, h_w, h_e and P_e, are constant.

Integrating the continuity equation yields

$$\rho v = \text{constant}$$

Since the density is not zero, application of the boundary condition (Eq. 6–5b) shows that $v = 0$ everywhere.

The x and y momentum equations are now

$$\frac{d}{dy}\left(\mu\frac{du}{dy}\right) = 0 \qquad (6-6)$$

and

$$\frac{\partial P}{\partial y} = 0 \qquad (6-7)$$

The energy equation is

$$\frac{1}{\Pr}\frac{d}{dy}\left(\mu\frac{dh}{dy}\right) + \mu\left(\frac{du}{dy}\right)^2 = 0 \qquad (6-8)$$

Before proceeding further, note that Eq. (6–6) can immediately be integrated to yield $\tau = \mu(du/dy) = $ constant, where τ is the shearing stress. This result shows that the shearing stress in compressible simple Couette flow is constant throughout the flow field, the same as in incompressible constant property simple Couette flow. However, here the temperature field, through the dependence of viscosity on temperature, and the velocity field interact in a precise manner to maintain constant shearing stress.

Integrating the y momentum equation and using the pressure boundary condition (Eq. 6–5g) yields

$$P = \text{constant} = P_e$$

In considering the remaining two equations it is convenient to introduce nondimensional variables

$$f = \frac{u}{U} \qquad g = \frac{h}{h_e} \qquad \bar{\mu} = \frac{\mu}{\mu_e} \qquad \eta = \frac{y}{L}$$

Equations (6–6) and (6–8) then become

$$(\bar{\mu} f')' = 0 \qquad (6-9)$$

and

$$(\bar{\mu} g')' + (\gamma - 1) \, \mathrm{M}_e^2 \, \mathrm{Pr} \, \bar{\mu} f'^2 = 0 \qquad (6-10)$$

where in deriving Eq. (6–10) we used the fact that $U^2/h_e = (\gamma - 1) \, \mathrm{M}_e^2$. The nondimensional boundary conditions are

$$f(0) = 0 \qquad g(0) = \frac{h_w}{h_e} = g_w \qquad (6-11a, b)$$

$$f(1) = 1 \qquad g(1) = 1 \qquad (6-11c, d)$$

Note that in contrast to our previous studies neither the transformations, the resulting governing equations nor the boundary conditions depend on the Reynolds number. Thus, the solutions for the velocity and enthalpy field depend on the Mach number, the viscosity law and g_w but not on the Reynolds number. Equation (6–9) can immediately be integrated to give $\bar{\mu} f' = \mathrm{constant}$ which, after using the boundary conditions, becomes

$$\bar{\mu} f' = f'(1) = \bar{\mu}(0) f'(0) \qquad (6-12)$$

Noting that $(\bar{\mu} f')' = 0$ and thus $(\bar{\mu} f f')' = \bar{\mu} f'^2$ we rewrite Eq. (6–10) as

$$\{\bar{\mu}[g' + (\gamma - 1) \, \mathrm{M}_e^2 \, \mathrm{Pr} \, \bar{\mu} f f']\}' = 0$$

Integrating once and using the boundary conditions yields

$$\bar{\mu}[g' + (\gamma - 1) \, \mathrm{M}_e^2 \, \mathrm{Pr} \, \bar{\mu} f f'] = \bar{\mu}(0) g'(0) = g'(1) + (\gamma - 1) \, \mathrm{M}_e^2 \, \mathrm{Pr} \, f'(1) \qquad (6-13)$$

Further, integration of the energy equation is achieved by dividing by Eq. (6–12) and subsequently multiplying by f' to yield

$$g' + (\gamma - 1) \, \mathrm{M}_e^2 \, \mathrm{Pr} \, f f' = \frac{g'(0)}{f'(0)} f'$$

Since $g'(0)$ and $f'(0)$ are constants (as yet unknown), integration yields

$$g + \frac{\gamma - 1}{2} \, \mathrm{M}_e^2 \, \mathrm{Pr} \, f^2 - \frac{g'(0)}{f'(0)} f = \mathrm{constant} \qquad (6-14)$$

The constant of integration is evaluated from the boundary conditions, i.e.

$$1 + \frac{\gamma - 1}{2} \, \mathrm{M}_e^2 \, \mathrm{Pr} - \frac{g'(0)}{f'(0)} = g(0) = g_w = \mathrm{constant}$$

202   Laminar Flow Analysis

Equation (6–14) then becomes

$$g + \frac{\gamma-1}{2} M_e^2 \Pr (f^2 - 1) - \frac{g'(0)}{f'(0)}(f-1) = 1 \qquad (6-15)$$

Integration of the momentum equation, Eq. (6–12), requires a knowledge of the viscosity law, i.e., $\mu = \mu(h)$. The integrated energy equation (Eq. 6–15) yields the enthalpy, h, as a function of the velocity, u. Thus, we consider the viscosity as a function of the velocity, i.e., $\mu = \mu(u)$, or in nondimensional variables $\bar{\mu} = \bar{\mu}(f)$. With the help of the boundary conditions, formally integrating Eq. (6–12) yields

$$\int_0^f \bar{\mu}\, df = f'(1)\eta \qquad (6-16)$$

The constant $f'(1)$ is determined by carrying out the integration across the flow field from the stationary plate, where $f(0) = 0$, to the moving plate, where $f(1) = 1$

$$\int_0^1 \bar{\mu}\, df = f'(1) \qquad (6-17)$$

Providing that $\bar{\mu} = \bar{\mu}(f)$ is known, Eqs. (6–12), (6–13), (6–15) and (6–16) represent four equations in the four unknown constants $f'(1)$, $g'(0)$, $f'(0)$ and $g'(1)$. In principle the problem is solved.

As a particular example, we assume that the viscosity law is given by

$$\bar{\mu} = g \qquad (6-18)$$

This corresponds to a Chapman-Rubesin viscosity law with $C = 1$ and the additional assumption that the specific heat at constant pressure, c_p, is constant. Although, as previously mentioned, no real gas exactly satisfies these conditions, the results serve to illustrate the fundamental characteristics of compressible simple Couette flow.

Substituting Eqs. (6–18) and (6–15) into Eq. (6–17) and integrating yields an explicit equation for

$$f'(1) = 1 + \frac{\gamma-1}{2} M_e^2 \Pr - \frac{g'(0)}{2f'(0)} \qquad (6-19)$$

Assuming that $\bar{\mu}(0) = g(0) = g_w$ is a known quantity from the boundary conditions, Eqs. (6–12), (6–13), (6–15) and (6–19) represent four explicit equations in the unknowns $f'(1)$, $g'(0)$, $f'(0)$ and $g'(1)$. Solution yields

$$f'(1) = \frac{1+g_w}{2} + \frac{1}{6}\left(\frac{\gamma-1}{2} M_e^2 \Pr\right) \qquad (6-20)$$

$$f'(0) = \frac{1+g_w}{2g_w} + \frac{1}{6}\left(\frac{\gamma-1}{2} M_e^2 \Pr\right) \qquad (6-21)$$

Compressible Couette Flow

$$g'(0) = \left((1+g_w) + \frac{\gamma-1}{2} M_e^2 \Pr\right)\left(\frac{1+g_w}{2g_w} + \frac{1}{6g_w}\frac{\gamma-1}{2} M_e^2 \Pr\right) \quad (6-22)$$

$$g'(1) = (1+g_w) - \frac{\gamma-1}{2} M_e^2 \Pr \quad (6-23)$$

From Eqs. (6–15), (6–16) and (6–18) to (6–22), the nondimensional velocity and enthalpy profiles are

$$\frac{\gamma-1}{2}\frac{M_e^2 \Pr}{3} \frac{f^3}{9} - \frac{1}{2}\left[(1-g_w) + \frac{\gamma-1}{2} M_e^2 \Pr\right] f^2$$

$$- g_w f + \frac{1}{2}\left[(1+g_w) + \frac{\gamma-1}{2}\frac{M_e^2 \Pr}{3}\right]\eta = 0 \quad (6-24)$$

and
$$g = g_w + \left(1 + \frac{\gamma-1}{2} M_e^2 \Pr - g_w\right) f - \frac{\gamma-1}{2} M_e^2 \Pr f^2 \quad (6-25)$$

The second term on the right hand side of Eq. (6–25) is the contribution of the heat transfer at the surface to the enthalpy of the fluid as it diffuses outward from the surface. The third term is the contribution from the internal heating or viscous dissipation of the fluid. These profiles are shown in Figure 6–5b for $[(\gamma-1)/2] M_e^2 \Pr = 5$ for various values of g_w.

When there is no heat transfer at the surface, $g'(0) = 0$. Equation (6–15) then gives the nondimensional adiabatic wall enthalpy at the stationary surface, i.e.

$$g_{aw} = 1 + \frac{\gamma-1}{2} M_e^2 \Pr \quad (6-26)$$

g_{aw} is sometimes called the recovery enthalpy. For constant specific heat at constant pressure, c_p, this is equivalent to a nondimensional recovery temperature. Here note that the adiabatic wall enthalpy is independent of the particular viscosity law. Further, g_{aw} is proportional to the Prandtl number and to the square of the Mach number. For large Mach numbers, a significant increase in the recovery enthalpy results. Since $[(\gamma-1)/2] M_e^2$ is related to the Eckert number based on a reference temperature T_1 (see Section 4–3), this is not surprising. This large increase in recovery enthalpy (temperature) for large Mach numbers is a result of the increased importance of viscous heating within the compressible fluid.

The velocity and enthalpy profiles for an adiabatic wall are

$$\frac{1}{3}\frac{\gamma-1}{2} M_e^2 \Pr f^3 - \left(1 + \frac{\gamma-1}{2} M_e^2 \Pr\right) f$$

$$+ \left(1 + \frac{2}{3}\frac{\gamma-1}{2} M_e^2 \Pr\right)\eta = 0 \quad (6-27)$$

and
$$g = 1 + \frac{\gamma-1}{2} M_e^2 \Pr (1 - f^2) \quad (6-28)$$

204 Laminar Flow Analysis

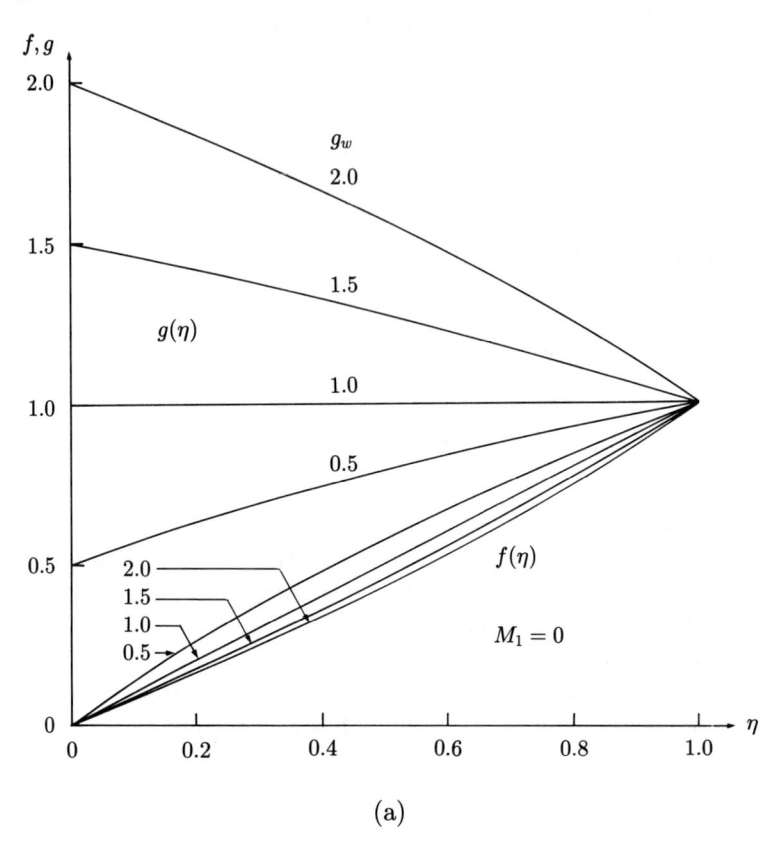

(a)

Figure 6–5. Velocity and enthalpy profiles for compressible Couette flow. (a) $M_1 = 0$; (b) $M_1 \geq 0$, $\bar{\mu} = g$, $[(\gamma - 1)/2] M_1^2 \Pr = 5$; (c) $M_1 \to \infty$, $\bar{\mu} = g$.

To further illustrate these results consider two special cases, $M_e \to 0$ and $M_e \to \infty$.

When $M_e \to 0$ the velocity and enthalpy profiles are given by

$$f = \frac{-g_w}{1 - g_w} + \frac{1}{1 - g_w}\left[\eta + g_w^2(1 - \eta)\right]^{1/2} \qquad g_w \neq 1 \qquad (6-29)$$

and
$$g = 1 + g_w(1 - f) \qquad (6-30)$$

Here we see that when $M_e \to 0$ viscous dissipation makes no contribution to the enthalpy of the fluid, i.e., no heat is added by viscous dissipation. This is the case of low speed compressible heat transfer.

For $M_e \to 0$, the adiabatic wall enthalpy reduces to $g_{aw} = 1$. For constant specific heats we write $T_r = T_w = T_e$, where T_r is the recovery temperature. g_w is then unity, and Eq. (6–30) shows that the temperature, and thus the viscosity,

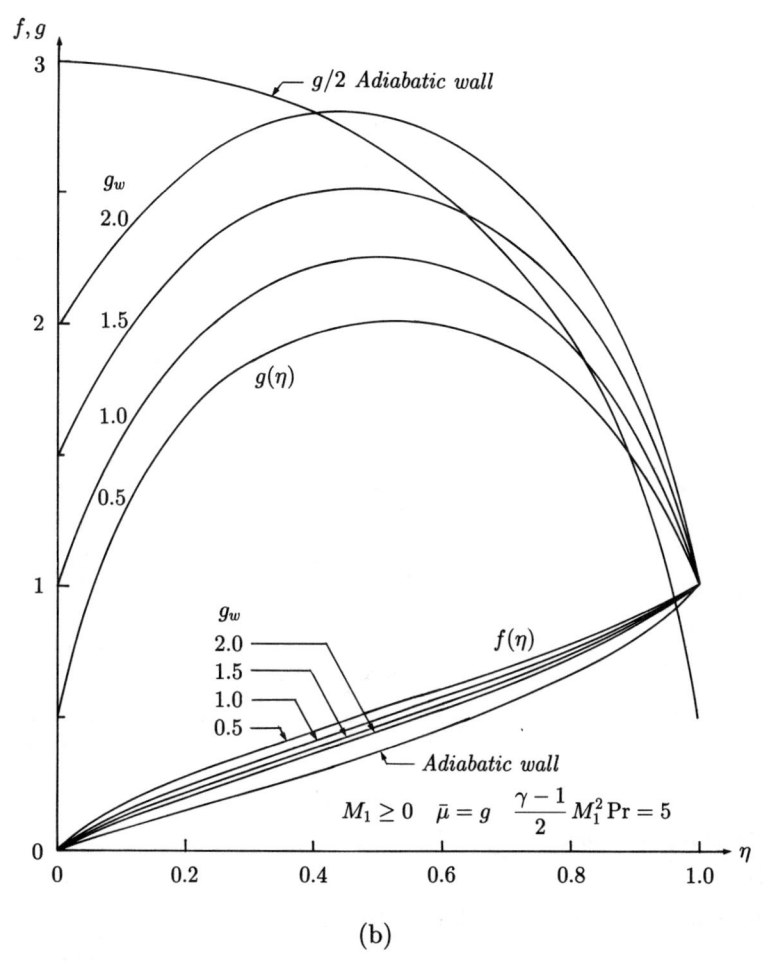

Figure 6–5. (cont.)

is constant throughout the flow field. In addition, Eq. (6–27) yields $f = \eta$, a linear velocity relation. Thus, when viscous dissipation and heat transfer from the surface by conduction are negligible, the flow can be considered as incompressible. Velocity and enthalpy profiles for $M_e \to 0$ are shown in Figure 6–5a for various values of g_w. Figure 6–5a shows that heat transfer by conduction at the surface significantly affects the enthalpy profile but has relatively little effect on the velocity profile.

When $M_e \to \infty$, i.e., in the hypersonic limit, with heat transfer at the surfaces it is convenient to introduce the variable

$$g^* = \frac{g}{\frac{\gamma-1}{2} M_e^2 \Pr}$$

206 Laminar Flow Analysis

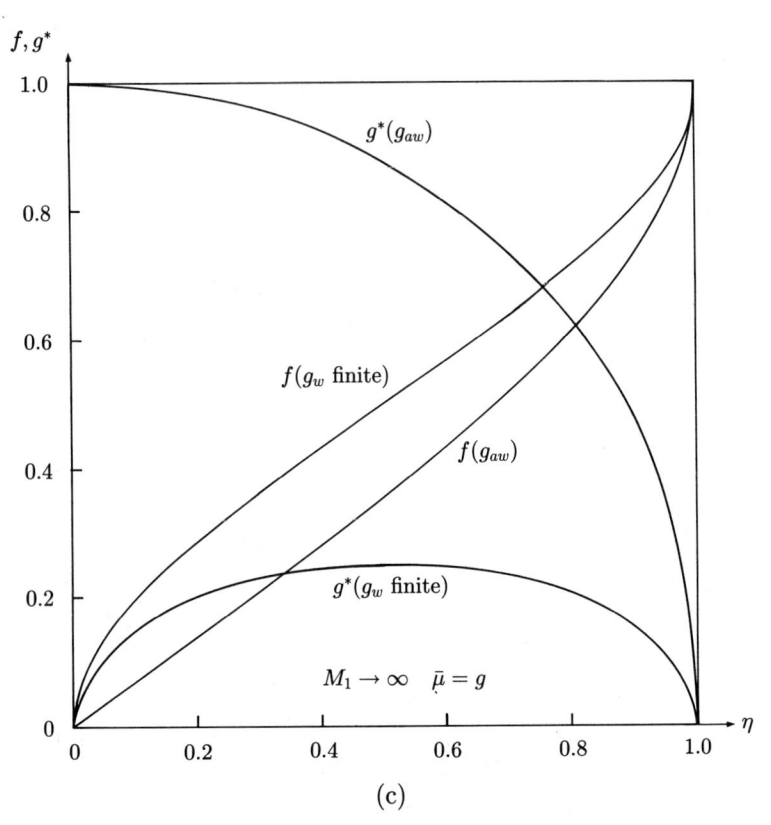

Figure 6–5. (cont.)

The velocity and enthalpy profiles are then given by

$$2f^3 - 3f^2 + \eta = 0 \tag{6-31}$$

and
$$g^* = f - f^2 \tag{6-32}$$

In deriving these results terms of $O(1/M_e^2)$ were neglected and g_w^* considered negligible for $g_w = O(1)$. These results show that, in the hypersonic limit, heat transfer to the fluid by conduction at the surface is negligible with respect to the internal heat generated by viscous dissipation within the fluid.

For an adiabatic wall, in the hypersonic limit Eq. (6–26) yields $g_{aw}^* = 1$, with the adiabatic wall velocity and enthalpy profiles given by

$$f^3 - 3f + 2\eta = 0 \tag{6-33}$$

and
$$g^* = 1 - f^2 \tag{6-34}$$

Velocity and enthalpy profiles with and without heat transfer are shown in Figure 6–5c.

Returning to the general case for $M_e > 0$, we see that the shearing stress at the lower surface, written in terms of nondimensional variables, is

$$C_f = \frac{\tau}{\dfrac{\rho_e U^2}{2}} = \frac{1}{\text{Re}_L} g f'(0)$$

where C_f is the skin friction coefficient. Substituting Eqs. (6–21) and (6–25) yields

$$C_f = \frac{1}{\text{Re}_L}\left[(1+g_w) + \frac{g_w}{3}\left(\frac{\gamma-1}{2} M_e^2 \Pr\right)\right] \quad (6-35)$$

Here we see that the skin friction coefficient at the stationary surface increases with increasing Mach number of the upper surface, increasing heat transfer to the fluid by conduction at the lower surface and increasing Prandtl number of the fluid. The skin friction coefficient decreases with decreasing Reynolds number of the upper surface.

The heat transfer rate at this surface is given by Fourier's law, i.e., $q = k(dT/dy)$. In the present nondimensional variables this becomes

$$q = \frac{k}{\mu_r c_p} \frac{h_e \mu_e}{L} g' = \frac{1}{\Pr} \frac{h_e}{L} \mu_e g'$$

Introducing a Nusselt number based on the temperature of the moving surface, T_e, and the distance between the plates, we have

$$\text{Nu}_e = \frac{L \dfrac{dT}{dy}}{T_e} = g'$$

For the stationary surface, substitution of Eq. (6–22) yields

$$\text{Nu}_e = \frac{(1+g_w)^2}{2g_w} + \frac{2}{3}\left(\frac{1+g_w}{g_w}\right)\left(\frac{\gamma-1}{2} M_e^2 \Pr\right) + \left(\frac{\gamma-1}{2} M_e^2 \Pr\right)^2 \quad (6-36)$$

while for the moving surface Eq. (6–23) yields

$$\text{Nu}_e = (1+g_w) - \frac{\gamma-1}{2} M_e^2 \Pr \quad (6-37)$$

From these results we see that the heat transfer by conduction at the surface is a function of the Mach number, wall enthalpy ratio and Prandtl number.

Introducing a Stanton number based on conditions at the moving surface yields

$$\text{St} = \frac{k \dfrac{dT}{dy}}{\rho_e c_p U T_e} = \frac{1}{\Pr \text{Re}_L} g' = \frac{\text{Nu}_e}{\Pr \text{Re}_L} = \frac{C_f}{2\Pr} \quad (6-38)$$

This result shows that Reynolds analogy (see Section 4–6) is applicable for simple compressible Couette flow.

In simple compressible Couette flow the flow field is bounded, hence the shear layer has fixed dimensions. Because of this fixed dimension, the effects of the variable viscosity of the fluid are relatively simple: The skin friction increases with increasing Mach number, increasing heat transfer and increasing Prandtl number. As we shall subsequently see, the effects of variable properties in compressible boundary layer theory are not so simple. In boundary layer flow the variations of density and viscosity with temperature result in two opposing effects. The increase in temperature results in an increase in the viscosity, μ, and hence in the shearing stress at the surface. Conversely, since the pressure is constant across the boundary layer the increase in temperature results in a decrease in density. This increases the boundary layer thickness and thus decreases the shearing stress at the surface.

6–4 Compressible Boundary Layer

We now turn our attention to laminar compressible boundary layer flows. Here, as in previous discussions, we concentrate on the similar solutions. These solutions serve to illustrate the fundamental properties and problems of compressible boundary layers.

The governing equations for a laminar compressible boundary layer were derived in Chapter 3. For a steady two-dimensional compressible boundary layer without body forces or bulk heat transfer, Eqs. (3–16), (3–22) and (3–23) become

continuity
$$\frac{\partial}{\partial x}(\rho u) + \frac{\partial}{\partial y}(\rho v) = 0 \tag{6-39}$$

x momentum
$$u\frac{\partial u}{\partial x} + v\frac{\partial u}{\partial y} = -\frac{1}{\rho}\frac{\partial P}{\partial x} + \frac{1}{\rho}\frac{\partial}{\partial y}\left(\mu\frac{\partial u}{\partial y}\right) \tag{6-40}$$

energy
$$u\frac{\partial h}{\partial x} + v\frac{\partial h}{\partial y} = \frac{u}{\rho}\frac{\partial P}{\partial x} + \frac{1}{\Pr}\frac{1}{\rho}\frac{\partial}{\partial y}\left(\mu\frac{\partial h}{\partial y}\right) + \nu\left(\frac{\partial u}{\partial y}\right)^2 \tag{6-41}$$

The appropriate coordinate system is shown in Figure 6–6.

Comparing these equations to those used in Chapter 4 (see Eqs. 4–55 to 4–57), we see that here we retain the compressive work term $(u/\rho)(\partial P/\partial x)$ in the energy equation. The second term on the right hand side of the energy equation represents the diffusion or spreading of heat transferred to the fluid or generated within the fluid. The third term represents heat generated due to viscous stresses within the fluid, i.e., viscous dissipation.

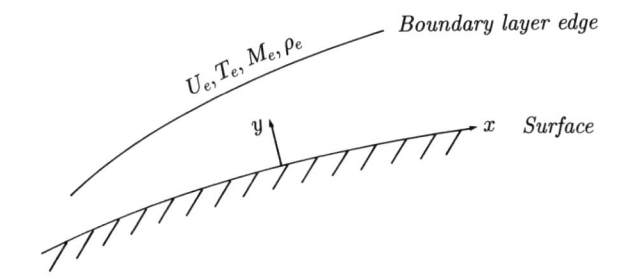

Figure 6-6. Boundary layer coordinate system.

The appropriate boundary conditions at the surface are given by the no-velocity-slip condition with or without mass transfer or heat transfer, i.e.

$$u(0) = 0 \quad v(0) = v(x) \quad h(0) = h(x) \quad \text{or} \quad \frac{\partial h}{\partial y} = 0 \qquad (6-42a,b,c,d)$$

The boundary conditions at the edge of the boundary layer require that the viscous flow interior to the boundary layer smoothly transition into the inviscid flow outside the boundary layer, i.e.

$$u(y \to \infty) \to U_e(x) \quad h(y \to \infty) \to h_e(x) \qquad (6-42e,f)$$

where the subscript e is used for conditions at the edge of the boundary layer.

Introducing nondimensional variables

$$\bar{u} = \frac{u}{U_e} \quad \bar{v} = \frac{v}{U_e} \quad \bar{y} = \frac{y}{L} \quad \bar{x} = \frac{x}{L}$$

$$\bar{h} = \frac{h}{h_e} \quad \bar{\mu} = \frac{\mu}{\mu_e} \quad \bar{P} = \frac{P}{\rho_e U_e^2} \quad \bar{\rho} = \frac{\rho}{\rho_e}$$

Equations (6-39) to (6-41) then become

$$\frac{\partial(\bar{\rho}\bar{u})}{\partial \bar{x}} + \frac{\partial(\bar{\rho}\bar{v})}{\partial \bar{y}} = 0$$

$$\bar{u}\frac{\partial \bar{u}}{\partial \bar{x}} + \bar{v}\frac{\partial \bar{u}}{\partial \bar{y}} = -\frac{1}{\bar{\rho}}\frac{\partial \bar{P}}{\partial \bar{x}} + \frac{1}{\text{Re}}\frac{\partial}{\partial \bar{y}}\left(\bar{\mu}\frac{\partial \bar{u}}{\partial \bar{y}}\right)$$

and

$$\bar{u}\frac{\partial \bar{h}}{\partial \bar{x}} + \bar{v}\frac{\partial \bar{h}}{\partial \bar{y}} = (\gamma - 1)\,\text{M}_e^2 \frac{\bar{u}}{\bar{\rho}}\frac{\partial \bar{P}}{\partial \bar{x}} + \frac{1}{\text{Pr}}\frac{1}{\text{Re}}\frac{\partial}{\partial \bar{y}}\left(\bar{\mu}\frac{\partial \bar{h}}{\partial \bar{y}}\right) + \frac{(\gamma - 1)\,\text{M}_e^2}{\text{Re}}\bar{\nu}\left(\frac{\partial \bar{u}}{\partial \bar{y}}\right)^2$$

where

$$\text{Re} = \frac{\rho_e U_e L}{\mu_e} \quad \text{and} \quad (\gamma - 1)\,\text{M}_e^2 = \frac{U_e^2}{h_e}$$

210 Laminar Flow Analysis

The nondimensional energy equation shows that the work due to compression and the heat generated by viscous dissipation become increasingly important as the Mach number of the external flow increases. However, as pointed out in our discussion oboth two- and three-equation models for the locally nonsimilar boundary layer method.[†]

A comparison of these results with the local similarity method and with

6–5 Transformation of the Compressible Boundary Layer Equations

Because of the complexity of the compressible boundary layer equations and associated boundary conditions, they have been extensively transformed into alternate forms. These transformations are in general of two types: those that use the stream function ψ or one of its derivatives, such as the tangential velocity, u, as an independent variable, and those of the similarity type. Of the former, the two transformations of most interest are the von Mises transformation and the Crocco transformation.

The von Mises transformation [vonM27] was originally derived for incompressible boundary layers. It was subsequently extended to compressible flow by von Karman and Tsien [vonK38]. Examination of the left hand side of the momentum and energy equations, Eqs. (6–40) and (6–41), helps in understanding the underlying philosophy of the von Mises transformation. In each case, the left hand side of the equation is the substantial derivative of either momentum or energy along a streamline. Since the changes to the flow occur along a streamline, it seems reasonable to use the stream function as an independent variable. Further, use of the stream function as an independent variable automatically satisfies the continuity equation. We therefore introduce the transformations

$$\xi = x \qquad \eta = \psi \qquad (6-43a,b)$$

where $\psi_y = \rho u$ and $\psi_x = -\rho v$. The transformation equations are then

$$\frac{\partial}{\partial x} = \frac{\partial}{\partial \xi} - \rho v \frac{\partial}{\partial \eta} \qquad \text{and} \qquad \frac{\partial}{\partial y} = \rho u \frac{\partial}{\partial \eta}$$

Using these relations, the continuity equation is automatically satisfied and the momentum and energy equations become

$$u \frac{\partial u}{\partial \xi} = -\frac{1}{\rho} \frac{\partial P}{\partial \xi} + u \frac{\partial}{\partial \eta}\left(\rho u \mu \frac{\partial u}{\partial \eta}\right) \qquad (6-44)$$

and

$$u \frac{\partial h}{\partial \xi} = \frac{u}{\rho} \frac{\partial P}{\partial \xi} + \frac{1}{\Pr} u \frac{\partial}{\partial \eta}\left(\rho u \mu \frac{\partial h}{\partial \eta}\right) + \mu \rho u^2 \left(\frac{\partial u}{\partial \eta}\right)^2 \qquad (6-45)$$

In obtaining this result we used the fact that $P = P(x) = P(\xi)$ only. In examining Eqs. (6–44) and (6–45) we see that the von Mises transformation eliminates the v term from the equations. Further, the number of equations is reduced from three to two and the order of the system of equations from five to four. These apparent simplifications are not without complication. Assuming no mass transfer at the surface and that $\psi = 0$ at the surface, the boundary conditions, Eq. (6–42), in von Mises coordinates are

$$\begin{array}{ccc} \text{at} & \psi = 0 & u = 0 & h = h_w(x) \\ \text{as} & \psi \to \infty & u \to u_e(x) & h \to h_e(x) \end{array}$$

However, because of the singular behavior of ψ near the surface, numerical integration of these equations is somewhat difficult. In order to see this we note that near the surface $\rho u = a_0(x)y + a_1(x)y^2 + \ldots$ Hence

$$\psi = b_0(x)y^2 + b_0(x)y^3 + \ldots \quad \text{and} \quad y = c_0(x)\psi^{1/2} + c_1(x)\psi + \ldots$$

Finally
$$u = d_0(x)\psi^{1/2} + d_1(x)\psi + \ldots$$

Thus, at the surface ($\psi = 0$) $\partial u/\partial \eta$ is infinite. Although these comments might lead one to conclude that the shearing stress at the surface is also infinite, this is, of course, not true physically, nor is it true mathematically. In order to see this we write the shearing stress as

$$\tau = \mu \frac{\partial u}{\partial y} = \rho u \mu \frac{\partial u}{\partial \eta} = \rho \mu \frac{\partial \left(\frac{u^2}{2}\right)}{\partial \eta} \tag{6 – 46}$$

Although $\partial u/\partial \eta$ is infinite at the surface, $\partial(u^2/2)/\partial \eta$ is finite. Similarly, the heat transfer at the surface is given by

$$q = -k\frac{\partial T}{\partial y} = -\frac{\mu}{\Pr}\frac{\partial h}{\partial y} = -\frac{\rho u \mu}{\Pr}\frac{\partial h}{\partial \eta} \tag{6 – 47}$$

Although $u = 0$ at the surface, $u(\partial h/\partial \eta)$ is generally finite. The heat transfer at the surface is obtained by evaluating Eq. (6–47) at $\psi = \eta = 0$. Once solutions of the transformed equations (Eq. 6–44 and 6–45) are obtained, the physical coordinate y is obtained by integration, i.e.

$$y = \int_0^\eta \frac{d\eta}{\rho u} = \int_0^\psi \frac{d\psi}{\rho u}$$

We now consider the Crocco transformation [Croc46]. The Crocco transformation uses the viscous shearing stress as a dependent variable. This has the effect of reducing the order of the governing differential equations. The longitudinal coordinate, x, and the tangential velocity component, u, are used as independent variables. Introducing the dependent and independent variable transformations

$$\tau = \mu \frac{\partial u}{\partial y} \tag{6 – 48a}$$

and
$$\xi = x \quad \eta = u(x, y) \tag{6 – 48b, c}$$

212   Laminar Flow Analysis

The appropriate transformation equations are

$$\frac{\partial}{\partial x} = \frac{\partial}{\partial \xi} - \left(\frac{\partial y}{\partial \xi}\right)\left(\frac{\partial y}{\partial \eta}\right)^{-1}\frac{\partial}{\partial \eta}$$

and

$$\frac{\partial}{\partial y} = \left(\frac{\partial y}{\partial \eta}\right)^{-1}\frac{\partial}{\partial \eta}$$

The continuity and momentum equations then become

$$\frac{\partial y}{\partial \eta}\frac{\partial(\rho\eta)}{\partial \xi} - \frac{\partial y}{\partial \xi}\frac{\partial(\rho\eta)}{\partial \eta} + \frac{\partial(\rho v)}{\partial \eta} = 0$$

and

$$-\rho\eta\frac{\partial v}{\partial \xi} + \rho v = \frac{\partial \tau}{\partial \eta} - \frac{\partial y}{\partial \eta}\frac{\partial P}{\partial \xi}$$

The v term is eliminated by differentiating the momentum equation with respect to η and subtracting the result from the continuity equation. Thus, after rearranging

$$\frac{\partial}{\partial \eta}\left(\frac{\mu}{\tau}\frac{\partial P}{\partial \xi}\right) - \frac{\partial}{\partial \xi}\left(\frac{\mu\rho u}{\tau}\right) = \frac{\partial^2 \tau}{\partial \eta^2} \qquad (6\text{–}49)$$

where we used the relation $\partial y/\partial \eta = \mu/\tau$ in deriving this result. Similarly, the energy equation becomes

$$\mu\rho\eta\frac{\partial h}{\partial \xi} - \rho\eta\tau\frac{\partial y}{\partial \xi}\frac{\partial h}{\partial \eta} + \rho v\tau\frac{\partial h}{\partial \eta} - \mu\eta\frac{\partial P}{\partial \xi} = \frac{\tau}{\Pr}\frac{\partial}{\partial \eta}\left(\tau\frac{\partial h}{\partial \eta}\right) + \tau^2$$

The term involving v in the energy equation is eliminated by using the momentum equation. Thus, after rearranging, the energy equation becomes

$$(1-\Pr)\tau\frac{\partial \tau}{\partial \eta}\frac{\partial h}{\partial \eta} + \tau^2\left(\frac{\partial^2 h}{\partial \eta^2} + \Pr\right) - \Pr\mu\rho\eta\frac{\partial h}{\partial \xi} + \Pr\mu\left(\frac{\partial h}{\partial \eta} + \eta\right)\frac{\partial P}{\partial \xi} = 0 \qquad (6\text{–}50)$$

Equations (6–49) and (6–50) are the Crocco equations for two-dimensional steady compressible boundary layer flow. Notice that, as is the case for the von Mises transformation, the v terms are eliminated from the equations, that the number of equations is reduced from three to two and that the order of the system of equations is reduced from five to four.

The appropriate transformed boundary conditions are obtained by noting that at $y = 0$, $u = 0$, and as $y \to \infty$, $u \to U_e(x)$ and $\tau \to 0$. From the momentum equation (Eq. 6–40) evaluated at the surface, we have

$$\rho v\frac{\tau}{\mu} + \frac{\partial P}{\partial x} = \frac{\partial \tau}{\partial y} = \frac{\partial \tau}{\partial \eta}\frac{\partial \eta}{\partial y} = \frac{\tau}{\mu}\frac{\partial \tau}{\partial \eta}$$

For the case of an impermeable wall, $v = 0$ and the boundary conditions become

at $u = 0$ $\quad\quad \tau \dfrac{\partial \tau}{\partial \eta} = \mu \dfrac{\partial P}{\partial \xi} \quad$ and $\quad h(\xi, 0) = h_w(\xi) \quad\quad$ (6 – 51a, b)

and as $u \to U_e(\xi) \quad\quad \tau \to 0 \quad$ and $\quad h(\xi, u \to U_e) \to h_e(\xi) \quad\quad$ (6 – 51c, d)

Generally, the surface enthalpy and hence the viscosity at the surface are known; however, for an adiabatic wall the surface temperature and thus the surface viscosity are unknown. Under these circumstances the boundary condition given in Eq. 6–51a depends on the unknown surface temperature.

Once the Crocco equations are solved, the physical coordinate, y, is obtained by integrating

$$y = \int_0^\eta \dfrac{\mu}{\tau} d\eta = \int_0^u \dfrac{\mu}{\tau} du$$

The tangential velocity, u, as a function of x and y, is then obtained by inverting this equation. Finally, knowing $u = u(x, y)$, the continuity equation is used to find v as a function of x and y. An important advantage of the Crocco transformation is that the skin friction and the enthalpy at the surface are obtained directly from the solution of the equations by application of the boundary conditions at the surface.

Before considering a generalized form of the similarity transformation for the compressible boundary layer, we investigate a restricted form of the transformation similar to the one due to Howarth [Howa48]. This transformation is derived by asking whether it is possible to reduce the compressible boundary layer equations to a form equivalent to the incompressible boundary layer equations. We begin by introducing the compressible stream function defined by

$$\dfrac{\partial \psi}{\partial y} = \dfrac{\rho u}{\rho_r} \quad \text{and} \quad \dfrac{\partial \psi}{\partial x} = -\dfrac{\rho v}{\rho_r}$$

where the subscript, r, indicates some reference condition. Independent variable transformations

$$\xi = \xi(x) \quad\quad \eta = \eta(x, y)$$

subject to the condition

$$\psi_\eta = u$$

are also introduced into the governing equations (Eqs. 6–39 to 6–41) and boundary conditions (Eq. 6–42). The particular functional forms chosen for the independent variable transformations are based on the equivalent forms for incompressible flow. The restriction that $\psi_\eta = u$ is also based on the incompressible results. Before proceeding with the formal transformation of the governing equations we note that

$$\psi_\eta = \dfrac{\rho_r}{\rho} \quad\quad \psi_y = \dfrac{\partial y}{\partial \eta} \psi_y$$

or

$$\dfrac{\partial \eta}{\partial y} = \dfrac{\rho}{\rho_r}$$

214   Laminar Flow Analysis

which yields the required independent variable transformation for η, i.e.

$$\eta = \int_0^y \frac{\rho}{\rho_r} dy \qquad (6-52)$$

Notice that, in contrast to the independent variable transformation used for the incompressible boundary layer, this is an integral relation. This is typical of similarity transformations for the compressible boundary layer equations.

The formal transformation equations are

$$\frac{\partial}{\partial y} = \frac{\rho}{\rho_r} \frac{\partial}{\partial \eta}$$

and

$$\frac{\partial}{\partial x} = \xi_x \frac{\partial}{\partial \xi} + \eta_x \frac{\partial}{\partial \eta}$$

Using these results the transformed momentum equation is

$$\xi_x(\psi_\eta \psi_{\xi\eta} - \psi_\xi \psi_{\eta\eta}) = -\frac{1}{\rho}\frac{\partial P}{\partial x} + \frac{1}{\rho_r^2}(\rho\mu\psi_{\eta\eta})_\eta \qquad (6-53)$$

Using the Chapman-Rubesin viscosity law (see Section 6–2) with $\omega = 1$, i.e.

$$\frac{\mu}{\mu_r} = C\frac{T}{T_r}$$

and recalling that $\partial P/\partial y = 0$, and using the equation of state, yields

$$\rho\mu = C\rho_r\mu_r$$

Equation (6–53) thus becomes

$$\xi_x(\psi_\eta \psi_{\xi\eta} - \psi_\xi \psi_{\eta\eta}) = -\frac{1}{\rho}\frac{\partial P}{\partial x} + \nu_r(C\psi_{\eta\eta})_\eta$$

If $C =$ constant, then with $\xi_x = C$ and thus $\xi = Cx$, we have

$$(\psi_\eta \psi_{\xi\eta} - \psi_\xi \psi_{\eta\eta}) = -\frac{1}{C}\frac{\rho_r}{\rho}\frac{1}{\rho_r}\frac{\partial P}{\partial x} + \nu_r\psi_{\eta\eta\eta} \qquad (6-54)$$

Except for the factor $(1/C)(\rho_r/\rho)$ in the pressure gradient term, this equation has the same form as the momentum equation governing incompressible constant property boundary layer flow (see Eq. 3–29). Here the fluid properties are evaluated at the reference condition. When the pressure gradient is zero, i.e., for a flat plate at zero incidence, Eq. (6–54) has exactly the same form as the incompressible constant property momentum equation. In the absence of a pressure gradient, the similarity transformations developed in Section 3–3 yield the Blasius equation, i.e.

$$\eta^* = \eta\left(\frac{U(\xi)}{2\nu\xi}\right)^{1/2} \qquad \psi(\xi,\eta) = (2\nu U(\xi)\xi)^{1/2} f^*(\eta^*)$$

yield

$$f^{*'''} + f^* f^{*''} = 0$$

The transformed boundary conditions for an impermeable surface are

$$f^*(0) = f^{*\prime}(0) = 0$$

and
$$f^{*\prime}(\eta^* \to \infty) \to 1$$

where the prime denotes differentiation with respect to η^*. Thus, the solution of the momentum equation for the compressible variable property boundary layer in the absence of pressure gradient is reduced to the solution of an equivalent incompressible constant property equation, i.e., the Blasius equation previously solved in Section 3–3. One interesting observation is that this implies that in the absence of a pressure gradient the momentum equation for compressible boundary layer flow is uncoupled from the energy equation. Formally this is true. However, notice that determining the physical coordinate, y, from the inverse of Eq. (6–52) requires a knowledge of the density distribution in the boundary layer and hence the solution of the energy equation. Thus, the momentum and energy equations for compressible boundary layer flow, even in the absence of a pressure gradient, are still technically coupled.

Transformation of the energy equation (Eq. 6–41) into ξ, η coordinates yields

$$\psi_\eta h_\xi - \psi_\xi h_\eta = \frac{1}{C}\frac{\rho_r}{\rho}\psi_\eta \frac{1}{P_r}\frac{\partial P}{\partial x} + \frac{1}{C \Pr \rho_r^2}(\rho\mu h_\eta)_\eta + \frac{\rho\mu}{C}\frac{1}{\rho_r^2}(\psi_{\eta\eta})^2 \quad (6-55)$$

Introducing the Chapman-Rubesin viscosity law and assuming that $C\rho_r\mu_r$ is constant yields

$$\psi_\eta h_\xi - \psi_\xi h_\eta = \frac{1}{C}\frac{\rho_r}{\rho}\psi_\eta \frac{1}{P_r}\frac{\partial P}{\partial x} + \frac{\nu_r}{\Pr}h_{\eta\eta} + \nu_r(\psi_{\eta\eta})^2 \quad (6-56)$$

In the absence of a pressure gradient, Eq. (6–56) is equivalent to the energy equation governing forced convection flow over a flat plate at zero incidence (see Section 4–7). The solution of the energy equation and the compressible flat plate boundary layer flow is discussed in more detail in the next section.

We now turn our attention to the derivation of a generalized similarity transformation for the steady two-dimensional compressible laminar boundary layer equations. We essentially follow the procedure adopted by Li and Nagamatsu [Li55]. For convenience we repeat the governing equations and boundary conditions here:

continuity
$$\frac{\partial}{\partial x}(\rho u) + \frac{\partial}{\partial y}(\rho v) = 0 \quad (6-39)$$

momentum
$$\rho u \frac{\partial u}{\partial x} + \rho v \frac{\partial u}{\partial y} = -\frac{\partial P}{\partial x} + \frac{\partial}{\partial y}\left(\mu \frac{\partial u}{\partial y}\right) \quad (6-40)$$

216   Laminar Flow Analysis

energy

$$\rho u \frac{\partial h}{\partial x} + \rho v \frac{\partial h}{\partial y} = u \frac{\partial P}{\partial x} + \mu \left(\frac{\partial u}{\partial y}\right)^2 + \frac{1}{\Pr} \frac{\partial}{\partial y}\left(\mu \frac{\partial h}{\partial y}\right) \qquad (6-41)$$

with boundary conditions

$$u(0) = 0 \qquad v(0) = v(x) \qquad h(0) = h(x) \quad \text{or} \quad \frac{\partial h}{\partial y} = 0 \qquad (6-42a,b,c,d)$$

and
$$u(y \to \infty) \to U_e(x) \qquad h(y \to \infty) \to h_e(x)$$

We first rewrite the energy equation in terms of the total enthalpy

$$H = h + \frac{u^2}{2}$$

Equation (6–41) then becomes

$$\rho u \frac{\partial H}{\partial x} + \rho v \frac{\partial H}{\partial y} - u\left(\rho u \frac{\partial u}{\partial x} + \rho v \frac{\partial u}{\partial y}\right) = u \frac{\partial P}{\partial x} + \mu \left(\frac{\partial u}{\partial y}\right)^2$$
$$+ \frac{1}{\Pr}\left[\frac{\partial}{\partial y}\left(\mu \frac{\partial H}{\partial y}\right) - \frac{\partial}{\partial y}\left(u\mu \frac{\partial u}{\partial y}\right)\right]$$

The pressure gradient term is eliminated by multiplying the momentum equation by u and adding the result to the energy equation. This yields

$$\rho u \frac{\partial H}{\partial x} + \rho v \frac{\partial H}{\partial y} = \left(1 - \frac{1}{\Pr}\right)\left[\mu\left(\frac{\partial u}{\partial y}\right)^2 + u \frac{\partial}{\partial y}\left(\mu \frac{\partial u}{\partial y}\right)\right] + \frac{1}{\Pr}\frac{\partial}{\partial y}\left(\mu \frac{\partial H}{\partial y}\right) \qquad (6-57)$$

We continue the search for the similarity transformations and the corresponding similar solutions by introducing the compressible stream function defined by

$$\frac{\partial \psi}{\partial y} = \rho u \qquad \frac{\partial \psi}{\partial x} = -\rho v \qquad (6-58)$$

which automatically satisfies the continuity equation. The momentum and energy equations (Eqs. 6–40 and 6–57) become

$$\frac{\partial \psi}{\partial y}\frac{\partial}{\partial x}\left(\frac{1}{\rho}\frac{\partial \psi}{\partial y}\right) - \frac{\partial \psi}{\partial x}\frac{\partial}{\partial y}\left(\frac{1}{\rho}\frac{\partial \psi}{\partial y}\right) = -\frac{\partial P}{\partial x} + \frac{\partial}{\partial y}\left[\mu \frac{\partial}{\partial y}\left(\frac{1}{\rho}\frac{\partial \psi}{\partial y}\right)\right] \qquad (6-59)$$

and
$$\frac{\partial \psi}{\partial y}\frac{\partial H}{\partial x} - \frac{\partial \psi}{\partial x}\frac{\partial H}{\partial y} = \left(1 - \frac{1}{\Pr}\right)\left\{\mu\left[\frac{\partial}{\partial y}\left(\frac{1}{\rho}\frac{\partial \psi}{\partial y}\right)\right]^2\right.$$

$$+ \frac{1}{\rho}\frac{\partial \psi}{\partial y}\frac{\partial}{\partial y}\left[\mu\frac{\partial}{\partial y}\left(\frac{1}{\rho}\frac{\partial \psi}{\partial y}\right)\right]\}$$

$$+ \frac{1}{\Pr}\frac{\partial}{\partial y}\left(\mu\frac{\partial H}{\partial y}\right) \qquad (6-60)$$

Based on our experience with the incompressible boundary layer equations (see Chapters 3 and 4) we introduce the dependent variable transformations

$$\psi(x,y) = N(x)f(\xi,\eta) \qquad (6-61a)$$
$$u(x,y) = U_e(x)f_\eta(\xi,\eta) \qquad (6-61b)$$
$$H(x,y) = H_e(x)g(\xi,\eta) \qquad (6-61c)$$

where the subscript, η, indicates partial differentiation. The form of the enthalpy transformation indicates that we expect the compressible boundary layer to be similar with respect to a nondimensional total enthalpy profile rather than the static enthalpy (temperature) profile, as is the case for the incompressible constant-property boundary layer. We further introduce independent variable transformations of the form

$$\xi = \xi(x) \qquad (6-61d)$$
$$\eta = \eta(x,y) \qquad (6-61e)$$

From the above definitions of the stream function, Eqs. (6–58) and (6–61a), we have

$$\frac{\partial \psi}{\partial y} = N(x)\frac{\partial \eta}{\partial y}f_\eta(\xi,\eta) = \rho u = \rho U_e(x)f_\eta(\xi,\eta)$$

which yields

$$\frac{\partial \eta}{\partial y} = \frac{U_e(x)}{N(x)}\rho$$

or integrating

$$\eta = \frac{U_e(x)}{N(x)}\int_0^y \rho\, dy \qquad (6-62)$$

$N(x)$ is determined from the transformed momentum and energy equations. Introducing Eqs. (6–61) and (6–62) into the momentum and energy equations yields

$$\frac{U^2}{N^2}(\rho\mu f_{\eta\eta})_\eta + \frac{N_x}{N}U f f_{\eta\eta} - U_x(f_\eta)^2$$

$$- \frac{1}{\rho U}\frac{dP}{dx} = U\xi_x(f_\eta f_{\xi\eta} - f_{\eta\eta}f_\xi) \qquad (6-63)$$

and
$$\frac{U}{N^2}H_e(\rho\mu g_\eta)_\eta + \frac{N_x}{N}H_e\Pr f g_\eta - \Pr H_{e_x}f_\eta g$$

$$+(\Pr - 1)\frac{U^3}{N^2}\left(\rho\mu f_\eta f_{\eta\eta}\right)_\eta = H_e\xi_x\left(f_\eta g_\xi - f_\xi g_\eta\right) \quad (6-64)$$

where here the subscripts η, ξ, x indicate partial differentiation.

Before continuing it is worthwhile to point out that in the transformed momentum and energy equations the density and the viscosity, except in the pressure gradient term, always appear in the combination $\rho\mu$. This is also true of the von Mises equations (see Eqs. 6–44 and 6–45), the Crocco equations (see Eqs. 6–49 and 6–50) and the modified Howarth equations (see Eqs. 6–53 and 6–55). This leads naturally to the assumption of a Chapman-Rubesin viscosity law, with $\omega = 1$. If we use conditions at the edge of the boundary layer for the reference condition, we have

$$\frac{\mu}{\mu_e} = C\frac{T}{T_e}$$

which upon recalling that $\partial P/\partial y = 0$ yields

$$\rho\mu = C\rho_e\mu_e$$

Substituting this result into Eqs. (6–63) and (6–64) yields

$$\frac{U^2}{N^2}\rho_e\mu_e\left(Cf_{\eta\eta}\right)_\eta + \frac{N_x}{N}Uff_{\eta\eta} - U_xf_\eta^2$$
$$- \frac{1}{\rho U}\frac{dP}{dx} = U\xi_x\left(f_\eta f_{\xi\eta} - f_{\eta\eta}f_\xi\right) \quad (6-65)$$

and

$$\frac{U}{N^2}H_e\rho_e\mu_e\left(Cg_\eta\right)_\eta + \frac{N_x}{N}H_e\Pr fg_\eta - \Pr H_{e_x}f_\eta g$$
$$+ (\Pr - 1)\frac{U^3}{N^2}\rho_e\mu_e\left(Cf_\eta f_{\eta\eta}\right)_\eta = H_e\xi_x\left(f_\eta g_\xi - f_\xi g_\eta\right) \quad (6-66)$$

where in this formulation we still allow the possibility that $C = C(\eta)$. Thus C can vary through the boundary layer. However, the simplifying assumption that C is a constant is frequently made. Except for the particularly simple results which occur when $C = 1$, the constant is generally evaluated at the surface conditions, e.g., using the Sutherland viscosity law. In this way the viscosity is accurately represented at the surface.

If we further assume that $f = f(\eta)$ and $g = g(\eta)$ only, i.e., that the flow is similar, the right hand sides of the momentum and energy equations are zero. Finally, if we assume that the total enthalpy at the boundary layer edge is constant, i.e., $H_{e_x} = 0$, Eqs. (6–65) and (6–66) become

$$f''' + \frac{NN_x}{C\rho_e\mu_e U_e}ff'' + \frac{N^2U_x}{C\rho_e\mu_e U_e^2}\left(\frac{\rho_e}{\rho} - f'^2\right) = 0 \quad (6-67)$$

$$g'' + \frac{NN_x}{C\rho_e\mu_e U_e}\Pr fg' = (1-\Pr)\frac{U_e^2}{H_e}(f'f'')' \qquad (6-68)$$

where the prime denotes differentiation with respect to η and the pressure gradient term in the momentum equation was rewritten using Euler's equation at the edge of the boundary layer, i.e., $U_x = -(1/\rho_e U_e)(dP/dx)$.

From these equations we note that the similarity conditions are

$$\frac{NN_x}{C\rho_e\mu_e U_e} = \text{constant} \qquad (6-69a)$$

$$\frac{N^2 U_x}{C\rho_e\mu_e U_e^2}\left(\frac{\rho_e}{\rho} - f'^2\right) = \text{function of } \eta \text{ only} \qquad (6-69b)$$

and

$$\frac{U_e^2}{H_e} = \text{constant} \quad \text{or} \quad \Pr = 1 \qquad (6-69c)$$

If we choose the constant in Eq. (6–69a) as unity, then, in the absence of a pressure gradient, the momentum equation reduces to the Blasius equation. Further, comparing Eq. (6–68) with the energy equation for forced convection (Eq. 4–111), we see that by choosing the constant in Eq. (6–69a) as unity the differential equation for the compressible boundary layer with unit Prandtl number has the same form as that for the incompressible boundary layer with an isothermal wall. Thus, we choose

$$\frac{NN_x}{C\rho_e\mu_e U_e} = 1$$

Rearranging and integrating yields

$$N(x) = \left(2\int_0^x C\rho_e\mu_e U_e \, dx\right)^{1/2} \qquad (6-70)$$

Using this result, Eq. (6–62) for η becomes

$$\eta = \frac{U_e(x)}{\sqrt{2\int_0^x C\rho_e\mu_e U_e \, dx}} \int_0^y \rho \, dy$$

Since $\xi = \xi(x)$, and noting that

$$\xi = \int_0^x C\rho_e\mu_e U_e \, dx \qquad (6-71)$$

a particularly simple form for η results, i.e.

$$\eta = \frac{U_e(x)}{\sqrt{2\xi}} \int_0^y \rho \, dy \qquad (6-72)$$

220 Laminar Flow Analysis

The transformations given in Eqs. (6–71) and (6–72) are frequently called the Illingworth-Levy transformations (see [Illi49] and [Li55]).

Examination of Eq. (6–69b) shows that for the particular case of constant stagnation enthalpy at the edge of the boundary layer, i.e., $H_e(x) = $ constant, this term becomes

$$\frac{N^2 U_e(x)}{C \rho_e \mu_e U_e^2} \frac{H_e}{h_e} (g - f'^2) = \hat{\beta}(g - f'^2)$$

To accomplish this we use the definition of $g(\eta)$ and of the stagnation enthalpy to obtain

$$g(\eta) = \frac{H}{H_e} = \frac{h + \dfrac{u^2}{2}}{h_e + \dfrac{U_e^2}{2}} = \frac{\left(\dfrac{h}{h_e} + \dfrac{u^2}{2h_e}\right)}{\left(1 + \dfrac{u^2}{2h_e}\right)}$$

Since $u/U_e = f'$, this is rewritten as

$$\frac{h}{h_e} = \left(1 + \frac{U_e^2}{2h_e}\right) g - \frac{U_e^2}{2h_e} f'^2$$

Finally, by virtue of the constant pressure assumption across the boundary layer we have

$$\frac{\rho_e}{\rho} - f'^2 = \frac{h}{h_e} - f'^2 = \left(1 + \frac{U_e^2}{2h_e}\right)(g - f'^2) = \frac{H_e}{h_e}(g - f'^2) \qquad (6-73)$$

Assuming that the inviscid flow at the edge of the boundary layer is isoenergetic, i.e., $H_e(x) = $ constant, is not as restrictive as it might first appear. Since $H_e = h_e + U_e^2/2$, both the static enthalpy and the velocity can vary along the edge of the boundary layer. Recalling that the stagnation enthalpy along any streamline is constant across a shock wave, we see that the isoenergetic flow assumption is reasonable when the shock wave is not significantly curved.

The last of the similarity conditions, Eq. 6–69c), is written as

$$\frac{U_e^2}{H_e} = \frac{U_e^2}{h_e + \dfrac{U_e^2}{2}} = \frac{(\gamma - 1) M_e^2}{1 + \dfrac{\gamma - 1}{2} M_e^2}$$

The governing equations (Eqs. 6–67 and 6–68) are now

$$f''' + f f'' + \hat{\beta}(g - f'^2) = 0 \qquad (6-74)$$

and
$$g'' + \Pr f g' = \bar{\sigma}(1 - \Pr)(f' f'')' \qquad (6-75)$$

where
$$\bar{\sigma} = \frac{(\gamma - 1) M_e^2}{1 + \dfrac{\gamma - 1}{2} M_e^2}$$

Similar solutions of the equations exist if

$$\hat{\beta} = \frac{2\xi}{U_e}\frac{dU_e}{d\xi}\frac{H_e}{h_e} = \frac{2\xi}{U_e}\frac{dU_e}{d\xi}\left(1 + \frac{\gamma-1}{2}M_e^2\right) \qquad (6-76)$$

is constant and if one of the following conditions is satisfied:

$\gamma = 1$, the ratio of specific heats is unity;
$M_e = 0$, the external Mach number is zero;
$\Pr = 1$, the Prandtl number is unity;
$M_e = $ constant other than zero;
$$\bar{\sigma} = \frac{(\gamma-1)M_e^2}{1 + \frac{\gamma-1}{2}M_e^2} = 2 \text{ i.e, for hypersonic flow, } M_e \to \infty.$$

Before considering the implications of each of the latter similarity requirements we rewrite $\hat{\beta}$ in terms of the external Mach number. This is accomplished by differentiating $[(\gamma-1)/2]M_e^2 = U_e^2/2h_e$ and evaluating $dh_e/d\xi$ by recalling that the stagnation enthalpy is constant at the edge of the boundary layer. Thus

$$(\gamma-1)M_e\frac{dM_e}{d\xi} = \frac{U_e}{h_e}\frac{dU_e}{d\xi} - \frac{U_e^2}{2h_e^2}\frac{dh_e}{d\xi} = (\gamma-1)M_e^2\left(\frac{1}{U_e}\frac{dU_e}{d\xi}\frac{H_e}{h_e}\right)$$

and
$$\frac{1}{U_e}\frac{dU_e}{d\xi}\frac{H_e}{h_e} = \frac{1}{M_e}\frac{dM_e}{d\xi}$$

Hence we write $\hat{\beta}$ as

$$\hat{\beta} = \frac{2\xi}{M_e}\frac{dM_e}{d\xi} = 2\left(\frac{d\ln M_e}{d\ln \xi}\right) = \text{constant} \qquad (6-77)$$

Integrating this equation yields

$$M_e = (\text{constant})\,\xi^{\frac{\hat{\beta}}{2}} \qquad (6-78)$$

Here we see that the similarity requirement for the momentum equation is satisfied by a power law variation of the Mach number *in the transformed plane*. This is analogous to the required power law velocity variation for similarity to exist for the incompressible boundary layer (see Section 3–4). Further, Li and Nagamatsu [Li55] and Cohen [Cohe54] showed that the similarity requirement for the compressible boundary layer is also satisfied by an exponential Mach number variation in the transformed plane. This particular case is not discussed here.

Returning to the similarity requirements imposed by the energy equation, we now consider each of these in detail. The assumption that $\gamma = 1$ is unrealistic for most gases. Further, it does not simplify the equations any more than the more reasonable assumption of unit Prandtl number. We therefore do not consider this assumption further. Letting the Mach number, M_e, equal zero neglects both the viscous dissipation and the compressive work terms in the energy equation. If we further assume that there is no heat transfer at the surface, the $M_e = 0$ assumption shows that the static temperature through the boundary layer is constant. However, since the static temperature in the boundary layer should vary from the surface or wall temperature to the static temperature at the boundary layer edge, the $M_e = 0$ assumption is less realistic than the unit Prandtl number assumption. However, it is still useful. When a unit Prandtl number is assumed, the stagnation enthalpy (temperature) for zero heat transfer at the surface is constant through the boundary layer. This result is close to the true adiabatic wall stagnation enthalpy variation, which is slight. The case of constant Mach number, M_e, strictly corresponds only to the flat plate ($\hat{\beta} = 0$). However, it may be sufficient to assume that the external Mach number is constant in the energy equation for small values of the pressure gradient parameter, $\hat{\beta}$. Finally, the assumption that the external Mach number is large, i.e., $M_e \to \infty$, leads to the hypersonic flow assumption that $\bar{\sigma} = 2$. This approximation is less than five percent in error at an external Mach number of ten. In addition, it allows the investigation of the effects of constant but nonunit Prandtl number on the heat transfer at the surface.

We turn now to the boundary conditions for Eqs. (6–74) and (6–75). Noting that at $y = 0$, $\eta = 0$ and, similarly, as $y \to \infty$, $\eta \to \infty$, the boundary conditions required at the surface for similar solutions to exist are

$$f(0) = -\frac{\sqrt{2\xi}\, v(x)}{\mu_w U_e(x)} = \text{constant} \qquad f'(0) = 0 \qquad (6-79a,b)$$

$$g(0) = g_w = \text{constant} \quad \text{or} \quad g'(0) = 0 \qquad (6-79c,d)$$

The outer boundary conditions become

$$f'(\eta \to \infty) \to 1 \qquad g(\eta \to \infty) \to 1 \qquad (6-79e,f)$$

Of particular interest is Eq. (6–79a), which represents the mass transfer normal to the surface. Equation (6–79a) is obtained by differentiating the stream function, Eq. (6–61a), with respect to x, using the definition of $N(x)$ given in Eq. (6–70) to evaluate N_x and the Chapman-Rubesin viscosity law to rearrange the result. For similar solutions to exist, $f(0)$ must be constant. This implies that either

$$v(x) = 0 \qquad (6-80a)$$

or

$$v(x) = \frac{\mu_w U_e(x)}{\sqrt{2\xi}} \qquad (6-80b)$$

Here, as for the Falkner-Skan equation (see Chapter 3), we note that negative values of $f(0)$ correspond to mass transfer from the surface to the fluid, i.e., injection or blowing, and positive values of $f(0)$ correspond to mass transfer from the fluid into the surface, i.e., suction. Implicit in our development are the assumptions that the injected fluid is the same as the main stream fluid and is at the same static temperature as the main stream fluid. Also of particular interest is the fact that an isothermal boundary condition is assumed in Eq. (6–79c) in order to satisfy similarity conditions.

In subsequent sections we discuss the similar solutions for Eqs. (6–74) and (6–75) subject to Eqs. (6–79) for the particular cases of

$\hat{\beta}$ = constant, $M_e = 0$, Pr = constant;
$\hat{\beta} = 0$, M_e = constant, Pr = constant;
$\hat{\beta}$ = constant, M_e = (constant) $\xi^{\hat{\beta}/2}$, Pr = 1;
$\hat{\beta}$ = constant, $M_e \to \infty$, Pr $\neq 1$.

The first case corresponds to the low speed compressible boundary layer with variable properties, the second to the compressible boundary layer on a flat plate, the third to the general similar compressible boundary layer with unit Prandtl number and the fourth to the similar hypersonic compressible boundary layer.

6–6 The Low Speed Compressible Boundary Layer

Assuming that the Mach number of the external flow approaches zero, i.e., $M_e \to 0$, amounts to neglecting the viscous dissipation and compressive work terms in the energy equation, Eq. (6–75). This is an acceptable approximation provided the term $\bar{\sigma}(1 - \text{Pr})(f'f'')'$ is small compared to the terms on the left hand side of Eq. (6–75). Since at $\eta = 0$ the left hand side of Eq. (6–75) is zero, this term cannot be neglected unless the Prandtl number is close to unity or the Mach number is small.

The boundary value problem for the low speed compressible boundary layer with variable properties is then given by

$$f''' + ff'' + \hat{\beta}(g - f'^2) = 0 \qquad (6-74)$$

and

$$g'' + \text{Pr}\, fg' = 0 \qquad (6-81)$$

with boundary conditions given by

$$f(0) = f_w = \text{constant} \qquad f'(0) = 0 \qquad (6-79a,b)$$
$$g(0) = g_w = \text{constant} \quad \text{or} \quad g'(0) = 0 \qquad (6-79c,d)$$
$$f'(\eta \to \infty) \to 1 \qquad g(\eta \to \infty) \to 1 \qquad (6-79e,f)$$

Since

$$\frac{h(\eta)}{h_e} = g(\eta) + \frac{\gamma - 1}{2} M_e^2 \left(g - f'^2\right)$$

or for $M_e = 0$

$$g(\eta) = \frac{h(\eta)}{h_e}$$

solutions of Eq. (6–81) represent nondimensional static enthalpy profiles through the boundary layer. For a constant specific heat at constant pressure, c_p, they represent nondimensional temperature profiles through the boundary layer. Further, Eq. (6–76) shows that $\hat{\beta} = (2\xi/U_e)(dU_e/d\xi)$. Using Eq. (6–71) yields $d\xi = C\rho_e \mu_e U_e \, dx$. Hence, integration yields

$$U_e = (\text{constant}) \, (2\xi)^{\frac{\hat{\beta}}{2}} = (\text{constant}) \, x^{\frac{\hat{\beta}}{(2-\hat{\beta})}} = (\text{constant}) \, x^m$$

where $m = \hat{\beta}/(2 - \hat{\beta})$. Note that $\hat{\beta}$ is the same as the Falkner-Skan pressure gradient parameter (see Chapter 3), i.e., $\hat{\beta} = \beta$.

Here, in contrast to the incompressible constant property solutions for forced convection, the momentum and energy equations are coupled. This is a result of the variable transport properties included in the solution. However, for a flat plate without mass transfer at the surface, i.e., for $\hat{\beta} = 0$, $f(0) = 0$, the nondimensional momentum and energy equations are uncoupled, and Eq. (6–74) reduces to the Blasius equation previously discussed in Section 3–3. Further, Eq. (6–81) has the same functional form as the energy equation (Eq. 4–82) governing the incompressible constant property forced convection thermal boundary layer without viscous dissipation. Thus, the nondimensional solutions of the Blasius equation obtained in Chapter 3 and of the energy equation obtained in Chapter 4 are applicable here. In particular, the solution of Eq. (6–81) for arbitrary but constant Prandtl number is

$$g(\eta) = 1 - (1 - g_w)\theta_1(\eta) \qquad (6-82)$$

where $\theta_1(\eta)$ is the nondimensional solution given by Eq. (4–86).

We now turn our attention to another case in which Eqs. (6–74) and (6–81) are uncoupled. Examination of Eq. (6–81) shows that for an adiabatic wall, i.e., $g'(0) = 0$, integrating twice and using the boundary conditions (Eqs. 6–79d, f) to evaluate the constants of integration yields

$$g(\eta) = 1 \qquad (6-83)$$

For zero Mach number, this result says that the static enthalpy is constant through the boundary layer. Equation (6–83) is analogous to the Busemann and Crocco integrals to be discussed later. However, the Busemann and Crocco integrals are restricted to unit Prandtl number. Using Eq. (6–83), we rewrite the momentum equation (Eq. 6–74) as

$$f''' + ff'' + \hat{\beta}(1 - f'^2) = 0$$

We immediately recognize this as the Falkner-Skan equation (see Chapter 3). Thus, for zero heat transfer at the surface, the nondimensional velocity field for the low speed ($M_e \to 0$) compressible boundary layer is governed by the Falkner-Skan equation. Although the nondimensional governing equations for the velocity and temperature fields are mathematically uncoupled, physically they are still coupled through the transport properties. Mathematically this is shown by considering the independent variable transformation for η. To determine the physical dimension, y, we invert Eq. (6–72) to yield

$$y = \int_0^\eta \frac{\left(2 \int_0^x C\rho_e\mu_e U_e\right)^{1/2}}{\rho U_e} d\eta = \frac{\sqrt{2\xi}}{\rho_e U_e} \int_0^\eta \frac{\rho_e}{\rho} d\eta$$

Explicit evaluation of this function is delayed until Section 6–10.

Further, we also delay discussion of the numerical integration of Eqs. (6–74) and (6–81) when they are coupled, i.e., for $\hat{\beta} \neq 0$, $g'(0) \neq 0$, until Section (6–12).

6–7 The Compressible Boundary Layer on a Flat Plate

We now turn our attention to the second case under consideration, i.e., $\hat{\beta} = 0$, M_e = constant, Pr = constant. Physically, these conditions correspond to the compressible boundary layer on a flat plate at zero incidence. Hence, in the external inviscid flow U_e, ρ_e and T_e are constant. Equation (6–74) reduces to the Blasius equation

$$f''' + ff'' = 0 \qquad (3-38)$$

The governing differential equations (Eq. 3–38 and 6–75) are thus uncoupled. The boundary equations are again given by Eqs. (6–79). Since the governing differential equations are uncoupled, the momentum and energy equations are integrated sequentially in a manner similar to that used for the incompressible constant property forced convection boundary layer (see Sections 4–9 to 4–11).

For the special case of unit Prandtl number (Pr = 1) Eq. (6–75) reduces to

$$g'' + fg' = 0 \qquad (6-84)$$

We previously showed that a specific integral of this form of the energy equation (see Eq. 6–83) for zero heat transfer at the surface is $g(\eta) = 1$. However, since here $M_e \neq 0$, $g(\eta)$ is the ratio of stagnation enthalpies. We thus conclude that for unit Prandtl number and zero heat transfer at the surface the stagnation enthalpy is constant through the boundary layer. Further, we note that the adiabatic wall condition is $g_{aw} = g(0) = 1$, i.e., for unit Prandtl number the stagnation enthalpy at the surface is equal to the stagnation enthalpy at the edge of the boundary layer. Since the velocity is zero at the surface, for constant specific heats the adiabatic wall temperature is equal to the stagnation temperature of the fluid at the boundary layer edge. Consequently, the recovery factor is unity.

226   Laminar Flow Analysis

Physically, the conversion of kinetic energy into thermal energy at the surface through viscous dissipation is as efficient as the conversion of kinetic energy into thermal energy through the action of pressure forces in the inviscid flow at the boundary layer edge. This particular integral of the energy equation, first discovered by Busemann [Buse31], is called the Busemann energy integral.

The restriction to an adiabatic wall is removed provided the surface is isothermal, i.e., g_w = constant. In order to show this we first multiply the Blasius equation (Eq. 3–38) by $(f' + A)$, where A is some constant. The result is then added to the energy equation (Eq. 6–84) to yield

$$(g'' + Af''') + f(g' + Af'') + f'(f''' + ff'') = 0$$

which is rewritten as

$$(g + Af')'' + f(g + Af')' = 0$$

Integrating once yields

$$(g + Af')' = \text{constant } e^{-\int_0^\eta f\,d\eta}$$

Using the boundary conditions (Eqs. 6–79) at the surface yields g'_w = constant. The constant of integration is zero if g_w = constant, i.e., if the surface is isothermal. Thus, after integrating again and using Eqs. (6–79b, c) to evaluate the constant of integration we have

$$g + Af' = \text{constant} = g_w$$

The constant A is evaluated from the boundary conditions at infinity, i.e., Eqs. (6–79e, f). Finally

$$g + (g_w - 1)f' = g_w \qquad (6-85)$$

This energy integral is called the Crocco integral (see [Croc32]). It is applicable for $\hat{\beta} = 0$, Pr = 1 and g_w = constant. Hence, for unit Prandtl number, $g_{aw} = 1$ and $g(\eta) = 1$ is a solution of the energy equation.

For nonunit Prandtl number the solution is obtained in a manner analogous to that for the incompressible constant property forced convection boundary layer on a flat plate (see Section 4–7). Since the solution of the Blasius equation is known, the energy equation is considered as a linear nonhomogeneous second-order ordinary differential equation with variable coefficients. The nonhomogeneous term is a known forcing-function that is physically attributed to heat addition due to viscous dissipation. Since the governing equation is linear, a solution is obtained as the sum of a complementary solution of the homogeneous equation and a particular solution of the nonhomogeneous equation, i.e.

$$g = \bar{K} + KG_1 + \bar{\sigma}G_2 \qquad (6-86)$$

where $G_1(\eta)$ is the solution of the homogeneous boundary value problem

$$G_1'' + \Pr f G_1' = 0 \tag{6-87}$$

$$G_1(0) = 1 \qquad G_1(\eta \to \infty) \to 0 \tag{6-88a, b}$$

and $G_2(\eta)$ is the solution of the nonhomogeneous boundary value problem

$$G_2'' + \Pr f G_2' = (1 - \Pr)(f' f'')' \tag{6-89}$$

$$G_2'(0) = 0 \qquad G_2(\eta \to \infty) \to 0 \tag{6-90a, b}$$

Physically, the homogeneous boundary value problem yields the solution with heat transfer at the surface, whereas the nonhomogeneous boundary value problem represents the adiabatic wall solution.

Comparison of Eqs. (4–82) and (4–83) and Eqs. (6–87) and (6–88) reveals that they are identical, i.e., $G_1(\eta) = \theta_1(\eta)$. The nondimensional solution for $G_1(\eta)$ is thus given by Eq. (4–86) and shown in Figure (4–9). Comparison of Eqs. (4–84) and (4–85) and Eqs. (6–89) and (6–90) shows that the nonhomogeneous terms are different. However, the similarities lead us to look for an analytical solution in a similar manner, i.e., by the method of variation of a parameter or by using an integrating factor. The solution is similar to that of Eqs. (4–84) and (4–85) and is given by

$$G_2(\eta) = \frac{f'^2}{2} + \Pr \int_{\xi=\eta}^{\infty} \left(f''(\xi)\right)^{\Pr} \left[\int_0^{\xi} \left(f''(\tau)\right)^{2-\Pr} d\tau\right] d\xi$$

$$= \frac{1}{2}(f'^2 - 1) + \theta_2(\eta) \tag{6-91}$$

where $\theta_2(\eta)$ is the nondimensional solution given by Eq. (4–96) and shown in Figure 4–11. The constants K and $\bar{K}$ in Eq. (6–86) are evaluated using the boundary conditions, Eqs. (6–79). Thus, the complete solution is

$$g(\eta) = 1 - (1 - g_w) G_1(\eta) + \bar{\sigma}\Big(G_2(\eta) - G_2(0) G_1(\eta)\Big) \tag{6-92}$$

Using Eq. (6–91) and the fact that $G_1(\eta) = \theta_1(\eta)$, we rewrite this result as

$$g(\eta) = 1 - (1 - g_w)\theta_1(\eta) + \bar{\sigma}\Big(\theta_2(\eta) - \theta_2(0)\theta_1(\eta)\Big) + \frac{\bar{\sigma}}{2}\Big(f'^2 + \theta_1(\eta)\Big) \tag{6-93}$$

When $\bar{\sigma} = 0$, Eq. (6–93) reduces to Eq. (6–82) obtained for $M_e = 0$. Recalling that $\bar{\sigma} = (\gamma-1) M_e^2 / [1 + ((\gamma-1)/2) M_e^2]$, we see that the effects of viscous dissipation on the enthalpy profile are significant when the Mach number in the external flow is significant. Selected results for the enthalpy profiles for $\hat{\beta} = 0$, $g_w = 0.6$ and $\Pr = 0.723$ are shown in Figure 6–7. Here we see that as $\bar{\sigma}$ increases the maximum enthalpy ratio in the boundary layer also increases. This is a result of the conversion of kinetic energy within the boundary layer into thermal energy

228   Laminar Flow Analysis

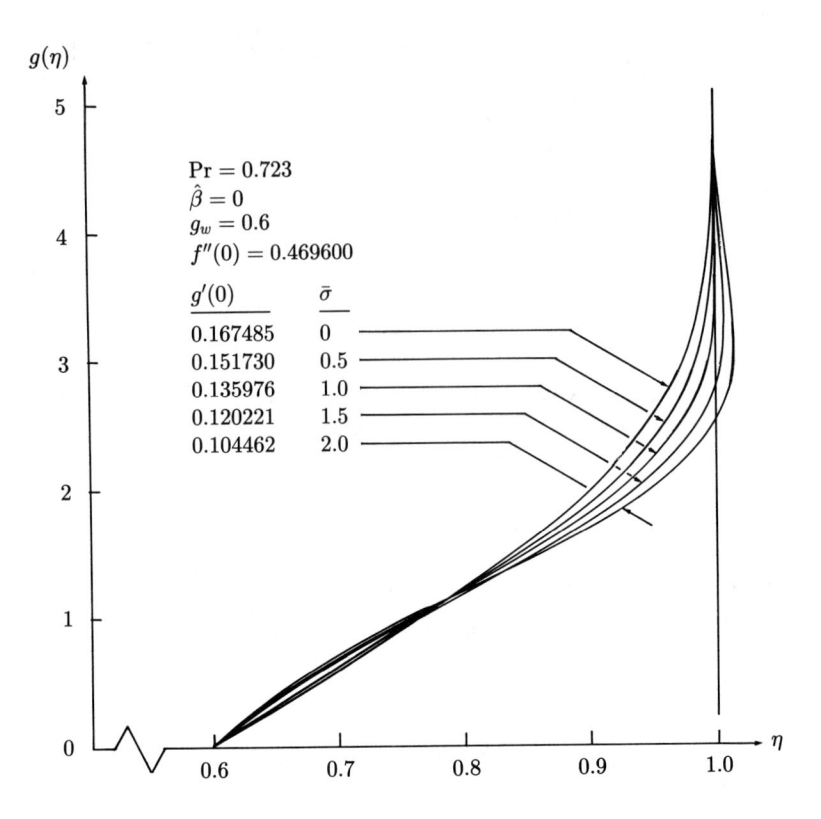

$g(\eta)$ vs η plot:

$\text{Pr} = 0.723$
$\hat{\beta} = 0$
$g_w = 0.6$
$f''(0) = 0.469600$

$g'(0)$	$\bar{\sigma}$
0.167485	0
0.151730	0.5
0.135976	1.0
0.120221	1.5
0.104462	2.0

Figure 6–7. The effect of Mach number on the enthalpy profiles for $\hat{\beta} = 0$, $g_w = 0.6$ and $\text{Pr} = 0.723$.

through viscous dissipation. Von Driest [VonD52], using a method due to Crocco [Croc32], studied this problem for $\text{Pr} = 0.75$ and a Sutherland viscosity law.

The adiabatic wall, or recovery, temperature is obtained by differentiating Eq. (6–93) and setting the result equal to zero, yielding

$$g'(0) = 0 = -\theta'_1(0)\left[(1 - g_{aw}) + \bar{\sigma}\left(\theta_2(0) - \frac{1}{2}\right)\right]$$

Since in general $\theta'_1(0) \neq 0$, the adiabatic wall enthalpy ratio is

$$g_{aw} = 1 + \bar{\sigma}\left(\theta_2(0) - \frac{1}{2}\right) \qquad (6-94)$$

Rewriting Eq. (6–94) in terms of a static recovery enthalpy at the surface yields

$$\frac{h_r}{h_e} = 1 + b(\text{Pr})\frac{\gamma - 1}{2}\text{M}_e^2 \qquad (6-95)$$

where $b(\mathrm{Pr}) = 2\theta_2(0,\mathrm{Pr})$. Recalling the results of Section 4–7 for unit Prandtl number, we have $b(1) = 1$. Further, near unit Prandtl number $b(\mathrm{Pr}) \approx \sqrt{\mathrm{Pr}}$.

Comparison of the theoretical results for the velocity profile on an adiabatic flat plate at a free stream Mach number of 2.41 and experimental measurements taken from von Driest [vonD52] is shown in Figure 6–8. The theory and experiment are in excellent agreement.

For the particular case of $\hat{\beta} = 0$, $M_e = $ constant, with the help of Eq. (6–61b) the shearing stress at the surface is written as

$$\tau = \mu \frac{\partial u}{\partial y} = \mu U_e \frac{\partial \eta}{\partial y} f''(0)$$

Using Eq. (6–72), and further assuming that $C = \rho_w \mu_w / \rho_e \mu_e$ so that the viscosity is correct at the wall, yields

$$\tau = \frac{\rho_w \mu_w U_e^2}{\sqrt{2\xi}}$$

Finally, recalling that for $\hat{\beta} = 0$ conditions are constant at the edge of the boundary layer, and using Eq. (6–71) allows the local skin friction coefficient to

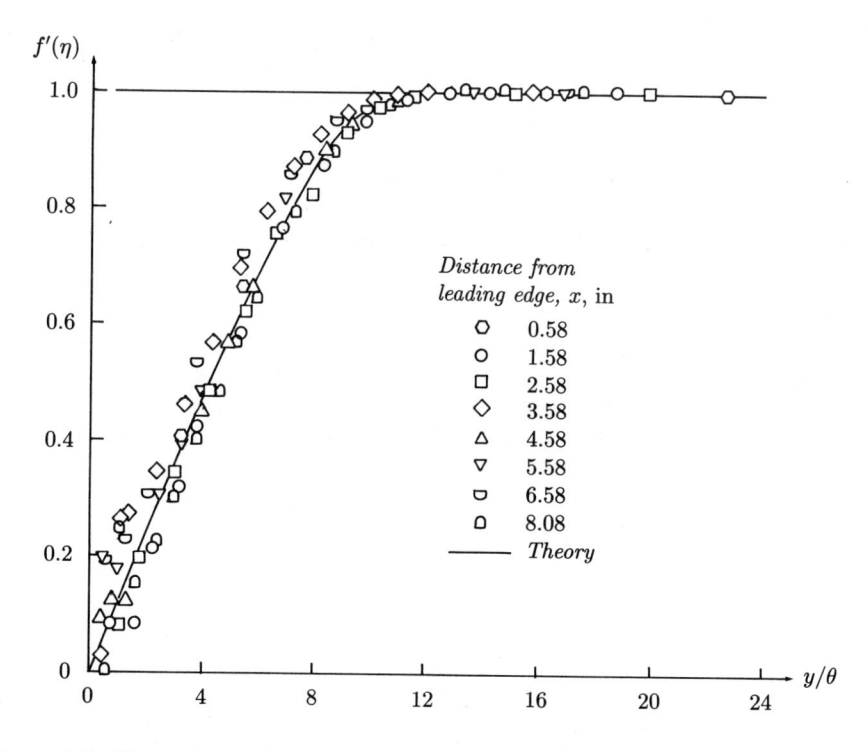

Figure 6–8. Comparison of experimental and theoretical velocity profiles for the compressible boundary layer on an adiabatic flat plate.

be written as

$$C_{f_x} = \frac{\tau}{\frac{\rho_w U_e^2}{2}} = \frac{\sqrt{2}}{\sqrt{Re_x}} f''(0)$$

where here $Re_x = (\rho_w U_e x)/\mu_w$.

Similarly, the heat transfer rate at the surface is

$$q = k\frac{\partial T}{\partial y} = \frac{1}{Pr}\left(\mu\frac{\partial h}{\partial y}\right)$$

From the definition of the stagnation enthalpy we have that $H(\partial g/\partial y) = \partial h/\partial y$. Thus, with the help of Eq. (6–72) we have

$$q = \frac{H_e}{Pr}\mu\frac{\partial \eta}{\partial y}\frac{\partial g}{\partial \eta} = \frac{\rho_w \mu_w}{Pr}\frac{U_e}{\sqrt{2\xi}} H_e g'(0)$$

Defining a local Nusselt number and using Eq. (6–71) allows us to write for $\hat{\beta} = 0$

$$Nu_x = \frac{C\rho_x}{k}\frac{q}{h_w - h_{aw}} = -\frac{1}{2}\sqrt{Re_x}\frac{g'(0)}{g_{aw} - g_w}$$

Finally, using Eqs. (6–22) and (4–89) yields

$$Nu_x = -\frac{1}{\sqrt{2}}\sqrt{Re_x}\,\theta_1'(0) = \frac{1}{\sqrt{2}}\sqrt{Re_x}\,\alpha_0\,(Pr)(f''(0))^{Pr}$$

Continuing, we write a form of Reynolds analogy as

$$Nu_x = \left\{\alpha_0(Pr)\left(f''(0)\right)^{Pr-1}\right\}\frac{Re_x C_f}{2} \qquad (6-96)$$

For unit Prandtl number this reduces to the familiar result for Reynolds analogy, i.e., $Nu_x = (1/2)Re_x C_f$ (see Eq. 4–68).

6–8 Similar Compressible Boundary Layer with Unit Prandtl Number

We now consider the general similar compressible boundary layer with unit Prandtl number, i.e., Eqs. (6–74), (6–75) and (6–79), with $\hat{\beta} = $ constant $\neq 0$. $M_e = $ (constant) $\xi^{\hat{\beta}/2}$ and $Pr = 1$. The governing boundary value problem is explicitly stated as

$$f''' + ff'' + \hat{\beta}(g - f'^2) = 0 \qquad (6-74)$$

$$g'' + fg' = 0 \qquad (6-81)$$

with
$$f(0) = f_w = \text{constant} \qquad f'(0) = 0 \qquad (6-79a,b)$$

$$g(0) = g_w = \text{constant} \quad \text{or} \quad g'(0) = 0 \qquad (6-79c,d)$$

and
$$f(\eta \to \infty) \to 1 \qquad g(\eta \to \infty) \to 1 \qquad (6-79e,f)$$

Similar Compressible Boundary Layer with Unit Prandtl Number

We recognize this as the same nondimensional boundary value problem governing the low speed ($M_e = 0$) compressible boundary layer. However, here the nondimensional dependent variable g is the ratio of stagnation enthalpies H/H_e. Since the use of H/H_e rather than h/h_e implicitly includes the effects of viscous dissipation, these effects are included in the unit Prandtl number solutions, whereas they are not included in the zero Mach number solutions.

Recall that for zero heat transfer at the surface an explicit integral of Eq. (6–81) subject to Eqs. (6–79d and f) is $g(\eta) = 1$. This shows that for an adiabatic wall the stagnation enthalpy is constant through the boundary layer. When the effects of viscous dissipation are neglected ($M_e = 0$), the static enthalpy is constant through the boundary layer. Thus, we see that for zero heat transfer at the surface the internal heat generated due to viscous dissipation in the velocity field and the heat transferred by diffusion and conduction in the temperature field interact in a precise manner to maintain the stagnation enthalpy constant throughout the boundary layer. This result is a consequence of the unit Prandtl number assumption.

Further analysis of the zero heat transfer solutions shows that, since $g(\eta) = 1$, the nondimensional momentum equation reduces to the Falkner-Skan equation. Hence the nondimensional momentum and energy equations are uncoupled. Thus, the nondimensional solutions of the Falkner-Skan equation discussed in Chapter 3 are applicable.

When there is heat transfer at the surface the boundary value problem specified by Eqs. (6–74), (6–79) and (6–81) has no known analytical solutions ($\hat{\beta} \neq 0$). This boundary value problem, and thus that associated with the $M_e = 0$ solutions, was studied numerically by Cohen [Cohe54], Levy [Levy54], Li and Nagamatsu [Li55], Cohen and Reshotko [Cohe56] and Rogers [Roge69b]. We defer consideration of the numerical solutions until Section 6–13.

Although the governing nondimensional boundary value problems are identical for the zero Mach number and unit Prandtl number similar compressible boundary layers, the shearing stress and heat transfer rate at the surface are not the same. Fundamentally, this is because of viscous dissipation. Recalling the development of the shearing stress at the surface given in Section 6–7, we have

$$\tau(x,0) = \frac{C \rho_e \mu_e U_e f''(0)}{\left(2 \int_0^x C \rho_e \mu_e U_e dx\right)^{1/2}} \qquad (6-97)$$

where we note that $\xi = \int_0^x C \rho_e \mu_e U_e \, dx$. Introducing $C = \rho_w \mu_w / \rho_e \mu_e$, to match the viscosity at the surface, the local skin friction coefficient becomes

$$C_f = \frac{\tau(x,0)}{\frac{\rho_w U_e^2}{2}} = 2 \left(\frac{\mu_w}{2 \int_0^x \rho_w U_e dx}\right)^{1/2} f''(0) \qquad (6-98)$$

232   Laminar Flow Analysis

Similarly, the heat transfer rate at the surface is

$$q(x,0) = \frac{C\rho_e\mu_e U_e g'(0)}{\Pr\left(2\int_0^x C\rho_e\mu_e U_e dx\right)^{1/2}} \qquad (6-99)$$

and the local Nusselt number is

$$\mathrm{Nu}_x = \frac{x\Pr q}{\mu(h_w - h_{aw})} = \frac{\rho_w U_e x}{\left(2\mu_w\int_0^x \rho_w U_e\, dx\right)^{1/2}} \frac{g'(0)}{g_w - g_{aw}} \qquad (6-100)$$

For the low speed compressible boundary layer ($M_e = 0$) we showed that

$$U_e(x) = (\text{constant})\; x^{\frac{\hat{\beta}}{(2-\hat{\beta})}}$$

whereas for the compressible boundary layer with unit Prandtl number

$$U_e(\xi) = \left(\frac{2H_e \frac{\gamma-1}{2}M_e^2}{1 + \frac{\gamma-1}{2}M_e^2}\right)^{1/2} \qquad (6-101)$$

with $M_e = (\text{constant})\; \xi^{\hat{\beta}/2}$. In order to obtain $U_e(x)$ for the unit Prandtl number solutions, the expression for $U_e(\xi)$ is substituted into $dx = d\xi/C\rho_e\mu_e U_e$ and the result integrated to yield $x(\xi)$. This result is then used in Eq. (6–101) to obtain $U_e(x)$. The final expression for $U_e(x)$ is considerably different than the simple result for $M_e = 0$. Substitution of the resulting expression for $U_e(x)$ and subsequent integration yield expressions for the skin friction coefficient and Nusselt numbers which are entirely different. However, Eqs. (6–97) and (6–99) show that for the same value of ξ the shearing stress and heat transfer rate at the surface are the same for dissipative and nondissipative flows. From Eq. (6–71) we see that as $x \to \infty$ the transformed variable $\xi \to$ constant. Consequently, the free stream velocity distribution for the dissipative flow approaches a constant as $\xi \to \infty$. However, the free stream velocity for the nondissipative flow increases without bound as $x \to \infty$. This is shown in Figure 6–9.

6-9 The Similar Hypersonic Compressible Boundary Layer with Nonunit Prandtl Number

Finally we consider the similar hypersonic compressible boundary layer with nonunit Prandtl number and nonzero pressure gradient. The hypersonic flow assumption implies that $M_e \to \infty$. Thus, the Mach number parameter is

$$\bar{\sigma} = \frac{(\gamma-1)M_e^2}{1 + \frac{\gamma-1}{2}M_e^2} = \frac{(\gamma-1)M_e^2}{\frac{\gamma-1}{2}M_e^2} = 2$$

Hypersonic Compressible Boundary Layer with Nonunit Prandtl Number

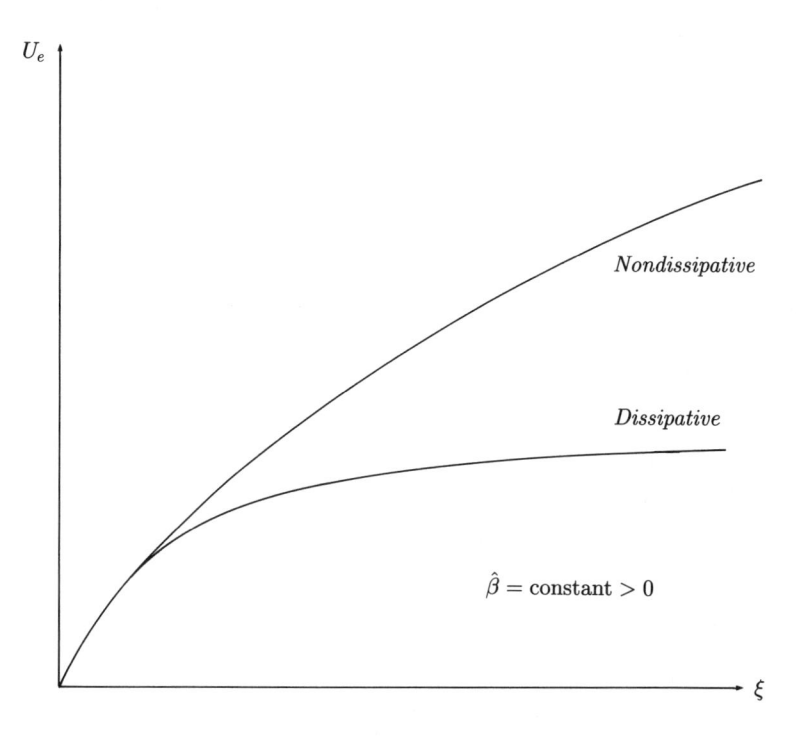

Figure 6–9. Free stream velocity variation for dissipative and nondissipative boundary layer flows.

The governing boundary value problem represented by Eqs. (6–74), (6–75) and (6–79) is then written as

$$f''' + ff'' + \hat{\beta}(g - f'^2) = 0 \qquad (6-74)$$

$$g'' + \Pr fg' = 2(1 - \Pr)(f'f'')' \qquad (6-102)$$

with boundary conditions

$$f(0) = f_w = \text{constant} \qquad f'(0) = 0 \qquad (6-79a,b)$$

$$g(0) = g_w = \text{constant} \quad \text{or} \quad g'(0) = 0 \qquad (6-79c,d)$$

$$f(\eta \to \infty) \to 1 \qquad g(\eta \to \infty) \to 1 \qquad (6-79e,f)$$

In Sections 6–6 to 6–8, under certain conditions we were able to reduce the boundary value problem for the compressible boundary layer to an equivalent incompressible boundary layer problem. However, this is not possible for the present problem. Fundamentally, this is because the stagnation enthalpy is not constant through the boundary layer even for an adiabatic wall, i.e., $g(\eta) = 1$ is *not* an integral of the energy equation. Thus, since $\hat{\beta} \neq 0$, the momentum equation (Eq. 6–74) cannot be reduced to the Falkner-Skan equation. Hence,

234   Laminar Flow Analysis

the functions $f(\eta)$, $f'(\eta)$ and $f''(\eta)$ required in the energy integrals, Eqs. (4–86) and (6–91), depend on $g(\eta)$, $g'(\eta)$ and $g''(\eta)$ because of the coupling between the momentum and energy equations, Eqs. (6–74) and (6–102). Because of this coupling, the energy integrals cannot be evaluated except by successive approximations using the incompressible Falkner-Skan solutions to begin the approximation (see, e.g., Levy [Levy54]). For an exact solution Eqs. (6–74) and (6–102), subject to the boundary conditions given in Eq. (6–79), must be considered as a pair of coupled ordinary differential equations. Rogers [Roge69b] performed extensive numerical integrations of this boundary value problem. These results are discussed in Section 6–14. The compressible boundary layer solutions discussed in Sections 6–6 to 6–8 are summarized in Table 6–1.

6–10 Displacement and Momentum Thickness in a Compressible Boundary Layer

Recalling the definitions of the boundary layer displacement and momentum thickness previously given in Chapter 3, we have

$$\delta^* = \int_0^\infty \left(1 - \frac{\rho}{\rho_e}\frac{u}{U_e}\right) dy \qquad (3-74)$$

and

$$\theta = \int_0^\infty \frac{\rho u}{\rho_e U_e}\left(1 - \frac{u}{U_e}\right) dy \qquad (3-75)$$

Looking first at the displacement thickness, with the help of Eqs. (6–72) and (6–73) we have

$$\delta^* = \frac{\sqrt{2\xi}}{\rho_e U_e}\left[\frac{H_e}{h_e}\int_0^\infty (g - f'^2)\, d\eta - \int_0^\infty f'(1 - f')\, d\eta\right] \qquad (6-103)$$

or

$$\delta^* = \frac{\sqrt{2\xi}}{\rho_e U_e}\left(\frac{H_e}{h_e} J_1 - J_2\right) \qquad (6-104)$$

where

$$J_1 = \int_0^\infty (g - f'^2)\, d\eta \qquad (6-105a)$$

and

$$J_2 = \int_0^\infty f'(1 - f')\, d\eta \qquad (6-105b)$$

Finally, we write

$$\delta^* = \frac{\sqrt{2\xi}}{\rho_e U_e}\left[\left(1 + \frac{\gamma - 1}{2} M_e^2\right)(J_4 - J_3) - J_2\right] \qquad (6-106)$$

where

$$J_3 = \int_0^\infty (1 - g)\, d\eta \qquad (6-107a)$$

Table 6-1. Summary of governing equations for similar compressible boundary layer.

	Governing Equations and Boundary Conditions	Adiabatic Wall Analytical Solutions	Isothermal Wall Analytical Solutions	Comments
$\hat{\beta} \neq 0$ $M_e = 0$	$f''' + ff'' + \hat{\beta}(g - f'^2) = 0$ $g'' + \Pr fg' = 0$	yes $f''' + ff'' + \hat{\beta}(1 - f'^2) = 0$	no	$g(\eta) = \dfrac{h}{h_e}$ Viscous dissipation terms neglected
$\hat{\beta} = 0$ $M_e = \text{constant}$	$f''' + ff'' = 0$ $g'' + fg'_1 =$ $\bar{\sigma}(1 - \Pr)(f'f''')'$	yes for $\Pr = 1$ $f''' + ff'' = 0$ $g(\eta) = 1$ (Buseman Integral)	yes for $\Pr = 1$ $f''' + ff'' = 0$ $g(\eta) = g_w - (g_w - 1)f'$ (Crocco Integral) yes for $\Pr \neq 1$ $g(\eta) = 1 - (1 - g_w)\theta_1(\eta)$ $+ \bar{\sigma}(\theta_2(\eta) - \theta_2(0)\theta_1(\eta))$ $+ \dfrac{\bar{\sigma}}{2}\left(f'^2 + \theta_1(\eta)\right)$	$g(\eta) = \dfrac{H}{H_e}$ Viscous dissipation terms included $g_{aw} = 1$ for $\Pr = 1$ $g_{aw} = 1 + \bar{\sigma}\left(\theta_2(0) - \dfrac{1}{2}\right)$ for $\Pr \neq 1$
$\hat{\beta} \neq 0$ $M_e = (\text{constant})\,\xi^{\hat{\beta}/2}$ $\Pr = 1$	$f''' + ff'' + \hat{\beta}(g - f'^2) = 0$ $g'' + fg' = 0$	yes $f''' + ff'' + \hat{\beta}(1 - f'^2) = 0$ $g(\eta) = 1$	no	No analytical solutions for $\Pr \neq 1$ Viscous dissipation terms included
$\hat{\beta} \neq 0$ $M_e \to \infty$	$f''' + ff'' + \hat{\beta}(g - f'^2) = 0$ $g'' + fg' = 2(1 - \Pr)(f'f''')'$	no	no	$M_e \to \infty$ yields $\bar{\sigma} = 2$ $g(\eta) \neq 1$ for the adiabatic wall

$$J_4 = \int_0^\infty (f1 - f'^2) \, d\eta \qquad (6-107b)$$

Similarly, the momentum thickness is written as

$$\theta = \frac{\sqrt{2\xi}}{\rho_e U_e} J_2 \qquad (6-108)$$

We now turn our attention to the evaluation of the physical distance y in terms of the known boundary layer parameters. Recalling that y is obtained by inverting Eq. (6–72), we have

$$y = \frac{\sqrt{2\xi}}{\rho_e U_e} \int_0^\eta \frac{\rho_e}{\rho} \, d\eta \qquad (6-109)$$

Using Eq. (6–73), this becomes

$$y = \frac{\sqrt{2\xi}}{\rho_e U_e} \int_0^\eta \left[1 - (1 - f'^2) + \frac{H_e}{h_e}(g - f'^2) \right] d\eta$$

and after integrating

$$y = \frac{\sqrt{2\xi}}{\rho_e U_e} \left(\eta - \bar{J}_4 + \frac{H_e}{h_e} \bar{J}_1 \right) \qquad (6-110)$$

where

$$\bar{J}_1 = \int_0^\eta (g - f'^2) \, d\eta$$

$$\bar{J}_4 = \int_0^\eta (1 - f'^2) \, d\eta$$

These integrals are analogous to the boundary layer displacement and momentum integrals except that they are evaluated progressively from zero to η rather than from zero to infinity. Thus, we see that at a specific location on the surface the physical distance, y, is given in terms of η and integrals of the boundary layer solutions.

6–11 Numerical Integration of the Compressible Boundary Layer Equations

We consider numerical integration of the general fifth-order coupled two-point asymptotic ordinary boundary value problem represented by Eqs. (6–74) and (6–75) subject to the boundary conditions given in Eqs. (6–79). These equations are an example of a Class 3 problem (see Section 1–10). We previously encountered

Class 3 problems in our discussion of the free convection similar boundary layer and of the locally nonsimilar boundary layer method. The program **comb**, which numerically integrates the governing two-point asymptotic boundary value problem for the compressible boundary layer equations, is similar in philosophy to the **freeb** program previously discussed in Section 5–3. The **comb** program allows for nonunit Prandtl number, similar mass transfer at the surface, variable values of the Mach number parameter $\bar{\sigma}$ and variable isothermal wall-to-stagnation temperature ratios, g_w. The integrals J_1, J_2, J_3 and J_4 required to evaluate the displacement and momentum thicknesses are also calculated.

Since the velocity and temperature (enthalpy) fields are coupled, the momentum and energy equations must be solved simultaneously. In addition, since two asymptotic boundary conditions, i.e., $f'(\eta \to \infty) \to 1$ and $g(\eta \to \infty) \to 1$, must be satisfied, it is necessary to estimate two unknown initial conditions at $y = \eta = 0$. These are $f''(0)$, the nondimensional shearing stress parameter, and $g'(0)$, the nondimensional heat transfer parameter. Recalling that the momentum and energy equations are coupled, we see that a change in $f''(0)$ affects both $f'(\eta_{\max})$ and $g(\eta_{\max})$. Similarly, a change in $g'(0)$ affects both $f'(\eta_{\max})$ and $g(\eta_{\max})$, i.e., there is a coupling between the effect of $f''(0)$ and $g'(0)$ on $f'(\eta_{\max})$ and $g(\eta_{\max})$. Here we again use the notation $\eta_{\max}$ for the numerical value used to represent infinity. Thus, the iteration scheme used to insure asymptotic convergence to the required outer boundary conditions must account for this coupling between $f''(0)$ and $g'(0)$. The Nachtsheim-Swigert iteration scheme incorporates this coupling effect; it is discussed in detail in Appendix B.

6–12 Results for the Low Speed Compressible Boundary Layer

In our discussion of the low speed compressible boundary layer (see Section 6–6) we pointed out that for the special case of a flat plate the momentum and energy equations are uncoupled. We further showed that for this particular case the solution of the momentum equation is given by the Blasius equation (Eq. 3–38) and that the solution of the energy equation is related to the analogous solution for the incompressible constant property forced convection problem, Eq. (4-78). The first run of the **comb** program shown below illustrates this solution. The details of the main program and the **eqmot** routine are given in Appendix D–16.

The solution is for a nonunit Prandtl number and for heat transfer at the surface, i.e., Pr = 0.7, g_w = 0.6, $\bar{\sigma}$ = 0, $f(0) = 0$, $\hat{\beta} = 0$. Recalling that $f''(0) = 0.469600$ for the Blasius solution (see Table C–1), we choose this value. The required value for $g'(0)$ is obtained using Eq. (6-82) and $\theta_1'(0) = -0.413912$ from Table C–13, i.e.

$$g'(0) = -(1 - g_w)\theta_1'(0) = 0.1655648$$

Recalling the solutions for the free convection boundary layer (see Chapter 5) we choose $\eta_{\max} = 8$.

238   Laminar Flow Analysis

```
Beta = 0
Prandtl Number = .7
Mach Number Parameter = 0
f''(0) = 0.469600
g'(0) = 0.1655648
```

η	f''	f'	f	g'	g
0	0.469600	0	0	0.1655648	0.6
8.	3.371572e−11	1.	6.78322	1.312928e−8	0.9999997

Convergence achieved

$J_1 = 1.132408 \qquad J_2 = 0.4695999 \qquad J_3 = 0.5539728 \qquad J_4 = 1.686380$

Convergence occurs immediately. It is interesting to note that the momentum thickness integral $J_2 = 0.469599$ in the limit as $\eta \to \infty$ equals $f''(0)$.

Recalling that (see Eq. 6–105)

$$J_2 = \int_0^\infty f'(1-f')\,d\eta = \int_0^\infty f'\,d\eta - \int_0^\infty f'f'\,d\eta \qquad (6-105)$$

and integrating the second integral by parts yields

$$J_2 = f - ff'\Big]_0^\infty + \int_0^\infty ff''\,d\eta$$

For $\beta = 0$ the Blasius equation yields $f = -f'''/f''$ and thus

$$J_2 = -\int_0^\infty f'''\,d\eta = f''(0)$$

since for asymptotic convergence $f''(\eta \to \infty) \to 0$.

An additional solution, in which the momentum and energy equations are uncoupled, is for an adiabatic wall. For this case the nondimensional momentum equation is the Falkner-Skan equation (see Section 3–3). The energy equation is then integrated to yield $g(\eta) = 1$. This solution is modeled in the **comb** run shown below, i.e., Pr = 0.7, $g_w = 1$, $\bar{\sigma} = 0$, $f(0) = 0$, $\hat{\beta} = 0.1$. Referring to our discussion in Chapter 3 and to Table C–1, we take $f''(0) = 0.5870354$. For an adiabatic wall $g'(0) = 0$. Convergence occurs immediately. Notice that $J_3 = \int_0^\infty (1-g)\,d\eta = 0$.

```
Beta = 0.1
Prandtl Number = .7
Mach Number Parameter = 0
f''(0) = 0.5870354
g'(0) = 0
```

η	f''	f'	f	g'	g
0	0.5870354	0	0	0	1
8.	1.220611e−8	1.	6.919682	0	1

Convergence achieved

$J_1 = 1.515773 \qquad J_2 = 0.4354552 \qquad J_3 = 0 \qquad J_4 = 1.515773$

Results for the Low Speed Compressible Boundary Layer

We now turn our attention to the low speed compressible boundary layer when the momentum and energy equations are coupled. A numerical solution is illustrated in a third run of the `comb` program. Here we choose Pr = 0.7, $g_w = 0.6$, $\bar{\sigma} = 0$, $f(0) = 0$, $\hat{\beta} = 0.1$, i.e., nonunit Prandtl number, heat transfer at the surface and a nonzero pressure gradient. Comparison with the first solution shows that these results serve to illustrate the effect of pressure gradient on the heat transfer and shearing stress at the surface. Comparison with the second solution illustrates the effect of heat transfer on the shearing stress at the surface. Expecting that these effects are small, we first take $f''(0) = 0.5870354$ and $g'(0) = 0.1655648$. Recalling that the first three trial integrations are used to determine the Nachtsheim-Swigert iteration derivatives and are for

$f''(0), g'(0);$
$f''(0) + \epsilon, g'(0);$
$f''(0), g'(0) + \epsilon,$

respectively, where ϵ is a small perturbation, we see that the results of the first two trial integrations shown below lead us to expect that $f''(0) < 0.5870354$, and those of the second and third trial integrations to expect $g'(0) > 0.1655648$.

```
Beta = 0.1
Prandtl Number = 0.7
Mach Number Parameter = 0
f''(0) = 0.5870354
g'(0) = 0.1655648
```

η	f''	f'	f	g'	g
0	0.5870354	0	0	0.1655648	0.6
8.	2.852534e−3	1.089444	7.414446	2.437444e−9	0.9788719

```
f''(0) = 0.5880354
g'(0) = 0.1655648
```

0	0.5880354	0	0	0.1655648	0.6
8.	2.9169118e−3	1.091667	7.428011	2.360933e−9	0.9786324

```
f''(0) = 0.5870354
g'(0) = 0.1665648
```

0	0.5870354	0	0	0.1665648	0.6
8.	2.802587e−3	1.088792	7.411793	2.463442e−9	0.9811686

```
f''(0) = 0.54832513
g'(0) = 0.17072791
```

0	0.54832513	0	0	0.17072791	0.6
8.	−7.77376e−5	0.9977903	6.866439	8.992490e−9	1.000829

```
f''(0) = 0.54924108
g'(0) = 0.1704624
```

0	0.54924108	0	0	0.1704624	0.6
8.	4.037124e−6	1.000112	6.880026	8.702357e−9	0.9999518

```
f"(0)  = 0.54919531
g'(0)  = 0.17047861
0       0.54919531      0            0           0.17047861    0.6
8.     -2.057939e-7    0.9999943    6.87934      8.7169e-9     1.000003

f"(0)  = 0.54919761
g'(0)  = 0.17047774
0       0.54919761      0            0           0.17047774    0.6
8.      1.057168e-8    1.000000     6.879374     8.716163e-9   0.9999999
Convergence achieved
J₁ = 1.026976     J₂ = 0.4464980     J₃ = 0.540148     J₄ = 1.567123
```

Convergence occurs in four iterations. The results show that $f''(0)$ and $g'(0)$ are increased as a result of the increase in the pressure gradient. Further, we see that the shearing stress at the surface, $f''(0)$, is decreased by heat transfer at the surface.

Extensive numerical investigations of the low speed compressible boundary layer were performed by Brown and Donoughe [Brow51], Low [Low53] and Levy [Levy54]. Since the energy equation for the low speed compressible boundary layer (see Section 6–6) and that for the unit Prandtl number solutions (see Section 6–8) are the same, the unit Prandtl number solutions of Cohen and Reshotko [Cohe56] and of Rogers [Roge69b] are applicable. Here $g(\eta)$ must be interpreted as a nondimensional static enthalpy (temperature) profile through the boundary layer, whereas for the unit Prandtl number solutions it is interpreted as a nondimensional stagnation enthalpy profile through the boundary layer. Selected values for $f''(0)$ and $g'(0)$ taken from Cohen and Reshotko [Cohe56] and Rogers [Roge69b] are shown in Figures 6–10 and 6–11, respectively. Numerical values are given in Table C–25.

Examination of Figure 6–10 shows that for $\hat{\beta} > 0$ the skin friction parameter $f''(0)$ is considerably increased from its incompressible value if the surface is heated and considerably decreased if it is cooled. In contrast, for retarded flows, i.e., $\hat{\beta} < 0$, $f''(0) > 0$, the skin friction parameter, $f''(0)$, is increased by surface cooling and decreased if the surface is heated. In the reverse flow region these trends tend to reverse themselves. Notice that for $\hat{\beta} = 0$, $f''(0) = 0.4696005$ for all values of the wall-to-edge temperature ratio, g_w. Figure 6–10 also shows that cooling the surface delays separation, whereas heating the surface hastens separation. Further, notice that the minimum value of $\hat{\beta}$ does not necessarily occur at the separation point. Figure 6–11 shows that for $\hat{\beta} > 0$ the heat transfer parameter, $g'(0)$, increases with increasing surface temperature, whereas for retarded flow, $\hat{\beta} < 0$, $f''(0) < 0$, these trends tend to reverse themselves

Figures 6–10 and 6–11 also show the effect of nonunit Prandtl number on the shearing stress and heat transfer at the surface. Results are shown for a Prandtl number of 0.723, which is representative of many gases. Selected numerical results are given in Table C–26. Here we see that the nonunit Prandtl number

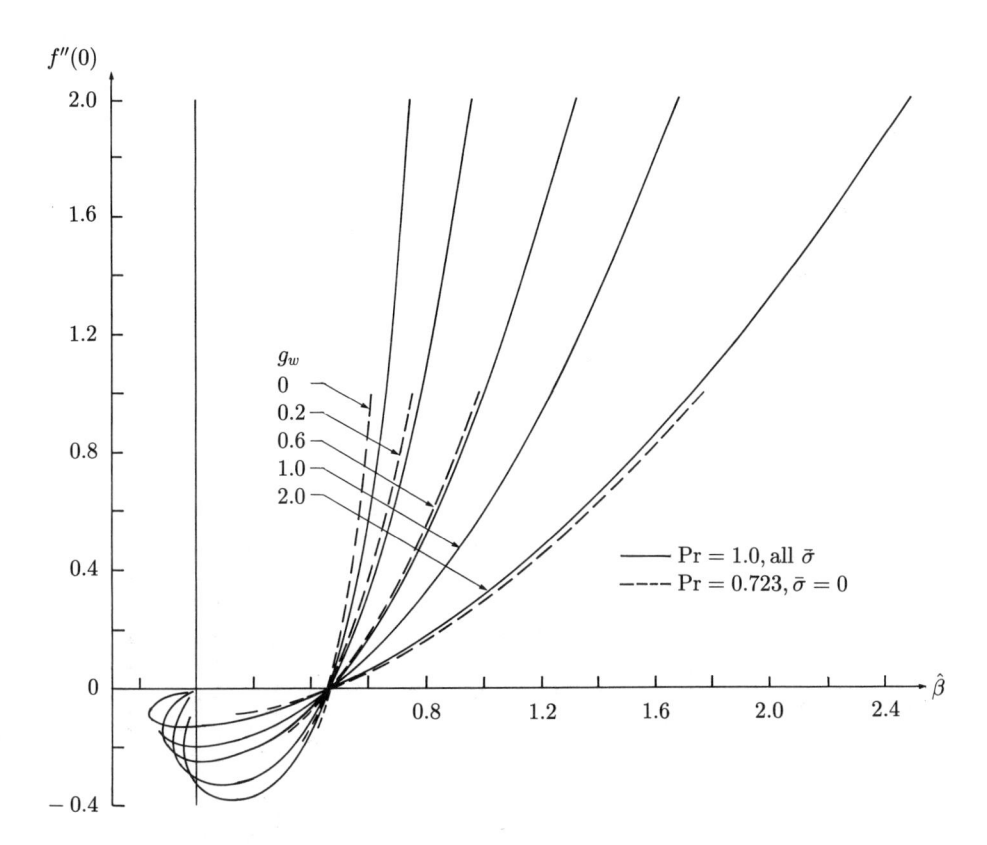

Figure 6-10. Variation of the wall shearing stress parameter, $f''(0)$, with the pressure gradient parameter, $\hat{\beta}$.

effect on the shearing stress is small. However, the heat transfer parameter, $g'(0)$, is significantly increased. Recalling Eqs. (6–97) and (6–99), we see that these observations are also true for the shearing stress and heat transfer rate at the surface. Figures 6–12 and 6–13 show the effect of wall-to-edge temperature ratio, g_w, i.e., heat transfer on the velocity and enthalpy profiles. The observed effects decrease with increasing value of $\hat{\beta}$.

6–13 Results for the Compressible Boundary Layer with Unit Prandtl Number

In our previous discussion of the unit Prandtl number solutions for the similar compressible boundary layer (see Section 6–8), we showed that the low speed compressible boundary layer ($M_e = 0$) and the compressible boundary layer ($M_e \neq 0$) were governed by the same nondimensional boundary value problem. Thus, the unit Prandtl number velocity and enthalpy profiles shown in Figures 6–12 and 6–13 are also solutions for the present problem. However, here

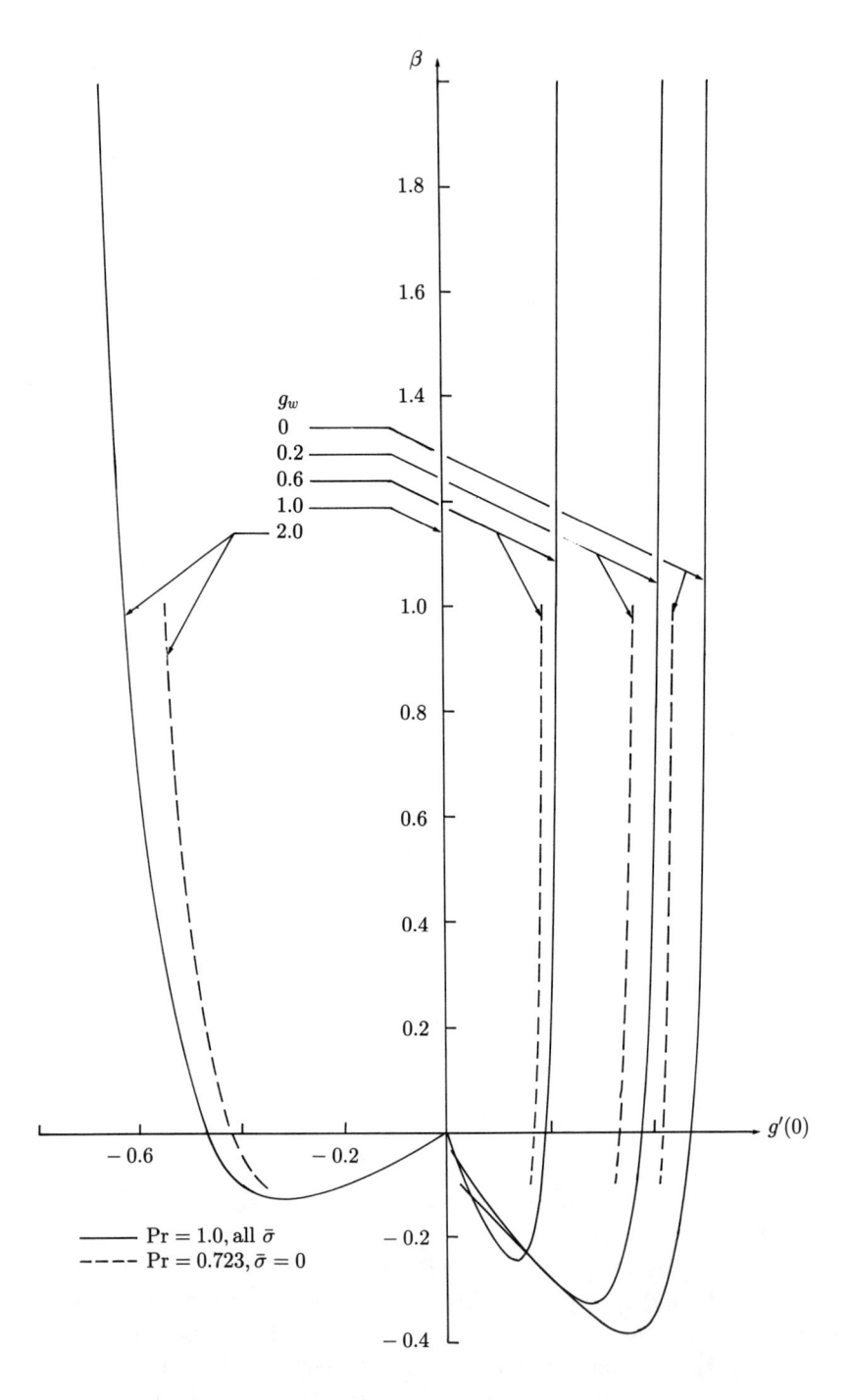

Figure 6–11. Variation of the wall heat transfer parameter, $g'(0)$, with pressure gradient parameter, $\hat{\beta}$.

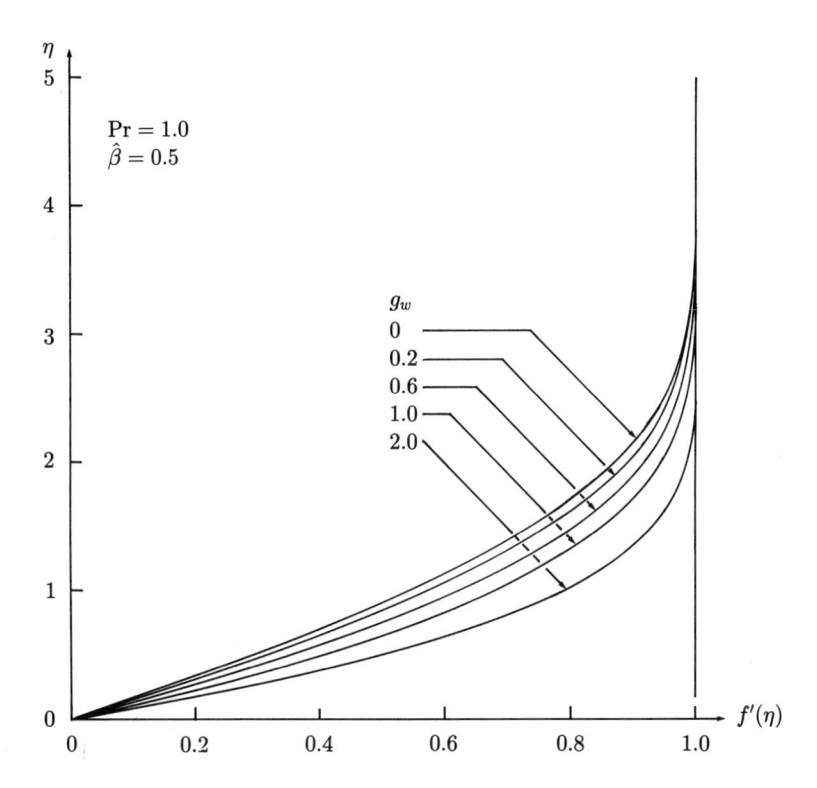

Figure 6–12. Effect of wall-to-edge temperature ratio on the nondimensional velocity profile, $\hat{\beta} = 0.5$.

the enthalpy profiles, $g(\eta)$, are interpreted as the ratio of the local stagnation enthalpy in the boundary layer to that in the inviscid flow at the edge of the boundary layer. In addition, the unit Prandtl number results for the shearing stress parameter, $f''(0)$, and the heat transfer parameter, $g'(0)$, shown in Figures 6–10 and 6–11 and given in Table C–25 are also solutions for the unit Prandtl number compressible boundary layer. However, because the external inviscid flows are different this does not mean that the physical shearing stress, τ, or the heat transfer rate, q, are the same for the two flows.

Before proceeding to a detailed discussion of the solutions we note that $\hat{\beta} = 1$ again corresponds to two-dimensional stagnation point flow [Stew50] and that the boundary layer equations for axisymmetric flow can be transformed to those for $\hat{\beta} = 1/2$ [Schl68]. Further, from Figures 6–10 and 6–11 we again note (see Section 3–4) that the lower branch solutions are not unique, i.e., for a given value of $\hat{\beta} < 0$ two values of $f''(0)$ and $g'(0)$ are possible. The uniqueness of the unit Prandtl number solution is discussed by Cohen and Reshotko [Cohe56].

Figures 6–14 show nondimensional velocity profiles for $g_w = 0.6$ and $g_w = 2$ for various values of $\hat{\beta}$. Here we notice that for the cooled surface ($g_w = 0.6$)

244   Laminar Flow Analysis

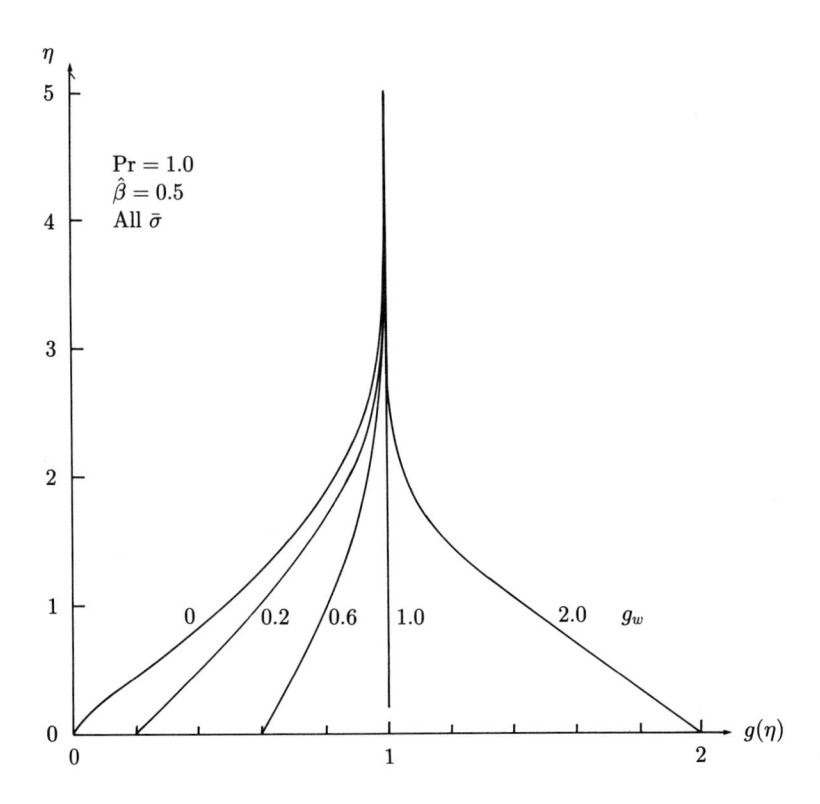

Figure 6–13. Effect of wall-to-edge temperature ratio on the nondimensional enthalpy profile, $\hat{\beta} = 0.5$.

with a favorable pressure gradient ($\hat{\beta} > 0$) the velocity increases monotonically from zero at the surface to the required value of one at the boundary layer edge. However, for the heated surface ($g_w = 2$) there is velocity overshoot ($f'(\eta) > 1$) in the boundary layer. Physically this is caused by the heat transfer from the surface to the fluid. The resulting increase in temperature and hence reduction in density allows the pressure forces to accelerate the flow to velocities in excess of that in the inviscid flow at the boundary layer edge in spite of the retarding effect of the viscous forces. For adverse pressure gradients ($\hat{\beta} < 0$) an inflection point exists in the velocity profiles for both heated and cooled surfaces. Further examination of the nondimensional velocity profiles shows that for $f''(0) > 0$ the boundary layer thickness increases as the shearing stress decreases while the separation point is approached. For the reverse flow solutions ($f''(0) < 0$) the boundary layer thickness continues to increase as the singular point (see Section 3–4) at $f'(0) = 0^-, \hat{\beta} = 0^-$ is approached. Physically, this increase in boundary layer thickness is a result of the decreased effectiveness of the external pressure field in accelerating the flow to the required inviscid velocity. Comparison of Figures 6–14a and b shows that in terms of η the boundary layer thickness for a given value of $\hat{\beta}$

increases as the wall temperature decreases. However, examination of Eqs. (6–72) and (6–110), along with the numerical results in Table C–25, shows that in terms of the physical distance, y, the *opposite* is true. Physically, a decrease in boundary layer thickness as the wall temperature increases is more realistic. This is because the increased heat transfer from the wall to the fluid for $g_w > 1$ decreases the density of the fluid near the surface. Thus, the pressure forces are more effective in accelerating the flow.

Examination of Figure 6–15 shows that the maximum value of the nondimensional shearing stress parameter, $f''(0)$, occurs at the surface for all favorable pressure gradients ($\hat{\beta} > 0$). However, for retarded and reverse flow the point of maximum shear moves increasingly away from the surface.

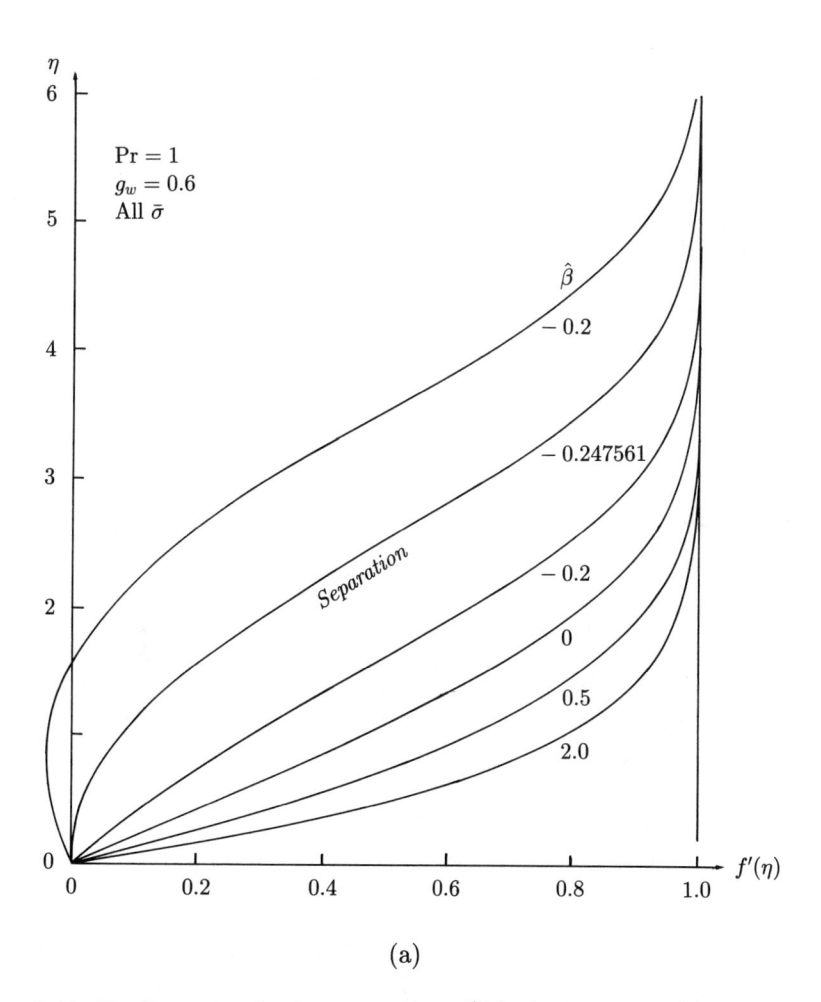

Figure 6–14. Nondimensional velocity profiles, $f'(\eta)$, for Pr $= 1$. (a) $g_w = 0.6$; (b) $g_w = 2$.

246   *Laminar Flow Analysis*

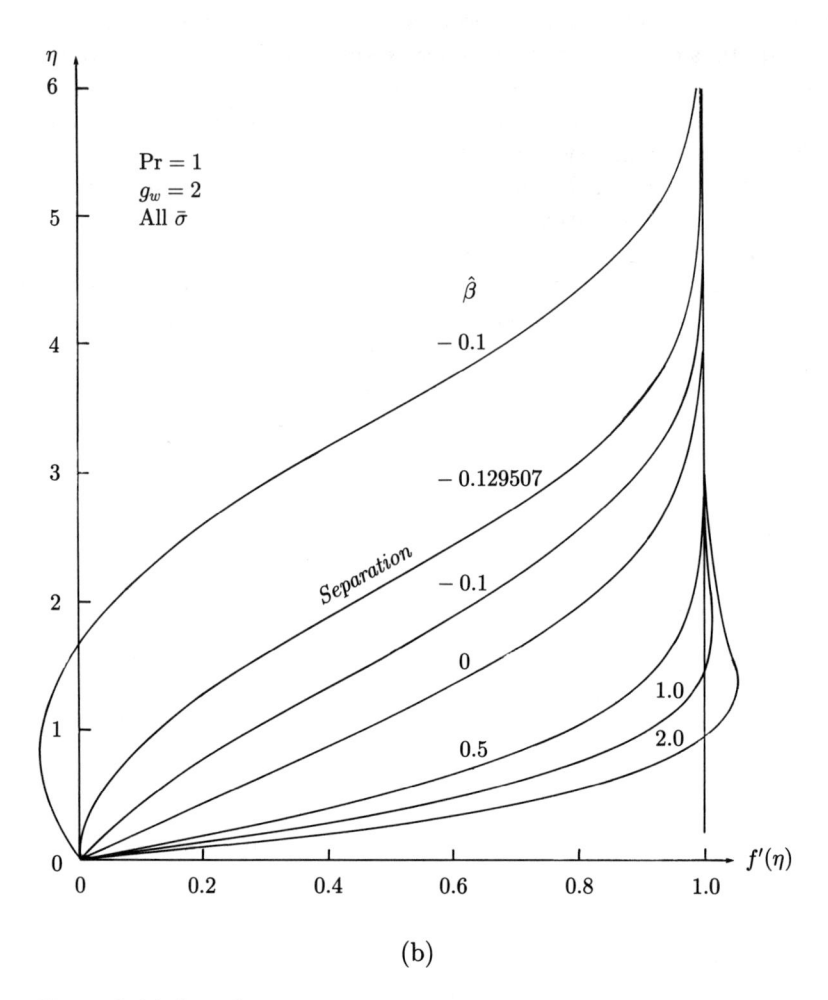

Figure 6–14. (*cont.*)

Recalling Figure 6–10, we note that the shearing stress parameter, $f''(0)$, is more sensitive to pressure gradient effects for heated walls than for cooled walls. Physically, this effect is caused by the decrease in density of the fluid in the boundary layer when the surface is heated. Thus, the pressure forces are more effective in accelerating the flow. Further, Figure 6–10 shows that estimating the shearing stress (see Eq. 6–97) on the basis of the flat plate results ($\hat{\beta} = 0$), or by estimates of $f''(0)$ obtained by linear extrapolations about $\hat{\beta} = 0$, leads to significant errors.

Of considerable interest in any boundary layer study is the value of $\hat{\beta}$ at the separation point. The comb program can be modified to calculate the adverse pressure gradient and heat transfer rate at the separation point. The necessary modifications to the iteration scheme are discussed in Appendix B.

Results for the Compressible Boundary Layer with Unit Prandtl Number 247

A typical run of comb incorporating these modifications is shown below for $Pr = 1$, $g_w = 0.6$, $f(0) = 0$. On the basis of incompressible solutions, we take $\hat{\beta} = -0.1988376$ as our first estimate of the separation value for the pressure gradient parameter. Based on the results of our previous run (see Section 6–12), we take $g'(0) = 0.1655648$ as our first estimate of the heat transfer parameter at the wall. Observing the results of the first integration, we anticipate that both $\hat{\beta}$ and $g'(0)$ are too large. After thirteen iterations our observations are confirmed with final results of $\hat{\beta} = -0.247561$ and $g'(0) = 0.125094$.

```
Beta = -0.1988376
Prandtl Number = 1
Mach Number Parameter = 0
f"(0) = 0
g'(0) = 0.1655648
```

η	f''	f'	f	g'	g
0	0	0	0	0.1655648	0.6
9.	3.922874e−3	1.016491	6.271214	2.770083e−10	1.152183

```
Beta = -0.1978376
g'(0) = 0.1655648
```

0	0	0	0	0.1655648	0.6
9.	3.999174e−3	1.015499	6.258592	2.924539e−10	1.152839

```
Beta = -0.1988376
g'(0) = 0.1665648
```

0	0	0	0	0.1665648	0.6
9.	3.938694e−3	1.017758	6.278702	2.722156e−10	1.155387

```
Beta = -0.233757
g'(0) = 0.12521709
```

0	0	0	0	0.12521709	0.6
9.	8.473501e−4	0.9918342	6.348089	1.002510e−10	1.005855

```
Beta = -0.24214339
g'(0) = 0.1251056
```

0	0	0	0	0.1251056	0.6
9.	3.254658e−4	0.9969078	6.428856	6.917563e−11	1.002135

⋮

9 iterations

⋮

```
Beta = -0.24756036
g'(0) = 0.12509448
```

0	0	0	0	0.12509448	0.6
9.	7.206177e−8	0.9999993	6.479478	5.474223e−11	1.000000

```
Beta = -0.24756105
g'(0) = 0.12509448
```

248   Laminar Flow Analysis

```
    0          0              0       0          0.12509448      0.6
    9.   3.163229e-8          1.      6.479484   5.474061e-11    1.000000
Convergence achieved

J₁ = 2.428472    J₂ = 0.6011969    J₃ = 0.6932411    J₄ = 3.121713
```

The adverse pressure gradient at separation is shown as a function of wall-to-stagnation temperature in Figure 6–16. Note that cooling the surface ($g_w < 1$) delays separation, whereas heating the surface ($g_w > 1$) hastens separation.

The nondimensional stagnation enthalpy profiles are shown in Figures 6–17. For positive values of the shearing stress parameter, i.e., $f''(0) > 0$, the stagnation enthalpy varies monotonically across the boundary layer. For the reverse flow solutions, $f''(0) < 0$, an inflection point is observed in the enthalpy profile. This implies that the maximum heat transfer rate does not occur at the surface. This effect is shown in Figures 6–18, where the nondimensional heat transfer parameter, $g'(\eta)$, is essentially constant. Heat transfer in this region occurs

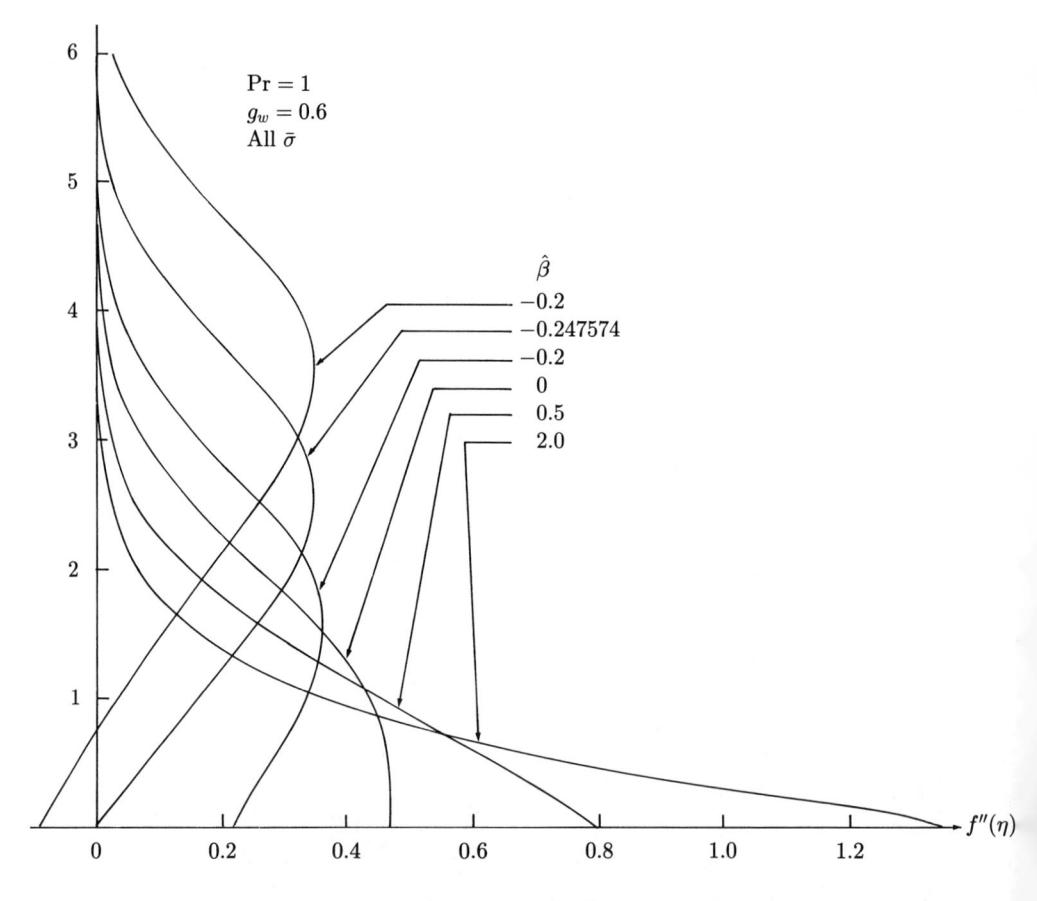

Figure 6–15. Nondimensional shearing stress profiles, $f''(\eta)$, for Pr = 1 and $g_w = 0.6$.

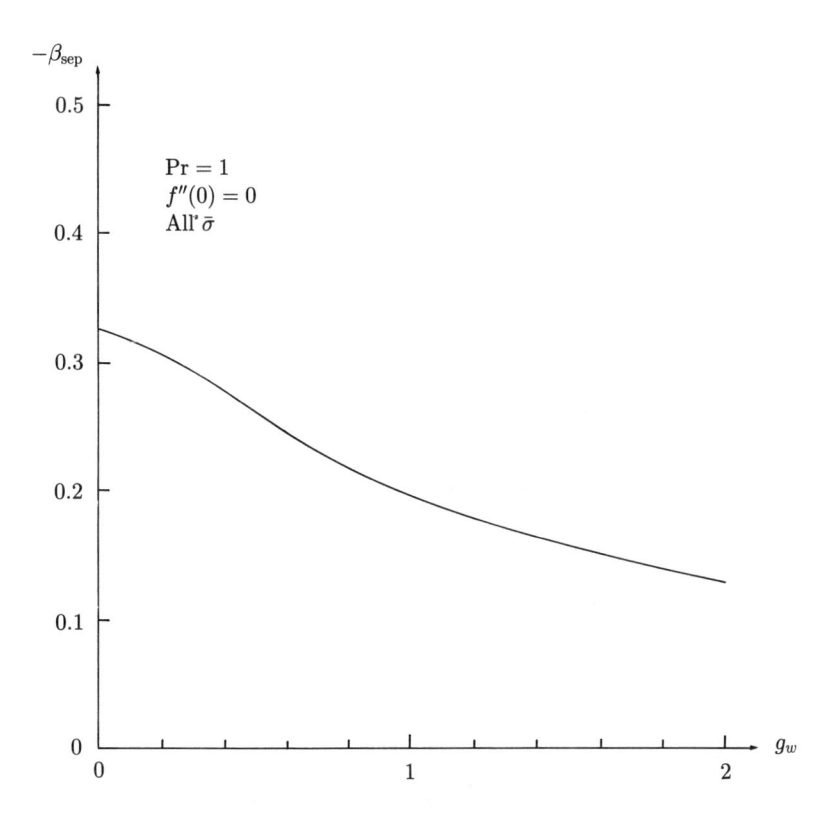

Figure 6–16. Effect of heat transfer on the adverse pressure gradient at separation, Pr = 1.

principally by conduction. Physically, this is because the nearly zero velocities in the region make convective heat transfer ineffective. Notice also that as the pressure gradient becomes more adverse the extent of the region increases. Thus, we conclude that in retarded flows heat transfer from or to the surface occurs primarily by conduction. Close comparison of Figures 6–14 and 6–17 shows that for favorable pressure gradients the thermal boundary layer is thicker than the velocity boundary layer, whereas for adverse pressure gradients the reverse is true, regardless of wall-to-stagnation temperature ratio.

Recalling Figure 6–11, we see that for favorable pressure gradients, i.e., $\hat{\beta} \geq 0$, and cooled walls, $g_w < 1$, the nondimensional heat transfer parameter, $g'(0)$, is almost constant. However, for heated walls the variation is significant. For retarded and reverse flows the variation of $g'(0)$ is quite distinct, with the effect increasing with increasing wall temperature. From these observations we conclude that estimation of the surface heat transfer rate on the basis of the flat plate results ($\hat{\beta} = 0$) is acceptable for cooled walls and favorable pressure gradients. For heated walls with favorable pressure gradients and retarded or reverse flows, this procedure yields significant errors.

250   Laminar Flow Analysis

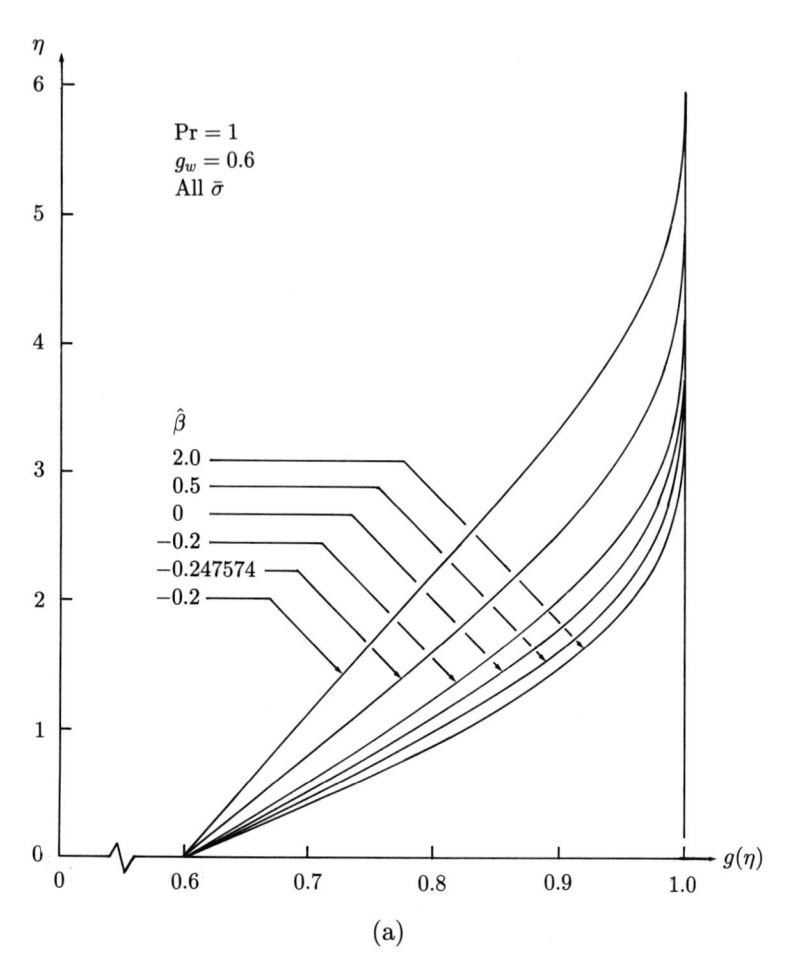

Figure 6–17. Nondimensional stagnation enthalpy profiles. (a) $g_w = 0.6$; (b) $g_w = 2$.

Recall that the skin friction coefficient is

$$C_f = \left(\frac{\mu_w}{2\int_0^x \rho_w U_e dx}\right)^{1/2} f''(0) \qquad (6-98)$$

and the Nusselt number is

$$\mathrm{Nu}_x = \frac{\rho_w U_e x}{\left(2\mu w \int_0^x \rho_w U_e dx\right)^{1/2}} \frac{g'(0)}{g_w - g_{aw}} \qquad (6-100)$$

Since for the unit Prandtl number solutions $g_{aw} = 1$, we now write Reynolds analogy as

Results for the Compressible Boundary Layer with Unit Prandtl Number

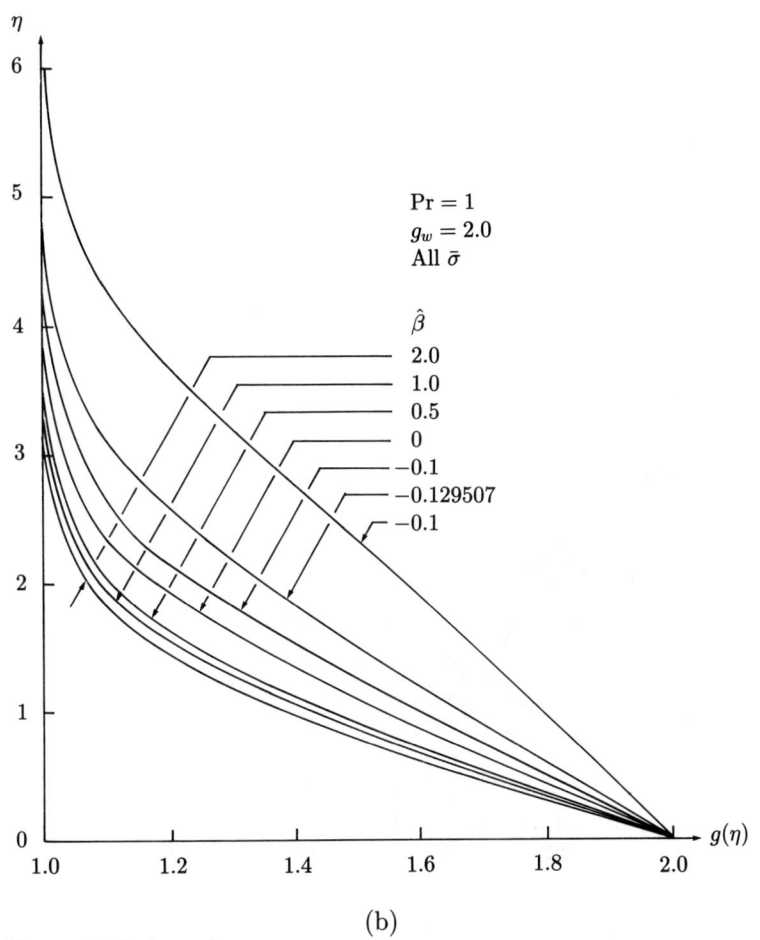

Pr = 1
$g_w = 2.0$
All $\bar{\sigma}$

$\hat{\beta}$
2.0
1.0
0.5
0
−0.1
−0.129507
−0.1

(b)

Figure 6–17. (cont.)

$$\frac{\mathrm{Nu}_x}{C_f \mathrm{Re}_x} = \frac{1}{2(1-g_w)} \frac{g'(0)}{f''(0)} \qquad (6-111)$$

In order to avoid infinite results when $f''(0) = 0$ at separation, the reciprocal of this parameter is shown in Figure 6–19. Examination of Figure 6–19 shows that the use of Reynolds analogy to estimate heat transfer results for nonzero pressure gradients also yields significant errors.

Numerical results for the boundary layer integrals J_1, J_2 and J_3 are given in Table C–25. J_1 and J_2 are shown in Figures 6–20. After recalling Eq. (6–103), i.e.

$$\delta^* = \frac{\sqrt{2\xi}}{\rho_e U_e}\left(\frac{H_e}{h_e} J_1 - J_2\right) \qquad (6-103)$$

252   *Laminar Flow Analysis*

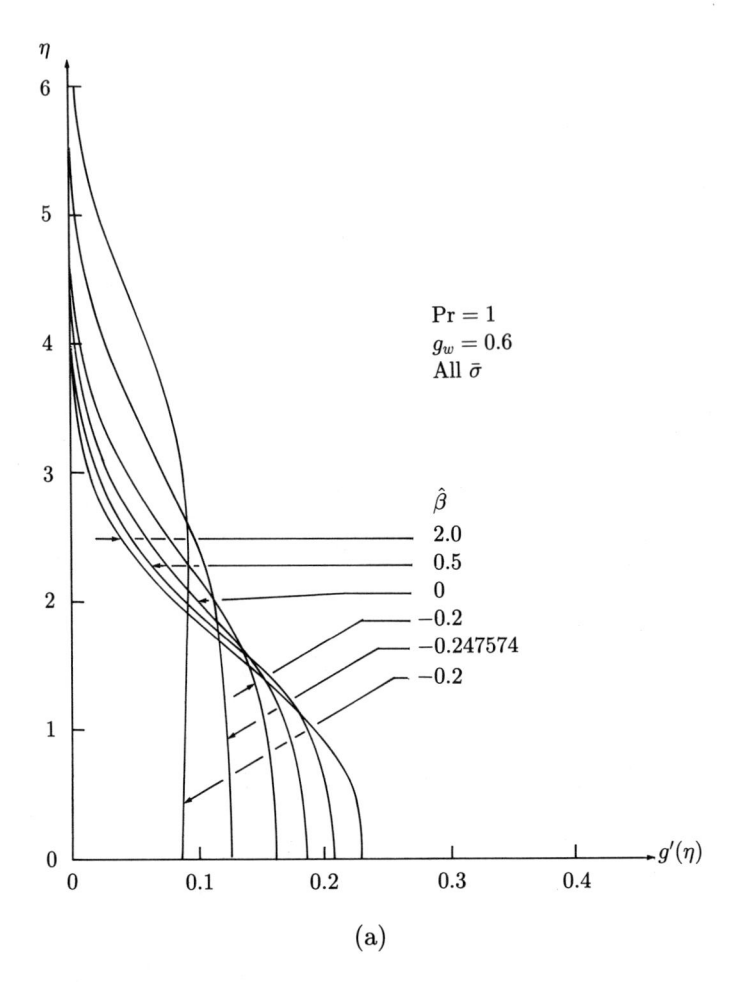

Pr = 1
$g_w = 0.6$
All $\bar{\sigma}$

$\hat{\beta}$
2.0
0.5
0
−0.2
−0.247574
−0.2

Figure 6–18. Nondimensional heat transfer parameter. (a) $g_w = 0.6$; (b) $g_w = 2$.

and examining the numerical results, we see that as $M_e \to 0$ the displacement thickness for highly cooled surfaces with favorable pressure gradient becomes negative. Physically, the increased density in the cool fluid near the surface results in more mass flow in the boundary layer than would exist in the equivalent inviscid flow. However, for larger Mach numbers and wall-to-stagnation temperature ratios the displacement thickness is positive, indicating a lower mass flow rate than would exist in the equivalent inviscid flow. Physically, for favorable pressure gradients this is due to the decreased density near the surface caused by heat transferred from the surface or generated internally by viscous dissipation reducing the mass flow in the boundary layer.

Further examination of Figure 6–20a shows that the displacement thickness increases with increasing wall temperature and decreasing pressure gradient. As

Results for the Compressible Boundary Layer with Unit Prandtl Number

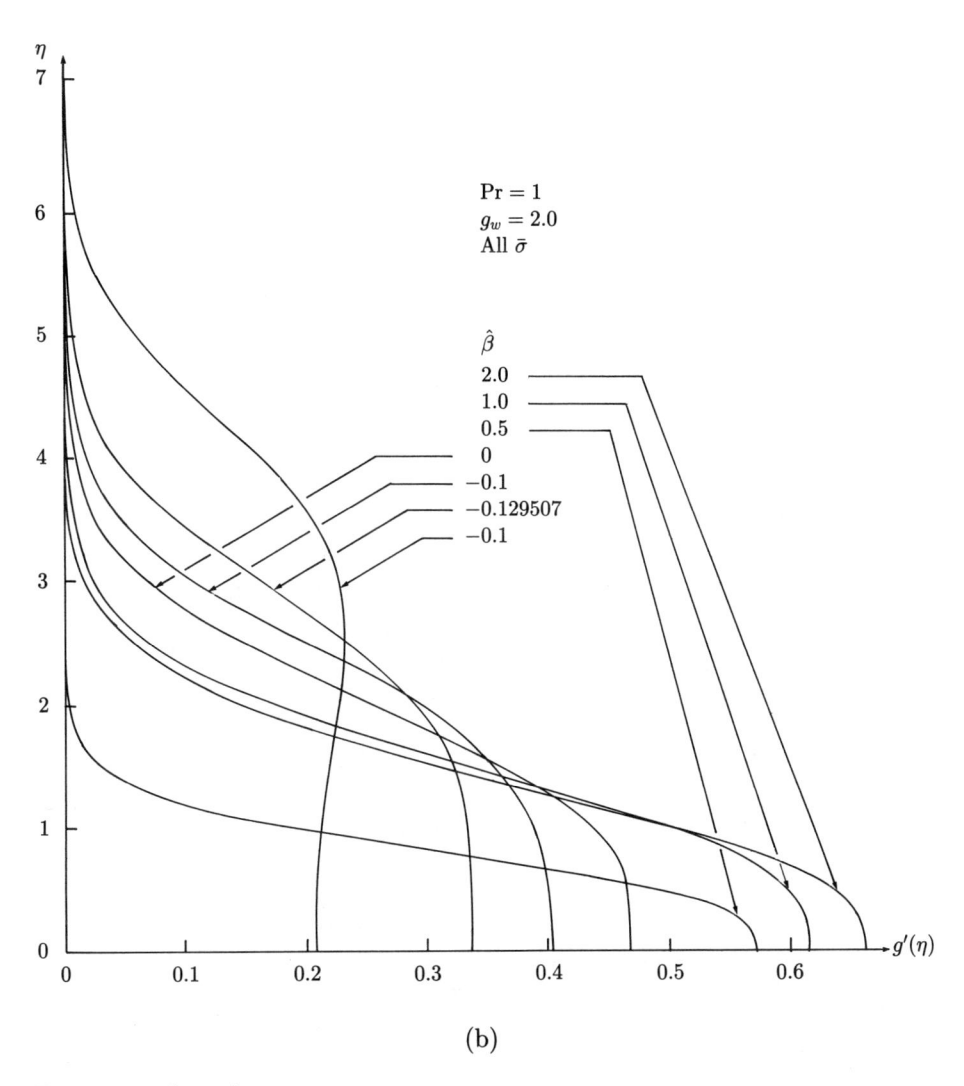

(b)

Figure 6–18. (cont.)

anticipated, for $g_w > 0$ the displacement thickness approaches infinity as the singular point at $\hat{\beta} = 0^-$, $f''(0) = 0^-$, $g = 0^\pm$ is approached along the lower branch. However, for $g_w = 0$ the displacement thickness 'appears' to reach a maximum and then decrease.

From Eq. (6–108) the boundary layer momentum thickness, θ, is

$$\theta = \frac{\sqrt{2\xi}}{\rho_e U_e} J_2 \qquad (6-108)$$

Examination of the momentum thickness, J_2, shown in Figure 6–20b reveals

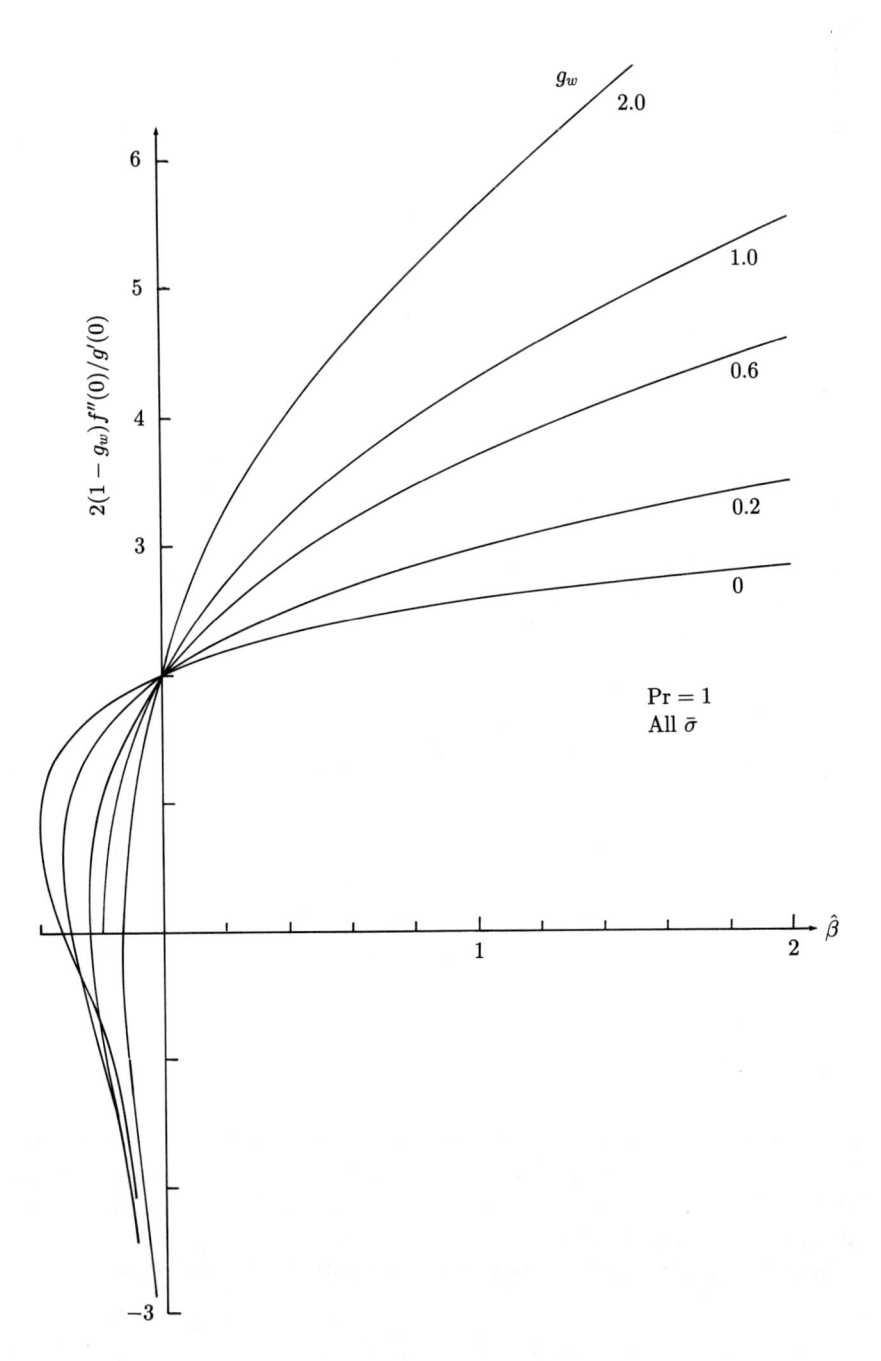

Figure 6–19. Reynolds analogy for the compressible boundary layer with nonzero pressure gradient.

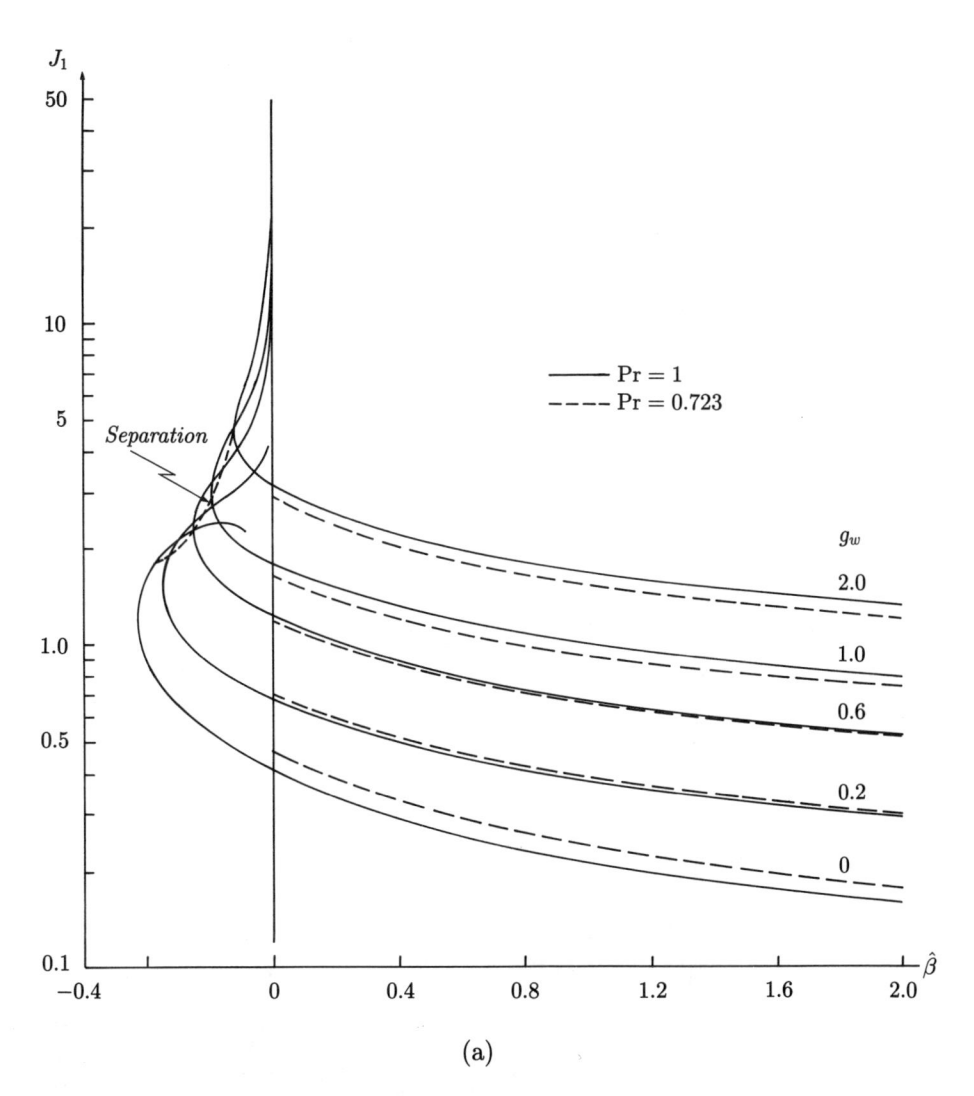

Figure 6–20. Boundary layer displacement and momentum integrals. (a) J_1; (b) J_2.

that for favorable pressure gradients the momentum thickness decreases with increasing wall temperature. Further, the momentum thickness reaches a maximum for the separation solution and then decreases toward zero for the reverse flow solutions. This is caused by the negative velocity found near the surface in the reverse flow solutions. Finally, we note that for large wall-to-stagnation temperature ratios the momentum thickness becomes negative, thus indicating more momentum flux in the boundary layer than in the equivalent inviscid flow. This is a result of the large velocity overshoot due to the severe heating of the fluid by the surface.

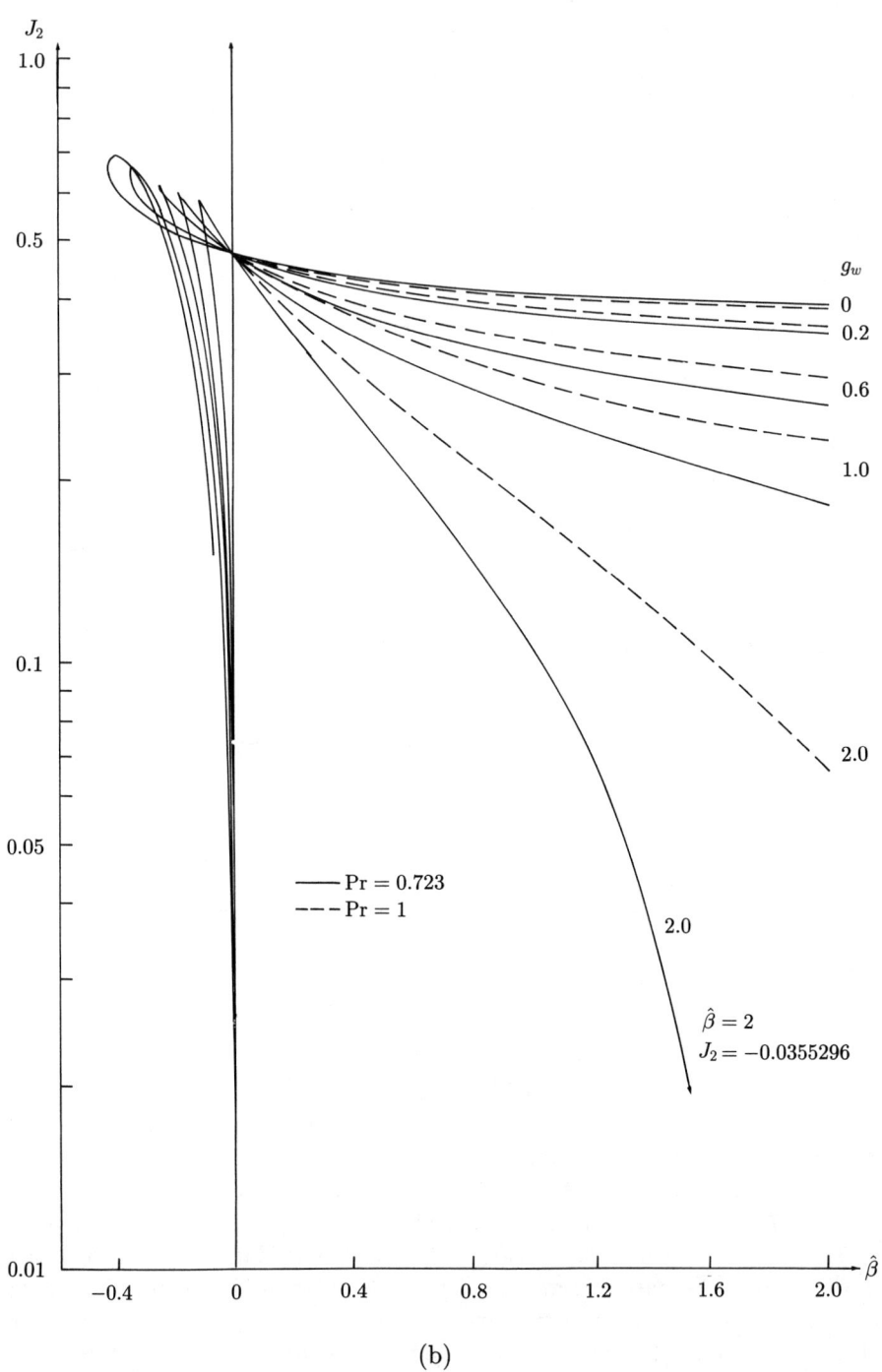

Figure 6–20. (cont.)

6-14 Results for the Similar Hypersonic Compressible Boundary Layer with Nonunit Prandtl Number

Our previous discussion of the similar hypersonic compressible boundary layer with nonunit Prandtl number showed that analytical solutions, even for an adiabatic wall, are not possible. Fundamentally this is because the stagnation enthalpy is not constant through the boundary layer. Rogers [Roge69b] numerically calculated solutions for the hypersonic compressible boundary layer with a Prandtl number of 0.723. Of particular interest are the effects of Prandtl number on the velocity and enthalpy profiles and the shearing stress and heat transfer rate at the surface.

Comparison of the numerical results for the $Pr = 1$ and $Pr = 0.723$ nondimensional velocity profiles indicates that the effect of the Prandtl number is small. Figure 6–21 shows the $Pr = 0.723$ nondimensional enthalpy profiles for a wall-to-stagnation temperature ratio of 0.6. Also shown for comparison are several unit Prandtl number profiles. As anticipated, the enthalpy profiles are significantly affected. Examination of Figure 6–21 and the numerical result shows that the effect of nonunit Prandtl number is to slightly increase the enthalpy ratio very near the wall. The enthalpy ratio is significantly decreased in the central portions but exhibits an enthalpy overshoot in order to maintain the total energy balance in the boundary layer.

The shearing stress parameter, $f''(0)$, and the heat transfer parameter, $g'(0)$, for a Prandtl number of 0.723 are shown in Figures 6–22 and 6–23. Selected numerical results are given in Table C–27. Selected unit Prandtl number solutions are also shown. Examination of Figures 6–22 and comparison of the numerical results in Tables C–25 and C–27 show that the effect of nonunit Prandtl number on the shearing stress parameter is small. Typically, for heat transfer from the surface to the fluid the reduction in Prandtl number increases the shearing stress parameter for favorable pressure gradients and decreases it for adverse pressure gradients. Examination of Figure 6–23 and the numerical results given in Tables C–25 and C–27 show that the effect of nonunit Prandtl number is to significantly decrease the *magnitude* of the wall heat transfer parameter, and hence the heat transfer rate to the surface, for all wall-to-stagnation enthalpy ratios. This effect is particularly significant for highly cooled surfaces, e.g., for a wall-to-stagnation enthalpy ratio, $g_w = 0.2$, on a flat plate ($\hat{\beta} = 0$) the reduction in $g'(0)$ is 26.6%.

Of particular interest in any heat transfer analysis is the adiabatic wall, or recovery temperature. For a unit Prandtl number we showed that the adiabatic wall-to-stagnation enthalpy is $g_{aw} = 1$. For a nonunit Prandtl number we anticipate from our previous discussions that it is less than one. The comb program can be modified to determine the adiabatic wall enthalpy ratio, g_{aw}. The required modifications to the iteration scheme are discussed in Appendix B.

A typical run of comb incorporating these modifications is shown below for $Pr = 0.723$, $\bar{\sigma} = 2$, $f(0) = 0$, $\hat{\beta} = 0$. Since the momentum and energy equations are uncoupled for $\hat{\beta} = 0$, the momentum equation reduces to the Blasius

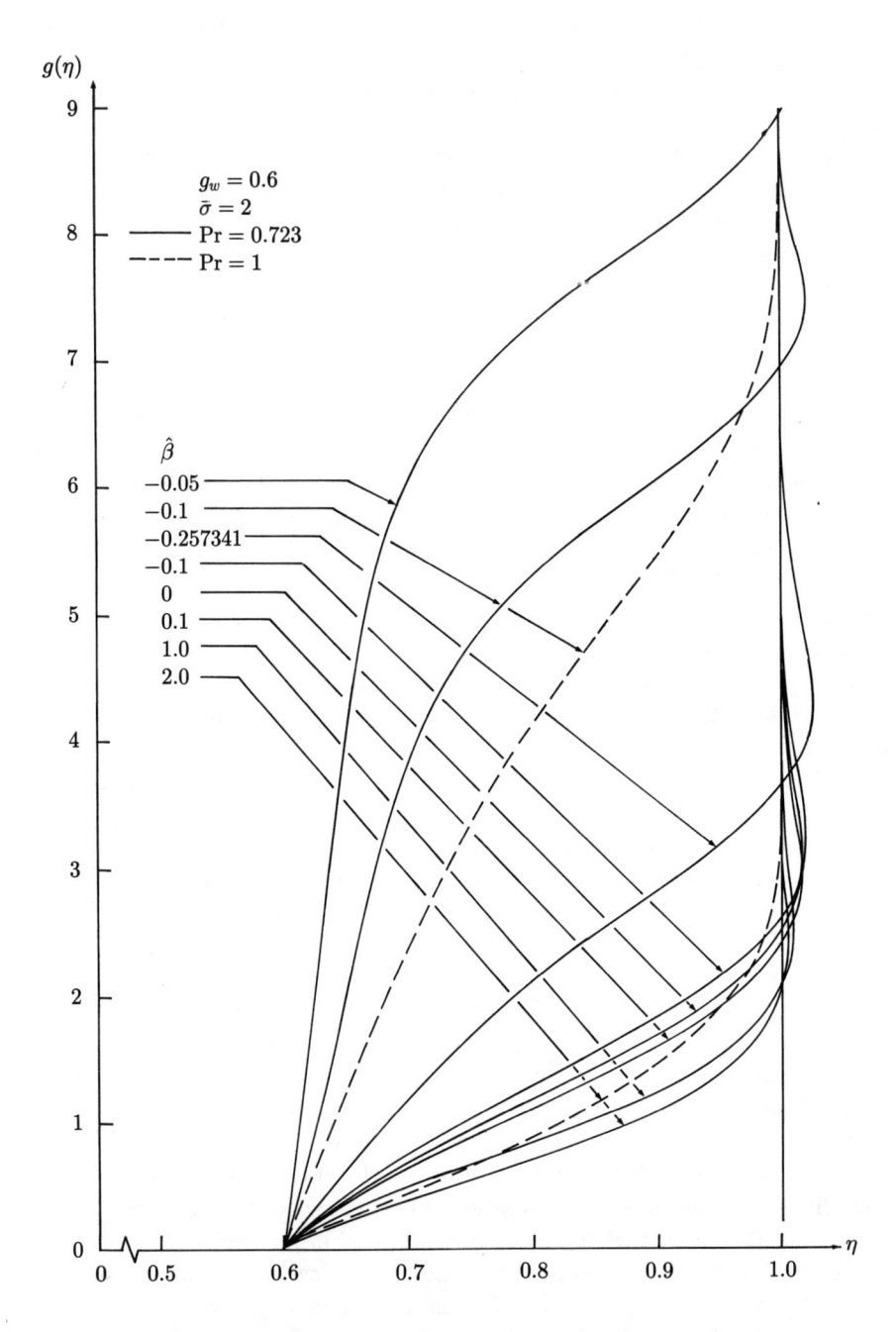

Figure 6-21. Nondimensional stagnation enthalpy profiles, Pr = 0.723, g_w = 0.6.

Nonunit Prandtl Number Hypersonic Compressible Boundary Layer 259

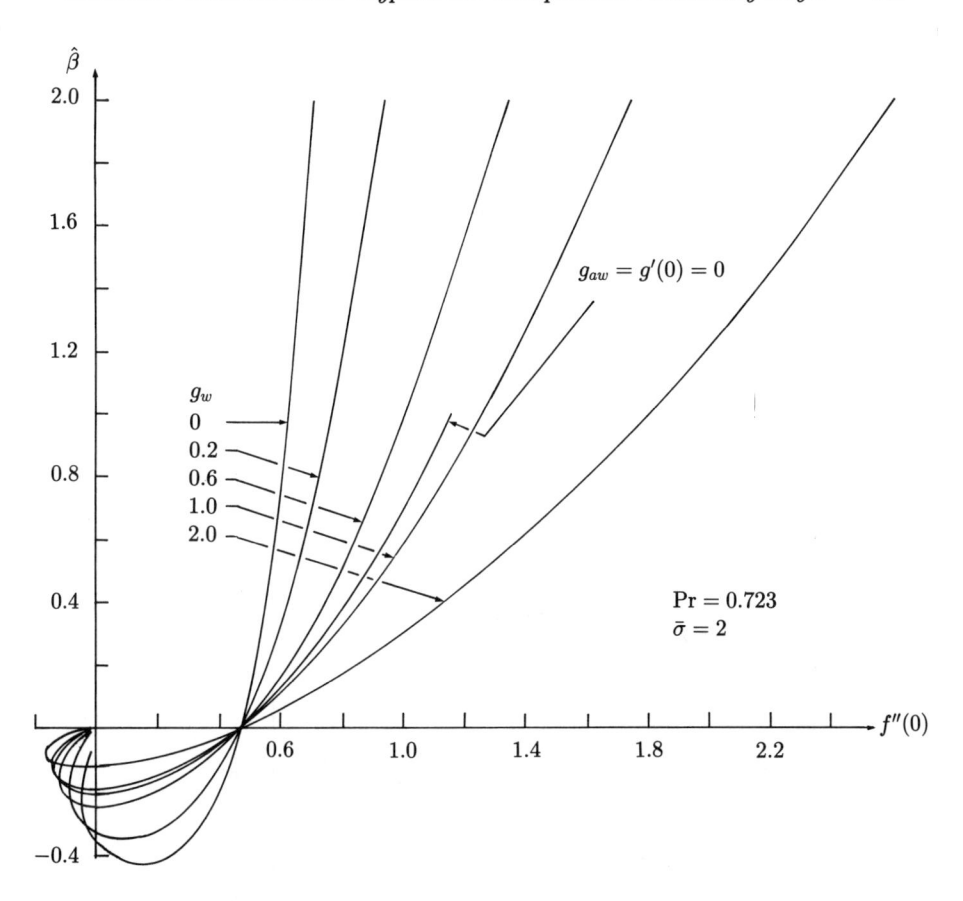

Figure 6–22. Variation of the wall shearing stress parameter, $f''(0)$, with the pressure gradient parameter, $\hat{\beta}$, Pr $= 0.723$, $\bar{\sigma} = 2$.

equation. Hence we choose $f''(0) = 0.4696005$. From our previous experience with the forced convection boundary layer (see Section 4–8) we anticipate that $g_{aw} = \sqrt{\text{Pr}} = \sqrt{0.723} \approx 0.85$. We therefore choose this value for our initial estimate of g_{aw}. The first integration shows that this is a reasonably good approximation; the final result after one iteration is 0.849495.

```
Beta = 0
Prandtl Number = 0.723
Mach Number Parameter = 2
f"(0) = 0.4696005
g'(0) = 0
gaw = 0.85
    η       f"           f'         f          g'             g
    0    0.4696005       0          0          0            0.85
    7.   1.805485e-8  1.000001   5.783224  -2.678094e-6   1.000505
```

260　*Laminar Flow Analysis*

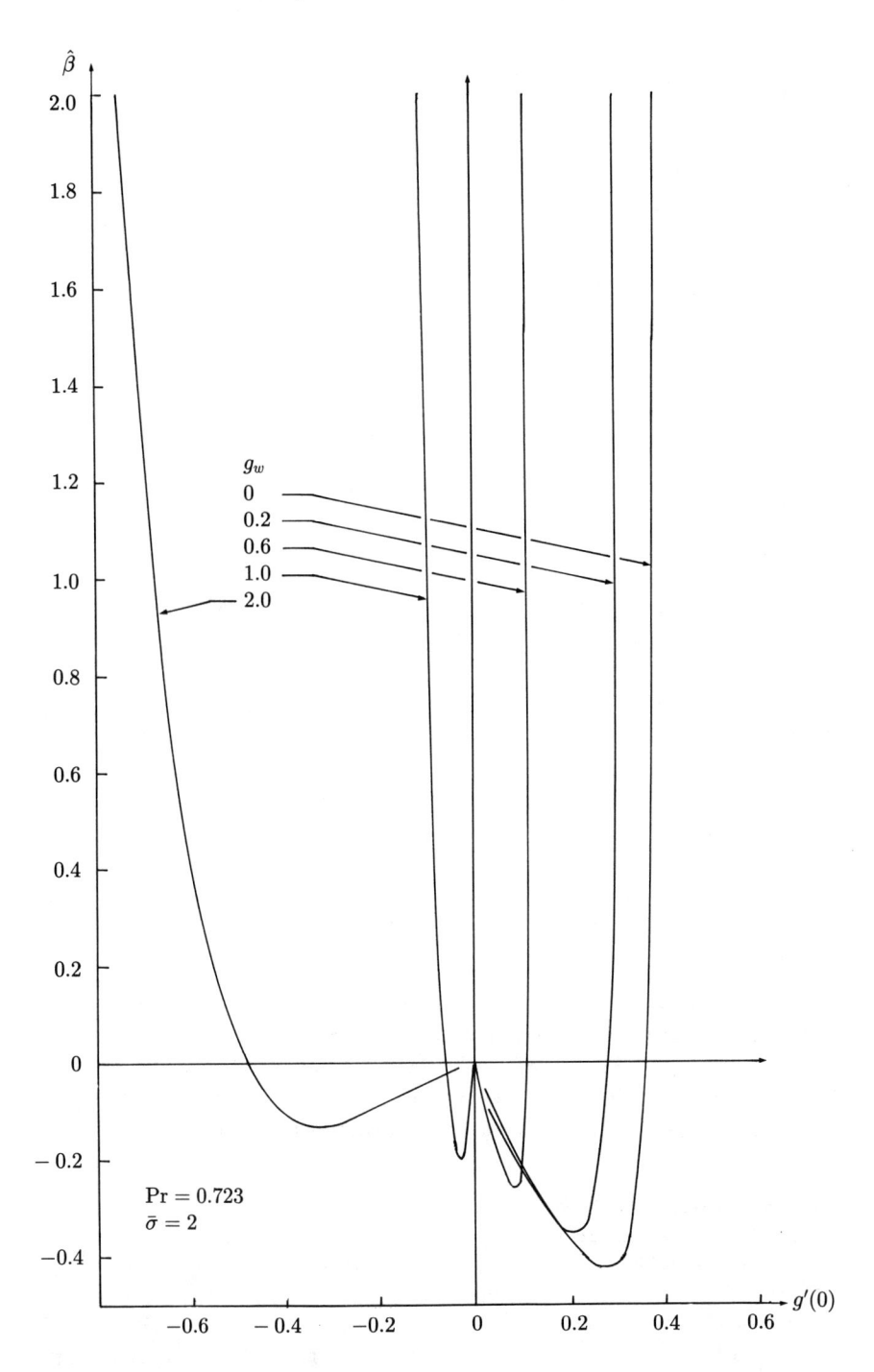

Figure 6–23. Variation of the wall heat transfer parameter, $g'(0)$, with the pressure gradient parameter, $\hat{\beta}$, Pr $= 0.723$, $\bar{\sigma} = 2$.

```
f"(0) = 0.4696005
gaw = 0.851
 0      0.4696005    0           0          0             0.851
 7.     1.805485e-8  1.000001    5.783224  -2.678094e-6   1.001505

f"(0) = 0.4706005
gaw = 0.85
 0      0.470601     0           0          0             0.85
 7.     1.758092e-8  1.001420    5.792295  -2.632641e-6   1.0009325

f"(0) = 0.4695999
gaw = 0.84949518
 0      0.46959999   0           0          0             0.8494952
 7.     1.805510e-8  1.          5.783219  -2.678117e-6   1.
Convergence achieved
J₁ = 1.570395    J₂ = 0.4696    J₃ = 0.115985    J₄ = 1.686381
```

The adiabatic wall-to-stagnation temperature ratio, g_{aw}, as well as the shearing stress parameter, $f''(0)$, for these solutions is shown in Figure 6–24 (see Rogers [Roge70]). Numerical results are given in Table C–28. Here we see that the maximum value of g_{aw} occurs at the separation point, where $f''(0) = 0$. Physically, this is caused by the increased heat conduction to the surface as a result of the near zero velocity close to the wall. The approximation $g_{aw} \approx \sqrt{\text{Pr}}$ overestimates g_{aw} for favorable pressure gradients and underestimates g_{aw} for retarded flows ($\hat{\beta} < 0$, $f''(0) > 0$). Finally, we observe that the adiabatic wall-to-stagnation enthalpy ratio is particularly sensitive to the pressure gradient in the reverse flow region. For Prandtl numbers from 0.5 to 1, the maximum adiabatic wall-to-stagnation enthalpy ratio, which occurs at the separation point, is approximated by (see Rogers [Roge70])

$$(g_{aw})_{\text{Pr} \neq 1} = (\text{Pr})^{0.423} (g_{aw})_{\text{Pr}=1} \qquad 0.5 \leq \text{Pr} \leq 1 \qquad (6-112)$$

Having completed our discussion of the compressible boundary layer without mass transfer, we now turn to the solution with mass transfer at the surface.

6–15 The Similar Compressible Boundary Layer with Mass Transfer

In Sections 6–6 to 6–14 we discussed the similar compressible boundary layer without mass transfer at the surface. In our previous discussion of the incompressible constant property forced convection boundary layer, we considered the effects of mass transfer at the surface (see Sections 3–10 and 4–10). We now consider the effects of mass transfer at the surface for the similar compressible boundary layer. We assume that mass is transferred either into (suction) or out of (injection or blowing) the surface, that the mass transfer occurs normal to the surface, that in the case of injection the fluid is injected normal to the surface and that the injected fluid is the same as the main stream fluid. Finally, we assume that the injected fluid has the same temperature as the surface. For these conditions the governing similar boundary value problem is given by Eqs. (6–74), (6–75) and (6–79), which we repeat here for convenience

$$f''' + ff'' + \bar{\beta}(g - f'^2) = 0 \qquad (6-74)$$

$$g'' + \Pr fg' = \bar{\sigma}(1 - \Pr)(f'f'')' \qquad (6-75)$$

with boundary conditions

$$f(0) = \frac{\sqrt{2\xi}\, v(x)}{\mu_w U_e(x)} = \text{constant} = f_0 \qquad f'(0) = 0 \quad (6-79a,b)$$

$$g(0) = g_w = \text{constant} \quad \text{or} \quad g'(0) = 0 \qquad (6-79c,d)$$

$$f'(\eta \to \infty) \to 1 \qquad g(\eta \to \infty) \to 1 \qquad (6-79e,f)$$

From Eq. (6–79a) we see that to satisfy similarity requirements the normal velocity at the surface given by

$$v(x) = \frac{\mu_w U_e(x) f(0)}{\sqrt{2\xi}}$$

must vary inversely as the square root of the similarity variable ξ. For a flat plate this result reduces to inversely as the square root of the distance from the leading edge x. Here we see that negative values of f_0 correspond to blowing or injection and positive values to suction. Although most practical applications of suction or blowing at the surface do not exhibit the required similar normal velocity distribution, the effects are qualitatively correct and in many cases represent acceptable quantitative approximations for the more practical situation of uniform suction or blowing (see, e.g., Lew and Fanucci [Lew55] or Libby and Chen [Libb65]).

Each of the previously discussed compressible boundary layers without mass transfer can be considered with mass transfer at the surface. However, we confine our discussions to the unit Prandtl number and hypersonic nonunit Prandtl number solutions. Recall that the low speed compressible boundary layer is included in the unit Prandtl number solutions. The nonunit Prandtl number low speed similar compressible boundary layer with mass transfer at the surface was investigated by Brown and Donoughe [Brow51] and Levy [Levy54]. These solutions can be obtained by using the comb program with surface mass transfer (see Section 6–12). The flat plate similar compressible boundary layer with injection at the surface was studied by Low [Low53] using a method similar to that discussed in Section 6–7.

6–16 Results for the Similar Compressible Boundary Layer with Mass Transfer

Solutions to the unit Prandtl number similar compressible boundary layer with mass transfer were obtained by Levy [Levy54] for $f_0 = -0.5$, by Beckwith

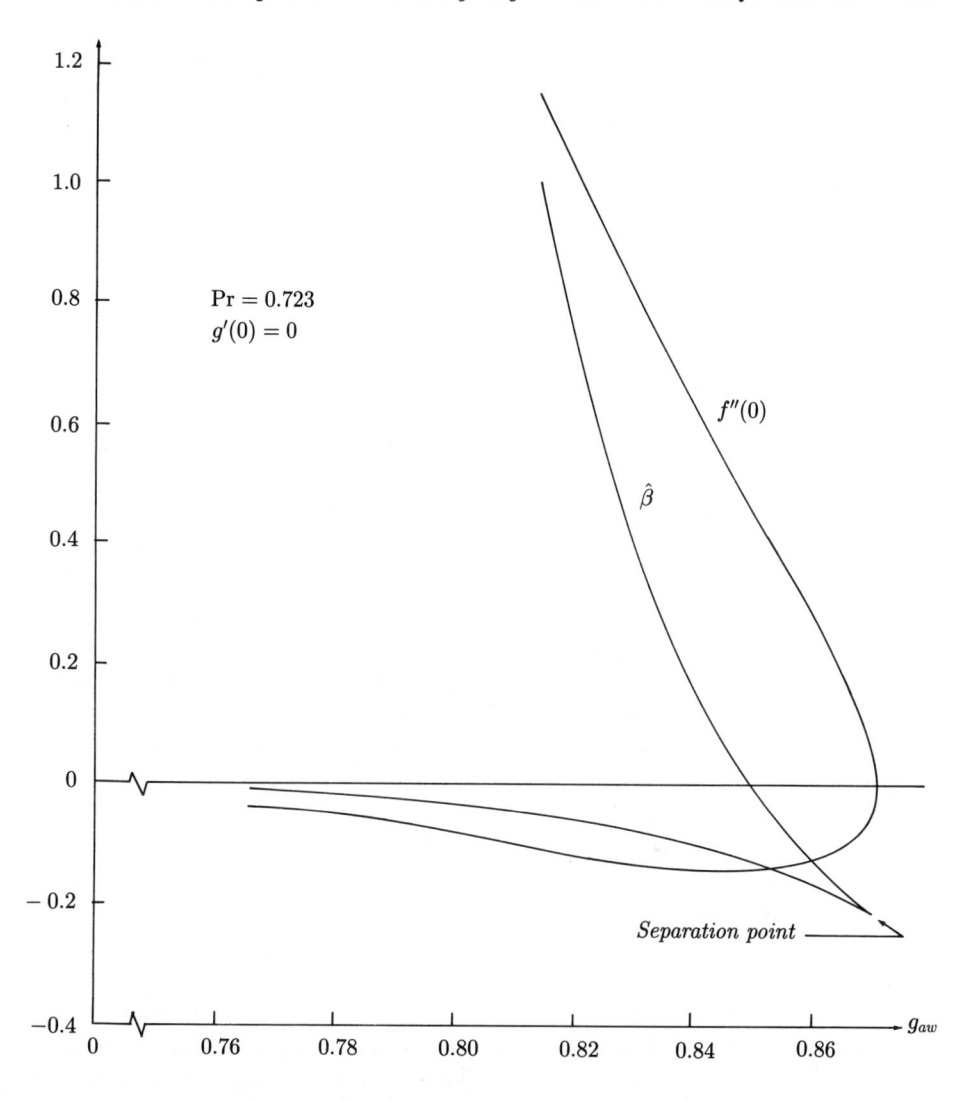

Figure 6–24. Variation of wall shearing stress and pressure gradient parameters, with the adiabatic wall temperature ratio Pr = 0.723, $\bar{\sigma} = 2$.

[Beck59] for $\bar{\beta} = 1$ and $f_0 = -0.5, -1$ and by Ball [Ball67] for reverse flow solutions with $f_0 = 0.5$ and 1. Tables C–29 to C–32 contain numerical results for the unit Prandtl number mass transfer solutions for $f_0 = -1, -0.5, 0.5, 1$.

The effects of mass transfer at the surface on the velocity profiles for the compressible boundary layer are similar to those for the incompressible boundary layer (see Section 3–10). With suction the boundary layer thickness decreases, and consequently the shearing stress at the surface increases. These results are similar to those for a favorable pressure gradient. With blowing, the thickness of the boundary layer increases and the shearing stress decreases. Further, the velocity profiles exhibit a characteristic inflection point which is similar to that for the retarded flow solutions.

Figure 6–25 shows the effect of mass transfer on the enthalpy profiles for a wall-to-stagnation enthalpy ratio of 0.6 and a favorable pressure gradient parameter, $\hat{\beta} = 0.5$. Here we see that suction ($f_0 > 0$) decreases the thickness of the thermal boundary layer and increases the heat transfer rate at the surface. Physically, the increased heat transfer rate is attributed to the increased effectiveness of convection caused by the larger mean velocities near the surface in the thinner suction boundary layer. Again, these effects are similar to those exhibited by increasingly favorable pressure gradients. Injection ($f_0 < 0$) increases the thickness of the thermal boundary layer and decreases the heat transfer rate at the surface. For larger injection rates than shown in Figure 6–25, a region of almost constant enthalpy develops near the surface. In this case heat transfer from the fluid to the surface takes place almost exclusively by conduction. With large rates of injection most of the fluid near the surface originated at the surface. It has the temperature of the surface and a velocity normal to the surface. With large injection rates, the available viscous mechanisms which realign the velocity with the inviscid flow at the boundary layer edge are relatively weak. This results in an increase in the distance above the surface required to turn the injected fluid, hence the observed increase in boundary layer thickness. Accompanying this is a reduction in heat transfer rate within the boundary layer, as well as reduced viscous dissipation due to the lower mean velocities near the surface. These effects result in an increase in the observed thermal boundary layer thickness. Eventually the injection rate becomes sufficiently large that boundary layer 'blow-off' occurs (see Section 3–10). Again, these results are similar to those for retarded flow.

Figures 6–26 and 6–27 show the shearing stress parameter, $f''(0)$, and the heat transfer parameter, $g'(0)$, variation with pressure gradient parameter, $\hat{\beta}$, and mass-transfer parameter, f_0. The wall-to-stagnation temperature is 0.6. Examination of Figure 6–26 shows that the shearing stress parameter, $f''(0)$, is increased by suction at the surface for all values of the pressure gradient parameter, $\hat{\beta}$. In contrast, mass injection at the surface decreases the shearing stress parameter, $f''(0)$, for all values of the pressure gradient parameter, $\hat{\beta}$. Further, notice that the shearing stress parameter is quite sensitive to pressure gradient. Hence, extrapolation of the zero pressure gradient results (except for small values of $\hat{\beta}$) to estimate the shearing stress for either favorable or adverse pressure gradients yields significant errors. In fact, the zero pressure gradients results indicate that there are no solutions with finite boundary layer thickness for f_0 less than the boundary layer 'blow-off' value of -0.875745 (see Section 3–10). However, provided $\hat{\beta} > 0$, solutions do exist. Examination of the numerical results for $f_0 = -1$ given in Table C–29 shows that the displacement thickness is finite. When the pressure gradient is zero, viscous shearing stress within the boundary layer is the only mechanism available to align the injected fluid with the inviscid

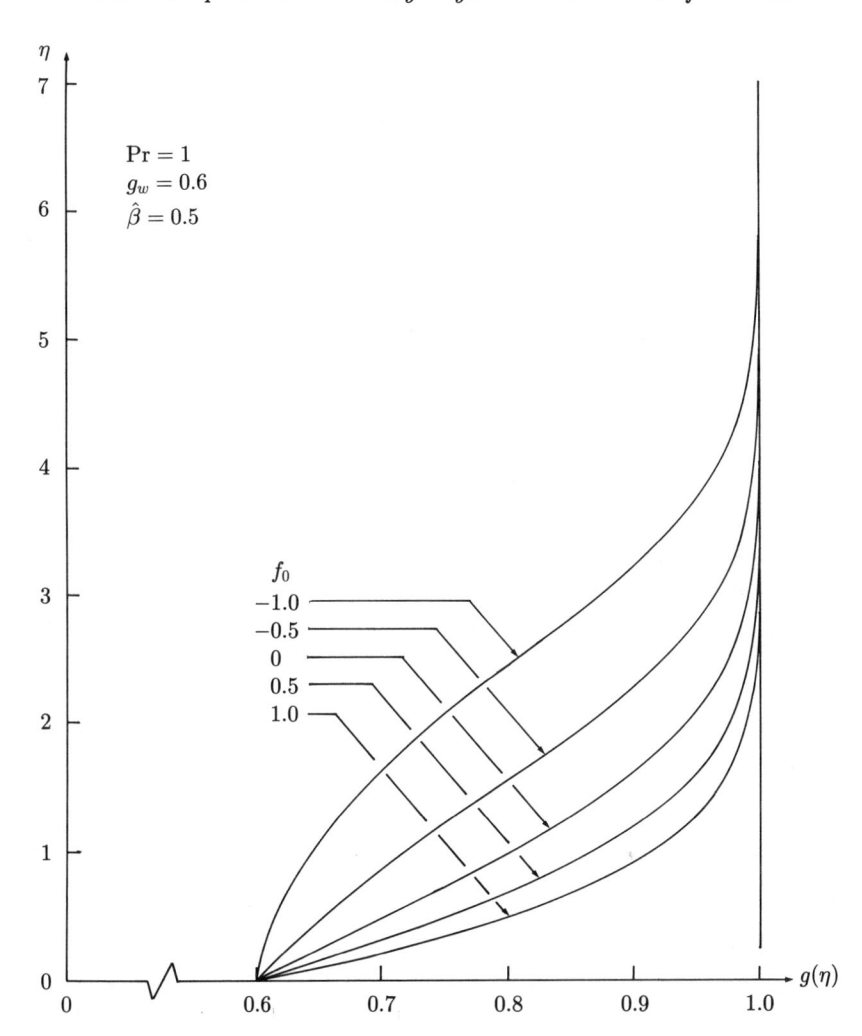

Figure 6–25. Effect of mass transfer on the nondimensional stagnation enthalpy profiles, $g_w = 0.6$.

flow. However, with a favorable pressure gradient ($\hat{\beta} > 0$) the pressure forces due to the external inviscid flow pressure gradient assist the viscous shearing stress in aligning the injected fluid with the inviscid flow. The larger the favorable pressure gradient the more easily the injected fluid is realigned and the thinner the boundary layer. Notice that there are no retarded flow solutions ($\hat{\beta} < 0$) for $f_0 > -0.875745$. This is because the adverse pressure gradient associated with the retarded flow solutions decreases the effectiveness of the viscous shearing stress in realigning the injected fluid. Finally, we observe that blowing decreases the adverse pressure gradient that the similar boundary layer can sustain before separation occurs, whereas suction increases it.

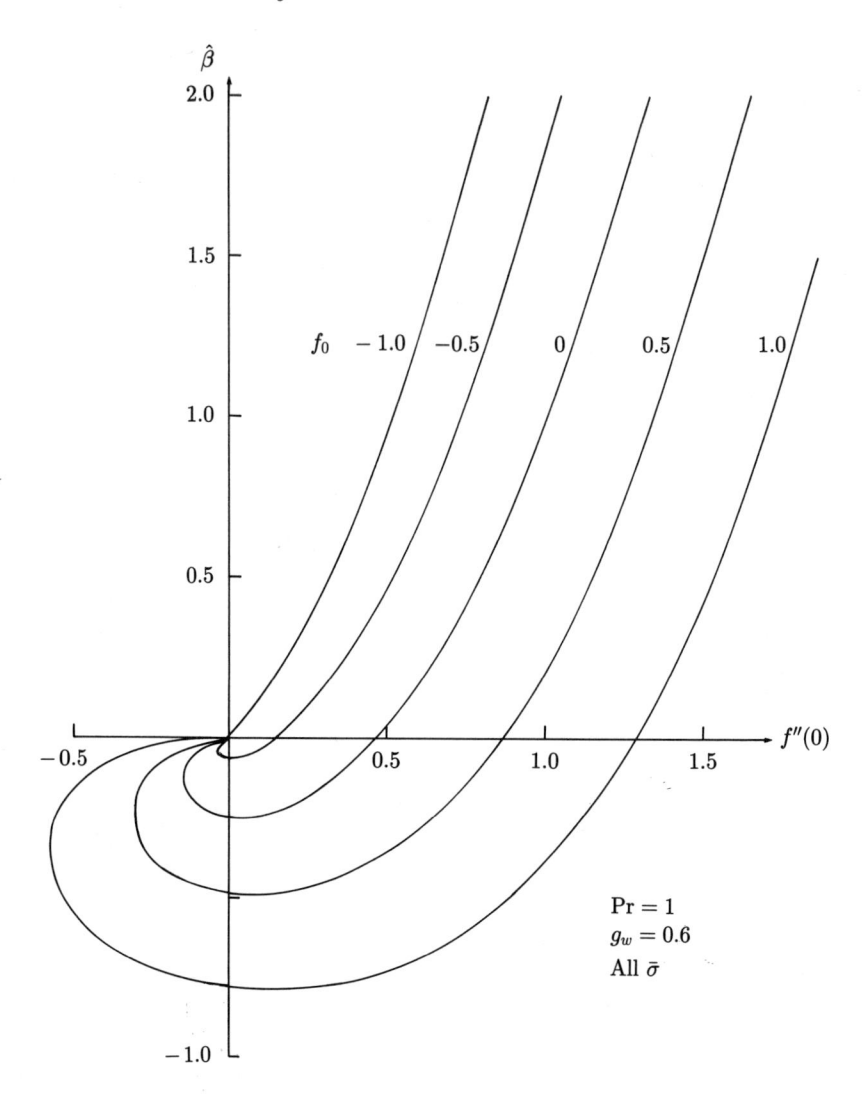

Figure 6–26. Effect of mass transfer on the shear stress parameter, $f''(0)$, $g_w = 0.6$.

Examination of Figures 6–27 and 6–28 shows that the heat transfer parameter, $g'(0)$, is increased by suction at the surface, whereas it is decreased by fluid injection at the surface for all accelerated ($\hat{\beta} > 0$) and retarded flows ($\hat{\beta} < 0$, $f''(0) > 0$). However, for reverse flows with suction the heat transfer parameter, $g'(0)$, can decrease below the value without suction. Again we notice (see Section 6–13) that for the cooled wall shown in Figure 6–28 the heat transfer parameter is almost independent of $\hat{\beta}$ except for the reverse flow solutions. However, this is again not the case for heated walls with favorable pressure gradients. Hence, the use of the zero pressure gradient results with mass transfer to estimate

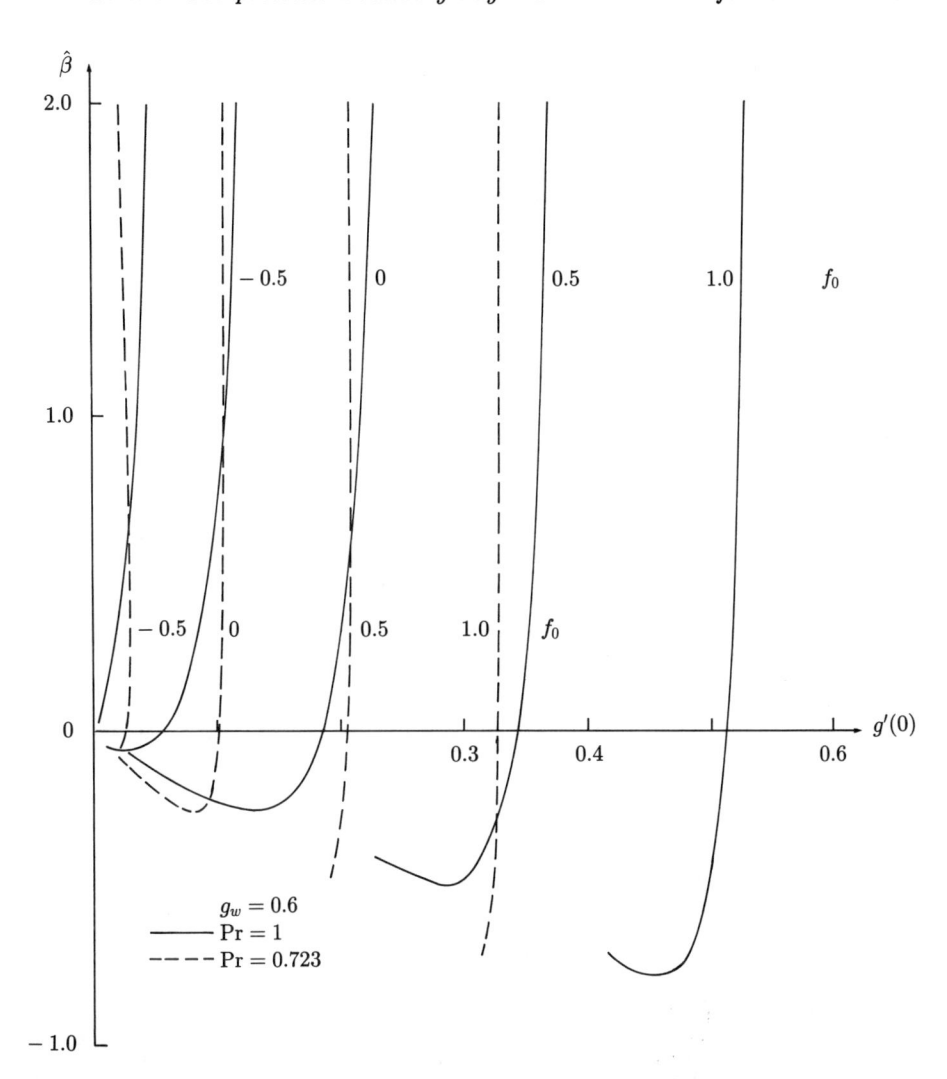

Figure 6–27. Effect of mass transfer on the heat transfer parameter, $g'(0)$, $g_w = 0.6$.

nonzero pressure gradient heat transfer results with mass transfer for heated surfaces with favorable pressure gradients or reverse flow yields erroneous results.

Figure 6–29 shows the effect of wall-to-stagnation enthalpy ratio on the shearing stress parameter, $f''(0)$, and the heat transfer parameter, $g'(0)$, for various mass transfer parameters, f_0. Here we see that for a favorable pressure gradient ($\hat{\beta} = 0.5$) the shearing stress parameter, $f''(0)$, and the heat transfer parameter, $g'(0)$, vary almost linearly with the wall-to-stagnation enthalpy ratio, g_w. However, Figure 6–28, which shows $\hat{\beta}$ and $g'(0)$ at the separation point, i.e., where $f''(0) = 0$, indicates that the variation of $g'(0)$ with wall-to-stagnation enthalpy ratio is not linear in the retarded or reverse flow regions.

268   Laminar Flow Analysis

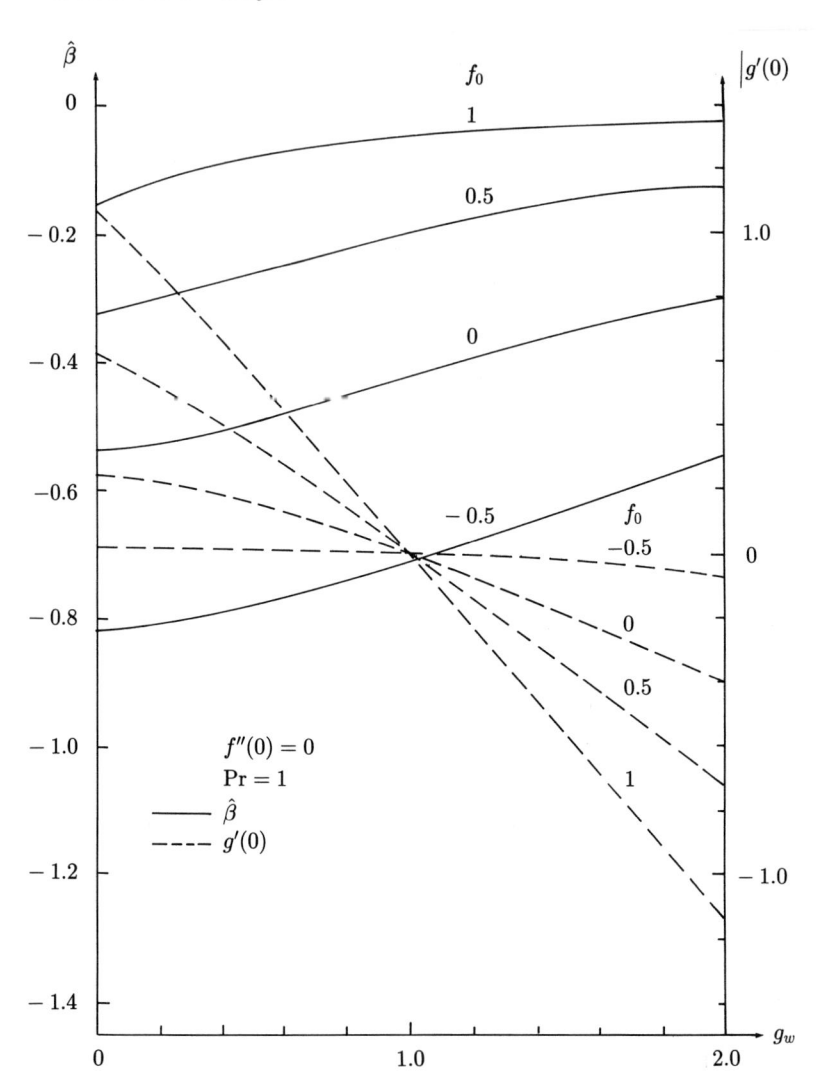

Figure 6–28. Effect of heat and mass transfer on the pressure gradient and heat transfer parameters, $\hat{\beta}$ and $g'(0)$, at the separation point, i.e., at $f''(0) = 0$.

The effects of nonunit Prandtl number (Pr < 1) on the compressible boundary layer with mass transfer are similar to those without mass transfer. Comparison of the numerical results given in Tables C–29 to C–32 for Pr = 1 and those given in Table C–33 for Pr = 0.723 and a favorable pressure gradient, $\hat{\beta} = 0.5$, show that for very cold surfaces ($g_w \to 0$) the shearing stress parameter is slightly decreased, while for moderately cool and heated surfaces it is increased. This characteristic is exhibited for both suction and injection with a favorable

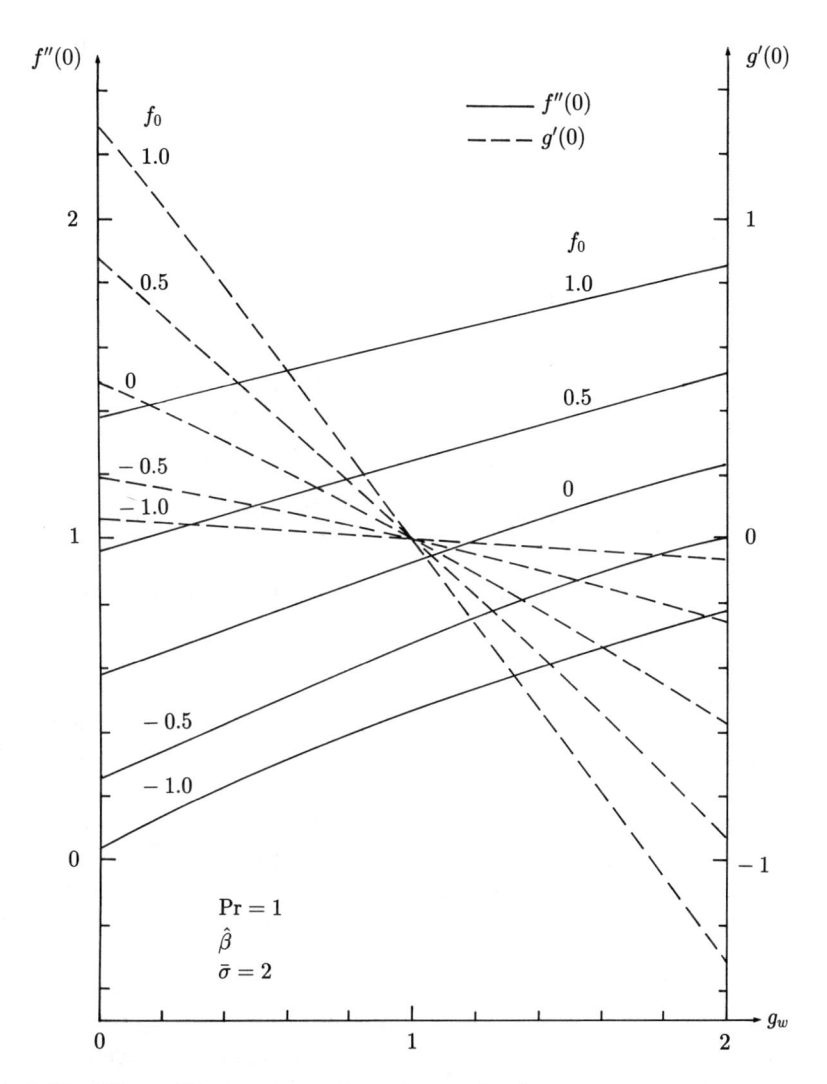

Figure 6–29. Effect of heat and mass transfer on the shearing stress and heat transfer parameters, $f''(0)$ and $g'(0)$, for $\hat{\beta} = 0.5$.

pressure gradient. Again, the effect of the reduced Prandtl number is to decrease the magnitude of the heat transfer parameter, $g'(0)$, for all wall-to-stagnation enthalpy ratios. For both suction and injection the adiabatic wall-to-stagnation temperature ratio, g_{aw}, occurs for $g_w < 1$.

Having discussed the similar compressible boundary layer in some detail, we now consider an application of these results, i.e., we consider the hypersonic shock-wave boundary layer interaction which occurs near the leading edge of a sharp flat plate.

6-17 Hypersonic Shock Wave-Boundary Layer Interaction

We now turn our attention to a classical problem involving the compressible boundary layer — the interaction of the shock wave and the boundary layer near the leading edge of a sharp flat plate in hypersonic flow. For a mathematically thin flat plate aligned with the flow direction, inviscid flow theory shows that even for supersonic flow there is no disturbance generated by the plate. However, a boundary layer grows on the plate. Examination of Eq. (6–106) shows that for hypersonic flow (i.e., $M_e \gg 1$) the boundary layer displacement thickness is significant. Recalling the definition of the displacement thickness (see Section 3–7) as the distance a streamline at the outer edge of the boundary layer is displaced with respect to its position in the corresponding potential flow, we see that the flat plate appears as a pseudobody to the oncoming flow. The pseudobody is considered to have thickness, $\delta^*(x)$, and a slope, $d\delta^*/dx$. Because of the existence of the boundary layer a shock wave forms at the leading edge of the plate, as shown in Figure 6–30. The shape of the shock wave depends on the growth of the boundary layer, i.e., $d\delta^*/dx$. However, the growth of the boundary layer depends on the pressure distribution in the inviscid flow between the shock wave and the boundary layer and, hence, upon the shock shape and the expansion behind the shock wave. Thus, the inviscid and viscous flows are coupled. Consequently, a complete solution involves the simultaneous solution of the coupled inviscid-viscous equations. The above description of the flow near the leading edge of a sharp flat plate is considerably simplified. The complete physical description of the flow encompasses flow regimes from free molecular near the leading edge to that for the compressible flat plate boundary layer far downstream. For a more complete description of the flow field the reader can consult McCroskey et al. [McCr66], where one finds additional references on this problem. However, for the purpose of the following discussion we consider the simplified flow model described above sufficient.

We begin our analysis by considering the inviscid flow across the shock wave and between the shock wave and the viscous boundary layer which grows along the plate surface. Our objective is to develop an expression for the pressure distribution as a function of the distance along the plate. The local pressure at any position behind an oblique shock wave is given by (see Liepmann and Roshko [Liep57])

$$\frac{P_s}{P_\infty} = \frac{2\gamma M_\infty^2 \sin\theta_s}{\gamma+1} - \frac{\gamma-1}{\gamma+1} \qquad (6-113)$$

where the subscripts ∞ and s indicate conditions immediately ahead of and immediately behind the shock wave. Here θ_s is the local inclination of the shock wave. Since for hypersonic flow $M_\infty \gg 1$, $(M_\infty \sin\theta_s)^2 \gg (\gamma-1)/2\gamma$ and Eq. (6–113) becomes

$$\frac{P_s}{P_\infty} = \frac{2\gamma}{\gamma+1} K_s^2 \qquad (6-114)$$

where $K_s = M_\infty \sin\theta_s$. For small angles $K_s \approx M_\infty \theta_s$.

For a two-dimensional shock wave, the relationship between the local flow direction or the body inclination, θ_b, and the local shock wave angle, θ_s, is given by (see Liepmann and Roshko [Liep57])

$$M_\infty^2 \sin^2 \theta_s - 1 = \frac{\gamma + 1}{2} \frac{M_\infty^2 \sin \theta_s \sin \theta_b}{\cos(\theta_s - \theta_b)} \quad (6-115)$$

Recalling our previous discussions of the displacement thickness in a compressible boundary layer and anticipating our subsequent results, we assume θ_b is small. Equation (6–115) is then written as

$$K_s^2 - 1 = \frac{\gamma + 1}{2} M_\infty^2 \theta_b \tan \theta_s \quad (6-116)$$

For hypersonic flow $M_\infty \gg 1$ and $\theta_s \ll 1$, but $K_s \gg 1$. Hence, Eq. (6–116) is written as

$$\theta_s \approx \frac{\gamma + 1}{2} \theta_b \quad (6-117)$$

Substituting into Eq. (6–114) yields

$$\frac{P_s}{P_\infty} = \gamma \left(\frac{\gamma + 1}{2}\right) K_b^2 \quad (6-118)$$

where $K_b = M_\infty \theta_b$. Although Eq. (6–118) gives the local pressure behind an oblique shock as a function of the local body or flow inclination, we require the pressure at the edge of the pseudobody (boundary layer). In order to accomplish this we utilize the tangent wedge approximation (see Hayes and Probstein [Haye59]). The tangent wedge approximation assumes that the pressure at any point on a body is equal to the equivalent pressure on the surface of a wedge with a half-angle equal to the local inclination of the body. Since the surface pressure on a two-dimensional wedge is the same as the pressure immediately behind the shock wave (see Liepmann and Roshko [Liep57]), the pressure at the edge of the boundary layer is

$$\frac{P_e}{P_\infty} = \gamma \left(\frac{\gamma + 1}{2}\right) \left(M_\infty \frac{d\delta^*}{dx}\right)^2 \quad (6-119)$$

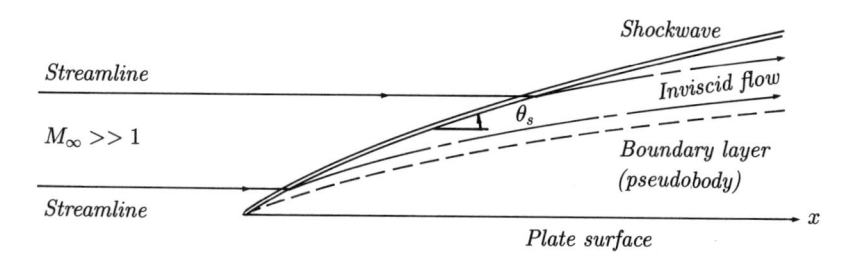

Figure 6–30. Hypersonic flow model for the sharp flat plate with strong interaction.

where in arriving at this result we used the fact that the local inclination of the pseudobody is $d\delta^*/dx$. Equation (6–119) represents the coupling between the inviscid flow which must yield $P_e(x)$, and the viscous boundary layer flow, which must yield $d\delta^*/dx$. However, Eq. (6–119) is one equation in two unknowns, $P_e(x)$ and $d\delta^*/dx$. Our subsequent analysis must develop a second equation in these two unknowns.

Before attempting to do this we consider the functional form of $P_e(x)$ and $\delta^*(x)$. Recalling our original derivation of the boundary layer equations (see Eq. 3–15), we have

$$\delta^{*2} \sim \delta^2 \sim \frac{\bar{\mu}x}{\bar{\rho}U_e}$$

where $\bar{\mu}$ and $\bar{\rho}$ are some average dynamic viscosity and density in the boundary layer. Using a linear viscosity law, $\bar{\mu}/\mu_\infty = C(\bar{T}/T_\infty)$ and the y-momentum equation, i.e., $\bar{P} = P_e$, yields

$$\delta^{*2} \sim C\mu_\infty \frac{\bar{T}}{T_\infty} \frac{\bar{T}}{P_e} \frac{x}{U_e} = \left(\frac{\bar{T}}{T_\infty}\right)^2 C\frac{\mu_\infty x}{\rho_e U_e}$$

Recalling that the temperature in the boundary layer is $\bar{T}/T_e \sim ((\gamma-1)/2)M_e^2$, we have

$$\left(\frac{\delta^*}{x}\right)^2 \sim M_e^4 \frac{T_e}{T_\infty} \frac{C\mu_\infty}{x} \frac{T_e}{P_e} \frac{1}{U_e}$$

Since $M_e^4 = M_\infty^4 (T_\infty/T_e)^2$ we have

$$\left(\frac{\delta^*}{x}\right)^2 \sim C \frac{M_\infty^4}{Re_{x\infty}} \frac{P_\infty}{P_e} \frac{U_\infty}{U_e}$$

Analysis of the oblique shock relation (see Liepmann and Roshko [Liep57]) shows that for slender bodies in hypersonic flow $U_e \approx U_\infty$ as $M_\infty \to \infty$ and as $\theta_s \to \theta_b \to 0$. We thus write

$$\left(\frac{\delta^*}{x}\right)^2 \sim C \frac{M_\infty^4}{Re_{x\infty}} \frac{P_\infty}{P_e}$$

For a power law variation of δ^* with x, $d\delta^*/dx \sim \delta^*/x$. Using Eq. (6–119) we write

$$\frac{P_e}{P_\infty} \sim C \frac{M_\infty^3}{(Re_{x\infty})^{1/2}} \sim x^{-1/2} = A\chi \qquad (6-120)$$

$$\frac{\delta^*}{x} \sim C \frac{M_\infty^{1/2}}{(Re_{x\infty})^{1/4}} \sim x^{-1/4} = \frac{B(\chi)^{1/2}}{M_\infty} \qquad (6-121)$$

where A and B are unknown constants. The parameter $\chi = CM_\infty^3/\sqrt{Re_{x\infty}}$ is called the hypersonic interaction parameter.

These two relations establish the functional variation for the pressure along the boundary layer edge and the growth of the boundary layer displacement thickness. A knowledge of these functional relationships allows us to complete the coupling of the inviscid and viscous flows. To this end, we proceed by recalling that one of the similarity requirements for the compressible boundary layer is that the pressure gradient parameter, $\hat{\beta}$, be constant. We now attempt to evaluate $\hat{\beta}$. From Eqs. (6–76) and (6–71) we have

$$\hat{\beta} = \frac{2\xi}{U_e} \frac{dU_e}{dx} \frac{1}{C\rho_e\mu_e U_e} \frac{H_e}{h_e} \qquad (6-76)$$

and

$$\xi = \int_0^x C\rho_e\mu_e U_e \, dx \qquad (6-71)$$

Introducing the Chapman-Rubesin viscosity relationship and using the equation of state allows Eq. (6–71) to be written as

$$\xi = C\rho_\infty \mu_\infty U_\infty \int_0^x \frac{P_e}{P_\infty} dx$$

After using Eq. (6–120) and integrating once we have

$$\xi = 2C\rho_\infty \mu_\infty U_\infty \frac{P_e}{P_\infty} x \qquad (6-123)$$

Using Euler's equation in the inviscid flow at the boundary layer edge, i.e., $(1/U_e)(dU_e/dx) = -(1/\rho_e U_e^2)(dP_e/dx)$, $\hat{\beta}$ is written as

$$\hat{\beta} = -\frac{2\xi}{C\rho_e\mu_e U_e^3} \frac{P_\infty}{\rho_e} \frac{d\left(\frac{P_e}{P_\infty}\right)}{dx} \frac{H_e}{h_e}$$

Using Eq. (6–123) and $U_e \approx U_e$, $H_e/h_e \approx U_e^2/2c_p T_e \approx U_\infty^2/c_p T_e$ for hypersonic flow along with a linear viscosity-temperature relation (Chapman-Rubesin constant C=1) yields

$$\hat{\beta} = -2x \frac{R}{c_p} \frac{P_\infty}{P_e} \frac{d\left(\frac{P_e}{P_\infty}\right)}{dx}$$

Using Eq. (6–120) yields

$$\hat{\beta} = \frac{\gamma - 1}{\gamma} = \text{constant} \qquad (6-124)$$

From this result we conclude that the compressible boundary layer which develops on a sharp flat plate in hypersonic flow is similar.[†] Thus, it is governed

---

[†]The hypersonic strong interaction problem is similar to zeroth-order only. The nonsimilar character of the self-induced pressure gradient can be accounted for by using an expansion for the pressure. This procedure results in a series of systems of ordinary differential equations, of which the present system is the zeroth-order. For additional discussion see Hayes and Probstein [Haye59].

274   Laminar Flow Analysis

by Eqs. (6–74), (6–75) and (6–79). Further, this fact provides the additional required coupling relation between the inviscid and viscous flow (see Eq. 6–119).

From Eq. (6–103) we see that the displacement thickness in hypersonic flow is given by

$$\delta^* = \frac{\sqrt{2\xi}}{\rho_e U_e} \frac{H_e}{h_e} J_1 \qquad (6-125)$$

Using Eq. (6–123), we have

$$\left(\frac{\delta^*}{x}\right)\left(\frac{P_e}{P_\infty}\right)^{1/2} = \frac{\gamma-1}{M_\infty} \chi J_1 \qquad (6-126)$$

which is the second required coupling relation.

Since J_1 is known from the boundary layer solution, Eqs. (6–119) and (6–126) represent two equations in two unknowns. Further, Eqs. (6–120) and (6–121) give the functional relations for P_e/P_∞ and δ^*/x to within an arbitrary constant. Thus, we can now determine the unknown constants A and B in terms of the boundary layer displacement integral, J_1. We proceed by substituting Eqs. (6–120) and (6–121) into Eq. (6–119) to yield

$$A = \frac{9}{32}\gamma(\gamma+1)B^2 \qquad (6-127)$$

Similarly, substituting Eqs. (6–120) and (6–121) into Eq. (6–126) yields

$$\sqrt{A}\, B = (\gamma-1)J_1 \qquad (6-128)$$

Solving Eqs. (6–127) and (6–128) for A and B, we have

$$A = \frac{3}{4}(\gamma-1)\left[\gamma\left(\frac{\gamma+1}{2}\right)\right]^{1/2} J_1 \qquad (6-129)$$

and

$$B = 2\left[2\gamma(\gamma+1)\right]^{1/4}\left[\frac{2}{3}(\gamma-1)J_1\right]^{1/2} \qquad (6-130)$$

These results, in combination with Eqs. (6–120) and (6–121), give the pressure distribution and the growth of the boundary layer displacement thickness on a sharp flat plate in hypersonic strong interaction once J_1 is known from the boundary layer solution. The shearing stress and heat transfer at the surface are evaluated using Eqs. (6–97) and (6–99) once $f''(0)$ and $g'(0)$ are known. For air, $\gamma = 1.4$ and $\hat{\beta} = 0.286$, whereas for a monatomic gas such as helium $\gamma = 1.67$ and $\hat{\beta} = 0.4$. Results for A, B and J_1 obtained using the comb program are shown in Figure 6–31 as functions of the wall-to-stagnation temperature ratio, g_w. Similarly, the results for $f''(0)$ and $g'(0)$ are shown in Figure 6–32. Numerical results are given in Table C–34. These results compare favorably with those of Lees [Lees53] and Li and Nagamatsu [Li55], who originally investigated this problem. From Figure 6–31 we see that the pressure distribution and the growth of the displacement thickness increase significantly with increasing pressure gradient, $\hat{\beta}$, and wall-to-stagnation temperature ratio, g_w. From Figure 6–32 we see that the shearing stress parameter, $f''(0)$, increases with increasing pressure gradient parameter, $\hat{\beta}$, and wall-to-stagnation temperature ratio, g_w, whereas the heat transfer parameter, $g'(0)$, decreases.

Figure 6–33 shows a comparison of hypersonic strong interaction theory and the experiments of Bertram [Bert57] on an insulated plate. Here we see that the zeroth-order strong interaction theory presented above for $\Pr = 1$, $g_w = 1$, i.e.,

$$\frac{P_e}{P_\infty} = 0.514\chi$$

predicts the correct slope for the experimental pressure distribution but does not give the correct magnitude. However, also shown is the first-order correction

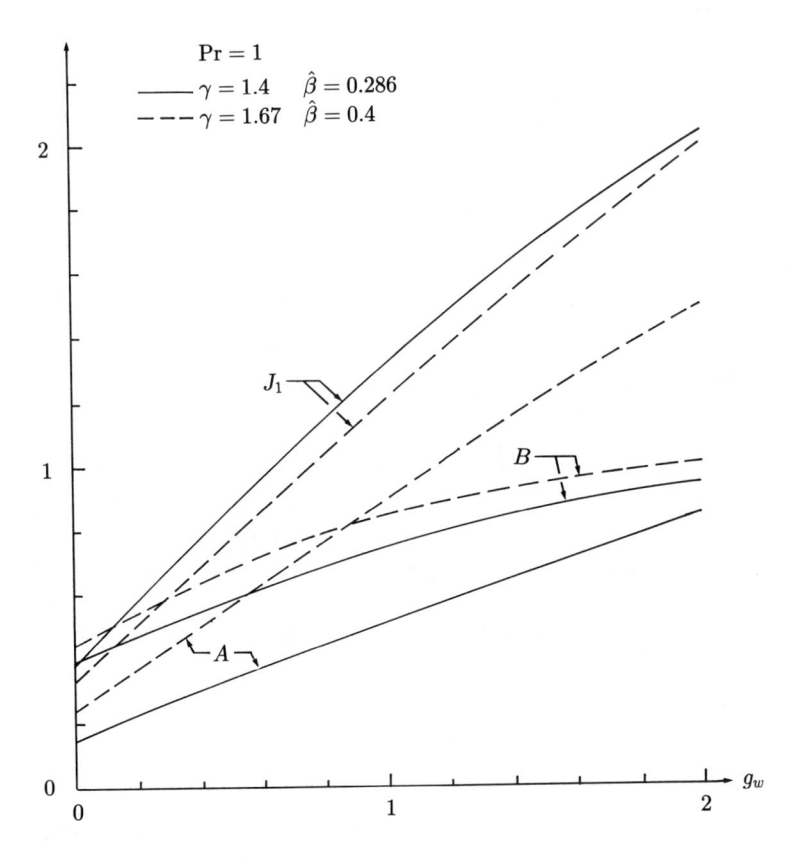

Figure 6–31. Results for hypersonic strong interaction on a sharp flat plate — A, B, J_1.

276 *Laminar Flow Analysis*

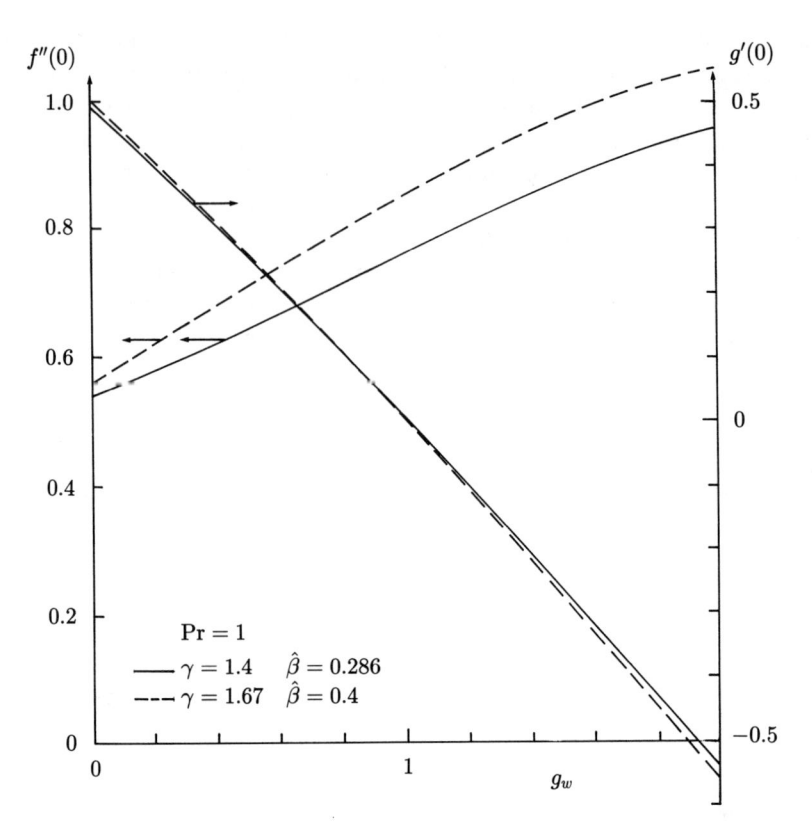

Figure 6-32. Results for hypersonic strong interaction on a sharp flat plate $f''(0)$, $g'(0)$.

determined by Nagakura and Naruse for $\text{Pr} = 1$, $g_w = 1$ [Naga57], i.e.

$$\frac{P_e}{P_\infty} = 0.514\chi + 0.759$$

Here we see that the theory and the experimental results are in good agreement.

Before leaving this topic we remark that the effects of nonunit Prandtl number and mass transfer can be investigated using the comb program. Further, hypersonic strong interaction implies $(M_\infty d\delta^*/dx) \gg 1$, whereas if $(M_\infty d\delta^*/dx) \ll 1$ then one has the hypersonic weak interaction regime. For a discussion of this problem the reader is directed to Hayes and Probstein [Haye59].

6–18 The Nonsimilar Compressible Boundary Layer

In Sections 6–6 through 6–17 we extensively discussed the similar laminar compressible boundary layer. In Section 6–17 the hypersonic strong interaction problem, which is a classical application of the compressible similar boundary layer, was discussed. However, many compressible boundary layer problems of interest are nonsimilar. In particular, the compressible boundary layer on most 'practical' bodies is nonsimilar because of the velocity distribution in the external flow. The practical case of constant mass transfer at the surface also yields a nonsimilar boundary layer. The particularly interesting problem of axisymmetric viscous flow over a long slender cylinder aligned with the flow direction is nonsimilar due to the transverse curvature terms which appear in the equations. The methods for solving the nonsimilar compressible boundary layer parallel those for the nonsimilar incompressible boundary layer previously discussed in Section 3–11. Here, we develop the locally nonsimilar boundary layer method for the compressible boundary layer.

Rather than repeat the analysis here, we begin by recalling the transformed forms of the momentum and energy equations given in Eqs. (6–65) and (6–66). In deriving these equations a Chapman-Rubesin viscosity law with $\omega = 1$ is assumed, i.e., $\rho\mu = C\rho_e\mu_e$. For convenience C is assumed constant. Using Euler's equation to evaluate conditions at the edge of the boundary layer yields

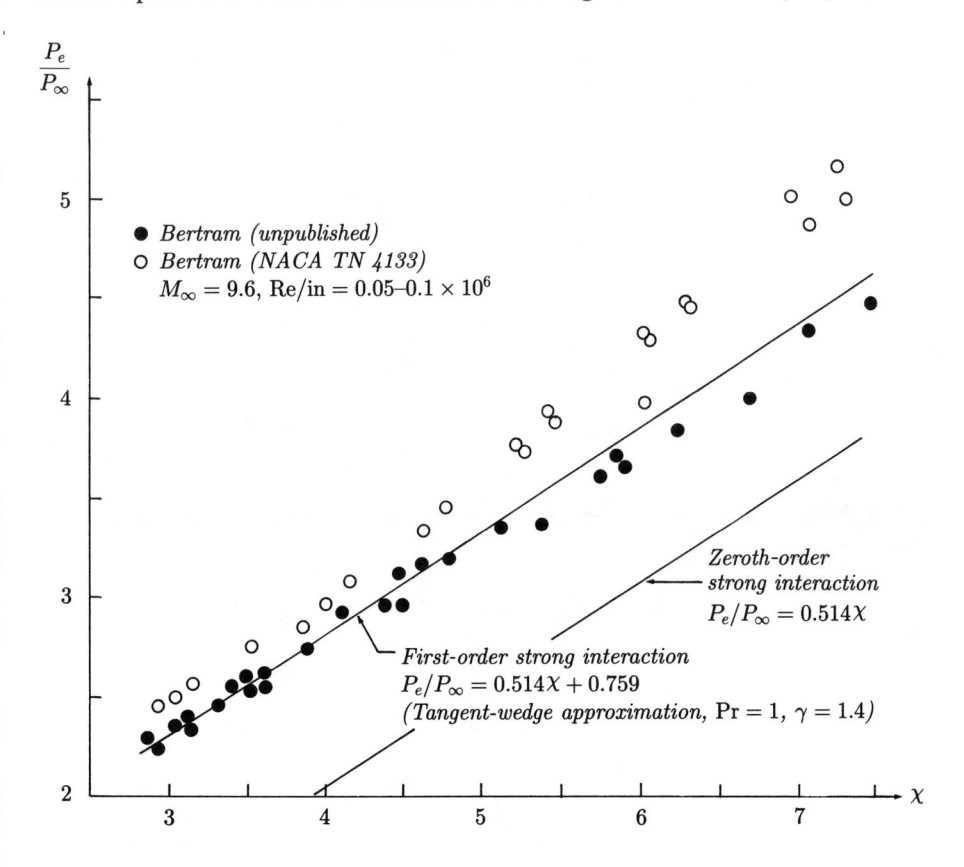

Figure 6–33. Comparison of hypersonic strong interaction theory and experiment.

278   Laminar Flow Analysis

$(\rho_e/\rho)(dU/dx) = -(dP/dx)/(\rho U)$. Substituting into the momentum equation, Eq. (6–65), and dividing by $C\rho_e\mu_e(N^2/U^2)$, yields

$$f_{\eta\eta\eta} + \frac{NN_x}{C\rho_e\mu_e U}ff_{\eta\eta} - N^2\frac{U_x}{U}\frac{1}{C\rho_e\mu U}(f_\eta)^2$$
$$+ \frac{N^2}{C\rho_e\mu_e U}\frac{U_x}{U}\frac{\rho_e}{\rho} = \frac{N^2}{C\rho_e\mu_e U}\xi_x(f_\eta f_{\xi\eta} - f_{\eta\eta}f_\xi) \qquad (6-131)$$

Dividing Eq. (6–66) by $H_e(C\rho_e\mu_e)(U/N^2)$ yields the form of the energy equation given by

$$g_{\eta\eta} + \frac{N_\perp N}{C\rho_e\mu_e U}\Pr f g_\eta - \frac{N^2}{C\rho_e\mu_e U}\frac{H_{e_x}}{H_e}\Pr g f_\eta$$
$$+ (\Pr - 1)\frac{U^2}{H_e}(f_\eta f_{\eta\eta})_\eta = \frac{N^2\xi_x}{C\rho_e\mu_e U}(f_\eta g_\xi - f_\xi g_\eta) \qquad (6-132)$$

Recalling that

$$\xi = \int_0^x C\rho_e\mu_e U\,dx \qquad (6-71)$$

$$N(x) = \left(2\int_0^x C\rho_e\mu_e U\,dx\right)^{1/2} = \sqrt{2\xi} \qquad (6-70)$$

and

$$\hat{\beta} = \frac{N^2}{C\rho_e\mu_e U}\frac{U_x}{U}\frac{H_e}{h_e}$$

we note that

$$\frac{N_x N}{C\rho_e\mu_e U} = 1$$

Finally, we note that

$$\left(\frac{\rho_e}{\rho} - f_\eta^2\right) = \frac{H_e}{h_e}(g - f_\eta^2) \qquad (6-73)$$

Using these results, the momentum and energy equations, including the nonsimilar terms, are written as

$$f_{\eta\eta\eta} + ff_{\eta\eta} + \hat{\beta}(g - f_\eta^2) = 2\xi(f_\eta f_{\xi\eta} - f_{\eta\eta}f_\xi) \qquad (6-133)$$

and
$$g_{\eta\eta} + \Pr f g_\eta + (\Pr - 1)\bar{\sigma}(f_\eta f_{\eta\eta})_\eta =$$
$$2\xi\bar{\alpha}\Pr f_\eta g + 2\xi(f_\eta g_\xi - f_\xi g_\eta) \qquad (6-134)$$

where

$$\bar{\alpha} = \frac{1}{C\rho_e\mu_e U}\frac{H_{e_x}}{H_e} \qquad (6-135)$$

and

$$\bar{\sigma} = \frac{U_e^2}{H_e} \qquad (6-136)$$

The appropriate surface boundary conditions are

$$f(\xi, 0) + 2\xi f_\xi = -\frac{v(x)\sqrt{2\bar{\sigma}}}{U(x)\mu_w(x)} \qquad f_\eta(\xi, 0) = 0 \qquad (6-137a,b)$$

$$g(0) = g_w(x) \qquad (6-137c)$$

and in the far field

$$f_\eta(\xi, \eta \to \infty) \to 1 \qquad g(\xi, \eta \to \infty) \to 1 \qquad (6-137d,e)$$

In our previous similarity analysis we neglected the term $2\xi(f_\eta f_{\xi\eta} - f_{\eta\eta}f_\xi)$ in the momentum equation and $2\xi(f_\eta g_\xi - f_\xi g_\eta)$ in the energy equation. Further, $\hat{\beta}$, $\bar{\sigma}$ and g_w were assumed constant, and $\bar{\alpha} = 0$. In the mass transfer boundary condition the term $2\xi f_\xi$ was neglected, and it was required that $f_0 = $ constant to satisfy the similarity requirements. The locally nonsimilar boundary layer method developed below eliminates each of these requirements.

The analysis is begun by introducing auxiliary functions

$$S(\xi, \eta) = f_\xi \qquad t(\xi, \eta) = g_\xi \qquad (6-138a,b)$$

The momentum and energy equations are now written as

$$f''' + ff'' + \hat{\beta}(g - f'^2) = 2\xi(f'S' - f''S) \qquad (6-139)$$

and

$$g'' + \Pr fg' + (\Pr - 1)\bar{\sigma}(f'f'')' =$$
$$2\xi\bar{\alpha}\Pr f'g + 2\xi(f't - Sg') \qquad (6-140)$$

where the primes denote differentiation with respect to η. The necessary equations needed to evaluate $S(\xi, \eta)$ and $t(\xi, \eta)$ are obtained by differentiating the momentum and energy equations with respect to ξ. Differentiating the momentum equation yields

$$S''' + f''S + fS'' + \frac{d\hat{\beta}}{d\xi}(g - f'^2) + \hat{\beta}(t - 2f'S') = 2(f'S' - f''S)$$
$$+ 2\xi(S'^2 - SS''') + 2\xi(f'S'_\xi - f''S_\xi) \qquad (6-141)$$

Differentiating the energy equation yields

$$t'' + \Pr(Sg' + ft') - 2\Pr\bar{\alpha}f'g$$
$$+ (\Pr - 1)\left(\frac{d\bar{\sigma}}{d\xi}(f'f'')' + \bar{\sigma}(S'f''' + 2f''S'' + f'S''')\right)$$
$$+ 2(Sg' - f't) = 2\xi\bar{\alpha}\Pr\left(S'g + f't + \frac{1}{2}\frac{d\bar{\alpha}}{d\xi}\right)$$
$$+ 2\xi(S't - St') + 2\xi(f't_\xi - S_\xi g') \qquad (6-142)$$

280   Laminar Flow Analysis

Examination of the system of four equations given by Eqs. (6–139) to (6–142) reveals that they could be considered as four simultaneous ordinary differential equations in the dependent variables $f(\xi,\eta)$, $g(\xi,\eta)$, $S(\xi,\eta)$, $t(\xi,\eta)$, with known coefficients except for the appearance of the term $2\xi(f'S'_\xi - f''S_\xi)$ in the first auxiliary momentum equation (Eq. 6–141) and the term $2\xi(f't_\xi - S_\xi g')$ in the first auxiliary energy equation (Eq. 6–142). It is now argued that these terms are negligibly small either because ξ itself is small or because the term in the brackets is small. The first auxiliary momentum equation thus becomes

$$S''' + f''S + fS'' + \frac{d\hat{\beta}}{d\xi}(g - f'^2) + \hat{\beta}(t - 2f'S') =$$
$$2(f'S' - f''S) + 2\xi(S'^2 - SS'') \qquad (6-143)$$

and the first auxiliary energy equation becomes

$$t'' + \Pr(Sg' + ft') - 2\Pr\bar{\alpha}fg'$$
$$+ (\Pr - 1)\frac{d\bar{\sigma}}{d\xi}\left((f'f'')' + \bar{\sigma}(S'f''' + 2f''S'' + f'S''')\right)$$
$$+ 2(Sg' - f't) = 2\xi\bar{\alpha}\Pr\left(S'g + f't + \frac{1}{\bar{\alpha}}\frac{d\bar{\alpha}}{d\xi}\right) + 2\xi(S't - St') \qquad (6-144)$$

Equations (6–139), (6–140), (6–143) and (6–144) are the governing equations for the two-equation locally nonsimilar boundary layer method for the compressible boundary layer. Note that, as is the case for the nonsimilar free convection and combined forced and free convection boundary layers, the four equations are coupled. Thus they must be solved simultaneously.

The appropriate boundary conditions for the first auxiliary momentum and energy equations are obtained by differentiating Eqs. (6–137) with respect to ξ. Differentiating the mass transfer boundary condition yields

$$3S(\xi,0) + 2\xi S_\xi(\xi,0) = -(2\xi)^{-1/2}\left(\frac{v(x)}{U_e\mu_w}\right) - (2\xi)^{1/2}\frac{\partial}{\partial\xi}\left(\frac{v(x)}{U_e\mu_w}\right)$$

Consistent with the two-equation locally nonsimilar boundary layer analysis, the term $2\xi S_\xi(\xi,0)$ is assumed negligibly small. The boundary conditions for the first auxiliary momentum and energy equations are then

$$S(\xi,0) = -\frac{1}{3}\left[(2\xi)^{-1/2}\frac{v(x)}{U_e\mu_w} + (2\xi)^{1/2}\frac{\partial}{\partial\xi}\left(\frac{v(x)}{U_e\mu_w}\right)\right] \qquad (6-145a)$$

$$S'(\xi,0) = 0 \qquad t(\xi,0) = \frac{\partial g_w}{\partial\xi} \qquad (6-145b,c)$$

and $\qquad S'(\xi,\eta\to\infty)\to 0 \qquad t(\xi,\eta\to\infty)\to 0 \qquad (6-145d,e)$

Nonsimilar Compressible Boundary Layer

The complete system of boundary conditions for the two-equation model is given by Eqs. (6–137) and (6–145). Note that satisfaction of four asymptotic boundary conditions is required.

The governing boundary value problem for the two-equation locally nonsimilar boundary layer method for the compressible boundary layer is numerically integrated by making suitable modifications and additions to the comb program. The required Nachtsheim-Swigert iteration technique for four asymptotic boundary conditions is discussed in Appendix B.

The three-equation locally nonsimilar model for a laminar compressible boundary layer is developed similarly by introducing additional auxiliary functions $h(\xi, \eta) = S_\xi(\xi, \eta) = f_{\xi\xi}(\xi, \eta)$ and $j(\xi, \eta) = t_\xi(\xi, \eta) = g_{\xi\xi}(\xi, \eta)$ into Eqs. (6–141) and (6–142) and subsequently differentiating with respect to ξ to obtain the equations necessary to evaluate $h(\xi, \eta)$ and $j(\xi, \eta)$ by evaluating those terms involving h_ξ and j_ξ. The boundary conditions are also developed in this manner. The resulting system of six equations is of fifteenth-order, with six asymptotic boundary conditions.

REFERENCES

[Adam68] Adams, J.A., and Lowell, R.L., Jr., Free convection organic sublimation on a vertical semi-infinite plate, *Int. J. Heat Mass Trans.*, Vol. 11, pp. 1215–1224, 1968.

[Ball67] Ball, K.O.W., Similarity solutions for the compressible laminar boundary layer with heat and mass transfer, *Phys. of Fluids*, Vol. 10, pp. 1823–1826, 1967.

[Beck59] Beckwith, Z.E., Similar solutions for the compressible boundary layer on a yawed cylinder with transpiration cooling, NASA TR R-42, 1959.

[Bert57] Bertram, M.H., Boundary-layer displacement effects in air at Mach numbers of 6.8 and 9.6, NACA TN 4133, 1957.

[Blas08] Blasius, H., Grenzschichten in Flüssigkeiten mit kleiner Reinbung, *Zeit. Math. Phys.*, Vol. 56, pp. 1–37 (English transl. in NACA TM1256), 1908.

[Brow51] Brown, W.B., and Donoughe, P.L., Tables of exact laminar-boundary solutions when the wall is porous and fluid properties are variable, NACA TN 2479, 1951.

[Buse31] Busemann, A., Gasdynamik, *Handbuch der Experimentalphysik*, Vol. 4, Part I, pp. 341–460, Akad. Verlag. M.B.H., Leipzig, 1931.

[Cath65] Catherall, D., Stewartson, K., and Williams, P., Viscous flow past a flat plate with uniform injection, *Proc. Roy. Soc. London Ser. A*, Vol. 284, pp. 370–396, 1965.

[Chap39] Chapman, S., and Cowling, T.G., *The Mathematical Theory of Nonuniform Gases*, Cambridge, UK: Cambridge Univ. Press, (Chap. 16) 1939.

[Chap49] Chapman, D.R., and Rubesin, M.W., Temperature and velocity profiles in the compressible laminar boundary layer with arbitrary distribution of surface temperature, *JAS*, Vol. 16, pp. 547–565, 1949.

[Cohe54] Cohen, C., Similar solutions of compressible laminar boundary layer equations, *JAS*, Vol. 21, pp. 281–282, 1954.

[Cohe56] Cohen, C.B., and Reshotko, E., Similar solutions for the compressible laminar boundary layer with heat transfer and pressure gradient, NACA TR 1293, 1956.

[Croc32] Crocco, L., Transmissione del calore da una lamina piona a un fluidoscorrente ad alte velocita, *L'Aerotecnica*, Vol. 12, pp. 181–197, 1932 (available in translation as NACA TM690, 1932).

[Croc46] Crocco, L,. Lo strato limite laminare nei gas, *Monografie Sci. Aeronaut.*, Vol. 3, pp. 3–78, December 1946.

[Dewe67] Dewey, C.F., and Gross, J.F., Exact similar solutions of the laminar boundary layer equations, *Advances in Heat Transfer*, Vol. 4, pp. 317–446, New York: Academic Press, 1967.

[Eich60] Eichhorn, R., The effect of mass transfer on free convection, *Trans. ASME Series C, J. Heat Transfer*, Vol. 82, pp. 260–263, 1960.

[Elzy67] Elzy, E., and Sisson, R.M., Tables of similar solutions to the equations of momentum, heat and mass transfer in laminar boundary layer flow, Bulletin No. 49, Engineering Experiment Station, Oregon State University, Corvallis, Oregon, February 1967.

[Emmo53] Emmons, H.W., and Leigh, D., Tabulation of the Blasius function with blowing and suction, Harvard Univ. Div. Applied Sci. Combustion Aerodynamics Lab. Interim Tech. Report No. 9 (DDC Report No. AD27068), 1953.

[Evan62] Evans, H.L., Mass transfer through laminar boundary layers—8: Further solutions to the velocity equation, *Int. Jour. Heat Transfer*, Vol. 5, pp. 373–407, 1962.

[Evan63] Evans, H.L., Integration of the velocity equation of the laminar boundary layer including the effect of mass transfer, *AIAA Jour.*, Vol. 1, p. 1677, 1963.

[Evan66] Evans, H.L., Laminar boundary layers with uniform fluid properties—similar solutions to the velocity equations involving mass transfer, Ministry of Aviation Aeronautical Research Council C.P. No. 857 (DDC Document No. AD800220), 1966.

[Evan68] Evans, H.L., *Laminar Boundary Layer Theory*, Reading, MA: Addison-Wesley, 1968.

[Falk31] Falkner, V.M., and Skan, S.W., Some approximations of the boundary layer equations, *Phil. Mag.*, Vol. 12, pp. 865–896, 1931.

[Fett54] Fettis, H.E., On a differential equation occurring in the theory of heat flow in boundary layers with Hartree's velocity profiles, *JAS*, Vol. 21, pp. 132–133, 1954.

[Flug62] Flügge-Lotz, I., and Baxter, D.C., Computation of the compressible laminar boundary-layer flow including displacement-thickness interaction using finite-difference methods, TR 131 (AFOSR2206), Stanford Univ., Stanford, California, 1962.

[Gall69] Gallo, W.F., Marvin, J.G., and Gnos, U.S., A study of the nonsimilar nature of the laminar boundary layer in a region of adverse pressure gradient, AIAA paper 69-35 presented at the 7th Aerospace Sciences, New York, January 20–22, 1969.

[Gold30] Goldstein, S., Concerning some solutions of the boundary layer equations in hydrodynamics, *Proc. Camb. Phil. Soc.*, Vol. 26, pp. 1–30, 1930.

[Gold38] Goldstein, S., Ed., *Modern Developments in Fluid Dynamics*, Oxford, UK: Oxford University Press, 1938.

[Gort52] Görtler, H., A new series for the calculation of steady laminar boundary layer flows, *Jour. Math. and Mechanics*, Vol. 6, pp. 1–66, 1957; see also original German paper: *ZAMM*, Vol. 32, pp. 270–271, 1952.

[Grie63] Grieser, D.R., and Goldthwaite, W.H., Experimental determination of the viscosity of air in the gaseous state at low temperatures and pressures, Arnold Engineering Development Center TDR 63-143, June 1963.

[Hans64] Hansen, A.G., *Similarity Analysis of Boundary Value Problems in Engineering*, Englewood Cliffs, NJ: Prentice-Hall, 1964.

[Hart37] Hartree, D.R., On the equation occurring in Falkner and Skan's approximate treatment of the equations of the boundary layer, *Proc. Camb. Phil. Soc.*, Vol. 33, Part II, pp. 223–239, 1937.

[Haye59] Hayes, W.D., and Probstein, R.F., *Hypersonic Flow Theory*, New York: Academic Press, 1959.

[Homa36] Homann, F., Der Einfluss grosser Zähigkeit bei der Strömung um den Zylinder und um die Kugel., *Zeitschr. f. Angew. Math. u. Mech.*, Vol. 16, pp. 153–164, 1936.

[Howa39] Howarth, L., On the solution of the laminar boundary layer equations, *Proc. Roy. Soc. London Ser. A*, Vol. 164, pp. 547–579, 1939.

[Howa48] Howarth, L., Concerning the effect of compressibility on laminar boundary layers and their separation, *Proc. Roy. Soc. London Ser. A*, Vol. 194, pp. 16–42, 1948.

[Howa53] Howarth, L., *Modern Developments in Fluid Dynamics High Speed Flow*, Vol. 1, Oxford, UK: Clarendon Press, 1953.

[Illi49] Illingworth, C.R., Steady flow in the laminar boundary layer of a gas, *Proc. Roy. Soc. London Ser. A*, Vol. 199, pp. 533–558, 1949.

[Illi50] Illingworth, C.R., Some solutions of the equations of flow of a viscous compressible fluid, *Proc. Cambridge Phil. Soc.*, Vol. 46, pp. 469–478, 1950.

[Koh61] Koh, J.C.Y., and Hartnett, J.P., Skin friction and heat transfer for incompressible laminar flow over porous wedges with suction and variable wall temperature, *Int. J. Heat Mass Transfer*, Vol. 2, pp. 185–198, 1961.

[Kuet59] Kuethe, A.M., and Schetzer, J.D., *Foundations of Aerodynamics*, 2nd ed., New York: Wiley, 1959.

[Lees53] Lees, L., On the boundary-layer equations in hypersonic flow and their approximate solutions, *JAS*, Vol. 20, pp. 143–145, 1953.

[Lees64] Lees, L., and Reeves, B.L., Supersonic separated and reattaching laminar flows, *AIAA Jour.*, Vol. 2, pp. 1907–1920, 1964.

[Levy52] Levy, S., Heat transfer to constant-property laminar boundary-layer flows with power-function free-stream velocity and wall temperature variation, *JAS*, Vol. 19, pp. 341–348, 1952.

[Levy54] Levy, S., Effect of large temperature changes (including viscous heating) upon laminar boundary layers with variable free-stream velocity, *JAS*, Vol. 21, pp. 459–474, 1954.

[Lew55] Lew, H.G., and Fanucci, J.B., On the laminar compressible boundary layer over a flat plate with suction or injection, *JAS*, Vol. 22, pp. 589–597, 1955.

[Li55] Li, T.Y., and Nagamatsu, H.T., Similar solutions of compressible boundary layer equations, *JAS*, Vol. 22, pp. 607–616, 1955.

[Libb65] Libby, P.A., and Chen, K., Laminar boundary layer with uniform injection, *Phys. of Fluids*, Vol. 8, pp. 568–574, 1965.

[Libb68] Libby, P.A., and Liu, T.M., Some similar laminar flows obtained by quasi-linearization, *AIAA Jour.*, Vol. 6, p. 1541–1548, 1968.

[Liep57] Liepmann, H.W., and Roshko, A., *Elements of Gas Dynamics*, New York: John Wiley & Sons, (Chap. 11) 1957.

[Ligh50] Lighthill, M.J., Contributions to the theory of heat ransfer through a laminar boundary layer, *Proc. Roy. Soc. London Ser. A*, Vol. 202, pp. 359–377, 1950.

[Lock51] Lock, R.C., The velocity distribution in the laminar boundary layer between parallel streams, *Quart. Jour. Mech. and Appl. Maths.*, Vol. 4, Part I, p. 42, 1951.

[Low53] Low, G.M., The compressible laminar boundary layer with heat transfer and small pressure gradient, NACA TN 3028, 1953.

[Mang48a] Mangler, W., Zusammenhang zwischen ebenen und rotations—symmetrischen grenzschichten in kompressiblen Flüssigkeiten, *ZAMM*, Vol. 28, pp. 97–103, 1948.

[Mang48b] Mangler, W., Special exact solutions, Section 1.2 of *Boundary Layers*, Tollmien, W., Ed., Monograph Aerodyn. VersAnst, Gottingen B., 1948.

[Marv67] Marvin, J.G., and Sinclair, R.A., Convective heating in regions of large favorable pressure gradients, *AIAA Jour.*, Vol. 5, pp. 1940–1948, 1967.

[McCr66] McCroskey, W.S., Bogdonoff, S.M., and McDougall, J.G., An experimental model for the sharp flat plate in rarefied hypersonic flow, *AIAA Jour.*, Vol. 4, pp. 1580–1587, 1966.

[Meks61] Meksyn, D., *New Methods in Laminar Boundary Layer Theory*, Oxford, UK: Pergamon Press, 1961.

[Merk59] Merk, H.J., Rapid calculations for boundary-layer transfer using wedge solutions and asymptotic expansions, *Jour. Fluid Mechs.*, Vol. 5, pp. 460–480, 1959.

[Moor64] Moore, F.K., Ed., *Theory of Laminar Flows*, Vol. 4, Princeton Series on High Speed Aerodynamics and Jet Propulsion, Princeton, NJ: Princeton Univ. Press, 1964.

[Murp60] Murphy, G.M., *Ordinary Differential Equations and Their Solutions*, Princeton, NJ: Van Nostrand, 1960.

[Nach65] Nachtsheim, P.R., and Swigert, P., Satisfaction of asymptotic boundary conditions in numerical solution of systems of nonlinear equations of boundary layer type, NACA TN D-3004, October 1965.

[Naga57] Nagakura, T., and Naruse, H., An approximate solution of the hypersonic laminar boundary layer equations and its application, *Jour. Phys. Soc. Japan*, Vol. 12, pp. 1298–1304, 1957.

[Niku42] Nikuradse, J., *Laminare Reibungsschichten an der längsangeträmten Platte*, Monograph, Zentrale f. wiss, Berichtswesen, Berlin, 1942.

[Odon54] O'Donnell, R.M., Experimental investigation at a Mach number of 2.41 of average skin friction coefficients and velocity profiles for laminar and turbulent boundary layers and an assessment of probe effects, NACA TN 3122, 1954.

References 287

[Ostr53] Ostrach, S., An analysis of laminar free-convection flow and heat transfer about a flat plate parallel to the direction of the generating body force. NACA TR-1111, 1953.

[Pohl12] Pohlhausen, K., Zur naherungsweisen integration der differentialgleichund der grenzschight, *ZAMM*, Vol. 1, pp. 252–268, 1912.

[Pohl21] Pohlhausen, E., Der Wärmeaustausch zwischen festen Körpern und Flüssigkeiten mit kleiner Reibung und kleiner Wärmeleitung, *ZAMM*, Vol. 1, p. 115, 1921.

[Pran28] Prandtl, L., Uber Flüssigheitbewegung bein sehr kleiner Reibung, *Proc. 3rd Inter. Math. Congress*, Heidelberg 1904, p. 484–491, also NACA TM452, 1928.

[Punn53] Punnis, B., A remark on 'On asymptotic solutions for the heat transfer at varying wall temperatures in a laminar boundary layer with Hartree's velocity profiles', *JAS*, Vol. 20, p. 505, 1953.

[Roge69a] Rogers, D.F., Further similar laminar flow solutions, *AIAA Jour.*, Vol. 7, pp. 967–978, 1969.

[Roge69b] Rogers, D.F, Reverse flow solutions for compressible laminar boundary layer equations, *Phys. of Fluids*, Vol. 12, pp. 517–523, 1969.

[Roge70] Rogers, D.F., Adiabatic wall solutions for the hypersonic compressible laminar boundary-layer equations, *Phys. of Fluids*, Vol. 13, pp. 1142–1145, 1970.

[Roge90] Rogers, D.F., and Adams, A.J., *Mathematical Elements for Computer Graphics*, 2nd ed., New York: McGraw-Hill, 1990.

[Rose63] Rosenhead, L., Ed., *Laminar Boundary Layers*, Oxford, UK: Oxford University Press, 1963.

[Scha58] Schaaf, S.A., and Chambre, P.L., Flow of Rarefied Gases in *Fundamentals of Gasdynamics*, Emmons, H.W., Ed., Vol. III High Speed Aerodynamics and Jet Propulsion, Princeton, 1958.

[Schl43] Schlichting, H., and Bussmann, K., Exaklelosungen für die laminare grenzschicht mit absangung und ausblasen, Schriften der Dentschen Akademie der Luftahrtforschung, Bd. 7b, Heft 2, 1943.

[Schl60] Schlichting, H., *Boundary Layer Theory*, 4th ed., New York: McGraw Hill, 1960.

[Schl68] Schlichting, H., *Boundary-Layer Theory*, 6th ed., New York: McGraw-Hill, 1968.

[Schm30] Schmidt, E., und Beckmann, W., Das Temperatur und Geschwindig-keitsfeld vor einer Warmeabgebenden senkrechter Platte bei naturlicher Konvektion. Tech. Mech u. Thermodynamik, Bd. I, pp. 341–349, Oct. 1930), cont, Bd. 1, pp. 391–406, November 1930.

[Scho64] Schönauer, W., Ein differenzenverfahren zur Lösung der Grenzschicht gleichuing für stationäre, laminare, inkompressible Strömung, *Ingen. Archiv*, Vol. 33, pp. 173–189, 1964.

[Schu53] Schuh, H., On asymptotic solutions for the heat transfer at varying wall temperatures in a laminar boundary layer with Hartree's velocity profiles, *JAS*, Vol. 20, pp. 146–147, 1953.

[Sing58] Singh, S.N., Heat transfer by laminar flow in cylindrical tubes, *Applied Sci. Res. Serial A*, Vol. 7, p. 325, 1958.

[Smit63] Smith, A.M.O., and Clutter, D.W., Solution of the incompressible laminar boundary layer equations, *AIAA Jour.*, Vol. 1, pp. 2062–2071, 1963.

[Spal61a] Spalding, D.B., Mass transfer through laminar boundary layers—1. The velocity boundary layer, *Int. Jour. Heat Transfer*, Vol. 2, pp. 15–32, 1961.

[Spal61b] Spalding, D.B., and Evans, H.L., Mass transfer through laminar boundary layers—2. Auxiliary functions for the velocity boundary layer, *Int. Jour. Heat Transfer*, Vol. 2, pp. 199–221, 1961.

[Spal62] Spalding, D.B., and Pun, W.M., A review of methods for predicting heat-transfer coefficients for laminar uniform-property boundary layer flows, *Int. J. Heat Mass Trans.*, Vol. 5, pp. 239–249, 1962.

[Spar58] Sparrow, E.M., and Gregg, J.L., Similar solutions for free convection from a non-isothermal vertical plate, *Trans. ASME 80*, pp. 379–396, 1958.

[Spar59] Sparrow, E.M., Eichhorn, R., and Gregg, J.L., Combined forced and free convection in boundary layer flow, *The Physics of Fluids*, Vol. 2, pp. 319–328, 1959.

[Spar70] Sparrow, E.M., Quack, H., and Boerner, C.J., Local non-similar boundary-layer solutions, *AIAA Jour.*, Vol. 8, pp. 1936–1942, 1970.

[Spie58] Spiegel, M.R., *Applied Differential Equations*, Englewood Cliffs, NJ: Prentice Hall, p. 168, 1958.

[Stew50] Stewartson, K., Correlated incompressible and compressible boundary layers, *Proc. Roy. Soc. London Ser. A*, Vol. 200, pp. 84–100, 1950.

[Stew54] Stewartson, H., Further solutions of the Falkner-Skan equations, *Proc. Camb. Phil. Soc.*, Vol. 50, pp. 454–465, 1954.

[Stew64] Stewartson, K., *The Theory of Laminar Boundary Layers in Compressible Fluids*, Oxford, UK: Clarendon Press, 1964.

[Stre58] Streeter, L. (ed.), *Fluid Mechanics*, New York: McGraw-Hill, 1958.

[Terr60] Terrill, R.M., Laminar boundary layer flow near separation with and without suction, *Phil. Trans. Roy. Soc.*, Ser. A, Vol. 253, pp. 55–100, 1960.

[Topf12] Töpfer, C., Bemerkung zu dem Aufsatz von H. Blasius Grenzschichten in Flüssigkeiten mit kleiner Reitung, *Zeit. fur Math. Phys.*, Vol. 60, pp. 397–398, 1912.

[VanD64] Van Dyke, M., *Perturbation Methods in Fluid Mechanics*, New York: Academic Press, 1964.

[vonD52] von Driest, E.R, Investigation of laminar boundary layer in compressible fluids using the Crocco method, NACA TN 2597, 1952.

[vonK38] von Karman, Th., and Tsien, H.S., Boundary layer in compressible fluids, *JAS*, Vol. 5, pp. 227–232, 1938.

[vonM27] von Mises, R., Bermenkungen zur Rydromechanik, ZAMM 7, p. 425, 1927.

[Weas87] Weast, R.C., Ed., *Handbook of Chemistry and Physics*, 68th Edition, The Chemical Rubber Publishing Co., Cleveland, Ohio, 1987.

[Yang60] Yang, K.T., Possible similarity solutions for laminar free convection on vertical plates and cylinders, *Trans. ASME J. Appl. Mech.*, Vol. 27, pp. 230–236, 1960.

APPENDIX A

RUNGE-KUTTA INTEGRATION SCHEME

A–1 Derivation of Equivalent First-order Equations

The Runge-Kutta method is used to determine the solution of an nth-order, ordinary differential equation of the form

$$\frac{d^n y}{dx^n} = f\left(x, y, \frac{dy}{dx}, \ldots \frac{d^{n-1}y}{dx^{n-1}}\right) \qquad (A-1)$$

However, it is usually more convenient to work with the equivalent set of first-order differential equations obtained from Eq. (A–1). The mathematical transformation of variables leading to the equivalent set of equations is derived by letting

$$Y_1 = y$$
$$Y_2 = \frac{dy}{dx}$$
$$\vdots$$
$$Y_n = \frac{d^{n-1}y}{dx^{n-1}} \qquad (A-2)$$

and then defining

$$F_1 = \frac{dY_1}{dx} = \frac{dy}{dx} = Y_2$$
$$F_2 = \frac{dY_2}{dx} = \frac{d^2 y}{dx^2} = Y_3$$
$$\vdots$$
$$F_n = \frac{dY_n}{dx} = \frac{d^n y}{dx^n} = f(x, Y_1, Y_2, \ldots Y_n) \qquad (A-3)$$

For example, consider a third-order differential equation given by

$$\frac{d^3 y}{dx^3} = \left(\frac{dy}{dx}\right)^2 - \left(\frac{dy}{dx}\right)\left(\frac{d^2 y}{dx^2}\right) - 1 \qquad (A-4)$$

290   Laminar Flow Analysis

with initial conditions at $x = 0$ specified by

$$\frac{d^2y}{dx^2} = C \qquad \frac{dy}{dx} = 0 \qquad y = 0 \qquad \text{(A-5)}$$

Let
$$Y_1 = y$$
$$Y_2 = \frac{dy}{dx}$$
$$Y_3 = \frac{d^2y}{dx^2} \qquad \text{(A-6)}$$

and then define

$$F_1 = \frac{dY_1}{dx} = Y_2$$
$$F_2 = \frac{dY_2}{dx} = Y_3$$
$$F_3 = \frac{dY_3}{dx} = (Y_2)^2 - Y_2 Y_3 - 1 \qquad \text{(A-7)}$$

with initial conditions at $x = 0$ given by

$$Y_1 = 0 \qquad Y_2 = 0 \qquad Y_3 = C \qquad \text{(A-8)}$$

The above three equations for F_1, F_2 and F_3 are three first-order differential equations equivalent to Eq. (A-4). There are now three dependent variables, Y_1, Y_2 and Y_3, instead of one.

A-2 Derivation of Recursion Formula

We now focus our attention on obtaining a numerical solution to a first-order differential equation of the form

$$F = \frac{dy}{dx} = f(x, y) \qquad \text{(A-9)}$$

The objective of a numerical solution is to approximate the solution $y(x)$ at many discrete values of the independent variable, x, over a specified interval, $a \leq x \leq b$. When the value of $y(x)$ is known at some starting point, e.g., $y = y_i$ at $x = x_i$, the value of y_{i+1} at $x_i + \Delta x$ is calculated by use of the Runge-Kutta recurrence formula given by

$$y_{i+1} = y_i + w_1 k_1 + w_2 k_2 + \ldots w_n k_n \qquad \text{(A-10)}$$

The weighting functions $w_1, w_2, \ldots w_n$ are determined by matching the terms in the recurrence formula to an nth-order Taylor series expansion of y about y_i. The k values depend upon derivatives (dy/dx) given by the differential equation, Eq. (A-9), evaluated within the increment Δx. For an nth-order approximation the k's are given by

$$k_1 = f(x_i, y_i) \Delta x \qquad \text{(A-11a)}$$

$$k_2 = f(x_i + p_1 \Delta x, y_i + q_{11} k_1) \Delta x \qquad (A-11b)$$

$$k_3 = f(x_i + p_2 \Delta x, y_i + q_{21} k_1 + q_{22} k_2) \Delta x \qquad (A-11c)$$

$$\vdots$$

$$k_n = f(x_i + p_{n-1} \Delta x, y_i + q_{n-1,1} k_1 + q_{n-1,2} k_2$$
$$+ q_{n-1,3} k_3 + \ldots q_{n-1,n-1} k_{n-1}) \Delta x \qquad (A-11n)$$

As shown below, the p's and q's are also determined by matching Eq. (A–11) with the terms of a Taylor series expansion.

The recurrence formula for the second-order Runge-Kutta technique requires an equation of the form

$$y_{i+1} = y_i + w_1 k_1 + w_2 k_2 \qquad (A-12)$$

with
$$k_1 = f(x_i, y_i) \Delta x \qquad (A-13a)$$

$$k_2 = f(x_i + p \Delta x, y_i + q k_1) \Delta x \qquad (A-13b)$$

The motivation of the Runge-Kutta method is to evaluate y_{i+1} in terms of y_i and the first derivative, dy/dx. This has advantages over a second-order Taylor series given by

$$y_{i+1} = y_i + \frac{dy_i}{dx} \Delta x + \frac{1}{2!} \frac{d^2 y_i}{dx^2} \Delta x^2 \qquad (A-14)$$

since the second derivative is not necessarily known or well-behaved. The Taylor series expansion leaves room for improvement, since the terms on the right side of Eq. (A–14) are all evaluated at the beginning of the ith increment of x.

Using Eq. (A–9), we write $(dy_i/dx) = f(x_i, y_i)$ and $d^2 y_i/dx^2 = dF_i/dx$. Since F is a point function of x and y, the chain rule of calculus gives

$$dF = \left(\frac{\partial f}{\partial x}\right) dx + \left(\frac{\partial f}{\partial y}\right) dy \qquad (A-15)$$

It follows that

$$\frac{dF}{dx} = \frac{\partial f}{\partial x}(1) + \frac{\partial f}{\partial y}\frac{dy}{dx} \qquad (A-16)$$

Combining the above relationships with Eq. (A–14) and using Eq. (A–9) yields

$$y_{i+1} = y_i + f(x_i, y_i) \Delta x + \frac{1}{2!} \left(\frac{\partial f(x_i, y_i)}{\partial x} + \frac{\partial f(x_i, y_i)}{\partial y} f(x_i, y_i) \right) (\Delta x)^2 \qquad (A-17)$$

This expression for a second-order Taylor series expansion serves as the standard of accuracy.

By defining k_1 and k_2 in terms of the function $f(x, y)$ in Eq. (A–13), we force these coefficients to be functions of the first derivative (dy/dx). The value of k_1 is based upon the derivative at the beginning of each interval in x, dy_i/dx. The value of k_2 is based upon the derivative at some unknown location within the increment Δx, since $0 < p < 1$.

292   Laminar Flow Analysis

We now expand the derivative $dy/dx = f(x,y)$ about $f(x_i, y_i)$ by use of a Taylor series expansion for two variables. This is given by

$$f(x_i + \Delta x, y_i + \Delta y) = f(x_i, y_i) + \frac{\partial f(x_i, y_i)}{\partial x}\Delta x + \frac{\partial f(x_i, y_i)}{\partial y}\Delta y$$
$$+ \frac{\partial^2 f(x_i, y_i)}{\partial x^2}\frac{(\Delta x)^2}{2} + \frac{\partial^2 f(x_i, y_i)}{\partial x \partial y}\Delta x \Delta y$$
$$+ \frac{\partial^2 f(x_i, y_i)}{\partial y^2}\frac{(\Delta y)^2}{2} + \cdots \qquad (A-18)$$

Second-order terms and higher are neglected. However, consistent with the definition of k_2 given by Eq. (A–13b), we want an expansion for $f(x_i + p\Delta x, y_i + qk_1)$. This is

$$f(x_i + p\Delta x, y_i + qk_1) = f(x_i, y_i) + \frac{\partial f(x_i, y_i)}{\partial x}p\Delta x + \frac{\partial f(x_i, y_i)}{\partial y}qk_1 \qquad (A-19)$$

Multiplying each side of this equation by Δx and using the definition of k_1 given by Eq. (A–13a) allows us to write

$$k_2 = f(x_i + p\Delta x, y_i + qk_1)\,\Delta x$$
$$= f(x_i, y_i)\Delta x + \frac{\partial f(x_i, y_i)}{\partial x}p(\Delta x)^2 + \frac{\partial f(x_i, y_i)}{\partial y}qf(x_i, y_i)(\Delta x)^2 \qquad (A-20)$$

Using the derived expressions for k_1 and k_2 in Eq. (A–12) gives

$$y_{i+1} = y_i + w_1 f(x_i, y_i)\Delta x + w_2 f(x_i, y_i)\Delta x$$
$$+ w_2 \left[\frac{\partial f(x_i, y_i)}{\partial x}p(\Delta x)^2 + \frac{\partial f(x_i, y_i)}{\partial y}qf(x_i, y_i)(\Delta x)^2\right] \qquad (A-21)$$

To match the accuracy of a second-order Taylor series expansion about y_i we choose values of w_1 and w_2 such that Eq. (A–21) agrees with Eq. (A–17). Equating coefficients of like terms requires that

$$w_1 + w_2 = 1 \qquad (A-22a)$$
$$w_2 p = \frac{1}{2} \qquad (A-22b)$$
$$w_2 q = \frac{1}{2} \qquad (A-22c)$$

The above three equations have four unknowns. Looking at the form of Eq. (A–13b), we see that a choice of $p = 1$ requires that k_1 be evaluated at the convenient location $x = x_i + \Delta x$. Letting $p = 1$, it follows from Eq. (A–22) that $w_2 = 1/2$, $w_1 = 1/2$ and $q = 1$. Equations (A–12) and (A–13) are now written as

$$y_{i+1} = y_i + \frac{1}{2}(k_1 + k_2) \qquad (A-23)$$

where
$$k_1 = f(x_i, y_i)\Delta x \qquad (A-24a)$$

and
$$k_2 = f(x_i + \Delta x, y_i + k_1) \Delta x \qquad (A-24b)$$

The second-order Runge-Kutta method is used as follows. First, the initial value of y_i at $x = x_i$ is assumed known. Then $f(x_i, y_i)$ gives the initial value of k_1 from Eq. (A-24a) when the value of Δx is specified. Then the value of $y_i + k_1$ is calculated. Note that $k_1 = (dy/dx)_i \Delta x = \Delta y$, the increment in y assuming a linear variation across Δx and a slope of k_1. Next, $f(x, y)$ is evaluated at $x_i + \Delta x$ and $y_i + k_1$ to obtain k_2. Then Eq. (A-23) is used to calculate y_{i+1}. Finally, this value of y and the value of $x = x_i + \Delta x$ are used in Eq. (A-24a) to repeat the process for the next increment in x.

When the same procedure is followed for a third-order Runge-Kutta technique, the results are

$$y_{i+1} = y_i + \frac{k_1 + 4k_2 + k_3}{6} \qquad (A-25)$$

$$k_1 = f(x_i, y_i) \Delta x \qquad (A-26a)$$

$$k_2 = f\left(x_i + \frac{\Delta x}{2}, y_i + \frac{k_1}{2}\right) \Delta x \qquad (A-26b)$$

$$k_3 = f(x_i + \Delta x, y_i - k_1 + 2k_2) \Delta x \qquad (A-26c)$$

The results for a fourth-order technique suggested by Runge are

$$y_{i+1} = y_i + \frac{k_1 + 2k_2 + 2k_3 + k_4}{6} \qquad (A-27)$$

$$k_1 = f(x_i, y_i) \Delta x \qquad (A-28a)$$

$$k_2 = f\left(x_i + \frac{\Delta x}{2}, y_i + \frac{k_1}{2}\right) \Delta x \qquad (A-28b)$$

$$k_3 = f\left(x_i + \frac{\Delta x}{2}, y_i + \frac{k_2}{2}\right) \Delta x \qquad (A-28c)$$

$$k_4 = f(x_i + \Delta x, y_i + k_3) \Delta x \qquad (A-28d)$$

This set of equations forms the algorithm that is used to obtain numerical solutions to ordinary differential equations.

The fourth-order Runge-Kutta algorithm can also be explained from a graphical point of view. We illustrate this by considering the solution of a first-order differential equation using a fourth-order Runge-Kutta technique. Mathematically we represent this type of equation by writing

$$F = \frac{dy}{dx} = f(x, y)$$

Recalling the fourth-order Runge-Kutta recurrence formula (see Eq. A-28) and noting that $k_1 = f(x_i, y_i) \Delta x = f(\Delta x)$ illustrates the significance of k_1, as shown graphically in Figure A-1a. There, k_1 is equal to the shaded rectangular area with height $f(x_i, y_i)$ and width Δx. Note that Δx is greatly enlarged to show the areas. Since

$$y = \int_0^x \frac{dy}{dx} \Delta x$$

a point on the curve in Figure A–1b at $x = c_1$ (e.g., point A) has a value equal to the area under the curve in Figure A–1a between 0 and $x_i = c_1$, plus the initial value of y_i.

A geometric interpretation is applicable for the other three k values. Point B in Figure A–1b is formed by the intersection of the lines $x_i + \Delta x/2$ and $y_i(x) + k_1/2$. According to the mathematical equation for k_2, point B also corresponds to an area in Figure A–1a. The additional area is equal to the increase in y between points A and B in Figure A–1b. It also has width Δx. The height of the rectangle which produces the required area is the value $F_2 = f(x_i + \Delta x/2, y_i + k_1/2)$. The lightly shaded area in Figure A–1a represents k_2.

The geometric interpretation of k_3 and k_4 is similar. We first identify a point, such as D in Figure A–1b, formed by the intersection of $x_i + \Delta x/2$ and $y_i + k_2/2$. A third area equal to k_3 is then identified in Figure A–1b, with width Δx and height $F_3 = f(x_i + \Delta x/2, y_i + k_2/2)$. Finally, point E formed by $x_i + \Delta x$ and $y_i + k_3$ is identified. The area equal to k_4 is also identified in Figure A–1a, with width Δx and height $F_4 = f(x_i + \Delta x, y_i + k_3)$.

The geometric interpretation of the Runge-Kutta recurrence formula is now clear. The actual area under the curve of dy/dx vs x between x_i and $x_i + \Delta x$ is equal to $y(x_i + \Delta x - \theta(x))$. This area A is approximated by four rectangular areas, weighted according to the equation

$$A = \frac{k_1 + 2k_2 + 2k_3 + k_4}{6}$$

A–3 Simultaneous First-order Differential Equations

As indicated in Section A–1, higher-order differential equations can be recast into an equivalent set of first-order differential equations. The solution of a set of simultaneous first-order equations is then required.

Consider a second-order equation of the form

$$\frac{d^2y}{dx^2} = f\left(x, y, \frac{dy}{dx}\right) \qquad (A-29)$$

Letting
$$Y_1 = y \qquad (A-30a)$$

$$Y_2 = \frac{dy}{dx} \qquad (A-30b)$$

and defining
$$F_1 = Y_2 \qquad (A-31a)$$

$$F_2 = F(x, Y_1, Y_2) \qquad (A-31b)$$

yields two ordinary differential equations, where $F_1 = dY_1/dx$ and $F_2 = dY_2/dx$.

In general, we consider two coupled ordinary differential equations given by

$$\frac{dY_1}{dx} = f(x, Y_1(x), Y_1(x)) \qquad (A-32a)$$

$$\frac{dY_2}{dx} = g(x, Y_1(x), Y_2(x)) \qquad (A-32b)$$

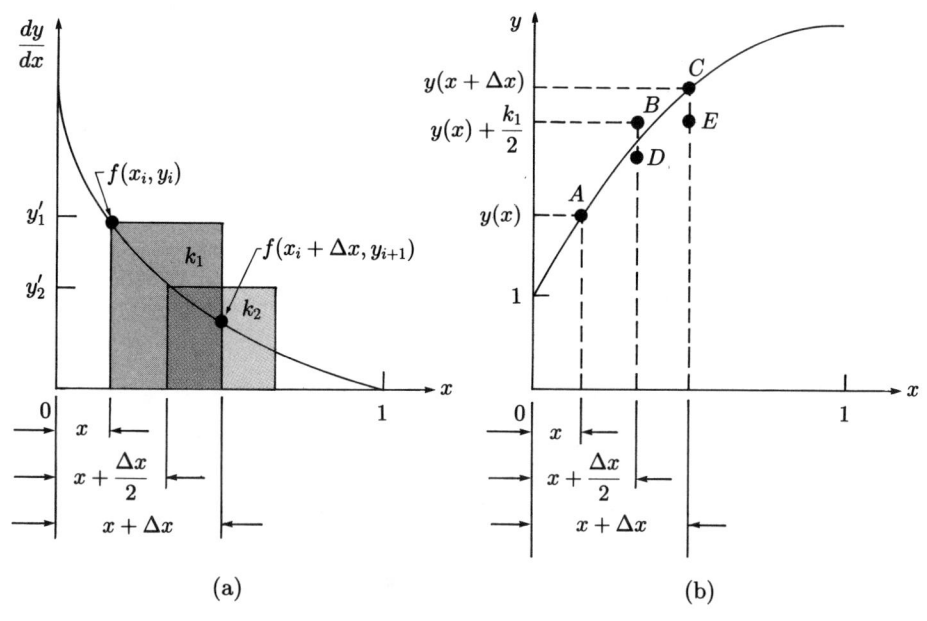

Figure A–1. Graphical interpretation of Runge-Kutta integration.

with initial conditions at $x = x_0$ given by $Y_1 = Y_1(0)$ and $Y_2 = Y_2(0)$. The fourth-order Runge-Kutta equations are then

$$Y_{1,i+1} = Y_{1,i} + \frac{k_1 + 2k_2 + 2k_3 + k_4}{6} \tag{A - 33}$$

$$k_1 = f(x_i, Y_{1,i}, Y_{2,i})\, \Delta x \tag{A - 34a}$$

$$k_2 = f\left(x_i + \frac{\Delta x}{2}, Y_{1,i} + \frac{k_1}{2}, Y_{2,i} + \frac{\ell_1}{2}\right) \Delta x \tag{A - 34b}$$

$$k_3 = f\left(x_i + \frac{\Delta x}{2}, Y_{1,i} + \frac{k_2}{2}, Y_{2,i} + \frac{\ell_2}{2}\right) \Delta x \tag{A - 34c}$$

$$k_4 = f(x_i + \Delta x, Y_{1,i} + k_3, Y_{2,i} + \ell_3)\, \Delta x \tag{A - 34d}$$

$$Y_{2,i+1} = Y_{2,i} + \frac{\ell_1 + 2\ell_2 + 2\ell_3 + \ell_4}{6} \tag{A - 35}$$

$$\ell_1 = g(x_i, Y_{1,i}, Y_{2,i})\, \Delta x \tag{A - 36a}$$

$$\ell_2 = g\left(x_i + \frac{\Delta x}{2}, Y_{1,i} + \frac{k_1}{2}, Y_{2,i} + \frac{\ell_1}{2}\right) \Delta x \tag{A - 36b}$$

$$\ell_3 = g\left(x_i + \frac{\Delta x}{2}, Y_{1,i} + \frac{k_2}{2}, Y_{2,i} + \frac{\ell_2}{2}\right) \Delta x \tag{A - 36c}$$

$$\ell_4 = g(x_i + \Delta x, Y_{1,i} + k_3, Y_{2,i} + \ell_3)\, \Delta x \tag{A - 36d}$$

A-4 Example

Consider the first-order equation given by

$$\frac{dx}{dt} = xt \tag{A-37}$$

with an initial condition $t_0 = 0$ at $x_0 = 1$. An analytical solution to this equation is given by

$$x = \exp\left(\frac{1}{2}t^2\right) \tag{A-38}$$

We now compare a numerical solution, obtained by using a second-order Runge-Kutta scheme, to the known analytical solution. We arbitrarily choose the independent variable increment $\Delta x = h = 0.1$. The recursion equations are

$$x(t_i + h) = x(t_i) + \frac{k_1 + k_2}{2} \tag{A-39}$$

$$k_1 = f(t_i, x_i)\Delta t \tag{A-40a}$$

$$k_2 = f(t_i + \Delta t, x_i + k_1)\Delta t \tag{A-40b}$$

Using the initial condition $x_0 = 1$, $t_0 = 0$, then $k_1 = 0$ and $k_2 = f(0+0.1, 1+0)\Delta t = (0.1)(0.1) = 0.01$. Then from Eq. (A-39)

$$x(t_0 + h) = x(t_0) + \frac{k_1 + k_2}{2} = 1 + \frac{0 + 0.01}{2} = 1.005$$

The value of $x(t) = 1.005$ at $t = 0.1$ is now used to calculate the value of $x(t)$ at $t = 0.2$

$$k_1 = f(0.1, 1.005)\Delta t = (0.1005)(0.1) = 0.01005$$

$$k_2 = f(0.2, 1.015)\Delta t = (0.203)(0.1) = 0.0203$$

Then, from Eq. (A-39)

$$x(0.2) = 1.005 + \frac{0.01005 + 0.0203}{2} = 1.0201$$

This procedure is repeated over the independent variable interval of interest. The analytical solution is compared with the numerical solution for $0 \leq t \leq 1$ in Table A-1. Higher-order Runge-Kutta schemes provide closer comparison between the analytical and numerical results. Alternatively, higher-order methods allow larger increments, $\Delta t = h$, to be used without sacrificing the accuracy of the numerical solution (see Table A-2). In general, for a given integration scheme the accuracy decreases as the increment in the independent variable increases. Table A-1 also shows the relative effect of increasing h when using a second-order Runge-Kutta technique on the example problem. The value of $x(t)$ at $t = 1$ is calculated using intervals of 0.1, 0.2, 0.5 and 1. As shown in Table A-1, the error is a strong function of the increment size.

Table A–1. Second-order accuracy solution of the differential equation, $dx/dt = xt$, using different size integration intervals.

t	Numerical Solution $x(t)$ $h = 0.1$	Numerical Solution $x(t)$ $h = 0.2$	Numerical Solution $x(t)$ $h = 0.5$	Numerical Solution $x(t)$ $h = 1$	Analytical Solution $x(t) = \exp(1/2\, t^2)$
0	1.0000	1.0000	1.0000	1.0000	1.0000
0.1	1.0050				
0.2	1.0202	1.0200			1.0202
0.3	1.0460				
0.4	1.0832	1.0828			1.0833
0.5	1.1330		1.1250		
0.6	1.1970	1.1963			1.1972
0.7	1.2773				
0.8	1.3767	1.3753			1.3771
0.9	1.4987				
1.0	1.6478	1.6449	1.6172	1.5000	1.6487

Table A–2. Higher-order accuracy.

Order	$x(1)$, with $h = 1$
Second	1.5000
Third	1.6667
Fourth	1.6458

A–5 Conclusions

Other numerical integration algorithms can be used to obtain solutions to ordinary differential equations. These are discussed in numerical analysis books. The Runge-Kutta method was chosen as the integration scheme, since it is self-starting when the required initial conditions and/or boundary conditions are specified. This leads to a simple computer program with wide applicability to engineering problems. Since the function $f(x,y)$ is evaluated several times between each independent variable increment, it is not necessarily the most computationally economic method. However, the slight increase in computer time is usually more than compensated by the ease in programming. The accuracy obtained by using the fourth-order Runge-Kutta recursion formula is well within engineering requirements when using appropriate incremental-step sizes for the independent variable.

APPENDIX B

SATISFACTION OF ASYMPTOTIC BOUNDARY CONDITIONS

B–1 Introduction

The boundary conditions associated with ordinary differential equations of the boundary layer type are of the two-point asymptotic class. Two-point boundary conditions have values of the dependent variable specified at two different values of the independent variable. Specification of an asymptotic boundary condition implies that the first derivative (and in the context of the boundary layer equations, all higher derivatives) of the dependent variable approaches zero as the specified value of the independent variable is approached.

One method of numerically integrating a two-point asymptotic boundary value problem of the boundary layer type, the initial value or 'shooting' method, requires that it be recast as an initial value problem. Thus, it is necessary to estimate as many boundary conditions at the surface as previously given at infinity. The governing differential equations are then integrated with these assumed surface boundary conditions. If the required outer boundary condition is satisfied, a solution is achieved. However, this is not generally the case. Hence, a method must be devised to logically estimate the new surface boundary conditions for the next trial integration. Asymptotic boundary value problems such as those governing the boundary layer equations are further complicated by the fact that the outer boundary condition is specified at infinity. In trial integrations infinity is numerically approximated by some large value of the independent variable. There is no a priori general method of estimating this value. Selecting too small a maximum value for the independent variable does not allow asymptotic convergence to the required accuracy. Selecting too large a value results in divergence of the trial integrations or in slow convergence to the surface boundary conditions required to satisfy the asymptotic outer boundary conditions. Selecting too large a value of the independent variable is expensive in terms of computer time.

In order to effectively illustrate the method used to overcome these problems, we apply it to the Falkner-Skan equation (see Section 3–4) and then develop the required analysis for more complex problems. For comparison purposes we first develop the Newton-Raphson iteration technique for estimating the surface boundary conditions for the Falkner-Skan equation. Subsequently, we develop the Nachtsheim-Swigert iteration technique (see Nachtsheim and Swigert [Nach65]) and indicate the modifications necessary to allow determination of the reverse flow solutions for the Falkner-Skan equation.

B–2 Newton-Raphson Iteration Technique — Falkner-Skan Equation, $f''(0)$

For clarity we develop the Newton-Raphson technique specifically for the boundary value problem associated with the Falkner-Skan equation. This boundary value problem is given by

$$f''' + f f'' + \beta(1 - f'^2) = 0 \qquad (3-50)$$

$$f(0) = f'(0) = 0 \qquad (3-51a,b)$$

and
$$f'(\eta \to \infty) \to 1 \qquad (3-51c)$$

To recast this as an initial value problem we estimate $f''(0)$ in order to perform the required trial numerical integration. Further, we specify the maximum value of the independent variable for the numerical integration. We designate this value as $\eta_{\max}$ and let it numerically represent infinity. Thus the boundary conditions are restated as

$$f(0) = f'(0) = 0 \qquad f''(0) = k_1 \qquad (B-1a,b)$$

The numerical solution is required to satisfy the asymptotic outer boundary condition

$$f'(\eta_{\max}) = 1 + \delta_1 \qquad (B-2)$$

where δ_1 is some small number.

All iteration methods used to estimate the values of the unknown surface boundary conditions required to satisfy the asymptotic boundary condition at $\eta_{\max}$, Eq. (B–2), depend on the integrals of the boundary layer equations at large values of the independent variable being functions of the unknown surface boundary conditions. For the Falkner-Skan equation this is mathematically stated as

$$f'(\eta_{\max}) = f'(f''(0)) = 1 + \delta_1 \qquad (B-3)$$

In order to obtain a correction equation for the value of $f''(0)$ required to satisfy the outer boundary condition at $\eta_{\max}$, Eq. (B–3), we expand $f'(\eta_{\max})$ in a first-order Taylor series, i.e.

$$f'(\eta_{\max}) = f'_c(\eta_{\max}) + \frac{\partial f'(\eta_{\max})}{\partial f''(0)} \Delta f''(0) + \ldots$$

where

$f'(\eta_{\max})$ is the required boundary value for f' at $\eta_{\max}$;

$f'_c(\eta_{\max})$ is the value of f' at $\eta_{\max}$ calculated using the assumed value of $f''(0)$;

$\partial f'/\partial f''(0)$ is the change in f' at $\eta_{\max}$ with a change in the assumed boundary condition at the surface $f''(0)$;

$\Delta f''(0)$ is the change in $f''(0)$ necessary to satisfy the boundary condition at $\eta_{\max}$.

Using a finite difference representation of the differentials and letting $f''(0) = x$, solving for Δx yields

$$\Delta x = \left.\frac{1 - f'_c}{f'_x}\right|_{\eta = \eta_{\max}} \quad (B-4)$$

where

$$f'_x = \frac{\partial f'(\eta_{\max})}{\partial f''(0)}$$

We can solve this algebraic equation for any given trial integration of the governing equations provided the derivative f'_x can be evaluated at $\eta_{\max}$. One method of determining this derivative (see [Nach65]) is to numerically integrate the perturbation equation obtained by differentiating the Falkner-Skan equation with respect to x, i.e., $f''(0)$. The resulting boundary value problem for the perturbation equation is an initial value problem, i.e.

$$f'''_x = -(f f''_x + f'' f_x + 2\beta f' f'_x)$$

with

$$f_x(0) = f'_x(0) = 0 \qquad f''_x(0) = 1$$

For an assumed value of $f''(0)$ the perturbation equation is integrated along with the Falkner-Skan equation, the derivative f'_x evaluated and the subsequent value of $f''(0)$ determined from Eq. (B-4). Since the perturbation equation is third-order and we require f'_x, integration of the perturbation equation along with the Falkner-Skan equation represents integration of a fifth-order system of equations.

Considerable simplification results if a finite difference representation of the derivative f'_x is used. In particular, we assume that

$$f'_x = \frac{\Delta f'(\eta_{\max})}{\Delta f''(0)} = \frac{f'_2(\eta_{\max}) - f'_1(\eta_{\max})}{f''_2(0) - f''_1(0)} \quad (B-5)$$

where

$f''_1(0) = k_1$, some assumed value;

$f''_2(0) = k_1 + \epsilon$, where ϵ is a small quantity, e.g., 1×10^{-5} to 1×10^{-2};

$f'_1(\eta_{\max})$ is the value of f' at $\eta_{\max}$ obtained by numerical integration of the Falkner-Skan equation using $f''_1(0)$;

$f'_2(\eta_{\max})$ is the value of f' at $\eta_{\max}$ obtained by numerical integration of the Falkner-Skan equation using $f''_2(0)$.

By performing two trial integrations with the assumed values of $f''_1(0)$ and $f''_2(0)$, the required convergence derivative, f'_x, is approximately determined using Eq. (B-5). Equation (B-4) then yields the required correction to $f''(0)$. For a simple one-parameter representation, such as is given in Eq. (B-3), the approximate value of f'_x is updated by using the current estimate of $f''(0)$ and the current value of $f'(\eta_{\max})$ to determine f'_x. However, this technique does not solve the problem of choosing $\eta_{\max}$, nor does the Newton-Raphson method assure asymptotic convergence to the required outer boundary condition. Figure B-1 is taken from Nachtsheim and Swigert [Nach65]. Examination of Figure B-1a shows that the boundary condition, $f'(\eta_{\max}) = 1$, is approximately satisfied at two values of $f''(0)$, namely $f''(0) = 0.85$ and 1.23.

302   Laminar Flow Analysis

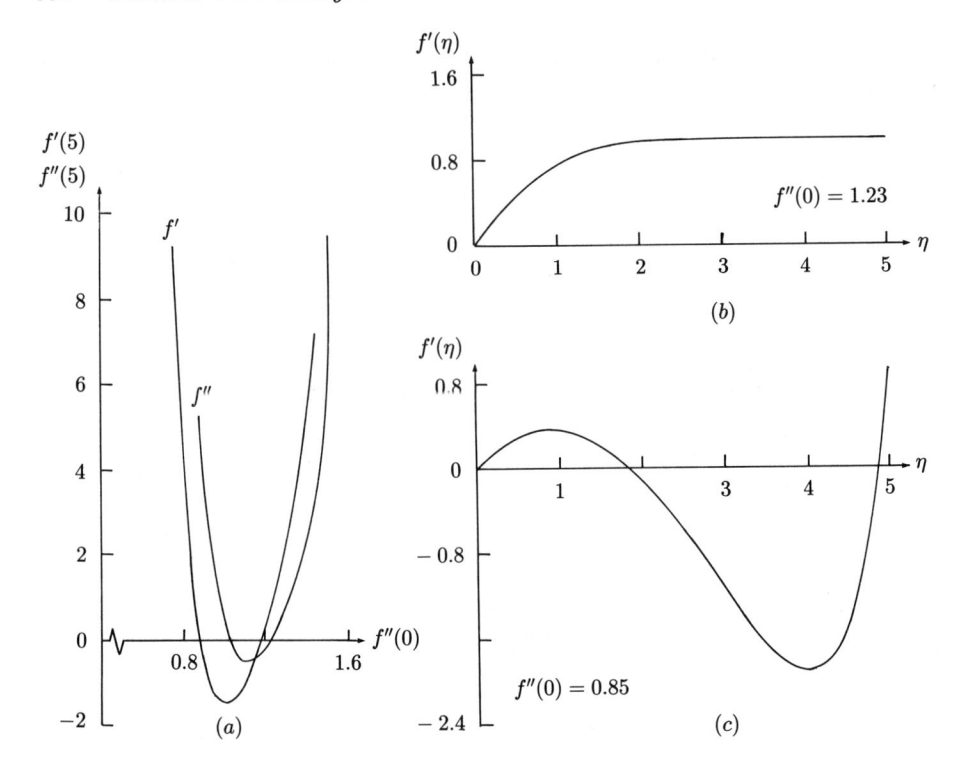

Figure B–1. Asymptotic convergence requirement for the Falkner-Skan equation, $\beta = 1$.

However, further examination of Figure B–1a shows that $f''(\eta_{max})$ is very large for $f''(0) = 0.85$, whereas it is quite small for $f''(0) = 1.23$. These observations are confirmed by Figures B–1b and c. However, if a Newton-Raphson iteration scheme is used, convergence to $f'(\eta_{max}) = 1 \pm \epsilon$ occurs for two values of $f''(0)$, i.e., approximately 0.85 and 1.23, with $\eta_{max} = 5$. Here we see that the Newton-Raphson iteration scheme does not yield asymptotic convergence, i.e., it does not assure that $f''(\eta_{max}) \to 0$ but only that $f'(\eta_{max}) = 1 \pm \epsilon$ regardless of $f''(\eta_{max})$. Further, a Newton-Raphson iteration scheme does not yield a unique solution for $f''(0)$.

B–3 Nachtsheim-Swigert Iteration Scheme — Falkner-Skan Equation, $f''(0)$

Nachtsheim and Swigert [Nach65] developed an iteration method which overcomes these difficulties. It is presented below in a somewhat modified form. Satisfying the asymptotic boundary condition at infinity, i.e., $f'(\eta \to \infty)$, requires that both $f'(\eta \to \infty) \to 1$ and $f''(\eta \to \infty) \to 0$. However, since there is only one adjustable parameter, namely $f''(0)$, it is not possible to simultaneously satisfy both conditions. We thus consider that the required asymptotic boundary condition is satisfied at η_{max} if

$$f'(\eta_{max}) = 1 + \delta_1 \tag{B-2}$$

and

$$f''(\eta_{max}) = \delta_2 \tag{B-6}$$

where δ_1 and δ_2 are some small numbers. Again performing a first-order Taylor series expansion, we have

$$f'(\eta_{\max}) = f'_c(\eta_{\max}) + \frac{\partial f'(\eta_{\max})}{\partial f''(0)} \Delta f''(0)$$

and

$$f''(\eta_{\max}) = f''_c(\eta_{\max}) + \frac{\partial f''(\eta_{\max})}{\partial f''(0)} \Delta f''(0)$$

Adopting a finite-difference representation of the differentials yields

$$\delta_1 = f'_c + f'_x \Delta x - 1 \qquad (B-7)$$

$$\delta_2 = f''_c + f''_x \Delta x \qquad (B-8)$$

where we used the required values for boundary conditions at $\eta_{\max}$ (see Eqs. B–2 and B–6) to arrive at these results. Following Nachtsheim and Swigert [Nach65] we seek a least-squares solution of Eqs. (B–7) and (B–8), i.e., we seek to minimize $\delta_1^2 + \delta_2^2$ with respect to x by differentiating with respect to x and setting the result equal to zero. Solving the result for $\Delta x = \Delta f''(0)$ yields

$$\Delta x = \frac{f'_x(1 - f'_c) - f''_x f''_c}{f'^2_x + f''^2_x} \qquad \text{at } \eta_{\max} \qquad (B-9)$$

Here again, Nachtsheim and Swigert use a perturbation equation to determine the required convergence derivatives, i.e., f'_x and f''_x. However, the finite difference representation of the convergence derivatives described above is also successful. Examination of Eq. (B–9) gives the fundamental reason. If the finite difference representation of f'_x and f''_x yields values which are too large but with $f'_c(\eta_{\max})$ and $f''_c(\eta_{\max}) = O(1)$, then the denominator is proportionately larger than the numerator, and Δx is smaller. The next iteration yields improved values of $f''(0)$, $f'_c(\eta_{\max})$ and hence, with continuous upgrading of f'_x and f''_x, convergence to the required value of $f''(0)$ occurs. However, when using *either* the appropriate perturbation equations or a finite difference method to determine the convergence derivatives it must be possible to determine these derivatives in some meaningful manner. This is not possible if the initial estimate of $f''(0)$ is so poor that the solution diverges. Such a case is shown in Figure B–2. Figure B–2 shows that for $\beta = 1$ trial solutions of the Falkner-Skan equation for $f''(0) \approx 1.23$ behave in approximately the same manner as the actual solution. Determination of the required convergence derivatives using trial solutions with $f''(0) \approx 1.23$ yields acceptable values. Convergence to $f''(0) = 1.232588$, the value required to satisfy the asymptotic outer boundary condition, occurs rapidly. However, trial solutions with $f''(0) \approx 1$ or 1.5 diverge or oscillate badly. Attempts to determine the required convergence derivatives from these trial solutions do not yield meaningful results. Convergence to the required value of $f''(0)$ does not occur. Nachtsheim and Swigert [Nach65] present a technique for overcoming this difficulty. From Figure B–2 observe that for small values of η the trial solutions for even poor estimates of $f''(0)$ behave approximately in the same manner as the correct solution. Thus, by initially using a small value for $\eta_{\max}$, e.g., 2, the required convergence derivatives for the Nachtsheim-Swigert iteration technique are approximately determined. This allows determination of a nearly correct value of $f''(0)$. Using this nearly correct value of $f''(0)$ and a larger value of $\eta_{\max}$, e.g., 5 or 6, a Nachtsheim-Swigert iteration scheme rapidly yields the value of $f''(0)$, satisfying the required outer boundary conditions. Nachtsheim and Swigert studied the effect of η_{max} on convergence. Their results for $\beta = 1$ are shown in Table B-1.

Along with our discussion of the effect of η_{max} on convergence to the required solution, it is appropriate to consider the effect of η_{max} on acceptable errors in satisfying the asymptotic boundary condition, i.e., δ_1 and δ_2. Accepting the sum of the squares of the deviations, δ_1 and δ_2, as a measure of the error in the solution, the error

$$E(\text{error}) = \delta_1^2 + \delta_2^2 = (1 - f_c')^2 + f_c''^2 \quad \text{at } \eta_{max}$$

is determined. Figure B-3 shows that, unlike the Newton-Raphson technique, the Nachtsheim-Swigert least-squares technique yields a unique solution for $f''(0)$.

One further comment about the Nachtsheim-Swigert iteration technique is of significance. The proposed technique of initially using a small value of η_{max} to obtain an approximately correct value for $f''(0)$ and then increasing η_{max} to obtain the final value of $f''(0)$ is *not* successful when searching for solutions with reverse flow. As illustrated by the example given in Table B-2, if an initially small value of η_{max}, e.g., 2, is used, convergence to the solution with reverse flow does not occur even if the correct value for the solution with reverse flow is taken as the initial estimate.

However, the reverse flow solutions are successfully obtained using the following technique. As an initial estimate for η_{max}, take the value of η_{max} used to obtain the separation solution $(f''(0) = 0)$ plus 2. For appropriate values of $\beta > \beta_0$ (the separation value), search for the negative values of $f''(0)$ which yield the reverse flow solutions (see Figure 3-3).

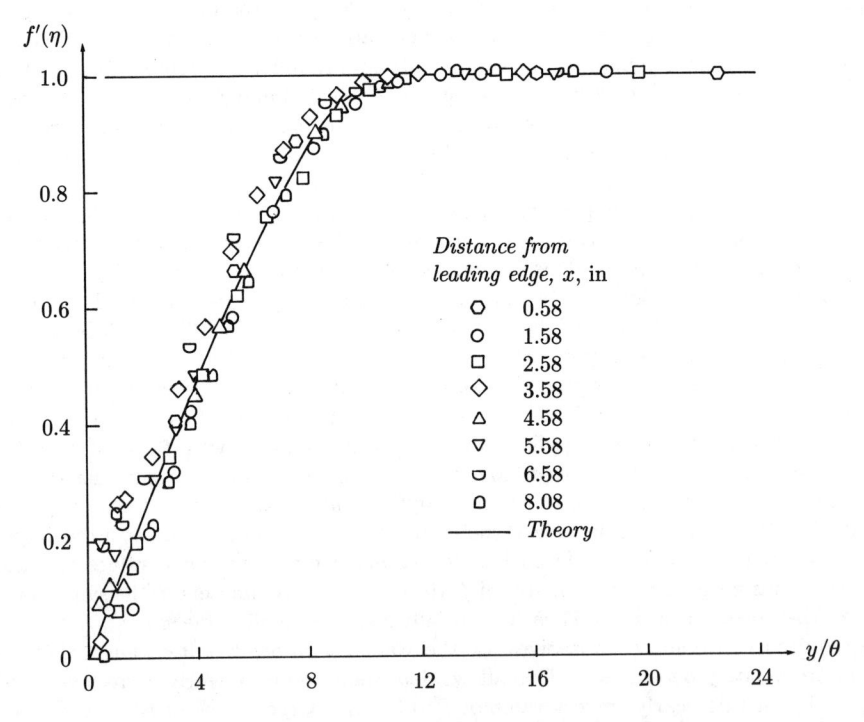

Figure B-2. Effect of $f''(0)$ on $f'(\eta)$ for $\beta = 1$.

Table B–1. Effect of η_{max} on convergence rate.

Initial guess $f''(0)$	$\eta_{max} = 2$ $f''(0)$ after two trials	Percent difference from needed value	$\eta_{max} = 5$ $f''(0)$ after two trials	Percent difference from needed value
0.25	1.2799034	3.83	0.26706180	−78.33
0.50	1.2449846	1.00	0.69205364	−43.85
0.75	1.2292604	−0.26	0.88486032	−28.21
1.00	1.2266764	−0.47	1.04137120	−15.51
1.25	1.2266282	−0.48	1.23258880	0.00
1.50	1.2266765	−0.47	1.24448130	0.96
1.75	1.2277231	−0.39	1.28907320	4.58
2.00	1.2312423	−0.10	1.35643890	10.04
2.25	1.2383960	0.47	1.88349300	52.80
2.50	1.2498889	1.40	2.49467760	102.39
2.75	1.2660459	2.71		
3.00	1.2869264	4.40		

Trial	Correction	η_{max}	$f''(0)$
1		2	1.0
2	1	2	1.2463981
3	2	5	1.2266764
4	3	5	1.2326729
5	4	5	1.2325878
	5		1.2325878

B–4 Nachtsheim-Swigert Iteration Scheme: Falkner-Skan Equation — Zero Shearing Stress Solutions, β_0

In determining the solutions for the Falkner-Skan equation it is necessary to determine the separation solution, i.e., the value of β for which $f''(0) = 0$ and for which the required asymptotic boundary conditions are satisfied. For this case we consider

$$f'(\eta_{max}) = f'(\beta) = 1 + \delta_1 \qquad (B-10)$$

and
$$f''(\eta_{max}) = f''(\beta) = \delta_2 \qquad (B-11)$$

where we assume that $f''(0) = 0$.

Again expanding in a first-order Taylor series and seeking a least-squares solution, we have

$$\Delta\beta = \frac{f'_\beta(1 - f'_c) + f''_\beta f''_c}{f'^2_\beta + f''^2_\beta} \qquad (B-12)$$

where
$$f'_\beta = \frac{\partial f'}{\partial \beta} \quad \text{and} \quad f''_\beta = \frac{\partial f''}{\partial \beta} \quad \text{at } \eta_{max}$$

306 Laminar Flow Analysis

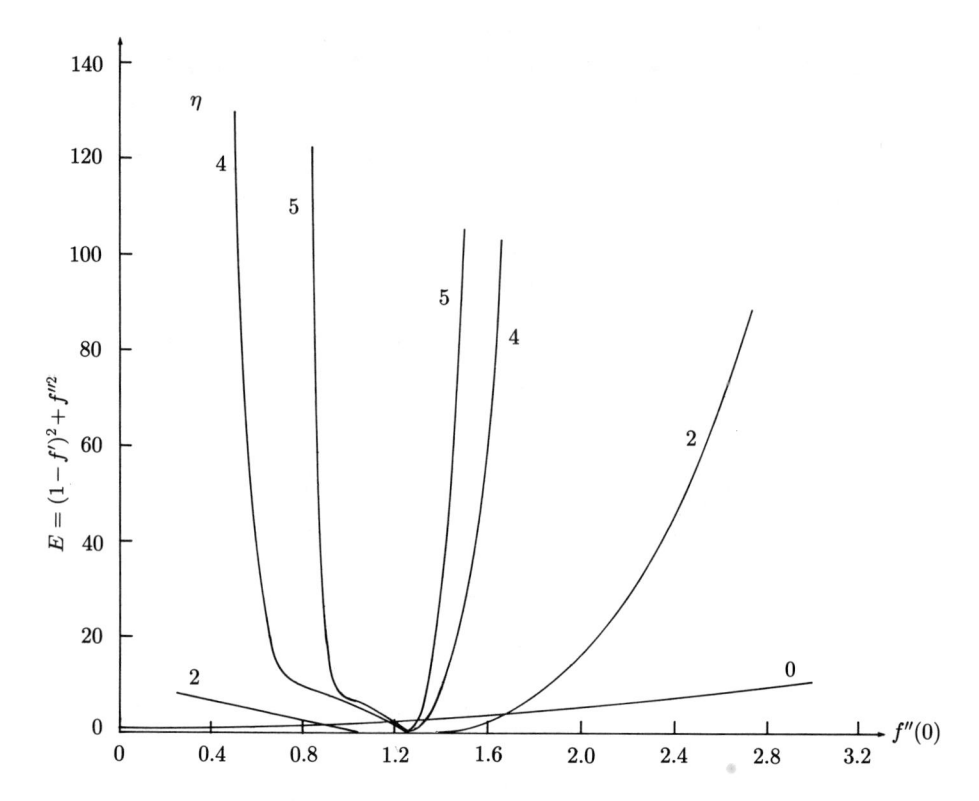

Figure B–3. Error as a function of $f''(0)$ for different values of $\eta_{\max}$, $\beta = 1$.

B–5 Nachtsheim-Swigert Iteration Scheme — Energy Equation, $\theta'(0)$

The boundary conditions for the incompressible energy equation (Eq. 4–78) are

$$\theta(0) = 1 \quad \text{and} \quad \theta(\eta \to \infty) \to 0 \quad (4-80a,b)$$

The asymptotic outer boundary conditions are functionally stated as

$$\theta(\eta_{\max}) = \theta(\theta'(0)) = \delta_1 \quad (B-13a)$$

$$\theta'(\eta_{\max}) = \theta(\theta'(0)) = \delta_2 \quad (B-13b)$$

Expanding in a first-order Taylor series and seeking a least-squares solution yields

$$\Delta\theta'(0) = \Delta y = \frac{\theta_y \theta_c - \theta'_y \theta'_c}{\theta_y^2 + \theta_y'^2} \quad \text{at } \eta_{\max} \quad (B-14)$$

where

$$\theta_y = \frac{\partial \theta(\eta_{\max})}{\partial \theta'(0)} \quad \text{and} \quad \theta'_y = \frac{\partial \theta'(\eta_{\max})}{\partial \theta'(0)}$$

If the energy equation is stated in its alternate form (see Eq. 4–109) with boundary conditions

$$\bar{\theta}(0) = 0 \quad \text{and} \quad \bar{\theta}(\eta \to \infty) \to 1 \quad (B-15a,b)$$

Table B–2. Convergence of the Falkner-Skan equation with unmodified Nachtsheim-Swigert iteration scheme.

$\beta = -0.1$

Trial	Correction	$\eta_{\max}$	$f''(0)$
1		2	−0.1405462
2	1	2	0.27164
3	2	2	0.492029
4	3	6	0.296224
5	4	6	0.319221
6	5	6	0.319272
7	6	6	0.319272
8	7	9	0.3192698

$\beta = -0.18$

Trial	Correction	$\eta_{\max}$	$f''(0)$
1		2	−0.09769212
2	1	2	0.404445
3	2	2	0.457885
4	3	6	0.170037
5	4	6	0.138257
6	5	6	0.129492
7	6	6	0.128708
8	7	6	0.128608
9	8	6	0.128677
10	9	6	0.128676
11	10	6	0.128676
12	11	9	0.128594
13	12	9	0.1286362

the required correction is

$$\Delta \bar{\theta}'(0) = \Delta y = \frac{\bar{\theta}_y(1 - \bar{\theta}_c) - \bar{\theta}'_y \bar{\theta}'_c}{\bar{\theta}_y^2 + \bar{\theta}_y'^2} \quad (B-16)$$

B–6 Nachtsheim-Swigert Iteration Scheme: Energy Equation — Adiabatic Wall Solution, θ_{aw}

When there is no heat transfer at the surface, the solution of the energy equation is the adiabatic wall solution. For the adiabatic wall solution, the appropriate boundary conditions for the energy equation (Eq. 4–109) are

$$\bar{\theta}'(0) = 0 \quad \bar{\theta}(\eta \to \infty) \to 1 \quad (B-17)$$

The asymptotic outer boundary conditions are then functionally expressed as

$$\bar{\theta}(\eta_{\max}) = \bar{\theta}(\bar{\theta}_{aw}) = 1 + \delta_1 \quad (B-18a)$$

$$\bar{\theta}'(\eta_{\max}) = \bar{\theta}'(\bar{\theta}_{aw}) = \delta_2 \qquad (B-18b)$$

The Nachtsheim-Swigert correction to $\bar{\theta}_{aw}$ is then

$$\Delta \bar{\theta}_{aw} = \frac{\bar{\theta}_a(1 - \bar{\theta}_c) - \bar{\theta}_a \bar{\theta}'_c}{\bar{\theta}_a^2 + \bar{\theta}_a'^2} \qquad (B-19)$$

where
$$\bar{\theta}_a = \frac{\partial \theta(\eta_{\max})}{\partial \theta_{aw}} \quad \text{and} \quad \bar{\theta}'_a = \frac{\partial \theta'(\eta_{\max})}{\partial \theta_{aw}}$$

B–7 Nachtsheim-Swigert Iteration Scheme: Systems of Equations — The Free Convection Boundary Layer, $f''(0)$, $\theta'(0)$

As shown by Nachtsheim and Swigert [Nach65], extension of the iteration scheme to systems of differential equations is straightforward. We illustrate this extension by considering the boundary value problem associated with free convection from a vertical flat plate (see Section 5–1), i.e.

$$f''' + (n+3)ff'' - 2(n+1)f'^2 + \theta = 0 \qquad (5-12)$$

$$\theta'' + \Pr[(n+3)f\theta' - 4nf'\theta] = 0 \qquad (5-13)$$

with boundary conditions

$$f(0) = f'(0) = 0 \qquad \theta(0) = 1 \qquad (5-14a,b,c)$$

and
$$f'(\eta \to \infty) \to 0 \qquad \theta(\eta \to \infty) \to 0 \qquad (5-14d,e)$$

Here we see that there are two asymptotic boundary conditions and hence two unknown surface conditions, $f''(0)$ and $\theta'(0)$. Within the context of the initial value method and the Nachtsheim-Swigert iteration technique, the outer boundary conditions (Eq. 5–14d, e) are functionally

$$f'(\eta_{\max}) = f'(f''(0), \theta'(0)) = \delta_1 \qquad (B-20)$$

$$\theta(\eta_{\max}) = \theta(f''(0), \theta'(0)) = \delta_2 \qquad (B-21)$$

with the asymptotic convergence criteria given by

$$f''(\eta_{\max}) = f''(f''(0), \theta'(0)) = \delta_3 \qquad (B-22)$$

$$\theta'(\eta_{\max}) = \theta'(f''(0), \theta'(0)) = \delta_4 \qquad (B-23)$$

Using Eqs. (B–20) to (B–23) and expanding in a first-order Taylor series yields

$$f'(\eta_{\max}) = f'_c(\eta_{\max}) + f'_x \Delta x + f'_y \Delta y = \delta_1 \qquad (B-24)$$

$$\theta(\eta_{\max}) = \theta_c(\eta_{\max}) + \theta_x \Delta x + \theta_y \Delta y = \delta_2 \qquad (B-25)$$

$$f''(\eta_{\max}) = f''_c(\eta_{\max}) + f''_x \Delta x + f''_y \Delta y = \delta_3 \qquad (B-26)$$

$$\theta'(\eta_{\max}) = \theta'_c(\eta_{\max}) + \theta'_x \Delta x + \theta'_y \Delta y = \delta_4 \qquad (B-27)$$

where we define $x = f''(0)$, $y = \theta'(0)$ and the x and y subscripts indicate partial differentiation, e.g., $f'_y = \partial f'(\eta_{\max})/\partial \theta'(0)$. The c subscript indicates the value of the function at $\eta_{\max}$ determined from the trial integration.

Solution of these equations in a least-squares sense requires determining the minimum value of $E = \delta_1^2 + \delta_2^2 + \delta_3^2 + \delta_4^2$ with respect to x and y. Differentiating E with respect to x yields

$$(f'^{2}_x + \theta'^{2}_x + f''^{2}_x + \theta'^{\prime 2}_x)\Delta x + (f'_x f'_y + \theta_x \theta_y + f''_x f''_y + \theta'_x \theta'_y)\Delta y$$
$$= -(f'_c f'_x + \theta_c \theta_x + f''_c f''_x + \theta'_c \theta'_x)$$

or
$$Q_{xx}\Delta x + Q_{xy}\Delta y = Q_{cx}$$

Similarly, differentiating E with respect to y yields

$$(f'_x f'_y + \theta_x \theta_y + f''_x f''_y + \theta'_x \theta'_y)\Delta x + (f'^{2}_y + \theta^2_y + f''^{2}_y + \theta'^{2}_y)\Delta y$$
$$= -(f'_c f'_y + \theta_c \theta_y + f''_c f''_y + \theta'_c \theta'_y)$$

or
$$Q_{xy}\Delta x + Q_{yy}\Delta y = Q_{cy}$$

The solutions for Δx and Δy, the corrections to $f''(0)$ and $\theta'(0)$ which yield asymptotic convergence to the required outer boundary conditions, are given by

$$\Delta x = \frac{Q_{cx}Q_{yy} - Q_{cy}Q_{xy}}{Q_{xx}Q_{yy} - Q_{xy}^2} \qquad (B-28)$$

and
$$\Delta y = \frac{Q_{xx}Q_{cy} - Q_{xy}Q_{cx}}{Q_{xx}Q_{yy} - Q_{xy}^2} \qquad (B-29)$$

The partial derivatives required above are determined from the appropriate perturbation equations (see Nachtsheim and Swigert [Nach65]), or by using the finite difference approximation previously discussed in Section B-2. The **freeb** program (see Section 5-3) uses the finite difference approximation. In order to implement the finite difference approximation for a two-parameter iteration it is necessary to perform three trial integrations with appropriate values of $f''(0)$ and $\theta'(0)$. These are

$$f''(0) = k_1 \qquad \theta'(0) = \ell_1$$
$$f''(0) = k_1 + \epsilon \qquad \theta'(0) = \ell_2$$
$$f''(0) = k_1 \qquad \theta'(0) = \ell_1 + \epsilon$$

where k_1 and ℓ_1 are some assumed constant value and ϵ is a small number, e.g., 1×10^{-5} to 1×10^{-2}. The first two integrations yield the change in f', f'', θ, θ' at $\eta_{\max}$, with a change in $f''(0)$ and hence the convergence derivatives f'_x, f''_x, θ_x, θ'_x. The first and third integrations yield the change in f', f'', θ, θ' at $\eta_{\max}$, with a change in $\theta'(0)$ and hence the convergence derivatives f'_y, f''_y, θ_y, θ'_y. Note that for a two-parameter iteration as described above continuous upgrading of the convergence derivatives is not possible when a finite difference approximation is used to estimate the derivatives. Since the arguments used in Section B-3 also apply in this case, this is not as important a restriction as it appears.

310   Laminar Flow Analysis

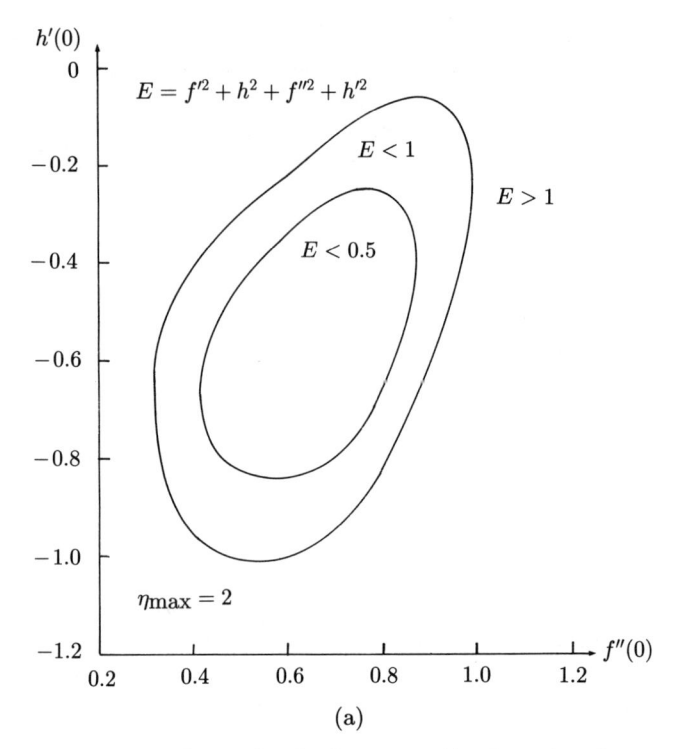

(a)

Figure B–4. Level curves of error for the free convection boundary layer equations. (a) $\eta_{max} = 2$; (b) $\eta_{max} = 5$.

The error in satisfying the asymptotic boundary conditions is

$$E = \delta_1^2 + \delta_2^2 + \delta_3^2 + \delta_4^2 = f_c'^2 + f_c''^2 + \theta_c^2 + \theta_c'^2 \quad \text{at } \eta_{max}$$

Figure B–4 shows the resulting level curves of error for the free convection problem (see Nachtsheim and Swigert [Nach65]). Here note that increasing η_{max} decreases the possible range of initial estimates of $f''(0)$ and $\theta'(0)$ which eventually converge within the required accuracy. In fact, for very large values of η_{max} or very small values of E the level curve of error reduces to a single point, i.e., one must guess the exact solution.

B–8 Nachtsheim-Swigert Iteration Scheme — Adiabatic Wall Solution, $f''(0)$, θ_{aw}

To determine the adiabatic wall solutions for the free convection boundary layer, i.e., the solution to Eqs. (5–12) and (5–13) subject to the boundary conditions

$$f(0) = f'(0) = 0 \quad \theta'(0) = 0 \qquad (B-30a,b,c)$$

and

$$f'(\eta \to \infty) \to 0 \quad \theta(\eta \to \infty) \to 0 \qquad (5-14d,e)$$

the Nachtsheim-Swigert iteration scheme is given by Eqs. (B–28) and (B–29) by considering y as $\theta(0)$ instead of $\theta'(0)$.

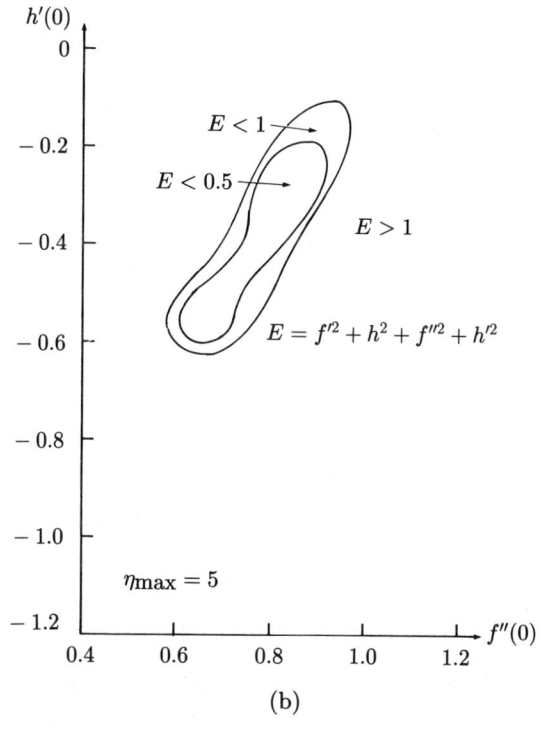

Figure B-4. (cont.)

By using the basic two-parameter Nachtsheim-Swigert iteration scheme given in Eqs. (B-28) and (B-29), various other solutions can be determined. For example, the value of n which, with surface heat transfer, yields zero shear at the surface, i.e., $f''(0) = 0$, and conceivably the nondimensional adiabatic wall temperature and the value of n which yields both zero heat transfer and zero shear at the surface, i.e., $f''(0) = \theta'(0) = 0$, can be determined.

B–9 Nachtsheim-Swigert Iteration Scheme — Compressible Boundary Layer, $f''(0)$, $g'(0)$

As a further illustration of the Nachtsheim-Swigert iteration technique, consider the boundary value problem associated with the hypersonic compressible boundary layer (see Section 6–9), i.e.

$$f''' + ff'' + \hat{\beta}(g - f'^2) = 0 \qquad (6-74)$$

$$g'' + \Pr fg' + 2(\Pr - 1)(ff'')' = 0 \qquad (6-102)$$

with boundary conditions

$$f(0) = f'(0) = 0 \qquad g(0) = g_w \qquad (6-79a,b,c)$$

and

$$f'(\eta \to \infty) \to 1 \qquad g(\eta \to \infty) \to 1 \qquad (6-79e,f)$$

312 Laminar Flow Analysis

Again we see that the two asymptotic boundary conditions yield two unknown surface conditions, $f''(0)$ and $g'(0)$. Within the context of the initial value method and the Nachtsheim-Swigert iteration scheme, the outer boundary conditions are functionally represented by

$$f'(\eta_{\max}) = f'(f''(0), g'(0)) = 1 + \delta_1 \qquad (B-31)$$

$$g(\eta_{\max}) = g(f''(0), g'(0)) = 1 + \delta_2 \qquad (B-32)$$

with the asymptotic convergence criteria given by

$$f''(\eta_{\max}) = f''(f''(0), g'(0)) = \delta_3 \qquad (B-33)$$

$$g'(\eta_{\max}) = g'(f''(0), g'(0)) = \delta_4 \qquad (B-34)$$

Using Eqs. (B–31) to (B–34) and expanding in a first-order Taylor series yields

$$f'(\eta_{\max}) = f'_c(\eta_{\max}) + f'_x \Delta x + f'_y \Delta y = 1 + \delta_1 \qquad (B-35)$$

$$g(\eta_{\max}) = g_c(\eta_{\max}) + g_x \Delta x + g_y \Delta y = 1 + \delta_2 \qquad (B-36)$$

$$f''(\eta_{\max}) = f''_c(\eta_{\max}) + f''_x \Delta x + f''_y \Delta y = \delta_3 \qquad (B-37)$$

$$g'(\eta_{\max}) = g'_c(\eta_{\max}) + g'_x \Delta x + g'_y \Delta y = \delta_4 \qquad (B-38)$$

where we define $x = f''(0)$, $y = g'(0)$ and the x and y subscripts indicate partial differentiation, e.g.

$$g'_y = \frac{\partial g'(\eta_{\max})}{\partial g'(0)}$$

The c subscript indicates the value of the function at $\eta_{\max}$ determined from a trial integration.

Solution of these equations in a least-squares sense requires determination of the minimum of $E = \delta_1^2 + \delta_2^2 + \delta_3^2 + \delta_4^2$ with respect to x and y. Differentiating E with respect to x and setting the result equal to zero yields

$$(f'^2_x + g^2_x + f''^2_x + g'^2_x)\Delta x + (f'_x f'_y + g_x g_y + f''_x f''_y + g'_x g'_y)\Delta y$$
$$= (1 - f'_c)f'_x + (1 - g_c)g_x - f''_c f''_x - g'_c g'_x \qquad (B-39a)$$

or $\qquad Q_{xx}\Delta x + Q_{xy}\Delta y = Q_{cx} \qquad (B-39b)$

Differentiating E with respect to y and setting the result equal to zero yields

$$(f'_x f'_y + g_x g_y + f''_x f''_y + g'_x g'_y)\Delta x + (f'^2_y + g^2_y + f''^2_y + g'^2_y)\Delta y$$
$$= (1 - f'_c)f'_y + (1 - g_c)g_y - f''_c f''_y - g'_c g'_y \qquad (B-40a)$$

or $\qquad Q_{xy}\Delta x + Q_{yy}\Delta y = Q_{cy} \qquad (B-40b)$

The solutions for Δx and Δy, the corrections for $f''(0)$ and $g'(0)$, are given by Eqs. (B–28) and (B–29) using these definitions of Q_{xx}, Q_{yy}, Q_{xy}, Q_{cx} and Q_{cy}.

B–10 Nachtsheim-Swigert Iteration Scheme — Compressible Boundary Layer, Zero Shearing Stress and Zero Heat Transfer Solutions

In determining the solution curves for the compressible boundary layer it is necessary to find the separation solution (see Figure 6–22). For the compressible boundary layer, the separation solution is defined as the solution for zero shearing stress at the surface, i.e., $f''(0) = 0$. Here we functionally write the asymptotic boundary conditions as

$$f'(\eta_{\max}) = f'(\hat{\beta}, g'(0)) = 1 + \delta_1$$

$$g(\eta_{\max}) = g(\hat{\beta}, g'(0)) = 1 + \delta_2$$

$$f''(\eta_{\max}) = f''(\hat{\beta}, g'(0)) = \delta_3$$

$$g'(\eta_{\max}) = g'(\hat{\beta}, g'(0)) = \delta_4$$

The Nachtsheim-Swigert corrections are given by Eqs. (B–28) and (B–29), using the definitions of Q_{xx}, Q_{yy}, Q_{xy}, Q_{cx} and Q_{cy} given in Eqs. (B–39) and (B–40), with $x = \hat{\beta}$. For example, $f'_x = \partial f'(\eta_{\max})/\partial \hat{\beta}$. Using a finite difference approximation for the convergence derivatives requires the following trial integration sequence to determine the convergence derivatives

$$\hat{\beta} = k_1 \qquad g'(0) = \ell_1$$

$$\hat{\beta} = k_1 + \epsilon \qquad g'(0) = \ell_1$$

$$\hat{\beta} = k_1 \qquad g'(0) = \ell_1 + \epsilon$$

with $f''(0) = 0$ and g_w a known constant value.

The iteration scheme for the zero heat transfer, i.e., the adiabatic wall, $g'(0) = 0$, solutions, is of the same form as that for zero shearing stress solutions. This is easily seen if the functional relationships for the asymptotic boundary conditions are written as

$$f'(\eta_{\max}) = f'(f''(0), g_{aw}) = 1 + \delta_1$$

$$g(\eta_{\max}) = g(f''(0), g_{aw}) = 1 + \delta_2$$

$$f''(\eta_{\max}) = f''(f''(0), g_{aw}) = \delta_3$$

$$g'(\eta_{\max}) = g'(f''(0), g_{aw}) = \delta_4$$

The Nachtsheim-Swigert corrections are then given by Eqs. (B–28) and (B–29), using the definitions of Q_{xx}, Q_{yy}, Q_{xy}, Q_{cx} and Q_{cy} given in Eqs. (B–39) and (B–40), with $y = g_{aw}$. The trial integration sequence necessary to determine the convergence derivatives is

$$f''(0) = k_1 \qquad g_{aw} = \ell_1$$

$$f''(0) = k_1 + \epsilon \qquad g_{aw} = \ell_1$$

$$f''(0) = k_1 \qquad g_{aw} = \ell_1 + \epsilon$$

with $g'(0) = 0$ and $\hat{\beta}$ a known constant value.

Further solutions of interest are those with zero heat transfer and zero shear at the surface, i.e., $f''(0) = 0$ and $g'(0) = 0$ (see Rogers [Roge70]). For this case the asymptotic boundary conditions are functionally represented by

$$f'(\eta_{\max}) = f'(\hat{\beta}, g_{aw}) = 1 + \delta_1$$

$$g(\eta_{\max}) = g(\hat{\beta}, g_{aw}) = 1 + \delta_2$$

$$f''(\eta_{\max}) = f''(\hat{\beta}, g_{aw}) = \delta_3$$

$$g'(\eta_{\max}) = g'(\hat{\beta}, g_{aw}) = \delta_4$$

The Nachtsheim Swigert corrections are then given by Eqs. (B–28) and (B–29), using the definitions of Q_{xx}, Q_{yy}, Q_{xy}, Q_{cx} and Q_{cy} given in Eqs. (B–39) and (B–40), with $x = \hat{\beta}$ and $y = g_{aw}$. The trial integration sequence necessary to determine the convergence derivatives is

$$\hat{\beta} = k_1 \qquad g_{aw} = \ell_1$$

$$\hat{\beta} = k_1 + \epsilon \qquad g_{aw} = \ell_1$$

$$\hat{\beta} = k_1 \qquad g_{aw} = \ell_1 + \epsilon$$

with $g'(0) = f''(0) = 0$.

The Nachtsheim-Swigert iteration scheme can be developed for three or more parameters using the methods described above. One particular three-parameter iteration scheme of some interest is described below.

B–11 Nachtsheim-Swigert Iteration Scheme for a Solution-Determined Parameter

Extension of the Nachtsheim-Swigert iteration scheme to a three-parameter iteration for the particular case of one of the parameters appearing in the governing differential equation being dependent upon the solution of the differential equation is given here. Again we assume that satisfaction of asymptotic boundary conditions is required. Here we functionally represent the asymptotic boundary conditions by

$$f'(\eta_{\max}) = f'(f''(0), \bar{\beta}) = 1 + \delta_1$$

$$g(\eta_{\max}) = g(f''(0), g'(0), \bar{\beta}) = 1 + \delta_2$$

where $\bar{\beta}$ is taken as the solution-dependent parameter. Since a value of $\bar{\beta}$ is not known at $\eta_{\max}$, and since the solution-determined value for $\bar{\beta}$ in general depends upon integrals of the governing equations, a solution subject to the appropriate boundary conditions is obtained by assuming a value of $\bar{\beta}$ (e.g., $\bar{\beta}_a$) performing the trial integrations with estimated values of $f''(0)$, $g'(0)$, iterating until the asymptotic boundary conditions are satisfied and determining the value of $\bar{\beta}$ (e.g., $\bar{\beta}_c$) from the 'solution'. If $\bar{\beta}_c = \bar{\beta}_a$, then a final solution results. This process is simplified by observing that in order to satisfy the solution conditions the difference between the assumed and calculated values of $\bar{\beta}$ at $\eta_{\max}$ must be zero. Thus, we consider $\Delta\bar{\beta} = \bar{\beta}_a - \bar{\beta}_c$ as the iteration parameter. Hence the additional functional requirement is written as

$$\Delta\bar{\beta}(\eta_{\max}) = \bar{\beta}(f''(0), g'(0), \bar{\beta}) = \delta_3$$

Satisfaction of Asymptotic Boundary Conditions

Finally, we write the functional asymptotic convergence conditions as

$$f''(\eta_{\max}) = f''(f''(0), g'(0), \bar{\beta}) = \delta_4$$

$$g'(\eta_{\max}) = g'(f''(0), g'(0), \bar{\beta}) = \delta_5$$

$$\Delta\bar{\beta}(\eta_{\max}) = \bar{\beta}(f''(0), g'(0), \bar{\beta}) = \delta_6$$

Expanding in a first-order Taylor series yields

$$f'(\eta_{\max}) = f'_c + f'_x \Delta x + f'_y \Delta y + f'_z \Delta z = 1 + \delta_1$$

$$g(\eta_{\max}) = g_c + g_x \Delta x + g_y \Delta y + g_z \Delta z = 1 + \delta_2$$

$$\Delta\bar{\beta}(\eta_{\max}) = \Delta\bar{\beta}_c + \Delta\bar{\beta}_x \Delta x + \Delta\bar{\beta}_y \Delta y + \Delta\bar{\beta}_z \Delta z = \delta_3$$

$$f''(\eta_{\max}) = f''_c + f''_x \Delta x + f''_y \Delta y + f''_z \Delta z = \delta_4$$

$$g'(\eta_{\max}) = g'_c + g'_x \Delta x + g'_y \Delta y + g'_z \Delta z = \delta_5$$

$$\Delta\bar{\beta}'(\eta_{\max}) = \Delta\bar{\beta}'_c + \Delta\bar{\beta}'_x \Delta x + \Delta\bar{\beta}'_y \Delta y + \Delta\bar{\beta}'_z \Delta z = \delta_6$$

where $x = f''(0)$, $y = g'(0)$ and $z = \bar{\beta}_a$. When used as subscripts, x, y and z indicate partial differentiation. The subscript c indicates the value of the function calculated at $\eta_{\max}$.

A least-squares solution of these six equations requires determining the minimum of $E = \delta_1^2 + \delta_2^2 + \delta_3^2 + \delta_4^2 + \delta_5^2 + \delta_6^2$ with respect to x, y and z, respectively. Differentiating with respect to x yields

$$(f'^2_x + g^2_x + \Delta\bar{\beta}^2_x + f''^2_x + g'^2_x + \Delta\bar{\beta}'^2_x)\Delta x$$

$$+ (f'_y f'_x + g_y g_x + \Delta\bar{\beta}_y \Delta\bar{\beta}_x + f''_y f''_x + g'_y g'_x + \Delta\bar{\beta}'_y \Delta\bar{\beta}'_x)\Delta y$$

$$+ (f'_z f'_x + g_z g_x + \Delta\bar{\beta}_z \Delta\bar{\beta}_x + f''_z f''_x + g'_z g'_x + \Delta\bar{\beta}'_z \Delta\bar{\beta}'_x)\Delta z$$

$$= (1 - f'_c)f'_x + (1 - g_c)g_x - \Delta\bar{\beta}_c \Delta\bar{\beta}_x - f''_c f''_x - g'_c g'_x - \Delta\bar{\beta}'_c \Delta\bar{\beta}'_x$$

or

$$Q_{xx}\Delta x + Q_{xy}\Delta y + Q_{xz}\Delta z = Q_{cx}$$

Differentiating with respect to y yields

$$(f'_x f'_y + g_x g_y + \Delta\bar{\beta}_x \Delta\bar{\beta}_y + f''_x f''_y + g'_x g'_y + \Delta\bar{\beta}'_x \Delta\bar{\beta}'_y)\Delta x$$

$$+ (f'^2_y + g^2_y + \Delta\bar{\beta}^2_y + f''^2_y + g'^2_y + \Delta\bar{\beta}'^2_y)\Delta y$$

$$+ (f'_z f'_y + g_z g_y + \Delta\bar{\beta}_z \Delta\bar{\beta}_y + f''_z f''_y + g'_z g'_y + \Delta\bar{\beta}'_z \Delta\bar{\beta}'_y)\Delta z$$

$$= (1 - f'_c)f'_y + (1 - g_c)g_y - \Delta\bar{\beta}_c \Delta\bar{\beta}_y - f''_c f''_y - g'_c g'_y - \Delta\bar{\beta}'_c \Delta\bar{\beta}'_y$$

or

$$Q_{xy}\Delta x + Q_{yy}\Delta y + Q_{yz}\Delta z = Q_{cy}$$

Finally, differentiating with respect to z yields

$$(f'_x f'_z + g_x g_z + \Delta\bar{\beta}_x \Delta\bar{\beta}_z + f''_x f''_z + g'_x g'_z + \Delta\bar{\beta}'_x \Delta\bar{\beta}'_z)\Delta x$$

316   Laminar Flow Analysis

$$+ (f_y'' f_z' + g_y g_z + \Delta\bar{\beta}_y \Delta\bar{\beta}_z + f_y'' f_z'' + g_y' g_z' + \Delta\bar{\beta}_y' \Delta\bar{\beta}_z')\Delta y$$

$$+ (f_z'^2 + g_z^2 + \Delta\bar{\beta}_z^2 + f_z''^2 + g_z'^2 + \Delta\bar{\beta}_z'^2)\Delta z$$

$$= (1 - f_c')f_z' + (1 - g_c)g_z - \Delta\bar{\beta}_c \Delta\bar{\beta}_z - f_c'' f_z'' - g_c' g_z' - \Delta\bar{\beta}_c' \Delta\bar{\beta}_z'$$

or
$$Q_{xz}\Delta x + Q_{yz}\Delta y + Q_{zz}\Delta z = Q_{cz}$$

Solving for Δx, Δy and Δz yields

$$\Delta x = \frac{Q_{cx}(Q_{yy}Q_{zz} - Q_{yz}^2) + Q_{cy}(Q_{xz}Q_{yz} - Q_{xy}Q_{zz}) + Q_{cz}(Q_{xy}Q_{yz} - Q_{yy}Q_{xz})}{Q_{xx}Q_{yy}Q_{zz} + 2Q_{xy}Q_{yz}Q_{xz} - Q_{yy}Q_{xz}^2 - Q_{xx}Q_{yz}^2 - Q_{zz}Q_{xy}^2}$$
(B – 41)

$$\Delta y = \frac{Q_{cx}(Q_{yz}Q_{xz} - Q_{xy}Q_{zz}) + Q_{cy}(Q_{xx}Q_{zz} - Q_{xz}^2) + Q_{cz}(Q_{xy}Q_{xz} - Q_{yz}Q_{xx})}{Q_{xx}Q_{yy}Q_{zz} + 2Q_{xy}Q_{yz}Q_{xz} - Q_{yy}Q_{xz}^2 - Q_{xx}Q_{yz}^2 - Q_{zz}Q_{xy}^2}$$
(B – 42)

$$\Delta z = \frac{Q_{cx}(Q_{yz}Q_{xy} - Q_{xz}Q_{yy}) + Q_{cy}(Q_{xy}Q_{xz} - Q_{yz}Q_{xx}) + Q_{cz}(Q_{yy}Q_{zz} - Q_{xy}^2)}{Q_{xx}Q_{yy}Q_{zz} + 2Q_{xy}Q_{yz}Q_{xz} - Q_{yy}Q_{xz}^2 - Q_{xx}Q_{yz}^2 - Q_{zz}Q_{xy}^2}$$
(B – 43)

Again, the required convergence derivatives are obtained using a finite difference approximation by performing the following four trial integrations

$$f''(0) = k_1 \qquad g'(0) = \ell_1 \qquad \bar{\beta}_a = m_1$$
$$f''(0) = k_1 + \epsilon \qquad g'(0) = \ell_1 \qquad \bar{\beta}_a = m_1$$
$$f''(0) = k_1 \qquad g'(0) = \ell_1 + \epsilon \qquad \bar{\beta}_a = m_1$$
$$f''(0) = k_1 \qquad g'(0) = \ell_1 \qquad \bar{\beta}_a = m_1 + \epsilon$$

where k_1, ℓ_1, m_1 are constants and ϵ is some small value, e.g., 0.0001–0.01.

B–12 Nachtsheim-Swigert Iteration Scheme for Systems of Equations — Satisfaction of Four Asymptotic Conditions

Extension of the Nachtsheim-Swigert iteration scheme to systems of differential equations requiring satisfaction of four asymptotic boundary conditions is straightforward. Satisfaction of four asymptotic boundary equations is required for the two-equation model for the locally nonsimilar method for the free convection and compressible boundary layers (see Sections 5–7 and 6–17). Here the equations are developed for the nonsimilar compressible boundary layer. To use the results for the free convection boundary layer the terms $(1-f_c')$ and $(1-g_c)$ are replaced by $-f_c'$ and $-g_c$, respectively. Within the context of the initial value method and the Nachtsheim-Swigert iteration scheme, the outer boundary conditions are functionally represented by

$$f'(\eta_{\max}) = f'(x, y, u, v) = 1 + \delta_1 \qquad (B – 44)$$

$$g(\eta_{\max}) = g(x, y, u, v) = 1 + \delta_2 \qquad (B – 45)$$

Satisfaction of Asymptotic Boundary Conditions 317

$$f''(\eta_{\max}) = f''(x, y, u, v) = \delta_3 \qquad (B-46)$$

$$g'(\eta_{\max}) = g'(x, y, u, v) = \delta_4 \qquad (B-47)$$

$$S'(\eta_{\max}) = S'(x, y, u, v) = \delta_5 \qquad (B-48)$$

$$t(\eta_{\max}) = t(x, y, u, v) = \delta_6 \qquad (B-49)$$

$$S''(\eta_{\max}) = S''(x, y, u, v) = \delta_7 \qquad (B-50)$$

$$t'(\eta_{\max}) = t'(x, y, u, v) = \delta_8 \qquad (B-51)$$

where $x = f''(0)$, $y = g'(0)$, $u = S''(0)$ and $v = t'(0)$, and Eqs. (B–44) to (B–47) arise from the asymptotic boundary conditions for the momentum and energy equations, and Eqs. (B–48) to (B–51) arise from the asymptotic boundary conditions for the first auxiliary momentum and energy equations for the two-equation locally nonsimilar model. Expanding in a first-order Taylor series yields

$$f'(\eta_{\max}) = f'_c + f'_x \Delta x + f'_y \Delta y + f'_u \Delta u + f'_v \Delta v = 1 + \delta_1 \qquad (B-52)$$

$$g(\eta_{\max}) = g_c + g_x \Delta x + g_y \Delta y + g_u \Delta u + g_v \Delta v = 1 + \delta_2 \qquad (B-53)$$

$$f''(\eta_{\max}) = f''_c + f''_x \Delta x + f''_y \Delta y + f''_u \Delta u + f''_v \Delta v = \delta_3 \qquad (B-54)$$

$$g'(\eta_{\max}) = g'_c + g'_x \Delta x + g'_y \Delta y + g'_u \Delta u + g'_v \Delta v = \delta_4 \qquad (B-55)$$

$$S'(\eta_{\max}) = S'_c + S'_x \Delta x + S'_y \Delta y + S'_u \Delta u + S'_v \Delta v = \delta_5 \qquad (B-56)$$

$$t(\eta_{\max}) = t_c + t_x \Delta x + t_y \Delta y + t_u \Delta u + t_v \Delta v = \delta_6 \qquad (B-57)$$

$$S''(\eta_{\max}) = S''_c + S''_x \Delta x + S''_y \Delta y + S''_u \Delta u + S''_v \Delta v = \delta_7 \qquad (B-58)$$

$$t'(\eta_{\max}) = t'_c + t'_x \Delta x + t'_y \Delta y + t'_u \Delta u + t'_v \Delta v = \delta_8 \qquad (B-59)$$

where the x, y, u, v subscripts indicate differentiation, e.g.,

$$f'_x = \frac{\partial f'(\eta_{\max})}{\partial f''(0)}$$

The c subscript indicates the value of the function at $\eta_{\max}$ determined from the trial integration.

Solution of the equations in a least-squares sense requires determination of the minimum of the error

$$E = \delta_1^2 + \delta_2^2 + \delta_3^2 + \delta_4^2 + \delta_5^2 + \delta_6^2 + \delta_7^2 + \delta_8^2$$

with respect to x, y, u and v. Differentiating E with respect to x and setting the result equal to zero yields

$$(f'^2_x + g^2_x + f''^2_x + g'^2_x + S'^2_x + t^2_x + S''^2_x + t'^2_x)\Delta x$$

$$+ (f'_x f'_y + g_x g_y + f''_x f''_y + g'_x g'_y + S'_x S'_y + t_x t_y + S''_x S''_y + t'_x t'_y)\Delta y$$

$$+ (f'_x f'_u + g_x g_u + f''_x f''_u + g'_x g'_u + S'_x S'_u + t_x t_u + S''_x S''_u + t'_x t'_u)\Delta u$$

$$+ (f'_x f'_v + g_x g_v + f''_x f''_v + g'_x g'_v + S'_x S'_v + t_x t_v + S''_x S''_v + t'_x t'_v)\Delta v$$

318   *Laminar Flow Analysis*

$$= (1-f'_c)f'_x + (1-g_c)g_x - f''_c f''_x - g'_c g'_x - S'_c S'_x - t_c t_x - S''_c S''_x - t'_c t'_x$$
(B – 60a)

or $\qquad Q_{xx}\Delta x + Q_{xy}\Delta y + Q_{xu}\Delta u + Q_{xv}\Delta v = Q_{cx}$ (B – 60b)

Differentiating E with respect to y and setting the result equal to zero yields

$$(f'_x f'_y + g_x g_y + f''_x f''_y + g'_x g'_y + S'_x S'_y + t_x t_y + S''_x S''_y + t'_x t'_y)\Delta x$$
$$+ (f'^2_y + g^2_y + f''^2_y + g'^2_y + S'^2_y + t^2_y + S''^2_y + t'^2_y)\Delta y$$
$$+ (f'_u f'_y + g_u g_y + f''_u f''_y + g'_u g'_y + S'_u S'_y + t_u t_y + S''_u S''_y + t'_u t'_y)\Delta u$$
$$+ (f'_v f'_y + g_v g_y + f''_v f''_y + g'_v g'_y + S'_v S'_y + t_v t_y + S''_v S''_y + t'_v t'_y)\Delta v$$
$$= (1-f'_c)f'_y + (1-g_c)g_y - f''_c f''_y - g'_c g'_y - S'_c S'_y - t_c t_y - S''_c S''_y - t'_c t'_y$$
(B – 61a)

or $\qquad Q_{xy}\Delta x + Q_{yy}\Delta y + Q_{yu}\Delta u + Q_{yv}\Delta v = Q_{cy}$ (B – 61b)

Differentiating E with respect to u and setting the result equal to zero yields

$$(f'_x f'_u + g_x g_u + f''_x f''_u + g'_x g'_u + S'_x S'_u + t_x t_u + S''_x S''_u + t'_x t'_u)\Delta x$$
$$+ (f'_y f'_u + g_y g_u + f''_y f''_u + g'_y g'_u + S'_y S'_u + t_y t_u + S''_y S''_u + t'_y t'_u)\Delta y$$
$$+ (f'^2_u + g^2_u + f''^2_u + g'^2_u + S'^2_u + t^2_u + S''^2_u + t'^2_u)\Delta u$$
$$+ (f'_v f'_u + g_v g_u + f''_v f''_u + g'_v g'_u + S'_v S'_u + t_v t_u + S''_v S''_u + t'_v t'_u)\Delta v$$
$$= (1-f'_c)f'_u + (1-g_c)g_u - f''_c f''_u - g'_c g'_u - S'_c S'_u - t_c t_u - S''_c S''_u - t'_c t'_u$$
(B – 62a)

or $\qquad Q_{xu}\Delta x + Q_{yu}\Delta y + Q_{uu}\Delta u + Q_{vu}\Delta v = Q_{cu}$ (B – 62b)

Differentiating E with respect to v and setting the result equal to zero yields

$$(f'_x f'_v + g_x g_v + f''_x f''_v + g'_x g'_v + S'_x S'_v + t_x t_v + S''_x S''_v + t'_x t'_v)\Delta x$$
$$+ (f'_y f'_v + g_y g_v + f''_y f''_v + g'_y g'_v + S'_y S'_v + t_y t_v + S''_y S''_v + t'_y t'_v)\Delta y$$
$$+ (f'_u f'_v + g_u g_v + f''_u f''_v + g'_u g'_v + S'_u S'_v + t_u t_v + S''_u S''_v + t'_u t'_v)\Delta u$$
$$+ (f'^2_v + g^2_v + f''^2_v + g'^2_v + S'^2_v + t^2_v + S''^2_v + t'^2_v)\Delta v$$
$$= (1-f'_c)f'_v + (1-g_c)g_v - f''_c f''_v - g'_c g'_v - S'_c S'_v - t_c t_v - S''_c S''_v - t'_c t'_v$$
(B – 63a)

or $\qquad Q_{xv}\Delta x + Q_{yv}\Delta y + Q_{uv}\Delta u + Q_{vv}\Delta v = Q_{cv}$ (B – 63b)

Gathering these results together we have

$$Q_{xx}\Delta x + Q_{xy}\Delta y + Q_{xu}\Delta u + Q_{xv}\Delta v = Q_{cx} \qquad \text{(B – 60b)}$$
$$Q_{xy}\Delta x + Q_{yy}\Delta y + Q_{yu}\Delta u + Q_{yv}\Delta v = Q_{cy} \qquad \text{(B – 61b)}$$

$$Q_{xu}\Delta x + Q_{yu}\Delta y + Q_{uu}\Delta u + Q_{vu}\Delta v = Q_{cu} \qquad (B-62b)$$
$$Q_{xv}\Delta x + Q_{yv}\Delta y + Q_{uv}\Delta u + Q_{vv}\Delta v = Q_{cv} \qquad (B-63b)$$

Defining

$$A = \begin{vmatrix} Q_{cx} & Q_{yx} & Q_{ux} & Q_{vx} \\ Q_{cy} & Q_{yy} & Q_{uy} & Q_{vy} \\ Q_{cu} & Q_{yu} & Q_{uu} & Q_{vu} \\ Q_{cv} & Q_{yv} & Q_{uv} & Q_{vv} \end{vmatrix} \qquad B = \begin{vmatrix} Q_{xx} & Q_{cx} & Q_{ux} & Q_{vx} \\ Q_{xy} & Q_{cy} & Q_{uy} & Q_{vy} \\ Q_{xu} & Q_{cu} & Q_{uu} & Q_{vu} \\ Q_{xv} & Q_{cv} & Q_{uv} & Q_{vv} \end{vmatrix}$$

$$C = \begin{vmatrix} Q_{xx} & Q_{yx} & Q_{cx} & Q_{vx} \\ Q_{xy} & Q_{yy} & Q_{cy} & Q_{vy} \\ Q_{xu} & Q_{yu} & Q_{cu} & Q_{vu} \\ Q_{xv} & Q_{yv} & Q_{cv} & Q_{vv} \end{vmatrix} \qquad D = \begin{vmatrix} Q_{xx} & Q_{yx} & Q_{ux} & Q_{cx} \\ Q_{xy} & Q_{yy} & Q_{uy} & Q_{cy} \\ Q_{xu} & Q_{yu} & Q_{uu} & Q_{cu} \\ Q_{xv} & Q_{yv} & Q_{uv} & Q_{cv} \end{vmatrix}$$

$$E = \begin{vmatrix} Q_{xx} & Q_{yx} & Q_{ux} & Q_{vx} \\ Q_{xy} & Q_{yy} & Q_{uy} & Q_{vy} \\ Q_{xu} & Q_{yu} & Q_{uu} & Q_{vu} \\ Q_{xv} & Q_{yv} & Q_{uv} & Q_{vv} \end{vmatrix}$$

the solutions for Δx, Δy, Δu and Δv, i.e., the corrections for $f''(0)$, $g'(0)$, $S''(0)$ and $t'(0)$, are given by

$$\Delta x = A/E \qquad (B-64a)$$
$$\Delta y = B/E \qquad (B-64b)$$
$$\Delta u = C/E \qquad (B-64c)$$
$$\Delta v = D/E \qquad (B-64d)$$

Alternately, consider $[Q]$ as the matrix of coefficients of Eqs. (B–60b) to (B–63b) and $[C]$ as the constant column matrix composed of the right hand side of these equations. The solution is then obtained as

$$[X] = [Q]^{-1}[C]$$

where $[X]$ is the column matrix

$$[X] = \begin{bmatrix} \Delta x \\ \Delta y \\ \Delta u \\ \Delta v \end{bmatrix}$$

APPENDIX C

NUMERICAL RESULTS FOR BOUNDARY LAYER FLOWS

C–1 Introduction

The numerical results given in this appendix represent one of the most extensive tabulations known to the author. For this reason it serves as a useful design and analysis tool. The results are generally given to six significant figures and are considered accurate to within $\pm 5 \times 10^{-5}$.

In using the tabulated results in conjunction with the computer programs given in this book (or others), it may be found that a converged solution is not obtained immediately. This is quite normal. The reasons are that neither all compilers nor all computers have the same accuracy. In addition, the stringent convergence criteria used in the boundary layer programs can require more accurate results than are given. This is particularly true for large values of β or $\bar{\beta}$. In these circumstances, convergence normally occurs within one or two iterations.

Although considerable care has been taken in generating the numerical results, some errors have undoubtedly resulted from the publishing process. The author would appreciate having these brought to his attention so that they can be corrected.

Laminar Flow Analysis

Table C–1. Results for the Falkner-Skan equation.

β	$f''(0)$	I_1	I_2	η_{max}	η_{conv}
2.00	1.687218	0.4974331	0.2307831	6.0	4.6
1.60	1.521514	0.5440214	0.2504147	6.0	4.8
1.20	1.335721	0.6068977	0.2761104	6.0	5.0
1.00	1.232588	0.6479002	0.2923434	6.0	5.2
0.80	1.120268	0.698680	0.3118461	6.0	5.2
0.60	0.9958365	0.7639711	0.3359076	6.0	5.4
0.50	0.9276801	0.804584	0.3502693	6.0	5.4
0.40	0.8544213	0.8526334	0.3666903	6.0	5.6
0.30	0.7747546	0.9109929	0.3857349	6.0	5.6
0.20	0.6867083	0.9841576	0.4082296	6.0	5.8
0.10	0.5870354	1.080319	0.4354562	6.0	5.8
0.05	0.5311299	1.141735	0.4514675	6.0	6.0
0	0.4696005	1.216778	0.469598	6.0	6.0
−0.05	0.4003233	1.312358	0.4904601	9.0	6.2
−0.10	0.3192698	1.442697	0.5150439	9.0	6.4
−0.14	0.239736	1.59902	0.5385607	9.0	6.6
−0.16	0.1907799	1.706649	0.5521947	9.0	6.8
−0.18	0.1286362	1.871575	0.567707	9.0	7.0
−0.19	0.08570037	2.006757	0.5765225	9.0	7.0
−0.1988376	0	2.358848	0.5854352	9.0	7.6
−0.19	−0.07133595	2.819803	0.5733665	9.0	8.0
−0.18	−0.09769212	3.083902	0.557817	9.0	8.2
−0.16	−0.1255678	3.527926	0.5224997	9.0	8.6
−0.10	−0.1405462	4.895107	0.3877379	12.0	9.8
−0.05	−0.1082711	6.688471	0.2380551	12.0	11.6
−0.025	−0.07436568	8.546494	0.142868	16.0	13.4

Table C–2. Results for the Falkner-Skan equation with mass transfer.

	$f_0 = -1.0$		
β	$f''(0)$	I_1	I_2
2.00	1.199996	0.65551	0.29624
1.60	1.038014	0.73868	0.32927
1.20	0.856849	0.85968	0.37510
1.00	0.756575	0.94498	0.40579
0.80	0.647461	1.0582	0.44466
0.60	0.526652	1.22148	0.49611
0.50	0.460428	1.33401	0.52894
0.40	0.389104	1.48187	0.56882
0.30	0.311187	1.68905	0.61882
0.20	0.224209	2.01310	0.68466
0.10	0.123327	2.65802	0.77957
0.05	0.0643559	3.43762	0.84998

Table C–3. Results for the Falkner-Skan equation with mass transfer.

	$f_0 = -0.5$		
β	$f''(0)$	I_1	I_2
2.00	1.425482	0.57064	0.26140
1.60	1.259542	0.63329	0.28704
1.20	1.073064	0.72125	0.32161
1.00	0.969230	0.78096	0.34414
0.80	0.855729	0.85781	0.37194
0.60	0.729201	0.96202	0.40749
0.50	0.659364	1.03037	0.42945
0.40	0.583678	1.11555	0.45533
0.30	0.503323	1.22625	0.48650
0.20	0.406219	1.37964	0.52524
0.10	0.295167	1.61788	0.57580
0.05	0.228856	1.80586	0.60816
0	0.148476	2.11187	0.64848
−0.05	0.00621227	3.28672	0.70584
−0.0501779	0	3.40214	0.70614
−0.05	−0.0576883	3.52601	0.70582
−0.04	−0.0305240	4.66797	0.68353
−0.03	−0.0320864	5.60980	0.65588
−0.02	−0.0270220	6.85387	0.62251
−0.01	−0.0166593	9.14773	0.58059

Table C–4. Results for the Falkner-Skan equation with mass transfer.

	$f_0 = 0.5$		
β	$f''(0)$	I_1	I_2
2.00	1.983124	0.43521	0.20423
1.60	1.821202	0.46962	0.21916
1.20	1.640968	0.51430	0.23809
1.00	1.541751	0.54233	0.24971
0.80	1.434611	0.57584	0.26330
0.60	1.317387	0.61693	0.27952
0.50	1.254023	0.64134	0.28890
0.40	1.186757	0.66913	0.29936
0.30	1.114818	0.70121	0.31112
0.20	1.037156	0.73886	0.32449
0.10	0.952276	0.78402	0.33989
0.05	0.906463	0.81034	0.34852
0	0.857916	0.83982	0.35792
−0.05	0.806123	0.87323	0.36819
−0.10	0.750402	0.91162	0.37952
−0.20	0.622914	1.01024	0.40620
−0.30	0.459619	1.16329	0.44087
−0.40	0.190189	1.51124	0.49114
−0.422018	0	1.87965	0.50737
−0.40	−0.167821	2.87506	0.48221
−0.30	−0.320577	3.48721	0.32226
−0.20	−0.339716	4.62752	0.10724
−0.10	−0.282671	6.33016	−0.16629

Table C-5. Results for the Falkner-Skan equation with mass transfer.

	$f_0 = 1.0$		
β	$f''(0)$	I_1	I_2
2.00	2.309817	0.38278	0.18142
1.60	2.154214	0.40819	0.19273
1.20	1.982722	0.44002	0.20668
1.00	1.889314	0.45932	0.21500
0.80	1.789455	0.48169	0.22450
0.60	1.681678	0.50808	0.23552
0.50	1.624199	0.52321	0.24173
0.40	1.563888	0.53992	0.24851
0.30	1.500310	0.55856	0.25596
0.20	1.432920	0.57953	0.26418
0.10	1.361010	0.60338	0.27334
0.05	1.323080	0.61662	0.27833
0	1.283635	0.63089	0.28364
-0.05	1.242510	0.64634	0.28929
-0.10	1.199489	0.66315	0.29534
-0.20	1.106691	0.70184	0.30882
-0.30	1.002351	0.74962	0.32463
-0.40	0.881661	0.81121	0.34357
-0.50	0.735461	0.89612	0.36704
-0.60	0.541438	1.02938	0.39766
-0.70	0.178834	1.80578	0.44288
-0.712041	0	1.58674	0.45085
-0.70	-0.173364	2.30473	0.43995
-0.60	-0.477762	2.98697	0.31442
-0.50	-0.591873	3.33398	0.15023
-0.40	-0.637487	4.02631	-0.04939
-0.30	-0.635429	4.82217	-0.26968
-0.20	-0.58884	5.84041	-0.52595

Table C-6. Results for the Falkner-Skan equation with mass transfer.

$f_0 = 1.5$

β	$f''(0)$	I_1	I_2
2.00	2.663174	0.33877	0.16193
1.60	2.515558	0.35763	0.17052
1.20	2.354448	0.38052	0.18083
1.00	2.267646	0.39399	0.18683
0.80	2.175753	0.40920	0.19355
0.60	2.077807	0.42661	0.20115
0.50	2.026176	0.43633	0.20534
0.40	1.972520	0.44682	0.20985
0.30	1.916588	0.45824	0.21471
0.20	1.858094	0.47073	0.21996
0.10	1.796680	0.48448	0.22567
0.05	1.764750	0.49190	0.22872
0	1.731913	0.49971	0.23191
-0.05	1.698100	0.50800	0.23527
-0.10	1.663237	0.51674	0.23879
-0.20	1.589938	0.53596	0.24641
-0.30	1.511070	0.55794	0.25493
-0.40	1.425355	0.58345	0.26456
-0.50	1.330969	0.61362	0.27556
-0.60	1.225199	0.65022	0.28833
-0.80	0.958771	0.75699	0.32183
-1.00	$1/2$[†]	1.00000	0.37673
-1.06489	0	1.38378	0.40730
-1.00	-0.499968	1.99997	0.34513
-0.80	-0.897156	3.01291	0.06587
-0.60	-1.02754	4.01226	-0.30046
-0.40	-1.02460	5.20024	-0.74085
-0.20	-0.895220	6.94147	-1.25866
-0.10	-0.747456	8.53987	-1.5483

[†] An analytical solution for $\beta = -1, f''(0) > 0$ exists (see [Evan66]).

Table C–7. Results for the Falkner-Skan equation with mass transfer.

	$f_0 = 2.0$		
β	$f''(0)$	I_1	I_2
2.00	3.039607	0.30185	0.14530
1.60	2.900391	0.31596	0.15187
1.20	2.750172	0.33263	0.15955
1.00	2.670056	0.34220	0.16393
0.80	2.585974	0.35279	0.16875
0.60	2.497306	0.36461	0.17409
0.50	2.451013	0.37106	0.17699
0.40	2.40326	0.37793	0.18006
0.30	2.35391	0.38527	0.18333
0.20	2.302800	0.39313	0.18681
0.10	2.249740	0.40161	0.19053
0.05	2.222410	0.40610	0.19248
0	2.194509	0.41077	0.19451
−0.05	2.166000	0.41563	0.19662
−0.10	2.136852	0.42070	0.19880
−0.20	2.076448	0.43155	0.20345
−0.30	2.012904	0.44347	0.20849
−0.40	1.945742	0.45665	0.21401
−0.50	1.874350	0.47137	0.22007
−0.60	1.797942	0.48795	0.22678
−0.80	1.625536	0.52876	0.24274
−1.00	$\sqrt{2}$[†]	$2 - \sqrt{2}$	0.26359
−1.20	1.13029	0.67594	0.29288
−1.40	0.642903	0.87213	0.34027
−1.47822	0	1.23253	0.37232
−1.40	−0.679757	1.82752	0.30307
−1.20	−1.18779	2.64802	0.050823
−1.00	−1.41421	3.41421	−0.26941
−0.80	−1.50996	4.22679	−0.64264
−0.60	−1.51640	5.15094	−1.06460
−0.40	−1.44174	6.30049	−1.53590
−0.20	−1.25298	8.03736	−2.05689
−0.10	−1.05973	9.68392	−2.32371

[†] An analytical solution for $\beta = -1, f''(0) > 0$ exists (see [Evan66]).

Table C-8. Results for the Falkner-Skan equation with mass transfer.

$f_0 = 2.5$

β	$f''(0)$	I_1	I_2
2.00	3.434967	0.27080	0.13112
1.60	3.304409	0.28148	0.13617
1.20	3.164876	0.29379	0.14197
1.00	3.091125	0.30072	0.14520
0.80	3.014297	0.30825	0.14872
0.60	2.933988	0.31650	0.15256
0.50	2.892374	0.32095	0.15460
0.40	2.849700	0.32561	0.15675
0.30	2.805880	0.33053	0.15902
0.20	2.760830	0.33572	0.16141
0.10	2.714434	0.34123	0.16392
0.05	2.690680	0.34413	0.16522
0	2.666567	0.34710	0.16657
−0.05	2.642050	0.35016	0.16796
−0.10	2.617113	0.35333	0.16938
−0.20	2.565908	0.35998	0.17238
−0.30	2.512763	0.36713	0.17557
−0.40	2.457468	0.37482	0.17899
−0.50	2.399765	0.38313	0.18266
−0.60	2.339345	0.39217	0.18661
−0.80	2.208776	0.41292	0.19555
−1.00	$\sqrt{17}/2$[†]	$(5 - \sqrt{17})/2$	0.20627
−1.20	1.89072	0.47115	0.21952
−1.60	1.41269	0.58215	0.25978
−1.80	0.994323	0.70532	0.29513
−1.95063	0	1.11428	0.34340
−1.80	−1.11723	1.91792	0.20622
−1.60	−1.60066	2.57956	−0.04402
−1.20	−1.99996	3.86399	−0.68417
−1.00	−2.06155	4.56155	−1.05872
−0.80	−2.06251	5.33632	−1.46726
−0.60	−2.00787	6.23948	−1.91045
−0.40	−1.88699	7.38415	−2.38888
−0.20	−1.6447	9.15588	−2.89190
−0.10	−1.39847	10.8931	−3.12128

[†]An analytical solution for $\beta = -1, f''(0) > 0$ exists (see [Evan66]).

Table C-9. Results for the Falkner-Skan equation with mass transfer.

	$f_0 = 3.0$		
β	$f''(0)$	I_1	I_2
2.00	3.846140	0.24458	0.11899
1.60	3.723984	0.25275	0.12292
1.20	3.594534	0.26200	0.12733
1.00	3.526640	0.26711	0.12977
0.80	3.456353	0.27258	0.13238
0.60	3.383400	0.27850	0.13519
0.50	3.345827	0.28165	0.13667
0.40	3.307470	0.28493	0.13821
0.30	3.268280	0.28834	0.13983
0.20	3.228200	0.29191	0.14152
0.10	3.187160	0.29566	0.14327
0.05	3.166260	0.29761	0.14417
0	3.145101	0.29959	0.14510
-0.05	3.123670	0.30163	0.14605
-0.10	3.101953	0.30372	0.14703
-0.20	3.057628	0.30807	0.14905
-0.30	3.012028	0.31266	0.15118
-0.40	2.965048	0.31752	0.15342
-0.50	2.916565	0.32267	0.15580
-0.60	2.866436	0.32815	0.15832
-0.80	2.760560	0.34026	0.16383
-1.00	$\sqrt{7}$[†]	$3 - \sqrt{7}$	0.17012
-1.20	2.51962	0.37074	0.17743
-1.40	2.37867	0.39065	0.18606
-1.60	2.21740	0.41543	0.19651
-2.00	1.78708	0.49307	0.22679
-2.40	0.861755	0.72129	0.29083
-2.48123	0	1.01866	0.31897
-2.40	-0.973767	1.50525	0.25798
-2.00	-2.14799	2.67752	-0.20706
-1.80	-2.39338	3.21784	-0.49842
-1.60	-2.54165	3.77037	-0.81824
-1.20	-2.65524	4.96850	-1.53484
-1.00	-2.64575	5.64575	-1.93071
-0.80	-2.59754	6.40880	-2.35254
-0.60	-2.50576	7.30893	-2.80100
-0.40	-2.35014	8.46503	-3.27355
-0.20	-2.05483	10.2942	-3.74499
-0.10	-1.74897	12.1424	-3.92749

[†]An analytical solution for $\beta = -1, f''(0) > 0$ exists (see [Evan66]).

Table C–10. Results for a circular cylinder in cross flow using the two-equation locally nonsimilar boundary layer method.

x/R	$f''(0)$	$g''(0)$	β
0	1.2326	−0.12346	1.00
0.10	1.2314	−0.12419	0.99750
0.25	1.2248	−0.12814	0.98421
0.50	1.2003	−0.14358	0.93480
0.75	1.1557	−0.17499	0.84506
1.00	1.0848	−0.23571	0.70155
1.25	0.98357	−0.36270	0.47946

Table C–11. Adiabatic wall solutions for the forced convection boundary layer.

$\beta = 0, \; \theta_2'(0) = 0$

Pr	$\theta_2(0)$
0.5	0.352153
0.723	0.424748
1.0	1/2
3.0	0.854390
1.0	1.26377
10.0	1.48098

Table C–12. Results for the energy equation for an isothermal wall without viscous dissipation or mass transfer.[†]

				$\theta'(0)$				
β\Pr	0.001	0.01	0.1	0.7	1.0	7.0	10.0	100.0
2.00	0.0249209	0.0768126	0.215477	0.523525	0.605197	1.277614	1.455750	3.286250
1.60	0.0248924	0.0765501	0.213589	0.514537	0.593857	1.244491	1.416603	3.184726
1.20	0.0248541	0.0761993	0.211105	0.503031	0.579417	1.203358	1.368148	3.060693
1.00	0.0248291	0.0759725	0.209522	0.495866	0.570465	1.178375	1.338797	2.986330
0.80	0.0247983	0.0756938	0.207598	0.487318	0.559823	1.149124	1.304497	2.900007
0.60	0.0247588	0.0753385	0.205179	0.476796	0.546773	1.113837	1.263203	2.797017
0.50	0.0247343	0.0751193	0.203705	0.470493	0.538979	1.093031	1.238892	2.736690
0.40	0.0247054	0.0748613	0.201984	0.463233	0.530022	1.069336	1.211234	2.668316
0.30	0.0246703	0.0745503	0.199933	0.454700	0.519518	1.041803	1.179130	2.589234
0.20	0.0246265	0.0741638	0.197413	0.444384	0.506854	1.008914	1.140816	2.495146
0.10	0.0245691	0.0736610	0.194183	0.431396	0.490949	0.967982	1.093171	2.378386
0.05	0.0245326	0.0733429	0.192165	0.423400	0.481177	0.942985	1.064086	2.307128
0	0.0244881	0.0729572	0.189743	0.413912	0.469600	0.913472	1.029747	2.222906
−0.05	0.0244316	0.0724706	0.186725	0.402236	0.455372	0.877291	0.987640	2.119318
−0.10	0.0243548	0.0718152	0.182721	0.386971	0.436798	0.830081	0.932653	1.983127
−0.14	0.0242651	0.0710562	0.178167	0.369883	0.416029	0.777134	0.870875	1.828105
−0.16	0.0242006	0.0705150	0.174971	0.358045	0.401651	0.740265	0.827756	1.718044
−0.18	0.0241050	0.0697203	0.170353	0.341138	0.381124	0.687156	0.765447	1.555102
−0.19	0.0240270	0.0690787	0.166687	0.327874	0.365021	0.644975	0.715750	1.420584

[†]Selected results from Elzy and Sisson [Elzy67] — $\theta = (T - T_w)/(T_\infty - T_w)$.

Table C–13. Nondimensional heat transfer parameter $\theta'(0)$ for forced convection from a nonisothermal surface without viscous dissipation.

			Pr = 1		
n\β	1.0	0.5	0.1	0	−0.1
−0.5	0.372983	0.216133	0.050670	0[†]	−0.056404
−0.1	0.537538	0.492470	0.438344	0.417351	0.319273
0	0.570465	0.538979	0.490950	0.469600	0.436800
0.1	0.601100	0.580600	0.536413	0.514294	0.478759
0.5	0.706260	0.714490	0.675433	0.649082	0.603241
1.0	0.811301	0.839037	0.798679	0.767132	0.710722
2.0	0.972394	1.02041	0.973088	0.933089	0.860656

[†]$\theta'(0) = 0$ for $n = -1/(2 - \beta)$.

Table C-14. Effect of viscous dissipation on the heat transfer parameter $\theta'(0)$ for a nonisothermal wall.[†]

	$\beta = 1$ $n = 2$	$\beta = -0.1$ $n = -0.0952381$	$\beta = 0.5$ $n = 2/3$
$E1$	$\theta'(0)$	$\theta'(0)$	$\theta'(0)$
-10	5.83437		
-1.0	1.45859	0.584567	1.13986
-0.1	1.02101		
0	0.972394	0.389712	0.759906
0.1	0.923775		
1.0	0.486197	0.194855	0.379953
2.0	0	0	0
10	-3.88958		

[†] These results are represented by the following linear relationships (see Figure 4-16):

$\beta = 1 \quad \theta'(0, E1) = 0.972394 - 0.4861965\, E1$

$\beta = 0.5 \quad \theta'(0, E1) = 0.389712 - 0.194856\, E1$

$\beta = -0.1 \quad \theta'(0, E1) = 0.759906 - 0.379953\, E1$

$\theta'(0, E1) = 0$ at $E1 = 2$.

Table C-15. Results for the energy equation for an isothermal wall with mass transfer, $\theta'(0)$.

		$f_0 = -1$		
β\Pr	0.1	0.7	1.0	7.0
2.00	0.162622	0.175427	0.148120	0.0008518
1.60	0.159578	0.166085	0.138125	0.0005782
1.20	0.155333	0.153668	0.125060	0.0003316
1.00	0.152463	0.145641	0.116752	0.0002249
0.80	0.148771	0.135721	0.106641	0.0001342
0.60	0.143744	0.122901	0.0938396	6.407×10^{-5}
0.50	0.140420	0.114829	0.0859452	3.830×10^{-5}
0.40	0.136225	0.105085	0.0765964	1.935×10^{-5}
0.30	0.130647	0.0928566	0.0651726	7.31×10^{-6}
0.20	0.122545	0.0765276	0.0505264	1.52×10^{-6}
0.10	0.108334	0.0518861	0.0300769	5.6×10^{-8}
0.05	0.0939365	0.0321076	0.0156629	7.5×10^{-10}

Table C–16. Results for the energy equation for an isothermal wall with mass transfer, $\theta'(0)$.

	$f_0 = -0.5$			
β\Pr	0.1	0.7	1.0	7.0
2.00	0.193296	0.324448	0.332386	0.104824
1.60	0.190754	0.314578	0.320441	0.0921826
1.20	0.187305	0.301611	0.304865	0.0772493
1.00	0.185036	0.293322	0.294975	0.0686278
0.80	0.182193	0.283181	0.282945	0.0589954
0.60	0.178464	0.270252	0.267710	0.0480818
0.50	0.176091	0.262224	0.258307	0.0420303
0.40	0.173207	0.252658	0.247158	0.0355037
0.30	0.169571	0.240877	0.233509	0.0284328
0.20	0.164724	0.225616	0.215959	0.0207519
0.10	0.157590	0.203985	0.191348	0.0124516
0.05	0.152262	0.188437	0.173858	0.0081359
0	0.144102	0.165569	0.148476	0.0038738
−0.05	0.117680	0.0995586	0.0783988	0.0001175

Table C–17. Results for the energy equation for an isothermal wall with mass transfer, $\theta'(0)$.

	$f_0 = 0.5$			
β\Pr	0.1	0.7	1.0	7.0
2.00	0.260467	0.763296	0.949707	3.958947
1.60	0.258825	0.755838	0.940318	3.938290
1.20	0.256739	0.746579	0.928725	3.913739
1.00	0.255453	0.740987	0.921757	3.899448
0.80	0.253939	0.734503	0.913703	3.883338
0.60	0.252114	0.726815	0.904193	3.864837
0.50	0.251045	0.722375	0.898717	3.854426
0.40	0.249840	0.717424	0.892624	3.843038
0.30	0.248466	0.711836	0.885765	3.830452
0.20	0.246874	0.705438	0.877929	3.816363
0.10	0.244993	0.697970	0.868809	3.800326
0.05	0.243910	0.693717	0.863625	3.791375
0	0.242708	0.689026	0.857916	3.781651
−0.05	0.241359	0.683802	0.851568	3.770991
−0.10	0.239825	0.677910	0.844420	3.759176
−0.20	0.235962	0.663261	0.826691	3.730671

Table C–18. Results for the energy equation for an isothermal wall with mass transfer, $\theta'(0)$.

	\multicolumn{4}{c}{$f_0 = 1$}
$\beta\backslash\mathrm{Pr}$	0.1	0.7	1.0	7.0
2.00	0.296461	1.033302	1.344664	7.211858
1.60	0.295169	1.027431	1.337423	7.206524
1.20	0.293577	1.020357	1.328752	7.194674
1.00	0.292626	1.016208	1.323691	7.188012
0.80	0.291536	1.011519	1.317992	7.180719
0.60	0.290266	1.006140	1.311481	7.172643
0.50	0.289545	1.003125	1.307844	7.168246
0.40	0.288755	0.999847	1.303899	7.163563
0.30	0.287881	0.996258	1.299589	7.158549
0.20	0.286906	0.992295	1.294844	7.153145
0.10	0.285808	0.987877	1.289569	7.147274
0.05	0.285203	0.985464	1.286694	7.144133
0	0.284554	0.982892	1.283635	7.140835
−0.05	0.283856	0.980140	1.280366	7.137359
−0.10	0.283101	0.977182	1.276859	7.133683
−0.20	0.281380	0.970511	1.268972	7.125612
−0.30	0.279286	0.962507	1.25954	7.11631
−0.40	0.276633	0.952527	1.24784	7.10521
−0.50	0.273052	0.939291	1.232392	7.091282

Table C-19. Results for the energy equation for an isothermal wall with mass transfer, $\theta'(0)$.

	$f_0 = 2$			
β\Pr	0.1	0.7	1.0	7.0
2.00	0.372303	1.631248	2.222521	14.086022
1.60	0.371508	1.627821	2.218502	14.082739
1.20	0.370580	1.623896	2.213927	14.079136
1.00	0.370052	1.621700	2.211379	14.077189
0.80	0.369472	1.619313	2.208620	14.075125
0.60	0.368829	1.616701	2.205610	14.072926
0.50	0.368480	1.615297	2.203997	14.071769
0.40	0.368110	1.613819	2.202303	14.070569
0.30	0.367716	1.612258	2.200518	14.069322
0.20	0.367296	1.610606	2.198632	14.068023
0.10	0.366847	1.608850	2.196634	14.066666
0.05	0.366610	1.607930	2.195588	14.065964
0	0.366363	1.606979	2.194509	14.065245
−0.05	0.366108	1.605995	2.193394	14.064508
−0.10	0.365842	1.604976	2.192240	14.063751
−0.20	0.365275	1.602821	2.189806	14.062176
−0.30	0.364656	1.60049	2.18718	14.0616
−0.40	0.363976	1.59795	2.18433	14.0598
−0.50	0.363223	1.595174	2.181222	14.056828
−0.60	0.362379	1.59210	2.17779	14.0558
−0.80	0.360331	1.58477	2.16966	14.0511
−1.00	0.357523	1.574999	2.158950	14.044213

Table C-20. Results for the energy equation for an isothermal wall with mass transfer, $\theta'(0)$.

	$f_0 = 3$			
$\beta\backslash\text{Pr}$	0.1	0.7	1.0	7.0
2.00	0.452349	2.272651	3.159198	21.050985
1.60	0.451852	2.270651	3.156945	21.049582
1.20	0.451294	2.268443	3.154469	21.048081
1.00	0.450988	2.267246	3.153132	21.047288
0.80	0.450661	2.265978	3.151720	21.046463
0.60	0.450310	2.264630	3.150224	21.045601
0.50	0.450124	2.263923	3.149440	21.045156
0.40	0.449931	2.263191	3.148631	21.044700
0.30	0.449730	2.262434	3.147794	21.044232
0.20	0.449520	2.261648	3.146929	21.043753
0.10	0.449301	2.260833	3.146032	21.043261
0.05	0.449188	2.260414	3.145571	21.043009
0	0.449072	2.259986	3.145101	21.042755
−0.05	0.448953	2.259549	3.144622	21.042496
−0.10	0.448832	2.259103	3.144134	21.042234
−0.20	0.448580	2.258183	3.143127	21.041697
−0.30	0.448315	2.25722	3.14208	21.0414
−0.40	0.448035	2.25621	3.14098	21.0408
−0.50	0.447740	2.255158	3.139830	21.039977
−0.60	0.447427	2.25405	3.13862	21.0396
−0.80	0.446740	2.25163	3.13602	21.0383
−1.00	0.445952	2.248918	3.133094	21.036625

Table C-21. Nondimensional wall shearing stress and heat transfer parameters for free convection flow from an isothermal vertical flat plate.[†]

Pr	$f''(0)$	$\theta'(0)$
0.01	0.9862	−0.0812
0.72	0.6760	−0.5046
0.733	0.6741	−0.5080
1	0.6421	−0.5671
2	0.5713	−0.7165
10	0.4192	−1.1694
100	0.2517	−2.191
1000	0.1450	−3.966

[†] Values from Sparrow and Gregg [Spar58].

Table C–22. Nondimensional shearing stress and heat transfer parameters for free convection from a nonisothermal vertical flat plate.

n	$f''(0)$	$\theta'(0)$	$f''(0)/f''_{n=0}(0)$	$\theta'(0)/\theta'_{n=0}(0)$
\multicolumn{5}{c}{Pr = 1}
3.0	0.432667	−1.087090	0.673760	1.916823
2.0	0.467445	−0.985803	0.727917	1.738228
1.5	0.491342	−0.921466	0.765130	1.624785
1.0	0.522892	−0.841558	0.814260	1.483886
0.5	0.567960	−0.734355	0.884441	1.294860
0.2	0.607000	−0.645462	0.945236	1.138118
0.1	0.623391	−0.608855	0.970760	1.073570
0.05	0.632441	−0.588719	0.984853	1.038065
0	0.642168	−0.567131	1.0	1.0
−0.05	0.652662	−0.543898	1.016342	0.959034
−0.1	0.664002	−0.518730	1.034000	0.914656
−0.2	0.689884	−0.461220	1.074305	0.813251
−0.3	0.721566	−0.390750	1.123641	0.688994
−0.4	0.761288	−0.300790	1.185497	0.530371
−0.5	0.813608	−0.179114	1.266971	0.315825
−0.6	0.887246	0	1.381642	0
\multicolumn{5}{c}{Pr = 0.7}
3.0	0.457771	−0.967339	0.674334	1.936622
2.0	0.494565	−0.876415	0.728535	1.754592
1.5	0.519865	−0.818637	0.765804	1.638919
1.0	0.553299	−0.746751	0.815055	1.495003
0.5	0.600814	−0.650161	0.885048	1.301629
0.2	0.641956	−0.570051	0.945654	1.141248
0.1	0.659164	−0.537054	0.971002	1.075187
0.05	0.668661	−0.518923	0.984992	1.038889
0	0.678849	−0.499498	1.0	1.0
−0.05	0.689885	−0.478596	1.016257	0.958154
−0.1	0.701752	−0.455978	1.033738	0.912873
−0.2	0.728800	−0.404374	1.073582	0.809561
−0.3	0.761584	−0.341375	1.121875	0.683436
−0.4	0.802531	−0.261459	1.182194	0.523444
−0.5	0.855868	−0.154521	1.260763	0.309353
−0.6	0.929670	0	1.369480	0

Table C-23. Nondimensional shearing stress and heat transfer parameters for free convection from an isothermal vertical flat plate with mass transfer.

$n = 0,\ \Pr = 0.73$[†]		
f_0	$f''(0)$	$-\theta'(0)$
1.0	0.409	2.192
0.9	0.446	1.994
0.8	0.488	1.798
0.7	0.530	1.606
0.6	0.573	1.420
0.5	0.611	1.241
0.4	0.644	1.072
0.3	0.668	0.912
0.2	0.681	0.764
0.1	0.684	0.629
0	0.674	0.507
−0.1	0.654	0.399
−0.2	0.627	0.305
−0.4	0.557	0.162
−0.6	0.476	0.0725
−0.8	0.399	0.0264
−1.0	0.336	0.00748

[†]After Eichhorn [Eich60].

Table C-24. Nondimensional shearing stress and heat transfer parameters for combined forced and free convection boundary layer flows on an isothermal surface.

$\beta = 2/3$, Pr $= 0.7$

Gr_x/Re_x^2	$f''(0)$	$-\theta'(0)$
100	35.389	1.3105
50	21.203	1.1115
20	10.893	0.90142
10	6.6935	0.77746
5	4.2297	0.68035
25/9	2.9705	0.61719
1	1.8130	0.54416
0.8	1.6685	0.53348
0.5	1.4434	0.51588
0.25	1.2466	0.49936
0.05	1.0815	0.48455
0.02	1.0560	0.48218
0	1.0389	0.48058
-0.02	1.0217	0.47895
-0.05	0.99579	0.47647
-0.10	0.95205	0.47223
-0.50	0.57308	0.43141
-0.80	0.22823	0.38495
-0.90	0.085249	0.36132
-0.94	0.018904	0.34900
-0.95	0.00097538	0.34549

Table C-25. Solutions of the compressible laminar boundary layer equations.

			Pr = 1			
g_w	$\hat{\beta}$	$f''(0)$	$g'(0)$	J_1	J_2	J_3
0	2.00	0.738646	0.520637	0.177507	0.383660	1.28810
	1.50	0.698714	0.514758	0.204855	0.391414	1.32633
	1.00	0.648858	0.506661	0.245559	0.403297	1.38265
	0.50	0.581143	0.494220	0.314630	0.423825	1.47691
	0.10	0.498666	0.476549	0.423879	0.456276	1.62464
	0	0.469600	0.469600	0.469598	0.469597	1.68638
	−0.10	0.433977	0.460534	0.530955	0.487091	1.76939
	−0.20	0.387508	0.447715	0.620116	0.511527	1.89070
	−0.30	0.318261	0.426305	0.772474	0.550002	2.10094
	−0.326419	0	0.247790	1.95750	0.638924	4.09547
	−0.30	−0.0201194	0.219819	2.11414	0.614006	4.47231
	−0.25	−0.0418397	0.170278	2.36227	0.548538	5.23478
	−0.20	−0.0485533	0.121835	2.55099	0.461489	6.19619
	−0.15	−0.0430414	0.0727043	2.63575	0.352215	7.63313
	−0.10	−0.0248272	0.0236310	2.41063	0.216142	10.8117
0.20	2.00	0.948323	0.433374	0.296497	0.355346	1.15564
	1.50	0.869491	0.426683	0.335632	0.366047	1.20614
	1.00	0.775537	0.417547	0.393393	0.382142	1.28023
	0.50	0.654961	0.403591	0.490937	0.409493	1.40442
	0.10	0.516777	0.383665	0.646479	0.452129	1.60181
	0	0.469600	0.375680	0.712953	0.469597	1.68637
	−0.10	0.412212	0.365003	0.804518	0.492664	1.80339
	−0.20	0.336409	0.349022	0.945504	0.525510	1.98531
	−0.30	0.209010	0.315893	1.24416	0.582260	2.38110
	−0.308622	0	0.226014	2.03446	0.627824	3.55511
	−0.30	−0.0147899	0.215398	2.12380	0.622345	3.70683
	−0.25	−0.0626109	0.166071	2.52592	0.568773	4.47551
	−0.20	−0.0804134	0.123730	2.85284	0.490069	5.27294
	−0.15	−0.0811743	0.0834549	3.14343	0.390220	6.26747
	−0.10	−0.0686004	0.0450293	3.39507	0.270823	7.73165
0.6	2.00	1.33335	0.230628	0.519429	0.294479	0.926629
	1.50	1.18536	0.225837	0.582101	0.312187	0.996556
	1.00	1.01219	0.219273	0.674252	0.337939	1.09927
	0.50	0.795193	0.209134	0.830082	0.380151	1.27288
	0.10	0.552320	0.194163	1.08513	0.443806	1.55779
	0	0.469600	0.187840	1.19966	0.469597	1.68637
	−0.10	0.367041	0.178822	1.36798	0.503838	1.87637
	−0.20	0.218417	0.162575	1.68107	0.554631	2.23409
	−0.247561	0	0.125095	2.42847	0.601196	3.12146
	−0.20	−0.100468	0.0876177	3.22051	0.543511	4.14312
	−0.15	−0.121171	0.0623341	3.82599	0.452631	5.00516
	−0.10	−0.113708	0.0392288	4.51552	0.337727	6.08870

Table C-25. (cont.)

g_w	$\hat{\beta}$	$f''(0)$	$g'(0)$	J_1	J_2	J_3
1.0	Given by the Falkner-Skan solutions, $g'(0) = 0$ — see Table C-1.					
2.0	2.00	2.48768	−0.661455	1.21019	0.0672736	0.307899
	1.50	2.14034	−0.642295	1.35175	0.112716	0.426665
	1.00	1.73668	−0.615585	1.56058	0.176103	0.601104
	0.50	1.23481	−0.572886	1.92036	0.274623	0.898878
	0.10	0.670713	−0.503755	2.56193	0.414519	1.41804
	0	0.469600	−0.469600	2.90315	0.469597	1.68638
	−0.10	0.181379	−0.403222	3.61429	0.542734	2.22478
	−0.129507	0	−0.338911	3.61429	0.542734	2.77261
	−0.10	−0.161671	−0.207312	6.44842	0.483171	4.08123
	−0.05	−0.151016	−0.105643	9.32033	0.314931	5.64387

Table C-26. Solutions of the low speed compressible laminar boundary layer equations.

$M_e = 0$, $\text{Pr} = 0.723$

g_w	$\hat{\beta}$	$f''(0)$	$g'(0)$	J_1	J_2	J_3
0	1.0	0.610911	0.442248	0.176773	0.434135	1.48255
	0.5	0.555599	0.433994	0.221500	0.444849	1.54831
	0.1	0.491300	0.422874	0.289930	0.462306	1.64679
	0.	0.469600	0.418711	0.317768	0.469600	1.68638
	−0.1	0.443805	0.413474	0.354308	0.479236	1.73811
0.2	1.0	0.746736	0.365466	0.339480	0.407252	1.35575
	0.5	0.635300	0.355072	0.417659	0.426468	1.45894
	0.1	0.510965	0.340609	0.540066	0.456958	1.61920
	0	0.469600	0.334969	0.591493	0.469600	1.68638
	−0.1	0.420249	0.327601	0.661195	0.486368	1.77754
0.6	1.0	0.998901	0.192617	0.648147	0.350753	1.13385
	0.5	0.785963	0.184493	0.794456	0.388746	1.29816
	0.1	0.549487	0.172516	1.03262	0.446225	1.56620
	0	0.469600	0.167485	1.13894	0.469600	1.68638
	−0.1	0.371286	0.160369	1.29399	0.500684	1.86247
1.0	Given by the Falkner-Skan Solutions, $g'(0) = 0$ — see Table C-1.					
2.0	1.0	1.76483	−0.543733	1.62132	0.143497	0.529297
	0.5	1.25480	−0.507821	2.00376	0.252921	0.845913
	0.1	0.677291	−0.448540	2.68830	0.408461	1.39897
	0	0.469600	−0.418711	3.05499	0.469599	1.68638
	−0.1	0.166218	−0.358628	3.84347	0.550566	2.28009

Table C-27. Solutions of the compressible laminar boundary layer equations.[†]

$\bar{\sigma} = 2.0$, $Pr = 0.723$, $f(0) = 0$

g_w	$\hat{\beta}$	$f''(0)$	$g'(0)$	J_1	J_2
0	2.00	0.703383[‡]	0.381550[‡]	0.158885	0.385545
	1.50	0.667808	0.378406	0.182897	0.393422
	1.00	0.623739	0.374173	0.218392	0.40531
	0.50	0.564591	0.367826	0.277945	0.425583
	0.10	0.493985	0.359069	0.370076	0.456948
	0	0.469600	0.355695	0.407694	0.469576
	−0.10	0.440236	0.351368	0.457202	0.485874
	−0.20	0.402921	0.345370	0.526534	0.508226
	−0.30	0.350892	0.336001	0.635111	0.541345
	−0.426121	0.20	0.299382	1.02892	0.638379
	−0.433244	0.15	0.282659	1.18867	0.664923
	−0.426845	0.10	0.262209	1.36846	0.684067
	−0.362483	0	0.198571	1.83816	0.666264
	−0.30	−0.0348202	0.153347	2.10194	0.595695
	−0.25	−0.0461998	0.120061	2.25817	0.518157
	−0.20	−0.0475812	0.0882142	2.36884	0.425941
	−0.15	−0.0409842	0.0576095	2.42059	0.321984
	−0.10	−0.0275362	0.0285845	2.36038	0.208501
0.2	2.00	0.935595	0.299591	0.294040	0.347458
	1.50	0.856740	0.296330	0.331515	0.359453
	1.00	0.763643	0.291938	0.386447	0.377149
	0.50	0.645825	0.285286	0.478214	0.406675
	0.10	0.513748	0.275785	0.621538	0.451563
	0	0.469600	0.271951	0.681389	0.469576
	−0.10	0.416813	0.266819	0.762222	0.492953
	−0.30	0.247150	0.244465	1.09999	0.577073
	−0.347875	0.15	0.225411	1.35982	0.622984
	−0.356406	0.10	0.212666	1.51969	0.641565
	−0.353555	0.05	0.196844	1.70503	0.652763
	−0.336754	0	0.176094	1.93018	0.649950
	−0.30	−0.0470249	0.147406	2.21581	0.617717
	−0.20	−0.0853451	0.0884386	2.74896	0.464273
	−0.15	−0.0822371	0.0626893	2.98188	0.364954
	−0.10	−0.0688031	0.0385981	3.23196	0.254318

Table C–27. (cont.)

g_w	$\hat{\beta}$	$f''(0)$	$g'(0)$	J_1	J_2
0.6	2.00	1.35704	0.107090	0.545626	0.265708
	1.50	1.20247	0.107365	0.610013	0.287392
	1.00	1.02262	0.107543	0.704201	0.318359
	0.50	0.799139	0.107306	0.862178	0.368005
	0.10	0.552454	0.105657	1.11660	0.440762
	0	0.469600	0.104462	1.22878	0.469576
	−0.10	0.368145	0.102284	1.39087	0.507147
	−0.20	0.225638	0.0970960	1.67787	0.561138
	−0.257071	0.05	0.0843375	2.18824	0.612477
	−0.257341	0	0.0782114	2.39411	0.616050
	−0.246783	−0.05	0.0697983	2.65975	0.606341
	−0.20	−0.111671	0.0509925	3.2424	0.536807
	−0.15	−0.125582	0.0359909	3.77005	0.439751
	−0.10	−0.114934	0.0228656	4.39482	0.324415
1.0	2.00	1.74124	−0.112440	0.780252	0.180671
	1.50	1.51946	−0.105971	0.871042	0.212828
	1.00	1.26207	−0.0971837	1.00405	0.257949
	0.50	0.943144	−0.0847436	1.22877	0.328707
	0.10	0.590174	−0.0688869	1.60211	0.429928
	0	0.469600	−0.0630259	1.77618	0.469576
	−0.10	0.316480	−0.05531570	2.04971	0.521370
	−0.196056	0.05	−0.0413344	2.77770	0.594526
	−0.200221	0	−0.0385233	2.98918	0.598439
	−0.1988	−0.0391949	−0.0361713	3.19303	0.595526
	−0.197477	−0.05	−0.0354832	3.25798	0.593325
	−0.15	−0.138730	−0.0270505	4.27206	0.501845
	−0.10	−0.142536	−0.0209630	5.23291	0.380581
2.0	2.00	2.60497	−0.743636	1.32019	−0.0355296
	1.50	2.23517	−0.713347	1.47437	0.0235425
	1.00	1.80639	−0.672821	1.70132	0.104997
	0.50	1.27508	−0.611937	2.09072	0.229650
	0.30	1.00927	−0.575473	2.35401	0.303011
	0.10	0.680817	−0.522479	2.77994	0.402781
	0	0.469600	−0.481746	3.14466	0.469575
	−0.10	0.164923	−0.406427	3.91549	0.556395
	−0.120701	0.05	−0.368047	4.36491	0.576789
	−0.125567	0	−0.347846	4.62019	0.580102
	−0.126950	−0.05	−0.324007	4.94089	0.577192
	−0.10	−0.160746	−0.228047	6.54058	0.493312
	−0.05	−0.151464	−0.129793	9.28577	0.312822

[†]Additional solutions are given in Rogers [Roge69].
[‡]For $\beta = 1.5$ and 2, the convergence criteria were relaxed to $(1 - f'|_{\eta_{\max}})$ and $(1 - g|_{\eta_{\max}}) < \pm 5 \times 10^{-6}$.

Table C–28. Solutions of the compressible laminar boundary layer equations — the adiabatic wall temperature branch $g'(0) = 0$.[†]

$\bar{\sigma} = 2.0, \ \Pr = 0.723, \ f(0) = 0$

$\hat{\beta}$	g_{aw}	$f''(0)$	J_1	J_2
1.00	0.814388	1.15295	0.866815	0.286134
0.50	0.826698	0.881723	1.07183	0.345808
0.10	0.843270	0.575496	1.412975	0.434199
0	0.849464	0.469600	1.570195	0.465967
−0.10	0.857375	0.335283	1.810908	0.516373
−0.15	0.862319	0.246432	2.002835	0.546857
−0.20	0.868479	0.114397	2.359336	0.586263
−0.215957	0.870973	0.01	2.745153	0.602836
−0.216103	0.871000	0	2.790413	0.603017
−0.215950	0.870971	−0.01	2.837782	0.602820
−0.20	0.867447	−0.0890432	3.33709	0.578377
−0.15	0.853340	−0.137194	4.117759	0.480471
−0.10	0.834879	−0.132742	4.91481	0.358742
−0.05	0.809091	−0.0970084	6.168873	0.211435

[†]Additional solutions are given in Rogers [Roge69].

Table C–29. Solutions of the compressible laminar boundary layer equations with mass transfer.

$\Pr = 1, \ f_0 = -1$

g_w	β	$f''(0)$	$g'(0)$	J_1	J_2
0	0.50	0.0197748	0.00923746	0.407631	0.815959
	0.40	0.0131445	0.00629761	0.453029	0.831936
	0.30	0.00627916	0.00305231	0.512063	0.852658
0.2	1.00	0.236298	0.0422767	0.552281	0.684008
	0.50	0.137559	0.0281596	0.785373	0.744874
	0.10	0.0282793	0.00550194	1.48431	0.879849
	0.05	0.0122991	0.00154728	1.86161	0.919218
0.6	1.00	0.516751	0.0360568	0.980262	0.536488
	0.50	0.311339	0.0258593	1.36413	0.629273
	0.10	0.0787323	0.00792748	3.54055	0.824676
	0.05	0.0392932	0.00365514	3.17527	0.880523
	0.025	0.0183326	0.00126853	3.95313	0.919500
2.0	1.00	1.27935	−0.165329	2.16234	0.117044
	0.50	0.786329	−0.125273	2.95172	0.310474
	0.10	0.222452	−0.0501511	5.37702	0.684751
	0.05	0.121260	−0.0292269	6.68866	0.786826

Table C-30. Solutions of the compressible laminar boundary layer equations with mass transfer.

		Pr $= 1$, $f_0 = -0.5$			
g_w	β	$f''(0)$	$g'(0)$	J_1	J_2
0	2.00	0.377051	0.219120	0.181436	0.514180
	1.50	0.346127	0.212528	0.214891	0.523788
	1.00	0.306724	0.203061	0.267464	0.539258
	0.50	0.251170	0.187340	0.365439	0.568450
	0.10	0.177582	0.161092	0.551970	0.622385
	0	0.148476	0.148476	0.648475	0.648475
	−0.10	0.107576	0.127764	0.812115	0.688787
	−0.15	0.0763459	0.108649	0.965589	0.721186
	−0.182218	0.04	0.0802620	1.19202	0.757211
	−0.166002	0	0.0240131	1.59577	0.764901
	−0.15	−0.00256549	0.0146463	1.65285	0.745361
0.2	2.00	0.626345	0.201476	0.330563	0.465221
	1.50	0.551444	0.193831	0.381684	0.478918
	1.00	0.461821	0.182975	0.461328	0.500489
	0.50	0.345091	0.165090	0.609624	0.540277
	0.10	0.203161	0.134598	0.903277	0.612833
	0	0.148476	0.118781	1.07085	0.648476
	−0.10	0.0636623	0.0859621	1.44225	0.707887
	−0.120805	0	0.0452002	1.95036	0.735612
	−0.10	−0.0178339	0.0217904	2.29923	0.712089
0.6	2.00	1.051122	0.119208	0.594015	0.363094
	1.50	0.903704	0.113976	0.678207	0.386395
	1.00	0.730545	0.106540	0.808801	0.421745
	0.50	0.510810	0.0942068	1.05317	0.484224
	0.10	0.250844	0.0722696	1.56755	0.594089
	0	0.148476	0.0593905	1.91560	0.648476
	−0.05	0.0736739	0.0467523	2.29678	0.688512
	−0.0713794	0	0.0277190	2.99585	0.713841
	−0.05	−0.0290882	0.00955724	4.10371	0.676097
	−0.025	−0.0222629	0.00211564	5.43463	0.613603
2.0	1.00	1.49713	−0.348265	1.83953	0.157594
	0.50	0.989668	−0.301578	2.38247	0.298430
	0.10	0.396349	−0.213301	3.64551	0.531798
	0	0.148476	−0.148476	4.87221	0.648476
	−0.02	0.0693751	−0.117301	5.63158	0.682006
	−0.0286743	0	−0.077054	6.94215	0.699062
	−0.02	−0.0354595	−0.0278466	10.0281	0.665104

Table C–31. Solutions of the compressible laminar boundary layer equations with mass transfer.

$\text{Pr} = 1, \quad f_0 = 0.5$

g_w	β	$f''(0)$	$g'(0)$	J_1	J_2
0	2.00	1.136615	0.892052	0.168165	0.300286
	1.50	1.09135	0.887425	0.189832	0.306601
	1.00	1.036201	0.881332	0.220457	0.315744
	0.50	0.964436	0.872611	0.268247	0.330314
	0.10	0.883910	0.861700	0.333862	0.350524
	0	0.857016	0.857016	0.357916	0.357916
	−0.10	0.828150	0.853424	0.387214	0.366873
	−0.30	0.750714	0.840936	0.472082	0.392339
	−0.50	0.618538	0.816881	0.644209	0.440644
	−0.5427613	0	0.628429	0.93867	0.552237
	−0.621854	0.4	0.768613	0.996522	0.519692
	−0.620874	0.2	0.711301	1.39939	0.528459
	−0.50	−0.0580294	0.593572	2.15119	0.517563
	−0.30	−0.179435	0.437474	3.01131	0.223959
	−0.10	−0.140492	0.216278	3.73296	−0.267195
0.2	2.00	1.316190	0.724152	0.266797	0.282594
	1.50	1.236073	0.718968	0.296601	0.291085
	1.00	1.141809	0.712190	0.338591	0.303217
	0.50	1.024118	0.702551	0.403668	0.322284
	0.10	0.897709	0.690520	0.492981	0.348411
	0	0.857916	0.686333	0.525879	0.357916
	−0.10	0.812802	0.681337	0.566168	0.369420
	−0.30	0.696659	0.667189	0.685196	0.402218
	−0.50	0.490555	0.636949	0.952949	0.467028
	−0.575864	0.20	0.579252	1.46603	0.544235
	−0.5334580	0	0.520358	1.96266	0.546995
	−0.5	−0.065789	0.493446	2.17953	0.523975
	−0.3	−0.219716	0.358243	3.19811	0.239716
	−0.1	−0.177347	0.183286	4.31699	−0.245648
0.6	2.00	1.658569	0.371524	0.456966	0.244639
	1.50	1.513586	0.367701	0.503598	0.258186
	1.00	1.345832	0.362689	0.568786	0.277042
	0.50	1.140713	0.355513	0.669784	0.303819
	0.10	0.925114	0.346403	0.809535	0.344160
	0	0.857916	0.343167	0.861806	0.357916
	−0.10	0.781810	0.339239	0.926770	0.374487
	−0.30	0.583034	0.327460	1.129132	0.421779
	−0.4825059	0	0.271434	2.12593	0.525775
	−0.30	−0.282062	0.189966	3.53498	−0.278432
	−0.10	−0.236124	0.101743	5.31843	−0.204281

Table C–31. (cont.)

g_w	β	$f''(0)$	$g'(0)$	J_1	J_2
2.0	2.00	2.738069	−0.993835	1.07064	0.0967862
	1.50	2.396962	−0.977086	1.17694	0.131548
	1.00	2.004030	−0.954629	1.32606	0.177970
	0.50	1.525341	−0.921236	1.56044	0.245124
	0.10	1.019156	−0.875587	1.90089	0.329066
	0	0.857916	−0.857916	2.03755	0.357916
	−0.10	0.669882	−0.834736	2.22004	0.391886
	−0.30	0.0391211	−0.723004	3.13687	0.480183
	−0.3046783	0	−0.713045	3.22133	0.481470
	−0.30	−0.263049	−0.621915	4.02421	0.444213
	−0.10	−0.369937	−0.324281	7.87534	−0.0824039
	−0.025	−0.213097	−0.164164	13.00709	−0.387920

Table C-32. Solutions of the compressible laminar boundary layer equations with mass transfer.

		Pr $= 1$, $f_0 = 1$			
g_w	β	$f''(0)$	$g'(0)$	J_1	J_2
0	2.00	1.557032	1.306320	0.156565	0.243901
	1.50	1.509139	1.302845	0.173438	0.248981
	1.00	1.452183	1.298444	0.196205	0.255980
	0.50	1.380977	1.292518	0.229267	0.266343
	0.10	1.306284	1.285783	0.269997	0.279284
	0	1.283635	1.283635	0.283635	0.283635
	-0.10	1.258712	1.281214	0.299346	0.288649
	-0.30	1.199570	1.275232	0.339576	0.301443
	-0.50	1.120479	1.266724	0.399679	0.320318
	-0.921910	0.4	1.163304	1.18659	0.493933
	-0.822459	0	1.075035	1.80952	0.488257
	-0.50	-0.384412	0.899611	2.94723	0.089205
	-0.30	-0.430368	0.786864	3.69633	-0.321470
	-0.10	-0.349190	0.614215	4.91190	-0.85800
0.2	2.00	1.713050	1.051690	0.240369	0.232318
	1.50	1.633658	1.047850	0.263091	0.239023
	1.00	1.541743	1.043016	0.293577	0.248165
	0.50	1.430367	1.036551	0.337622	0.261556
	0.10	1.317273	1.029273	0.391705	0.278103
	0	1.283635	1.026908	0.409812	0.283635
	-0.10	1.246925	1.024281	0.430691	0.289997
	-0.20	1.160888	1.017777	0.484330	0.306187
	-0.50	1.047345	1.008428	0.565420	0.330055
	-0.872426	0.4	0.937257	1.229074	0.472278
	-0.8162883	0	0.871091	1.81935	0.485111
	-0.50	-0.433597	0.720769	1.90951	0.475751
	-0.30	-0.474743	0.627535	3.86977	-0.313813
	-0.10	-0.377785	0.488044	5.26165	-0.851620
0.6	2.00	2.016454	0.532082	0.404387	0.207680
	1.50	1.876818	0.529211	0.439167	0.218067
	1.00	1.717543	0.525596	0.485609	0.231936
	0.50	1.528002	0.520747	0.552462	0.251771
	0.10	1.339185	0.515226	0.634575	0.275728
	0	1.283635	0.513454	0.662167	0.283635
	-0.10	1.22327	0.511446	0.694098	0.292676
	-0.30	1.082418	0.506407	0.777012	0.315522
	-0.50	0.895762	0.498894	0.906483	0.349004
	-0.7746315	0	0.444047	1.89620	0.468856
	-0.50	-0.521139	0.363115	3.28311	0.120414
	-0.30	-0.558717	0.313309	4.21486	-0.294260
	-0.10	-0.432379	0.242403	5.95612	-0.836768

Table C-32. (cont.)

g_w	β	$f''(0)$	$g'(0)$	J_1	J_2
2.0	2.00	3.007734	−1.376952	0.948607	0.110521
	1.50	2.678218	−1.363428	1.027705	0.136658
	1.00	2.303601	−1.346045	1.13345	0.170155
	0.50	1.858997	−1.322003	1.28670	0.215650
	0.10	1.415199	−1.293273	1.47901	0.267298
	0	1.283635	−1.283635	1.54540	0.283633
	−0.10	1.139516	−1.272385	1.62390	0.301906
	−0.30	0.794032	−1.241972	1.84040	0.346152
	−0.555091	0	−1.144072	2.55817	0.420017
	−0.573864	−0.3	−1.087995	2.97389	0.406611
	−0.30	−0.793845	−0.807186	5.33775	−0.192520
	−0.10	−0.594141	−0.615030	8.20257	−0.773883
	−0.025	−0.368227	−0.478004	12.63365	−1.05239

Table C–33. Mass transfer solutions for the hypersonic similar compressible boundary layer equations.

$\bar{\sigma} = 2.0, \ \bar{\beta} = 0.5, \ \mathrm{Pr} = 0.723$

g_w	f_0	$f''(0)$	$g'(0)$	J_1	J_2
0	−1.0	0.0198032	0.012234	0.399454	0.820076
	−0.5	0.237331	0.148286	0.328431	0.573034
	0	0.564591	0.367826	0.277945	0.425583
	0.5	0.949560	0.637478	0.238369	0.330301
	1.0	1.368889	0.937816	0.206484	0.265564
0.2	−1.0	0.133834	0.0203313	0.774758	0.746449
	−0.5	0.335770	0.117428	0.59251	0.539512
	0	0.645825	0.285286	0.478214	0.406675
	0.5	1.01795	0.497690	0.398204	0.318743
	1.0	1.42744	0.736992	0.338363	0.258167
0.6	−1.0	0.306946	−0.0077009	1.372826	0.620342
	−0.5	0.508761	0.0308990	1.072729	0.472347
	0	0.799139	0.107306	0.862178	0.368005
	0.5	1.15083	0.210864	0.711253	0.295116
	1.0	1.54273	0.331014	0.599168	0.243043
1.0	−1.0	0.456928	−0.0603066	1.89559	0.509138
	−0.5	0.663626	−0.0772162	1.51416	0.406541
	0	0.943144	−0.0847436	1.22877	0.328707
	0.5	1.27918	−0.0846900	1.01655	0.270914
	1.0	1.65576	−0.0804596	0.856358	0.227585
2.0	−1.0	0.787134	−0.252754	3.04730	0.263483
	−0.5	1.00801	−0.408147	2.51842	0.248804
	0	1.27508	−0.611937	2.09072	0.229650
	0.5	1.584273	−0.856165	1.751568	0.208380
	1.0	1.929838	−1.130914	1.484826	0.187340

Table C-34. Results for hypersonic strong interaction on a sharp flat plate.

			Pr = 1				
			$\gamma = 1.4$, $\hat{\beta} = 0.286$				
g_w	$f''(0)$	$g'(0)$	J_1	J_2	J_4	A	B
0	0.541983	0.486140	0.362838	0.438211	1.54221	0.141087	0.386392
0.2	0.588367	0.394581	0.559185	0.428441	1.49102	0.217436	0.479678
0.6	0.677625	0.202466	0.940457	0.408653	1.39600	0.365692	0.622073
1.0	0.762989	0	1.30886	0.388654	1.30886	0.508944	0.733870
2.0	0.963452	−0.543364	2.18601	0.338252	1.11592	0.850020	0.948415
			$\gamma = 1.67$, $\hat{\beta} = 0.4$				
0	0.563930	0.490771	0.335029	0.429917	1.50457	0.249716	0.446960
0.2	0.625440	0.399723	0.519772	0.417532	1.44101	0.387415	0.556716
0.6	0.742905	0.206287	0.876509	0.342300	1.32464	0.653311	0.722945
1.0	0.854421	0	1.21932	0.366691	1.21932	0.908831	0.852681
2.0	1.11422	−0.560461	2.03058	0.301985	0.989253	1.51351	1.10037

APPENDIX D

GENERAL BOUNDARY LAYER PROGRAM

D-1 Introduction

The General Boundary Layer Program (GBLP)[†] is designed to solve any two-point asymptotic boundary value problem of the boundary layer type. It may have application to more general asymptotic boundary value problems. A fixed-step size fourth-order Runge-Kutta numerical integration scheme and a Nachtsheim-Swigert iteration scheme are used. GBLP is modular in nature. Basically, the user need only reduce the governing equations to a system of first-order ordinary differential equations (see Appendix A) and write modules to describe the equations. If desired, a routine to provide an explanation for the specific problem is incorporated. Combined with a driver program for the specific problem, GBLP and these modules provide a complete system for solving two-point asymptotic boundary value problems.

D-2 Discussion of GBLPdriv Routine

The General Boundary Layer Program driver routine is shown below using pseudocode. For details of the pseudocode, see Rogers [Roge85] or Rogers and Adams [Roge90]. A complete listing of all the variables in the routine is given. The routine is designed to be called by a main program for the specific problem under consideration. The GBLP routine is quite simple, consisting mainly of calls to other routines. After initializing for systems of equations up to order 20, array variables are redimensioned for the specific problem at hand, the convergence and derivative flags set and the independent variable η initialized to zero. Routines that provide an explanation, get the parameters and initial conditions and output a summary of the conditions for the problem are called. The routine **header** outputs the table headings for the solution. The main integration and iteration loop is then entered. Integration and iteration continue until convergence is achieved (`cflag > 0`).

gblpdriv

    general driver program for solving two-point asymptotic boundary value problems

    routines called: cond, explain, header, initial, ns, parameter, rk

    variables:

    asymbc()   = *array containing the asymptotic boundary conditions*

[†] The general boundary layer program and all the subroutines implemented in True Basic, along with a limited version of the True Basic compiler, is available at modest cost from True Basic Inc., West Lebanon, NH 03784-9758, telephone 1-800-TR-BASIC.

base()	=	has dimensions asymbc(2*nasymbx) array containing the saved values of the dependent variables at the edge of the boundary layer has dimensions of base(m)
cflag	=	used to control calculation of new values for the unknown initial conditions 0 = convergence achieved 1 = convergence not achieved, calculate new values
con()	=	array containing control values relating dependent variables and boundary conditions has dimensions of con(m)
dperm	=	amount by which the first guess of the initial conditions is perturbed to find the derivatives of the dependent variables at the edge of the boundary layer w.r.t. changes in the unknown initial conditions — suggest 0.001
deriv(,)	=	array containing the derivatives of the dependent variables at the edge of the boundary layer w.r.t. changes in the unknown initial conditions has dimensions of deriv(nasymbc, 2*nasymbc)
dflag	=	used to control the calculation of the derivatives of the dependent variables at the edge of the boundary layer w.r.t. changes in the unknown initial conditions
err()	=	array containing the required error condition values has dimensions of err(m) — suggest 5×10^{-6}
etamax	=	the maximum value of the independent variable — the 'edge' of the boundary layer — suggest 6.0
h	=	Runge-Kutta fixed-step size — suggest 0.01
headernam()	=	character array containing the names of the dependent variables has dimensions of headernam(m)
init()	=	array containing the initial conditions has dimensions of init(m)
initnam()	=	character array containing the names of the initial conditions has dimensions of initnam(m)
interval	=	print interval for the dependent variables in multiples of h
m	=	order of the system of equations
nasymbc	=	number of asymptotic boundary conditions
nparam	=	number of parameters in the equations
oflag	=	output flag 1 = on 0 = off
param()	=	array containing the parameter values for the equations has dimensions of param(m)
paramnam()	=	character array containing the names of the parameter values has dimensions of paramnam(m)
x()	=	array containing the current values of the dependent variables as the integration proceeds. has dimensions of x(m)

subroutine gblpdriv(m,con(),err(),nasymbc,asymbc(),init(),initnam(),nparam, paramnam(),headnam(),h,dper,interval,oflag,etamax)

dimension for up to a system of order 20
dimension param(20),x(20),base(20),deriv(20,20)
redimension for the specific problem at hand
Mat param = **Zer**(m)
Mat x = **Zer**(m)
Mat base = **Zer**(m)
Mat deriv = **Zer**(nasymbc,2∗nasymbc)
set the initial values of the error and derivative flags
dflag = 1
cflag = 1
initialize the independent variable
eta = 0
call explain   *provide an explanation*
call parameter(nparam,paramnam,param)   *get the parameter values*
call initial(nasymbc,con,initnam,init)   *get the initial guesses*
call condition(nparam,paramnam,param,nasymbc,initnam,init,con)   *output*
call header(m,headernam,eta,x)   *output the header*
while cflag > 0
    call rk(m,h,etamax,param(),init(),oflag,interval,eta,x())   *integrate & iterate*
    call ns(m,nasymbc,asymbc,con,dper,cflag,dflag,err,base,deriv,x,initnam,init)
end while
return

D–3 The explain Routine

The **explain** routine shown below for illustrative purposes is very simple. More complex and complete explanations are constructed in a similar manner.

explain
provide an explanation for the specific problem for the general driver program for solving two-point asymptotic boundary value problems

the lcase function converts the input string to lower case
[1:1] extracts the first character from the string

subroutine explain
print "Do you want a program explanation ";
input achar
if lcase(achar[1:1]) = "y" **then**
    print "Falkner-Skan equation"
    print
end if
return

D–4 The parameter Routine

The purpose of the parameter routine is to obtain the parameters for the governing two-point asymptotic boundary value problem. Again, the illustrative routine here is quite simple. More elaborate routines are easily constructed.

parameter
> *to get the parameters for the general driver program
> for solving two-point asymptotic boundary value problems*
>
> *variables:*
>
> nparam   = *number of parameters in the equations*
> param()   = *array containing the parameter values for the equations
>   has dimensions of param(m)*
> paramnam() = *array containing the names of the parameter values
>   has dimensions of paramnam(m)*
>
> **subroutine** parameter(nparam, paramnam(), param())
>
> **for** i = 1 to nparam
>   Print "What is ";paramnam(i)
>   **input** param(i)
> **next** i
>
> **return**

D–5 The initial Routine

This routine acquires the initial conditions for the boundary value problem.

initial
> *to get the initial conditions for the general two-point asymptotic
> boundary value program*
>
> *variables:*
>
> con()   = *array containing control values relating dependent
>   variables and boundary conditions
>   has dimensions of con(m)*
> init()   = *array containing the initial conditions
>   has dimensions of init(m)*
> initnam() = *array containing the names of the initial conditions
>   has dimensions of initnam(m)*
> nasymbc   = *number of asymptotic boundary conditions*
>
> **subroutine** initial (nasymbc,con(),initnam(),init())
>
> **for** i = 1 to nasymbc
>   print "What is ";initnam(con(i))
>   **input** init(con(i))
> **next** i
>
> **return**

D–6 The condition Routine

The **condition** routine outputs the information from the **parameter** and **initial** routines to document the exact problem solved.

condition
> *outputs the problem conditions for the GBLP*

variables:

con() = array containing control values relating dependent variables and boundary conditions
has dimensions of con(m)
init() = array containing the initial conditions
has dimensions of init(m)
initnam() = array containing the names of the initial conditions
has dimensions of initnam(m)
nasymbc = number of asymptotic boundary conditions
nparam = number of parameters in the equations
param() = array containing the parameter values for the equations
has dimensions of param(m)
paramnam() = array containing the names of the parameter values
has dimensions of paramnam(m)

subroutine condition(nparam,paramnam(),param(),nasymbc,initnam(), init(),con())

print

for i = 1 to nparam
   print paramnam(i);" = ";param(i)
next i

for i = 1 to nasymbc
   print initnam(con(i));" = ";init(con(i))
next i

print

return

D–7 The header Routine

The **header** routine generates the table headings for the solution.

header

header routine for the general boundary layer program used to output a table header for the two-point asymptotic boundary value problem

variables:

eta = the independent variable
headernam() = character array containing the names
of the dependent variables
has dimensions of headernam(m)
m = order of the system of equations
x() = array containing the current values of the dependent variables as the integration proceeds.
has dimensions of x(m)

subroutine header(m,headernam(),eta,x())

for i = 1 to m+1
   print headernam(i),
next i

print

return

358   Laminar Flow Analysis

D-8 The Runge-Kutta rk Numerical Integration Routine

The **rk** routine implements the fourth-order fixed-step size Runge-Kutta numerical integration scheme discussed in Appendix A.

rk

    to perform fixed-step size fourth-order Runge-Kutta numerical integration on the set of simultaneous first-order equations described in the subroutine **eqmot**

    an output subroutine must also be supplied

    routines called: eqmot, output

    variables:

eta	=	*independent integration variable*
etamax	=	*maximum value of the independent variable — the 'edge' of the boundary layer — suggest 6.0*
h	=	*Runge-Kutta fixed-step size — suggest 0.01*
init()	=	*array containing the initial conditions has dimensions of init(m)*
interval	=	*output interval for the dependent variables in multiples of h*
m	=	*order of the system of equations*

oflag	=	*output flag*
		  0 = no output
		  1 = output

param()	=	*array containing the parameter values for the equations has dimensions of param(m)*
x()	=	*array containing the current values of the dependent variables as the integration proceeds has dimensions of x(m)*

subroutine rk(m,h,etamax,param(),init(),oflag,interval,eta,x())

dimension for up to a system of order 20

dimension k(20),l(20),f(20),r(20),n(20)

redimension for the specific problem at hand

Mat k = **Zer**(m)
Mat l = **Zer**(m)
Mat r = **Zer**(m)
Mat n = **Zer**(m)
Mat f = **Zer**(m)

eta = 0   *set the values of η etc.*
icount = 0
tcount = interval
for i = 1 to m   *set initial conditions*
    x(i) = init(i)
next i
call output(m,eta,x())   *output eta = 0 values*
while eta < (etamax − h)
    for ncount = 1 to 4   *fourth order Runge-Kutta integration*
        call eqmot(eta,x(),param(),f())
        if ncount = 1 **then**   *select the section for the Runge-Kutta*

                    for i = 1 to m   *calculate k's*
                       k(i) = h∗f(i)
                       x(i) = x(i) + 0.5∗k(i)
                    next i
                 else if ncount = 2 **then**
                    for i = 1 to m   *calculate l's*
                       l(i) = h∗f(i)
                       x(i) = x(i) − 0.5∗k(i) + 0.5∗l(i)
                    next i
                 else if ncount = 3 **then**
                    for i = 1 to m   *calculate r's*
                       r(i) = h∗f(i)
                       x(i) = x(i) − 0.5∗l(i) + r(i)
                    next i
                 else if ncount = 4 **then**
                    for i = 1 to m   *calculate n's & new dependent variables*
                       n(i) = h∗f(i)
                       x(i) = x(i) − r(i) + (k(i) + 2∗l(i) + 2∗r(i) + n(i))/6
                    next i
                 end if
              next ncount
              eta = eta + h   *step integration variable*
              icount = icount + 1
              if oflag = 1 **then**   *output ?*
                 if tcount = icount **then**
                    call output(m,eta,x())
                    tcount = icount + interval
                 end if
              end if
           end while
           if oflag = 0 **then call** output(m,eta,x())
           return

D–9 The Nachtsheim-Swigert ns Routine

The **ns** routine implements the Nachtsheim-Swigert iteration scheme for asymptotic boundary conditions as discussed in Appendix B.

ns

   Nachtsheim-Swigert iteration scheme to determine the unknown initial conditions for two-point asymptotic boundary value problems

   variables:

   asymbc() = *array containing the asymptotic boundary conditions*
               has dimensions asymbc(2∗nasymbx)
   base()   = *array containing the saved values of the dependent*
               variables at the edge of the boundary layer
               has dimensions of base(m)

cflag	=	used to control calculation of new values for the unknown initial conditions
		0 = convergence achieved
		1 = convergence not achieved calculate new values
con()	=	array containing control values relating dependent variables and boundary conditions
		has dimensions of con(m)
dperm	=	amount by which the first guess of the initial conditions is perturbed to find the derivatives of the dependent variables at the edge of the boundary layer w.r.t. changes in the unknown initial conditions — suggest 0.001
deriv(,)	=	array containing the derivatives of the dependent variables at the edge of the boundary layer w.r.t. changes in the unknown initial conditions
		has dimensions of deriv(nasymbc, 2*nasymbc)
dflag	=	used to control the calculation of the derivatives of the dependent variables at the edge of the boundary layer w.r.t. changes in the unknown initial conditions
err()	=	array containing the required error condition values
		has dimensions of err(m) — suggest 5×10^{-6}
init()	=	array containing the initial conditions
		has dimensions of init(m)
initnam()	=	array containing the names of the initial conditions
		has dimensions of initnam(m)
m	=	order of the system of equations
nasymbc	=	number of asymptotic boundary conditions
x()	=	array containing the current values of the dependent variables as the integration proceeds
		has dimensions of x(m)

subroutine ns(m,nasymbc,asymbc(),con(),dper,cflag,dflag,err(),base(),deriv(,), x(),initnam(),init())

dimension for up to a system of order 20

dimension q(20,20),invq(20,20),rhs(20,1),delta(20,1)

redimension for the specific problem at hand

Mat q = **Zer**(nasymbc,nasymbc)
Mat invq = **Zer**(nasymbc,nasymbc)
Mat rhs = **Zer**(nasymbc,1)
Mat delta = **Zer**(nasymbc,1)

cflag = 0
for i = 1 to 2*nasymbc   *check for convergence*
    if (abs(asymbc(i) − x(con(i)))) > err(i) **then** cflag = 1
next i

if cflag = 0 **then**
    print
    print "Convergence achieved"
else
    if dflag <= nasymbc **then**
        if dflag = 1 **then**

```
            for i = 1 to m
               base(i) = x(i)
            next i
            init(con(dflag)) = init(con(dflag)) + dper
         else
            for i = 1 to 2*nasymbc
               deriv(dflag−1,i) = (x(con(i)) − base(con(i)))/dper
            next i
            init(con(dflag−1)) = init(con(dflag−1))−dper
            init(con(dflag)) = init(con(dflag)) + dper
         end if
      else
         if dflag = nasymbc + 1 then
            for i = 1 to 2*nasymbc
               deriv(dflag−1,i) = (x(con(i)) − base(con(i)))/dper
            next i
            init(con(dflag−1)) = init(con(dflag−1)) − dper
            for i = 1 to m
               x(i) = base(i)
            next i
         end if
         if cflag = 1 then                           calculate corrections
            for i = 1 to nasymbc
               for j = 1 to nasymbc
                  for i1 = 1 to 2*nasymbc
                     q(i,j) = q(i,j) + deriv(i,i1)*deriv(j,i1)
                  next i1
               next j
            next i
            for i = 1 to nasymbc
               for i1 = 1 to 2*nasymbc
                  rhs(i,1) = rhs(i,1) + (asymbc(i1) − x(con(i1)))*deriv(i,i1)
               next i1
            next i
            Mat invq = inv(q)
            Mat delta = invq*rhs
            for i = 1 to nasymbc
               init(cons(i)) = init(con(i)) + delta(i,1)
            next i
         end if
      end if
      dflag = dflag + 1
      print
      for i = 1 to nasymbc
         print initnam(con(i));" = ";init(con(i))
      next i
      print
   end if
   return
```

D-10 The output Routine

The output routine generates a single entry for the solution table.

output

    create a table of the solution for a two-point asymptotic boundary value problem

    variables:

    eta = *the independent variable*
    m =   *order of the system of equations*
    x() = *array containing the current values of the dependent*
           variables as the integration proceeds
           has dimensions of x(m)

subroutine output(m,eta,x())

print eta,

for i = 1 to m

    print x(i),

next i

print

return

D-11 The General Form of the Main Program

The main program that calls the general boundary layer program driver is constructed using the following outline:

    Set the overall array dimensions.

    Set the order of the system of equations.

    Redimension the arrays for the specific problem at hand.

    Set the Runge-Kutta integration step size.

    Set the perturbation value for the unknown initial conditions used to determine the dependent variable derivatives at the boundary layer edge.

    Assign a finite value of the independent variable to specify the 'edge' of the boundary layer, η_{max}.

    Assign values to control array elements (see the discussion below for more details).

    Assign error criteria for the asymptotic dependent variables at the edge of the boundary layer.

    Fill in the names of the independent and dependent variables for the table header.

    Set the number of parameters and their names.

    Set the number of asymptotic boundary conditions and assign values at the 'edge' of the boundary layer.

    Fill in the names of the initial conditions.

    Set the output flag and the output interval.

    Call the GBLPdriv routine.

In the main program, the values assigned to the control array **con()** are critical.

The **rk** and **ns** routines assume that the highest-order derivative of the dependent variable for the first equation is assigned to **x(1)**, with each lower-order derivative assigned to succeeding elements of the array until the dependent variable itself is reached. The next element of the array is assigned to the highest-order derivative of the second equation in the coupled set of equations, with succeeding elements of the array assigned to successively lower-order derivatives until the dependent variable for the second equation is reached. Additional equations are handled in the same manner.

Since asymptotic boundary conditions can be associated with any of the dependent variables, a control array is used. The first element of the control array is associated with the first unknown initial condition for the first equation in the system of equations. Its value is the **x()** array element number for that dependent variable. The second element of the control array is associated with the second unknown initial condition (if any) for the first equation in the system of equations. Its value is the **x()** array element number for the appropriate dependent variable. Additional unknown initial conditions (if any) for the *first* equation are assigned to succeeding elements of the control array in like manner, until all are accounted for.

The next element in the control array is associated with the first unknown initial condition for the *second* equation in the system of equations. Its value is the **x()** array element number for that dependent variable. Additional unknown initial conditions (if any) for the second equation are assigned to succeeding elements of the control array in like manner, until all are accounted for. Unknown initial conditions for additional equations (if any) are handled similarly.

When all the unknown initial conditions are accounted for, the asymptotic boundary conditions at the edge of the boundary layer are considered. The next element in the control array is associated with the first asymptotic boundary condition for the first equation in the system of equations. Its value is the **x()** array element number for that dependent variable. Succeeding asymptotic boundary conditions for the first and all subsequent equations in the system are handled similarly.

The examples discussed in succeeding sections help to clarify the system.

D–12 Main Program for Two-Dimensional Stagnation Point Flow

The momentum equation for two-dimensional stagnation point flow is governed by (see Section 2–7), i.e.

$$f''' + f f'' + (1 - f'^2) = 0 \qquad (2-65)$$

with boundary conditions

$$\eta = 0; \quad f(0) = f'(0) = g(0) = 0 \qquad (2-67)$$

$$\eta \to \infty; \quad f'(\eta) \to 1 \qquad (2-68)$$

Writing Eq. (2–65) as three first-order simultaneous equations yields

$$x(1) = f''(\eta) \qquad (A-1a)$$

$$x(2) = f'(\eta) \qquad (A-1b)$$

$$x(3) = f(\eta) \qquad (A-1c)$$

364   Laminar Flow Analysis

The Runge-Kutta functions then become

$$\frac{dx(1)}{d\eta} = F(1) = -ff'' - (1 - f'^2)$$
$$= -x(3)x(1) - (1 - x(2)x(2)) \quad (A-2a)$$
$$\frac{dx(2)}{d\eta} = F(2) = x(1) \quad (A-2b)$$
$$\frac{dx(3)}{d\eta} = F(3) = x(2) \quad (A-2c)$$

There is one asymptotic boundary condition and hence one unknown initial condition, $f''(0) - \mathtt{x(1)}$. The system is third-order, and there are no parameters. Thus, $\mathtt{m} = 3$, $\mathtt{nasymbc} = 1$ and $\mathtt{nparms} = 0$. The asymptotic boundary condition $f'(\eta \to \infty) \to 1$ implies that $f''(\eta \to \infty) \to 0$. Hence, $\mathtt{con(1)} = 1$ and $\mathtt{con(2)} = 2$, since the dependent variables, f'' and f', correspond to $\mathtt{x(1)}$ and $\mathtt{x(2)}$, respectively. The asymptotic boundary conditions are imposed using $\mathtt{asymbc(1)} = 0$ and $\mathtt{asymbc(2)} = 1$. We choose the edge of the boundary layer as $\eta_{\max} = 6$ and impose error conditions on $f''(\eta \to \infty)$ of $\pm 5 \times 10^{-6}$ and on $f'(\eta \to \infty) - 1$ of $\pm 5 \times 10^{-6}$.

It is convenient to make a table of these values:

Variable	Control Array	Control Value	Asymptotic Value	Error Value
$f''(0)$	con(1)	1	asymbc(1)=0	err(1)= 5×10^{-6}
$f'(\eta_{\max})$	con(2)	2	asymbc(2)=1	err(2)= 5×10^{-6}

The complete main program for the solution of this boundary value problem is

stag2d

Main program for solution of two-dimensional stagnation point flow variables:

asymbc()	=	*array containing the asymptotic boundary conditions has dimensions asymbc(2∗nasymbx)*
base()	=	*array containing the saved values of the dependent variables at the edge of the boundary layer has dimensions of base(m)*
cflag	=	*used to control calculation of new values for the unknown initial conditions* *0 = convergence achieved* *1 = convergence not achieved calculate new values*
con()	=	*array containing control values relating dependent variables and boundary conditions has dimensions of con(m)*
dperm	=	*amount by which the first guess of the initial conditions is perturbed to find the derivatives of the dependent variables at the edge of the boundary layer w.r.t. changes in the unknown initial conditions — suggest 0.001*

General Boundary Layer Program   365

$deriv(,)$	=	array containing the derivatives of the dependent variables at the edge of the boundary layer w.r.t. changes in the unknown initial conditions has dimensions of deriv(nasymbc, 2∗nasymbc)
$dflag$	=	used to control the calculation of the derivatives of the dependent variables at the edge of the boundary layer w.r.t. changes in the unknown initial conditions
$err()$	=	array containing the required error condition values has dimensions of err(m) — suggest 5×10^{-6}
$etamax$	=	the maximum value of the independent variable — the 'edge' of the boundary layer — suggest 6.0
h	=	Runge-Kutta fixed-step size — suggest 0.01
$headernam()$	=	character array containing the names of the dependent variables has dimensions of headernam(m)
$init()$	=	array containing the initial conditions has dimensions of init(m)
$initnam()$	=	character array containing the names of the initial conditions has dimensions of initnam(m)
$interval$	=	print interval for the dependent variables in multiples of h
m	=	order of the system of equations
$nasymbc$	=	number of asymptotic boundary conditions
$nparam$	=	number of parameters in the equations
$oflag$	=	output flag 1 = on 0 = off
$param()$	=	array containing the parameter values for the equations has dimensions of param(m)
$paramnam()$	=	character array containing the names of the parameter values has dimensions of paramnam(m)
$x()$	=	array containing the current values of the dependent variables as the integration proceeds. has dimensions of x(m)

dimension the arrays for a general system of order 20
dimension con(20),err(20),asymbc(20),init(20)
dimension headnam(20),initnam(20),paramnam(20)

m=3   *the order of the system is 3*

redimension the arrays for the current problem

Mat con = **Zer**(m)
Mat err = **Zer**(m)
Mat asymbc = **Zer**(2∗m)
Mat init = **Zer**(m)
h=.01   *set the step size for the Runge-Kutta integration*
dper=.001   *set the perturbation value for unknown initial condition to calculate the b.l. edge derivatives*
etamax = 6   *set the value of the independent variable at the edge of the boundary layer*
con(1) = 1   *assign the control variables*
con(2) = 2

366 Laminar Flow Analysis

```
err(1)=.000005                    assign the error values for the dependent variables
err(2)=.000005                        at the edge of the boundary layer
headnam(1) = "eta"                fill in the names of the independent
headnam(2) = " f''"                   and dependent variables
headnam(3) = " f'"
headnam(4) = " f"
nparms=0                          set the number of parameters and assign their names
paramnam(1) = ""                  a null value is assigned for the parameter name
nasymbc=1                         set the number of asymptotic boundary conditions
asymbc(1)=0                           and assign their values
asymbc(2)=1
initnam(1)="f''(0)"               assign the names of the initial conditions
oflag = 0                         set the output flag and output interval
interval = 50
call the general boundary layer driver program
call gblpdriv(m,con,err,nasymbc,asymbc,init,initnam,nparms,paramnam,
                                  headnam,h,dper,interval,oflag,etamax)
```
finish

The **eqmot** routine is

eqmot2d

equations of motion routine for two-dimensional stagnation point flow
subroutine eqmot(eta, x(), param(), f())
f(1) = −x(3)∗x(1) − (1 − x(2)∗x(2))
f(2) = x(1)
f(3) = x(2)
return

To complete the program a simple explanation routine is given by

explain2d

simple explanation routine for two-dimensional stagnation point flow
the lcase function converts the input string to lower case
[1:1] extracts the first character from the string
subroutine explain
print "Do you want a program explanation ";
input achar
if lcase(achar[1:1]) = "y" **then**
   print "Two-dimensional stagnation point flow"
   print
end if
return

D–13 Main Program for the Falkner-Skan Equation

The incompressible steady two-dimensional constant property viscous boundary layer is governed by the Falkner-Skan equation (see Section 3–4)

$$f''' + ff'' + \beta(1 - f'^2) = 0 \qquad (3-50)$$

with $\alpha = 1$ and boundary conditions

$$f(0) = f'(0) = 0 \qquad (3-51a,b)$$

$$f(\eta \to \infty) \to 1 \qquad (3-51c)$$

Here we see that the two-dimensional stagnation point flow is a special case of the Falkner-Skan equation, with $\beta = 1$. The solution is thus very similar to that of the previous section. Equation (3–50) is written as three first-order simultaneous equations

$$x(1) = f''(\eta) \qquad (A-3a)$$

$$x(2) = f'(\eta) \qquad (A-3b)$$

$$x(3) = f(\eta) \qquad (A-3c)$$

The Runge-Kutta functions now become

$$\frac{dx(1)}{d\eta} = F(1) = -ff'' - \beta(1 - f'^2)$$

$$= -x(3)x(1) - param(1)(1 - x(2)x(2)) \qquad (A-4a)$$

$$\frac{dx(2)}{d\eta} = F(2) = x(1) \qquad (A-4b)$$

$$\frac{dx(3)}{d\eta} = F(3) = x(2) \qquad (A-4c)$$

where **param(1)** is the parameter, β.

Again, there is one asymptotic boundary condition and hence one unknown initial condition, $f''(0) = $ **x(1)**. The order of the system is three, but here there is one parameter, β. Thus, **m** = 3, **nasymbc** = 1 and **nparms** = 1. The asymptotic boundary condition $f'(\eta \to \infty) \to 1$ implies that $f''(\eta \to \infty) \to 0$. Thus, **con(1)** = 1 and **con(2)** = 2, since the dependent variables, f'' and f', correspond to **x(1)** and **x(2)**, respectively. The asymptotic boundary conditions are imposed using **asymbc(1)** = 0 and **asymbc(2)** = 1. We choose the edge of the boundary layer as $\eta_{max} = 6$ and impose error conditions on $f''(\eta \to \infty) = \pm 5 \times 10^{-6}$ and on $f'(\eta \to \infty) - 1 = \pm 5 \times 10^{-6}$.

Again, it is convenient to make a table of these values:

Variable	Control Array	Control Value	Asymptotic Value	Error Value
$f''(0)$	con(1)	1	asymbc(1)=0	err(1)= 5×10^{-6}
$f'(\eta_{max})$	con(2)	2	asymbc(2)=1	err(2)= 5×10^{-6}

The complete main program for the solution of the Falkner-Skan equation is:

faskan

Main program for solution of the Falkner-Skan equation
routines called: gblpdriv

variables:

asymbc()	=	array containing the asymptotic boundary conditions has dimensions asymbc(2∗nasymbx)
base()	=	array containing the saved values of the dependent variables at the edge of the boundary layer has dimensions of base(m)

cflag	=	used to control calculation of new values for the unknown initial conditions
		    0 = convergence achieved
		    1 = convergence not achieved calculate new values

con()	=	array containing control values relating dependent variables and boundary conditions has dimensions of con(m)
dperm	=	amount by which the first guess of the initial conditions is perturbed to find the derivatives of the dependent variables at the edge of the boundary layer w.r.t. changes in the unknown initial conditions — suggest 0.001
deriv(,)	=	array containing the derivatives of the dependent variables at the edge of the boundary layer w.r.t. changes in the unknown initial conditions has dimensions of deriv(nasymbc, 2∗nasymbc)
dflag	=	used to control the calculation of the derivatives of the dependent variables at the edge of the boundary layer w.r.t. changes in the unknown initial conditions
err()	=	array containing the required error condition values has dimensions of err(m) — suggest 5×10^{-6}
etamax	=	the maximum value of the independent variable — the 'edge' of the boundary layer — suggest 6.0
h	=	Runge-Kutta fixed step size — suggest 0.01
headernam()	=	character array containing the names of the dependent variables has dimensions of headernam(m)
init()	=	array containing the initial conditions has dimensions of init(m)
initnam()	=	character array containing the names of the initial conditions has dimensions of initnam(m)
interval	=	print interval for the dependent variables in multiples of h
m	=	order of the system of equations
nasymbc	=	number of asymptotic boundary conditions
nparam	=	number of parameters in the equations

oflag	=	output flag
		    1 = on
		    0 = off

param()	=	array containing the parameter values for the equations has dimensions of param(m)
paramnam()	=	character array containing the names of the parameter values has dimensions of paramnam(m)
x()	=	array containing the current values of the dependent variables as the integration proceeds. has dimensions of x(m)

General Boundary Layer Program

dimension the arrays for a general system of order 20
dimension con(20),err(20),asymbc(20),init(20)
dimension headnam(20),initnam(20),paramnam(20)

m=3   *the order of the system is 3*

redimension the arrays for the current problem
Mat con = **Zer**(m)
Mat err = **Zer**(m)
Mat asymbc = **Zer**(2*m)
Mat init = **Zer**(m)

h=.01   *set the step size for the Runge-Kutta integration*
dper=.001   *set the perturbation value for unknown initial*
                               condition to calculate the b.l. edge derivatives

etamax = 6   *set the value of the independent variable at the*
                               edge of the boundary layer

con(1) = 1   *assign the control variables*
con(2) = 2

err(1)=.000005   *assign the error values for the dependent variables*
err(2)=.0000005   *at the edge of the boundary layer*

headnam(1) = "eta"   *fill in the names of the independent*
headnam(2) = " f'''"   *and dependent variables*
headnam(3) = " f'"
headnam(4) = " f"

nparms=1   *set the number of parameters and assign their names*
paramnam(1) = "Beta"

nasymbc=1   *set the number of asymptotic boundary conditions*
asymbc(1)=0   *and assign their values*
asymbc(2)=1

initnam(1)="f''(0)"   *assign the names of the initial conditions*

oflag = 0   *set the output flag and output interval*
interval = 50

call the general boundary layer driver program
call gblpdriv(m,con,err,nasymbc,asymbc,init,initnam,nparms,paramnam,
                  headnam,h,dper,interval,oflag,etamax)
finish

The equation of motion routine is

eqmotfs

equations of motion routine for the Falkner-Skan equation
subroutine eqmot(eta, x(), param(), f())
f(1) = −x(3)*x(1) − param(1)*(1 − x(2)*x(2))
f(2) = x(1)
f(3) = x(2)
return

To complete the program a simple explanation routine is given by

explainfs
> *simple explanation routine for the Falkner-Skan equation*
> *the lcase function converts the input string to lower case*
> *[1:1] extracts the first character from the string*
> **subroutine** explain
> print "Do you want a program explanation "
> input achar
> **if** lcase(achar[1:1]) = "y" **then**
>   print "Falkner-Skan equation"
>   print
> **end if**
> **return**

D-14 Forced Convection Boundary Layer

The governing equations for the forced convection boundary layer are (see Eqs. 3–50 and 4–111, and Section 4–9)

$$f''' + ff'' + \beta(1 - f'^2) = 0 \qquad (3-50)$$

$$\theta'' + \Pr f\theta' + n\Pr(2-\beta)(1-\theta)f' = 0 \qquad (4-111)$$

with boundary conditions

$$f(0) = f'(0) = 0 \qquad (3-51a,b)$$

$$f(\eta \to \infty) \to 1 \qquad (3-51c)$$

$$\theta(0) = 0 \qquad \theta(\eta \to \infty) \to 1 \qquad (4-110a,b)$$

Here note that the momentum and energy equations are uncoupled. The momentum equation is the familiar Falkner-Skan equation, the numerical solutions for which are discussed in Section D–13. Once solutions for f' and f in the Falkner-Skan equation are obtained they are then used in obtaining the solution of the energy equation. Note that these values must be stored for each integration step so that they are available when the solution of the energy equation is attempted. Modification of the **GBLPdriv** is required.

The energy equation is second-order with two parameters, Pr and n, with a single asymptotic boundary condition. The set-up for the main program for solution of the energy equation is similar to that for the Falkner-Skan equation except that the previously obtained solution for the Falkner-Skan equation must be passed to the numerical integration routine. This, of course, requires modification of the **rk** routine. An outline of the required modifications is given here:

> Dimension array variables feta() and fpeta() in **gblpdriv** to hold the solutions for $f(\eta)$ and $f'(\eta)$.
>
> Pass these arrays to **rk**, where they are filled as the solution progresses.
>
> Modify **rk** to use either the **eqmot** routine for the Falkner-Skan equation or the energy equation, depending on the value of a new convergence flag.

Once convergence is obtained for the Falkner-Skan equation, guess the unknown initial condition and iterate to a solution of the energy equation.

The details of the modifications are left to the reader. This modified program is the one used for the discussion in Section 4–10.

An alternate method is to ignore the uncoupling of Eqs. (3–50) and (4–111) and consider them as a pair of *coupled* ordinary differential equations. **gblpdriv** can then be used directly. The main program and the **eqmot** routine are similar to those discussed in Section D–15 for the free convection boundary layer, and in Section D–16 for the compressible boundary layer. In this case, the known value of $f''(0)$ from the solution to the Falkner-Skan equation for the required value of β is used as the initial value for $f''(0)$, with an appropriate initial guess for $g'(0)$. Convergence for the Falkner-Skan equation is immediate. Iteration then takes place on $g'(0)$ only, until convergence to a solution results. Again, the details are left to the reader.

D–15 Free Convection Boundary Layer Equations

The governing equations for the free convection boundary layer are (see Section 5–2)

$$f''' + (n+3)ff'' - 2(n+1)f'^2 + \theta = 0 \qquad (5-12)$$
$$\theta'' + \Pr\left[(n+3)f\theta' - 4nf'\theta\right] = 0 \qquad (5-13)$$

with boundary conditions at $\eta = 0$

$$f(0) = f'(0) = 0 \qquad \theta(0) = 1 \qquad (5-14a,b,c)$$

and as $\eta \to \infty$

$$f'(\eta \to \infty) \to 0 \qquad \theta(\eta \to \infty) \to 0 \qquad (5-14d,e)$$

Writing Eqs. (5–12) and (5–13) as five first-order simultaneous equations yields

$$x(1) = f''(\eta) \qquad (A-5a)$$
$$x(2) = f'(\eta) \qquad (A-5b)$$
$$x(3) = f(\eta) \qquad (A-5c)$$
$$x(4) = \theta'(\eta) \qquad (A-5d)$$
$$x(5) = \theta(\eta) \qquad (A-5e)$$

The Runge-Kutta functions then become

$$\begin{aligned}\frac{dx(1)}{d\eta} &= F(1) = -(n+3)ff'' - 2(n+1)f'^2 + \theta \\ &= -(param(2)+3)x(3)x(1) \\ &\quad + 2(param(2)+1)x(2)x(2) - x(5)\end{aligned} \qquad (A-6a)$$

$$\frac{dx(2)}{d\eta} = F(2) = x(1) \qquad (A-6b)$$

$$\frac{dx(3)}{d\eta} = F(3) = x(2) \qquad (A-6c)$$

372   Laminar Flow Analysis

$$\frac{dx(4)}{d\eta} = F(4) = -\text{Pr}\left[(n+3)f\theta' - 4nf'\theta\right]$$
$$= -param(1)(param(2)+3)x(3)x(4)$$
$$\quad - 4\,param(2)x(2)x(5)) \tag{A-6d}$$

$$\frac{dx(5)}{d\eta} = F(5) = x(4) \tag{A-6e}$$

where **param(1)** is Pr and **param(2)** is n.

Here there are two asymptotic boundary conditions and hence two unknown initial conditions, $f''(0) = \mathbf{x}(1)$ and $\theta'(0) = \mathbf{x}(4)$. The order of the system is five, and there are two parameters. Thus, **m = 5**, **nasymbc = 2** and **nparms = 2**. The asymptotic boundary conditions $f'(\eta \to \infty) \to 1$ and $\theta(\eta \to \infty) \to 1$ imply that $f''(\eta \to \infty) \to 0$ and $\theta'(\eta \to \infty) \to 0$. Hence, **con(1) = 1** and **con(2) = 4**, since the dependent variables, f'' and θ', correspond to $\mathbf{x}(1)$ and $\mathbf{x}(4)$, respectively. In addition, **con(3) = 2** and **con(4) = 5**, since the dependent variables, f' and θ, correspond to $\mathbf{x}(2)$ and $\mathbf{x}(5)$. The asymptotic boundary conditions are imposed using **asymbc(1) = asymbc(2) = 0** and **asymbc(3) = asymbc(4) = 1**. We choose the edge of the boundary layer as $\eta_{\max} = 6$ and impose error conditions on $f''(\eta \to \infty)$ and $\theta'(\eta \to \infty)$ of $\pm 5 \times 10^{-6}$ and on $f'(\eta \to \infty) - 1$ and $\theta(\eta \to \infty) - 1$ of $\pm 5 \times 10^{-6}$. A table of these values is given below:

Variable	Control Array	Control Value	Asymptotic Value	Error Value
$f''(0)$	con(1)	1	asymbc(1)=0	err(1)= 5×10^{-6}
$\theta'(0)$	con(2)	4	asymbc(2)=0	err(2)= 5×10^{-6}
$f'(\eta_{\max})$	con(3)	2	asymbc(3)=1	err(3)= 5×10^{-7}
$\theta(\eta_{\max})$	con(4)	5	asymbc(4)=1	err(4)= 5×10^{-7}

The complete main program for the solution of the free convection boundary layer equations is

convbl

main program for solution of the free convection boundary layer
routines called: gblpdriv
variables:

asymbc()	=	*array containing the asymptotic boundary conditions*
		has dimensions asymbc(2∗nasymbx)
base()	=	*array containing the saved values of the dependent*
		variables at the edge of the boundary layer
		has dimensions of base(m)
cflag	=	*used to control calculation of new values*
		for the unknown initial conditions
		0 = convergence achieved
		1 = convergence not achieved calculate new values
con()	=	*array containing control values relating dependent*
		variables and boundary conditions
		has dimensions of con(m)
dperm	=	*amount by which the first guess of the initial conditions is*

General Boundary Layer Program 373

		perturbed to find the derivatives of the dependent variables at the edge of the boundary layer w.r.t. changes in the unknown initial conditions — suggest 0.001
deriv(,)	=	array containing the derivatives of the dependent variables at the edge of the boundary layer w.r.t. changes in the unknown initial conditions has dimensions of deriv(nasymbc, 2*nasymbc)
dflag	=	used to control the calculation of the derivatives of the dependent variables at the edge of the boundary layer w.r.t. changes in the unknown initial conditions
err()	=	array containing the required error condition values has dimensions of err(m) — suggest 5×10^{-6}
etamax	=	the maximum value of the independent variable — the 'edge' of the boundary layer — suggest 6.0
h	=	Runge-Kutta fixed-step size — suggest 0.01
headernam()	=	character array containing the names of the dependent variables has dimensions of headernam(m)
init()	=	array containing the initial conditions has dimensions of init(m)
initnam()	=	character array containing the names of the initial conditions has dimensions of initnam(m)
interval	=	print interval for the dependent variables in multiples of h
m	=	order of the system of equations
nasymbc	=	number of asymptotic boundary conditions
nparam	=	number of parameters in the equations
oflag	=	output flag 1 = on 0 = off
param()	=	array containing the parameter values for the equations has dimensions of param(m)
paramnam()	=	character array containing the names of the parameter values has dimensions of paramnam(m)
x()	=	array containing the current values of the dependent variables as the integration proceeds. has dimensions of x(m)

dimension the arrays for a general system of order 20
dimension con(20),err(20),asymbc(20),init(20)
dimension headnam(20),initnam(20),paramnam(20)

m=5　　　　　　　　　　　　　　　　　　　*the order of the system is 5*

redimension the arrays for the current problem
Mat con = **Zer**(m)
Mat err = **Zer**(m)
Mat asymbc = **Zer**(m)
Mat init = **Zer**(m)

h=.01　　　　　　　　　*set the step size for the Runge-Kutta integration*
dper=.001　　　　　　　*set the perturbation value for unknown initial condition to calculate the b.l. edge derivatives*

etamax = 6　　　　　　　*set the value of the independents variable at the*

```
              con(1) = 1                          edge of the boundary layer
              con(2) = 4                          assign the control variables
              con(3) = 2
              con(4) = 5
              err(1)=.00005          assign the error values for the dependent variables
              err(2)=.00005                  at the edge of the boundary layer
              err(3)=.00005
              err(4)=.00005
              headnam(1) = "eta"        fill in the names of the dependent variables
              headnam(2) = " f''"
              headnam(3) = " f'"
              headnam(4) = " f"
              headnam(5) = " theta'"
              headnam(6) = " theta"
              nparms=2              set the number of parameters and assign their names
              paramnam(1) = "Prandtl Number"
              paramnam(2) = "n"
              nasymbc=2             set the number of asymptotic boundary conditions
              asymbc(1)=0                          and assign their values
              asymbc(2)=0
              asymbc(3)=0
              asymbc(4)=0
              initnam(1)="f''(0)"        assign the names of the initial conditions
              initnam(2)="f'(0)"
              initnam(3)="f(0)"
              initnam(4)="theta'(0)"
              initnam(5)="theta(0)"
              init(5) = 1           set the value of the temperature at the wall
              oflag = 0             set the output flag and output interval
              interval = 50
              call the general boundary layer driver program
              call gblpdriv(m,con,err,nasymbc,asymbc,init,initnam,nparms,paramnam,
                                       headnam,h,dper,interval,oflag,etamax)
       finish
```

The corresponding **eqmot** routine is

eqmotcv

```
    equations of motion routine for the free convection boundary layer
    routines called: none
    subroutine eqmot(eta, x(), param(), f())
    f(1) = -(param(2) + 3)*x(3)*x(1) + 2*(param(2) + 1)*x(2)*x(2) - x(5)
    f(2) = x(1)
    f(3) = x(2)
    f(4) = -param(1)*((param(2) + 3)*x(3)*x(4) + 4*param(2)*x(2)*x(5))
    f(5) = x(4)
    return
```

To complete the program a simple explanation routine is given by

explaincv
  simple explanation routine for free convection boundary layer equations
  the lcase function converts the input string to lower case
  [1:1] extracts the first character from the string
  subroutine explain
  print "Do you want a program explanation ";
  input achar
  if lcase(achar[1:1]) = "y" **then**
    print "Free convection boundary layer equations"
    print
  end if
  return

D–16 Compressible Boundary Layer Equations

The steady two-dimensional compressible viscous similar boundary layer is governed by (see Section 6–5)

$$f''' + ff'' + \hat{\beta}(g - f'^2) = 0 \tag{6-74}$$

$$g'' + \Pr fg' = \bar{\sigma}(1 - \Pr)(f'f'')' \tag{6-75}$$

with boundary conditions given by

$$f(0) = \text{constant} \qquad f'(0) = 0 \qquad g(0) = \text{constant} \tag{6-79a, b, c}$$

$$f(\eta \to \infty) \to 1 \qquad g(\eta \to \infty) \to 1 \tag{6-79e, f}$$

Writing Eqs. (6–74) and (6–75) as five first-order simultaneous equations yields

$$x(1) = f''(\eta) \tag{A-7a}$$

$$x(2) = f'(\eta) \tag{A-7b}$$

$$x(3) = f(\eta) \tag{A-7c}$$

$$x(4) = g'(\eta) \tag{A-7d}$$

$$x(5) = g(\eta) \tag{A-7e}$$

The Runge-Kutta functions then become

$$\frac{dx(1)}{d\eta} = F(1) = -ff'' - \beta(g - f'^2)$$
$$= -x(3)x(1) - param(1)(x(5) - x(2)x(2)) \tag{A-8a}$$

$$\frac{dx(2)}{d\eta} = F(2) = x(1) \tag{A-8b}$$

$$\frac{dx(3)}{d\eta} = F(3) = x(2) \tag{A-8c}$$

376   Laminar Flow Analysis

$$\frac{dx(4)}{d\eta} = F(4) = -\Pr fg' - \bar{\sigma}(1-\Pr)(f'f'')'$$
$$= -\text{param}(2)x(3)x(4) + \text{param}(3)(1-\text{param}(2)) \times$$
$$(x(1)x(1) + x(2)F(1))   \quad (A-8d)$$

$$\frac{dx(5)}{d\eta} = F(5) = x(4)   \quad (A-8e)$$

where **param**(1) is β and **param**(2) is Pr and **param**(3) is $\bar{\sigma}$.

Here there are two asymptotic boundary conditions and hence two unknown initial conditions, $f''(0) = \mathtt{x(1)}$ and $g'(0) = \mathtt{x(4)}$. The order of the system is five, and there are three parameters. Thus, $\mathtt{m} = 5$, $\mathtt{nasymbc} = 2$ and $\mathtt{nparms} = 3$. The asymptotic boundary conditions $f'(\eta \to \infty) \to 1$ and $g(\eta \to \infty) \to 1$ imply that $f''(\eta \to \infty) \to 0$ and $g'(\eta \to \infty) \to 0$. Hence, $\mathtt{con(1)} = 1$ and $\mathtt{con(2)} = 4$, since the dependent variables, f'' and g', correspond to $\mathtt{x(1)}$ and $\mathtt{x(4)}$, respectively. In addition $\mathtt{con(3)} = 2$ and $\mathtt{con(4)} = 5$, since the dependent variables, f' and g, correspond to $\mathtt{x(2)}$ and $\mathtt{x(5)}$. The asymptotic boundary conditions are imposed using $\mathtt{asymbc(1)} = \mathtt{asymbc(2)} = 0$ and $\mathtt{asymbc(3)} = \mathtt{asymbc(4)} = 1$. We choose the edge of the boundary layer as $\eta_{\max} = 6$ and impose error conditions on $f''(\eta \to \infty)$ and $g'(\eta \to \infty)$ of $\pm 5 \times 10^{-6}$ and on $f'(\eta \to \infty) - 1$ and $g(\eta \to \infty) - 1$ of $\pm 5 \times 10^{-7}$. A table of these values is convenient.

Variable	Control Array	Control Value	Asymptotic Value	Error Value
$f''(0)$	con(1)	1	asymbc(1)=0	err(1)= 5×10^{-6}
$g'(0)$	con(2)	4	asymbc(2)=0	err(2)= 5×10^{-6}
$f'(\eta_{\max})$	con(3)	2	asymbc(3)=1	err(3)= 5×10^{-7}
$g(\eta_{\max})$	con(4)	5	asymbc(4)=1	err(4)= 5×10^{-7}

The complete main program for the solution of the compressible boundary layer equations is

combl

    main program for solution of the compressible boundary layer equations

    routines called: gblpdriv

    variables:

asymbc()	= *array containing the asymptotic boundary conditions has dimensions asymbc(2∗nasymbx)*
base()	= *array containing the saved values of the dependent variables at the edge of the boundary layer has dimensions of base(m)*

cflag	= *used to control calculation of new values for the unknown initial conditions*
	    0 = convergence achieved
	    1 = convergence not achieved calculate new values

con()	= *array containing control values relating dependent variables and boundary conditions has dimensions of con(m)*

General Boundary Layer Program 377

dperm	=	amount by which the first guess of the initial conditions is perturbed to find the derivatives of the dependent variables at the edge of the boundary layer w.r.t. changes in the unknown initial conditions — suggest 0.001
deriv(,)	=	array containing the derivatives of the dependent variables at the edge of the boundary layer w.r.t. changes in the unknown initial conditions has dimensions of deriv(nasymbc, 2∗nasymbc)
dflag	=	used to control the calculation of the derivatives of the dependent variables at the edge of the boundary layer w.r.t. changes in the unknown initial conditions
err()	=	array containing the required error condition values has dimensions of err(m) — suggest 5×10^{-6}
etamax	=	the maximum value of the independent variable — the 'edge' of the boundary layer — suggest 6.0
h	=	Runge-Kutta fixed step size — suggest 0.01
headernam()	=	character array containing the names of the dependent variables has dimensions of headernam(m)
init()	=	array containing the initial conditions has dimensions of init(m)
initnam()	=	character array containing the names of the initial conditions has dimensions of initnam(m)
interval	=	print interval for the dependent variables in multiples of h
m	=	order of the system of equations
nasymbc	=	number of asymptotic boundary conditions
nparam	=	number of parameters in the equations
oflag	=	output flag 1 = on 0 = off
param()	=	array containing the parameter values for the equations has dimensions of param(m)
paramnam()	=	character array containing the names of the parameter values has dimensions of paramnam(m)
x()	=	array containing the current values of the dependent variables as the integration proceeds. has dimensions of x(m)

dimension the arrays for a general system of order 20
dimension con(20),err(20),asymbc(20),init(20)
dimension headnam(20),initnam(20),paramnam(20)
m=5

redimension the arrays for the current problem
Mat con = **Zer**(m)
Mat err = **Zer**(m)
Mat asymbc = **Zer**(m)
Mat init = **Zer**(m)
h=.01   *set the step size for the Runge-Kutta integration*
dper=.001   *set the perturbation value for unknown initial condition to calculate the b.l. edge derivatives*

378 Laminar Flow Analysis

etamax = 6   *set the value of the independents variable at the edge of the boundary layer*

con(1) = 1   *assign the control variables*
con(2) = 4
con(3) = 2
con(4) = 5

err(1)=.000005   *assign the error values for the dependent variables at the edge of the boundary layer*
err(2)=.000005
err(3)=.0000005
err(4)=.0000005

headnam(1) = "eta"   *fill in the names of the dependent variables*
headnam(2) = " f''"
headnam(3) = " f'"
headnam(4) = " f"
headnam(5) = " g'"
headnam(6) = " f"

nparms=3   *set the number of parameters and assign their names*
paramnam(1) = "Beta"
paramnam(2) = "Prandtl Number"
paramnam(3) = "Mach Number Parameter"

nasymbc=2   *set the number of asymptotic boundary conditions and assign their values*
asymbc(1)=0
asymbc(2)=0
asymbc(3)=1
asymbc(4)=1

initnam(1)="f''(0)"   *assign the names of the initial conditions*
initnam(2)="f'(0)"
initnam(3)="f(0)"
initnam(4)="g'(0)"
initnam(5)="g(0)"

oflag = 0   *set the output flag and output interval*
interval = 50

call the general boundary layer driver program

call gblpdriv(m,con,err,nasymbc,asymbc,init,initnam,nparms,paramnam,
                             headnam,h,dper,interval,oflag,etamax)

finish

The corresponding **eqmot** routine is

eqmotcb

equations of motion routine for the compressible boundary layer

subroutine eqmot(eta, x(), param(), f())
f(1) = −x(3)∗x(1) − param(1)∗(x(5)−x(2)∗x(2))
f(2) = x(1)
f(3) = x(2)
f(4) = −param(2)∗x(3)∗x(4) + param(3)∗(1−param(2))∗(x(1)∗x(1) + x(2)∗f(1))
f(5) = x(4)
return

To complete the program a simple explanation routine is given by

explaincb

>*simple explanation routine for the compressible boundary layer equations*
>
>*the lcase function converts the input string to lower case*
>*[1:1] extracts the first character from the string*
>
>**subroutine** explain
>print "Do you want a program explanation ";
>input achar
>**if** lcase(achar[1:1]) = "y" **then**
>> print "Compressible boundary layer equations"
>> print
>
>**end if**
>**return**

APPENDIX E

PROBLEMS

The problems given here are generally of three types: numerical, parametric and directed analysis. The numerical problems are generally of the form — 'Given the following specific information, calculate a required result'. This is the type of problem generally found in textbooks. The parametric problems are of the form — 'Investigate the effect of the variation of a parameter on the numerical solution'. These problems generally are amenable to solutions using programs derived from the pseudocode algorithms and/or modification of these algorithms. They allow the student to ask the question 'What if —— ?' and lead to a better physical understanding of a mathematical model. Such problems are useful in teaching the concept of parametric analysis. Directed analysis problems seek to guide the student through an original analysis of a given problem. They are frequently skeletons of the analyses in classical papers. They serve to extend and diversify the material and to tailor the material to individual interests and needs. For example, there are several directed analysis problems on rotational flows which can serve as an introduction to that topic. Frequently, the result of a directed analysis problem is a differential equation or system of differential equations that require numerical solution. Here, the reader is shown how to put these equations in a form similar to that of the equations solved by one of the analysis algorithms. Consequently, the directed analysis problems serve as a unifying mechanism for various topics in fluid dynamics. Directed analysis problems generally require considerable effort. Problems that require or lead to computer solutions are marked with (c).

CHAPTER 1

1–1 Complete the details of the derivation of Eq. (1–1).

1–2 Show in detail the derivation of Eqs. (1–6b) and (1–6c).

1–3 Complete the details of the derivation of Eq. (1–19).

1–4 Show that the x and y components of the angular velocity of a fluid element are

$$\omega_x = \frac{1}{2}\left(\frac{\partial w}{\partial y} - \frac{\partial v}{\partial z}\right) \qquad \omega_y = \frac{1}{2}\left(\frac{\partial u}{\partial z} - \frac{\partial w}{\partial x}\right)$$

1–5 Show that $\dot{\epsilon}_{xy} = \dot{\epsilon}_{yx}$. This implies that the shear rate-of-strain tensor is symmetrical.

1-6 Two conditions with respect to angular velocity and shear rate-of-strain on a fluid element are of frequent interest:

  a. the angular velocity is zero;

  b. the shear rate of strain is zero.

Consider these two cases for a two-dimensional element. Draw the resulting shapes for the element, and discuss the implications.

1-7 Complete the details of the derivation of the rate at which the shearing stresses do work on the element used in the derivation of the energy equation.

CHAPTER 2

2-1(c) Numerically integrate the governing boundary value problem for the suddenly accelerated flat plate. (Hint: Modify the **stag2d** algorithm, and reduce the order of the system.) Compare the results with those for the analytical solution.

2-2 Using Eqs. (2-52) and (2-53), derive Eqs. (2-54) to (2-58).

2-3 Using Eq. (2-78), derive Eqs. (2-79) and (2-80).

2-4 Using Eqs. (2-81) and (2-82), derive Eqs. (2-83) and (2-84).

2-5(c) Modify the **stag2d** program to automatically increment η_{max} as the iteration proceeds (see Appendix B and Figure 2-8).

2-6(c) Modify the **ns** routine of Appendix D to dynamically update the Nachtsheim-Swigert iteration derivative at the edge of the boundary layer for the **stag2d** program.

2-7 Consider the starting process in simple Couette flow. The flow is considered to be two-dimensional, incompressible and without body forces. The governing equations are then

$$\frac{\partial u}{\partial t} = \nu \frac{\partial^2 u}{\partial y^2}$$

with boundary conditions

$$u = 0 \quad \text{for} \quad 0 \leq y \leq h \quad t \leq 0$$

$$\left. \begin{array}{ll} u = 0 & y = 0 \\ u = U & y = h \end{array} \right\} \quad t > 0$$

  a. Using a transformation of the form

$$u = Af(\eta) \qquad \eta = Bt^n y^m$$

  show that the governing equation reduces to the linear ordinary differential equation

$$C_1 f'' + C_2 \eta f' = 0$$

  with boundary conditions of the form

$$f(0) = C_3 \qquad f(C_4 t) = C_5 g(t)$$

where the C_n are constants and the prime denotes differentiation with respect to η.

b. Since the resulting governing equation is linear, use the method of superposition to obtain a solution. Hint: the resulting solution is a series solution of the form

$$f = \sum_{n=0}^{\infty} F_n(\eta + C_n \eta_1)$$

c. Plot the velocity profile $u(y,t)$ vs y for several values of time $t > 0$. Show that as $t \to \infty$ the profile approaches the linear profile associated with simple steady Couette flow.

2–8 Consider the flow between two infinite concentric rotating cylinders of radius r_1 and r_2, with angular velocities ω_1 and ω_2, respectively. Assume that the flow is steady, two-dimensional and the fluid viscous and incompressible.

a. Show that the incompressible Navier-Stokes equations in cylindrical coordinates, i.e.

$$\rho \left(\frac{\partial u_r}{\partial t} + u_r \frac{\partial u_r}{\partial r} + \frac{u_\theta}{r} \frac{\partial u_r}{\partial \theta} - \frac{u_\theta^2}{r} + u_z \frac{\partial u_r}{\partial z} \right) = F_r - \frac{\partial P}{\partial r}$$

$$+ \mu \left(\frac{\partial^2 u_r}{\partial r^2} + \frac{1}{r} \frac{\partial u_r}{\partial r} - \frac{u_r}{r^2} + \frac{1}{r^2} \frac{\partial^2 u_r}{\partial \theta^2} - \frac{2}{r^2} \frac{\partial u_\theta}{\partial \theta} + \frac{\partial^2 u_r}{\partial z^2} \right)$$

$$\rho \left(\frac{\partial u_\theta}{\partial t} + u_r \frac{\partial u_\theta}{\partial r} + \frac{u_\theta}{r} \frac{\partial u_\theta}{\partial \theta} - \frac{u_r u_\theta}{r} + u_z \frac{\partial u_\theta}{\partial z} \right) = F_\theta - \frac{1}{r} \frac{\partial P}{\partial \theta}$$

$$+ \mu \left(\frac{\partial^2 u_\theta}{\partial r^2} + \frac{1}{r} \frac{\partial u_\theta}{\partial r} - \frac{u_\theta}{r^2} + \frac{1}{r^2} \frac{\partial^2 u_\theta}{\partial \theta^2} + \frac{2}{r^2} \frac{\partial u_r}{\partial \theta} + \frac{\partial^2 u_\theta}{\partial z^2} \right)$$

$$\rho \left(\frac{\partial u_z}{\partial t} + u_r \frac{\partial u_z}{\partial r} + \frac{u_\theta}{r} \frac{\partial u_z}{\partial \theta} + u_z \frac{\partial u_z}{\partial z} \right) = F_z - \frac{\partial P}{\partial z}$$

$$+ \mu \left(\frac{\partial^2 u_z}{\partial r^2} + \frac{1}{r} \frac{\partial u_z}{\partial r} + \frac{1}{r^2} \frac{\partial^2 u_z}{\partial \theta^2} + \frac{\partial^2 u_z}{\partial z^2} \right)$$

with the continuity equation given as

$$\frac{\partial u_r}{\partial r} + \frac{u_r}{r} + \frac{1}{r} \frac{\partial u_\theta}{\partial \theta} + \frac{\partial u_z}{\partial z} = 0$$

where r, θ and z are the radial, azimuthal and axial coordinates, respectively; u_r, u_θ, u_z are the velocity components in the r, θ, z directions, respectively; and F_r, F_θ, F_z are the r, θ, z components of the body force/unit mass, respectively, reduce to

$$\rho \frac{u_\theta^2}{r} = \frac{dP}{dr} \tag{1}$$

and

$$\frac{d^2 u_\theta}{dr^2} + \frac{d}{ddr}\left(\frac{u_\theta}{r}\right) = 0 \tag{2}$$

384   Laminar Flow Analysis

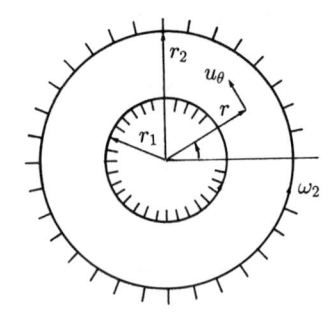

b. Assuming that the no-slip condition is valid at the surfaces of both cylinders, show that the velocity distribution is

$$u_\theta(r) = \frac{1}{r_2^2 - r_1^2}\left[r(\omega_2 r_2^2 - \omega_1 r_1^2) - \frac{r_1^2 r_2^2}{r}(\omega_2 - \omega_1)\right]$$

c. Use the radial velocity distribution to determine the radial pressure distribution.

d. Using Newton's law of friction, determine the shearing stress at the surface of each cylinder.

e. Assuming that the inner cylinder is at rest, i.e., $\omega_1 = 0$, determine the torque transmitted to the fluid by the rotation of the outer cylinder. Assume that the cylinders are of height h.

2-9   Consider the flow about an infinite flat plate undergoing harmonic oscillations in the plane of the plate. The fluid is viscous and incompressible.

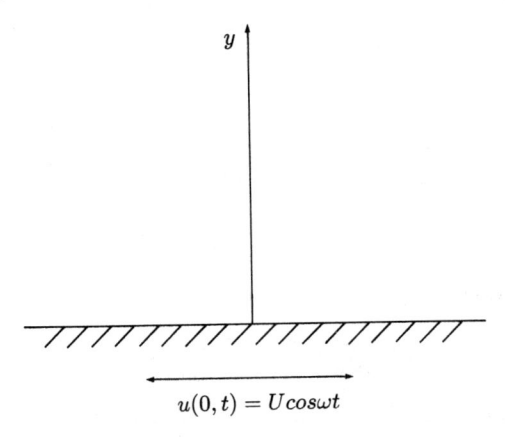

$u(0, t) = U \cos \omega t$

a. Show that the governing equation is

$$\frac{\partial u}{\partial t} = \nu \frac{\partial^2 u}{\partial y^2}$$

b. Assume that the no-slip boundary condition applies at the plate and that the motion of the plate is given by $u(0, t) = U \cos \omega t$. Since the motion of the plate is harmonic, it seems reasonable to expect the velocity of the fluid to vary harmonically with time. Since the fluid has viscosity, it is reasonable to expect the fluid motion to exhibit attenuation and phase shift with respect to the motion of the plate which logically depend on the distance above the plate. Thus, assume a solution of the form

$$u(y,t) = Ae^{a\eta}\cos(wt - \eta)$$

$$\eta = by$$

and determine the constants A, a, b such that the boundary conditions are satisfied.

c. Introduce appropriate nondimensional dependent and independent variables, f and η, and plot the velocity profile, $f(\eta, t)$, for several values of ωt, say 0, $\pi/2$, π, $3\pi/2$, 2π. Discuss the phase relationship between the fluid motion and the motion of the plate.

2-10 Consider the flow emerging from a point source into a quiescent fluid. Assume that the flow is viscous, incompressible, steady and axially symmetric. Using spherical polar coordinates (r, θ, ϕ), with θ measured from the axis of the jet, and u_r, u_θ, u_ϕ, the velocity components in the r, θ, ϕ directions, respectively, show that an exact solution of the Navier-Stokes equations can be obtained. Since the jet is axisymmetric, $u_\phi = 0$, and all quantities are independent of ϕ. Thus, the Navier-Stokes equations, i.e.

$$\rho\left[\frac{\partial u_r}{\partial t} + u_r\frac{\partial u_r}{\partial r} + \frac{u_\theta}{r}\frac{\partial u_r}{\partial \theta} + \frac{u_\phi}{r\sin\theta}\frac{\partial u_r}{\partial \phi} - \frac{u_\theta^2 + u_\phi^2}{r}\right] = F_r - \frac{\partial P}{\partial r}$$

$$+ \mu\left[\frac{1}{r^2}\frac{\partial}{\partial r}\left(r^2\frac{\partial u_r}{\partial r}\right) + \frac{1}{r^2\sin\theta}\frac{\partial}{\partial \theta}\left(\sin\theta\frac{\partial u_r}{\partial \theta}\right) + \frac{1}{r^2\sin^2\theta}\frac{\partial^2 u_r}{\partial \phi^2}\right.$$

$$\left. - \frac{2u_r}{r^2} - \frac{2}{r^2}\frac{\partial u_\theta}{\partial \theta} - \frac{2u_\theta\cot\theta}{r^2} - \frac{2}{r^2\sin\theta}\frac{\partial u_\phi}{\partial \phi}\right]$$

$$\rho\left[\frac{\partial u_\theta}{\partial t} + u_r\frac{\partial u_\theta}{\partial r} + \frac{u_\theta}{r}\frac{\partial u_\theta}{\partial \theta} + \frac{u_\phi}{r\sin\theta}\frac{\partial u_\theta}{\partial \phi} + \frac{u_r u_\theta}{r} - \frac{u_\phi^2\cot\theta}{r}\right] = F_\theta - \frac{1}{r}\frac{\partial P}{\partial \theta}$$

$$+ \mu\left[\frac{1}{r^2}\frac{\partial}{\partial r}\left(r^2\frac{\partial u_\theta}{\partial r}\right) + \frac{1}{r^2\sin\theta}\frac{\partial}{\partial \theta}\left(\sin\theta\frac{\partial u_\theta}{\partial \theta}\right) + \frac{1}{r^2\sin^2\theta}\frac{\partial^2 u_\theta}{\partial \phi^2}\right.$$

$$\left. + \frac{2}{r^2}\frac{\partial u_r}{\partial \theta} - \frac{u_\theta}{r^2\sin^2\theta} - \frac{2\cos\theta}{r^2\sin^2\theta}\frac{\partial u_\phi}{\partial \phi}\right]$$

$$\rho\left[\frac{\partial u_\phi}{\partial t} + u_r\frac{\partial u_\phi}{\partial r} + \frac{u_\theta}{r}\frac{\partial u_\phi}{\partial \theta} + \frac{u_\phi}{r\sin\theta}\frac{\partial u_\phi}{\partial \phi} + \frac{u_\phi u_r}{r} + \frac{u_\theta u_\phi\cot\theta}{r}\right] = F_\phi$$

$$- \frac{1}{r\sin\theta}\frac{\partial P}{\partial \phi} + \mu\left[\frac{1}{r^2}\frac{\partial}{\partial r}\left(r^2\frac{\partial u_\phi}{\partial r}\right) + \frac{1}{r^2\sin\theta}\frac{\partial}{\partial \theta}\left(\sin\theta\frac{\partial u_\phi}{\partial \theta}\right)\right.$$

$$\left. + \frac{1}{r^2\sin^2\theta}\frac{\partial^2 u_\phi}{\partial \phi^2} - \frac{u_\phi}{r^2\sin^2\theta} + \frac{2}{r^2\sin^2\theta}\frac{\partial u_r}{\partial \phi} + \frac{2\cos\theta}{r^2\sin^2\theta}\frac{\partial u_\theta}{\partial \phi}\right]$$

and the continuity equation

$$\frac{1}{r^2}\frac{\partial}{\partial r}(r^2 u_r) + \frac{1}{r\sin\theta}\frac{\partial}{\partial \theta}(u_\theta\sin\theta) + \frac{1}{r\sin\theta}\frac{\partial u_\phi}{\partial \phi} = 0$$

386   Laminar Flow Analysis

are considerably simplified.

a. Obtain the simplified governing equations.
b. Since there is no characteristic length in this flow, assume a solution of the form
$$\psi(r,\theta) = Ar^a f(\eta)$$
$$\eta = B(\cos\theta)^b$$
$$\frac{P - P_\infty}{\rho} = Cr^c P(\eta)$$

where P_∞ is the pressure at infinity where the velocity components vanish.

1. Show that the continuity equation is satisfied if we assume
$$u_r = +Ar^{-a} f'(\eta) \qquad u_\theta = -\frac{Ar^{-a} f(\eta)}{\sqrt{1-\eta^2}}$$

What are B and b?

2. Show that the θ momentum equation can be reduced to
$$P' = -\left[\frac{f^2}{2(1-\eta^2)}\right]' - f''$$

where the prime denotes differentiation with respect to η.

3. Show that the r momentum equation reduces to
$$P' = \frac{-f^2}{2(1-\eta^2)} - \frac{1}{2}[ff' - (1-\eta^2)f'']'$$

4. What are A, B, C and a, b, c?

c. First integrate the θ momentum equation, and then use this result to integrate the r momentum equation to give
$$f^2 = 2(1-\eta^2)f' + 4\eta f + C_1\eta^2 + C_2\eta + C_3$$

where C_1, C_2, C_3 are constants of integration.

d. Restricting the discussion to the case $C_1 = C_2 = C_3 = 0$, show that
$$f = \frac{2(1-\eta^2)}{\bar{a} + (1-\eta)}$$

where $\bar{a}$ is a constant of integration.

e. Plot the streamlines for $\bar{a} = 10$, 0.1 and 0.01, i.e., plot
$$\psi = Ar^a f(\eta) \quad \text{for} \quad \bar{a} = 10, 0.1 \text{ and } 0.01$$

and show that the stream tubes have a minimum throat area for
$$\eta = \frac{1}{1+\bar{a}}$$

2–11 Consider two-dimensional steady incompressible viscous flow without body forces in a channel with nonparallel straight walls, i.e., flow in a convergent or divergent channel.

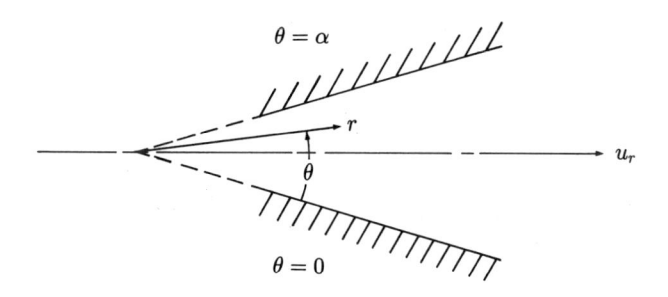

a. Using a substitution of the form

$$u(r,\theta) = Ar^a f(\eta)$$
$$\eta = Br^b \theta^t$$

determine the functional form of $u(r,\theta)$ which satisfies the continuity equation.

b. Show that the Navier-Stokes equations in cylindrical polar coordinates (see Problem 2–6) reduce to

$$u_r \frac{\partial u_r}{\partial r} = -\frac{1}{\rho}\frac{\partial P}{\partial r} + \nu\left(\frac{\partial^2 u_r}{\partial r^2} + \frac{1}{r}\frac{\partial u_r}{\partial r} + \frac{1}{r^2}\frac{\partial^2 u_r}{\partial \theta^2} - \frac{u_r}{r^2}\right)$$

and

$$0 = -\frac{1}{\rho}\frac{\partial P}{\partial \theta} + \frac{2\nu}{r}\frac{\partial u_r}{\partial \theta}$$

c. Eliminate the pressure from the Navier-Stokes equations, and use the results of (a) above to show that the problem is reduced to the solution of an ordinary differential equation of the form

$$f''' + C_1 f f' + C_2 f' = 0$$

where C_1 and C_2 are constants.

d. Show that the boundary conditions given by the no-slip condition at the surfaces, i.e., $u_r = 0$ at $\theta = 0$ and at α, plus the auxiliary condition that the velocity profiles be symmetrical about the center line, i.e., $\partial u_r / \partial r = 0$ at $\theta = \alpha/2$, is reduced to

$$f(0) = 0 \qquad f(2) = 0 \qquad f'(1) = 0$$

e. (c) This ordinary differential equation, subject to these boundary conditions, can be integrated analytically in terms of elliptic functions (see [Rose63], p. 144). However, the solution is somewhat involved. Therefore, consider a numerical solution. Noting that the velocity profile is symmetrical about $\eta = 1$, only the region $0 \leq \eta \leq 1$ need be considered. The integration problem is thus similar to that encountered for stagnation point flow. Modify the **stag2d** algorithm, and obtain solutions to the present governing equations. Plot and discuss these solutions.

2–12 Consider the viscous incompressible steady flow without body forces about a flat disk rotating in its own plane with a constant angular velocity, ω. The surrounding fluid is at rest.

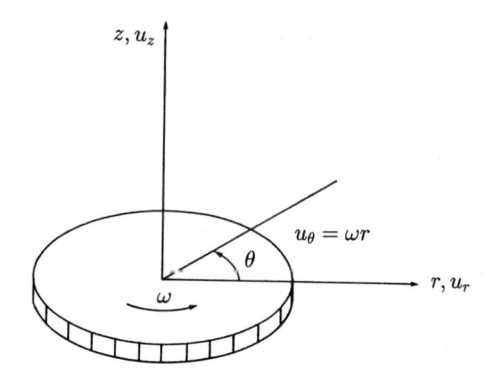

a. Show that the incompressible Navier-Stokes equations in cylindrical polar coordinates (see Problem 2–6) reduce to

$$u_r \frac{\partial u_r}{\partial r} - \frac{u_\theta^2}{r} + u_z \frac{\partial u_r}{\partial z} = -\frac{1}{\rho}\frac{\partial P}{\partial r} + \nu \left\{ \frac{\partial^2 u_r}{\partial r^2} + \frac{\partial}{\partial r}\left(\frac{u_r}{r}\right) + \frac{\partial^2 u_r}{\partial z^2} \right\}$$

$$u_r \frac{\partial u_\theta}{\partial r} + \frac{u_r u_\theta}{r} + u_z \frac{\partial u_\theta}{\partial z} = \nu \left\{ \frac{\partial^2 u_\theta}{\partial r^2} + \frac{\partial}{\partial r}\left(\frac{u_\theta}{r}\right) + \frac{\partial^2 u_\theta}{\partial z^2} \right\}$$

$$u_r \frac{\partial u_z}{\partial r} + u_z \frac{\partial u_z}{\partial z} = -\frac{1}{\rho}\frac{\partial P}{\partial z} + \nu \left\{ \frac{\partial^2 u_z}{\partial r^2} + \frac{1}{r}\frac{\partial u_z}{\partial r} + \frac{\partial^2 u_z}{\partial z^2} \right\}$$

with the continuity equation given by

$$\frac{\partial u_r}{\partial r} + \frac{u_r}{r} + \frac{\partial u_z}{\partial z} = 0$$

with the no-slip boundary conditions

$$z = 0 \qquad u_r = 0 \qquad u_\theta = \omega r \qquad u_z = 0$$
$$z \to \infty \qquad u_r \to 0 \qquad u_\theta \to 0$$

Note that this is a truly three-dimensional flow.

b. Using transformations of the form

$$u_r = A r^a \omega^m F(\eta)$$
$$u_\theta = B r^b \omega^n G(\eta)$$
$$u_z = C r^c \omega^q H(\eta)$$
$$P = D \rho r^d \omega^s P(\eta)$$

and

$$\eta = E r^e \omega^t z^i$$

show that the governing equations can be reduced to the system of simultaneous ordinary differential equations

$$C_1 F + C_2 H' = 0 \qquad (1)$$

$$C_3 F^2 + C_4 F' H + C_5 G^2 + C_6 F'' = 0 \qquad (2)$$

$$C_7 FG + C_8 HG' + C_9 G'' = 0 \qquad (3)$$

$$C_{10} P' + C_{11} HH' + C_{12} H'' = 0 \qquad (4)$$

with boundary conditions

$$\eta = 0 \quad F = 0 \quad G = 1 \quad H = 0 \quad P = 0$$
$$\eta \to \infty \quad F \to 0 \quad G \to 0$$

where the C_n are integer constants *generally* = ±1 or ±2, and the prime denotes differentiation with respect to η. Note: be careful in choosing a, b... etc., not to eliminate one of the independent variables.

c. Using Eq. (a), show that the solution can be reduced to the integration of a pair of coupled ordinary differential equations of the form

$$C_{13} H''' + C_{14} HH'' + C_{15} G^2 + C_{16} H'^2 = 0$$

and

$$C_{17} G'' + C_{18} (HG)' = 0$$

with boundary conditions

$$\eta = 0 \quad H = 0 \quad H' = 0 \quad G = 1$$
$$\eta \to \infty \quad H' \to 0 \quad G \to 0$$

The solutions for $P(\eta)$ and $F(\eta)$ are then obtained sequentially.

This problem is of a form similar to that governing the laminar free convection boundary layer discussed in Chapter 5. Its numerical integration is addressed in Problem 5–1.

CHAPTER 3

3–1 Show that the solution for steady, two-dimensional incompressible flow in a convergent channel, which is an exact solution of the incompressible Navier-Stokes equations (see Problem 2–9), is also an exact solution of the boundary layer equations.

3–2 Show that the solution for steady, two-dimensional incompressible stagnation point flow, which is an exact solution of the incompressible Navier-Stokes equations (see Section 2–7), is also an exact solution of the boundary layer equations.

3–3 Show that the solution for steady incompressible axisymmetric stagnation point flow, which is an exact solution of the incompressible Navier-Stokes equations (see Section 2–10), is also an exact solution of the boundary layer equations.

3–4 Using the results in Table 3–2, calculate and plot, at the appropriate position, x, the velocity profiles, $u(x, y)$, on a flat plate corresponding to Reynolds numbers of 5×10^4, 5×10^5, 5×10^6, 5×10^7. The free stream velocity is 100 ft/sec, and the kinematic viscosity is 2×10^{-4} ft^2/sec. Determine the displacement and momentum thickness for each Reynolds number, and indicate δ^* and θ on each velocity profile. Draw a line showing the growth of the displacement and momentum thickness with distance along the plate. On the same graph show the distribution of shearing stress, $\tau(x,0)$, along the plate.

3–5 Do Problem 3–4 for an included wedge angle of $\pi/6$.

3–6 Attempt to reduce the governing equations and boundary conditions for steady incompressible viscous boundary layer flow (see Eqs. 3–41 and 3–42) to a single ordinary nonlinear differential equation, using the free parameter method. That is, seek a similarity transformation.

First introduce a stream function of the form

$$u = \frac{\partial \psi}{\partial y} \qquad v = -\frac{\partial \psi}{\partial x}$$

and then assume that

$$u = U(x)f'(\eta) \qquad \text{and} \qquad \eta = \eta(x,y)$$

At this point $\eta = \eta(x,y)$ is maintained as a completely arbitrary function. Substitute these assumed forms into the governing equation. Reduce the resulting equation to an ordinary differential equation (Eq. 3–50) with appropriate boundary conditions (Eqs. 3–51 a, b) by postulating conditions on the coefficients of f and its derivatives. (Hint: Make the coefficient of f'^2 unity, and assume that the coefficient of f''' is an arbitrary function of η.)

What is the function $\eta = \eta(x,y)$?

What are the possible forms of $U(x)$?

3–7 Attempt to reduce the governing equations and boundary conditions for steady incompressible viscous boundary layer flow (Eqs. 3–41 and 3–42) to a single ordinary nonlinear differential equation, using a classical separation of variables technique. That is, seek a similarity transformation.

First introduce a stream function of the form

$$u = \frac{\partial \psi}{\partial y} \qquad v = -\frac{\partial \psi}{\partial x}$$

and then assume that

$$\psi(\xi, \eta) = H(\xi)f(\eta)$$

where $\qquad \xi = x \qquad \eta = \eta(x,y)$

What is the function $\eta = \eta(x,y)$?

What are the possible forms of $U(x)$?

3–8 Consider the efflux of an incompressible fluid from a two-dimensional jet. The jet is considered to be a long narrow slit from which the fluid emerges and mixes with the surrounding fluid. The jet flow is assumed to be laminar. Since the pressure in the surrounding fluid is constant, the flow in the jet is governed by the incompressible laminar boundary layer equations with zero pressure gradient (Eqs. 3–24 and 3–25). The boundary conditions are that as $y \to \pm\infty$, $u \to 0$ and, since the flow is symmetrical about the x-axis, $v = 0$ at $y = 0$. This implies that there is no-mass transfer across the $y = 0$ plane.

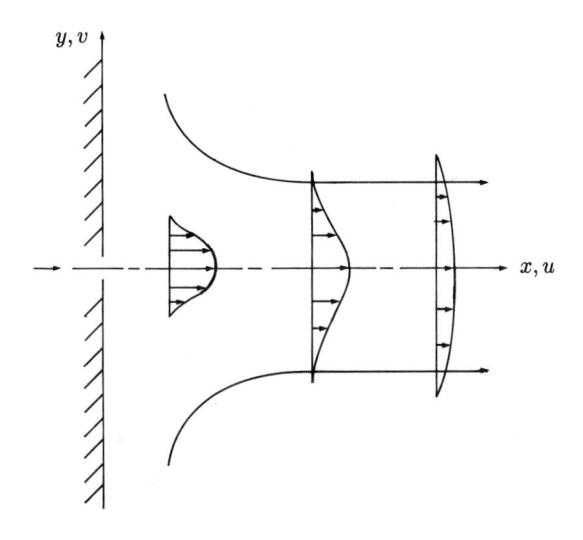

a. After introducing the two-dimensional stream function, assume a transformation of variables of the form

$$\psi(x,y) = Ax^a f(\eta) \qquad \eta = Bx^b y^c$$

and show that the governing equations reduce to

$$f''' + ff'' + f'^2 = 0$$

with homogeneous boundary conditions

$$\eta = 0 \qquad f = 0$$
$$\eta \to \pm\infty \qquad f' \to 0$$

b. Show that this equation can be analytically integrated by reduction to a Riccati-type differential equation with the solution

$$f(\eta) = 2\alpha \tanh(\alpha\eta)$$

where α is a constant.[†]

c. Use the physical condition that the rate of momentum transfer across any plane normal to the x-axis is a constant to evaluate α.

[†] In order to evaluate the hyperbolic tangent, recall that it can be written as

$$\tanh u = \frac{e^u - e^{-u}}{e^u + e^{-u}}$$

392   Laminar Flow Analysis

    d. Plot the velocity profiles for $u(x,y)$ and $v(x,y)$ for several positions along the x-axis, say $x = 1$, 5, 10 feet. Assume that the momentum transfer across any plane normal to the x-axis is 10 slug ft/sec^2, the kinematic viscosity is 1×10^{-5} ft^2/sec and the density is 2 slugs/ft^3.

    e. Determine the transverse velocity, v, at the edge of the jet.

    f. Determine the volume rate of discharge through the jet.

3-9   Consider the efflux of a viscous fluid from a small circular hole in a surface. As the fluid emerges from the hole it forms a jet which mixes with the surrounding stationary flow. Assume that the flow is steady, axisymmetric and incompressible.

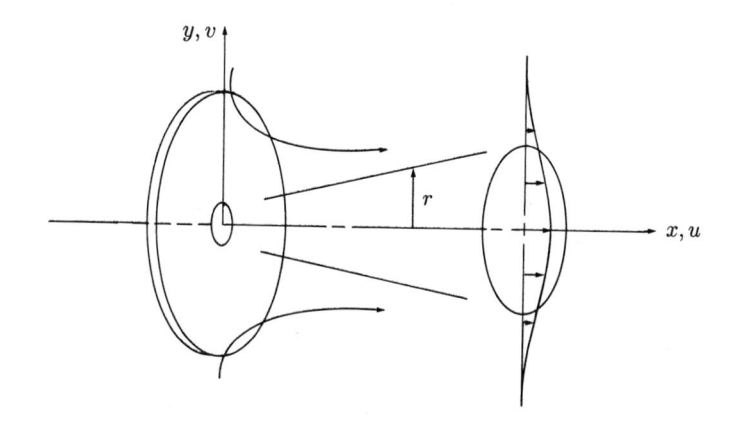

Since the pressure in the surrounding fluid is constant, the flow in the jet is governed by the axisymmetric laminar boundary layer equations with zero pressure gradient (Eq. 3–83). The boundary conditions are: that since the surrounding fluid is at rest $u \to 0$ as $y \to \infty$; that since the axis of the jet is a streamline of the flow no mass is transferred across the axis hence $v = 0$ at $y = 0$; and that since the flow is symmetrical about the axis, $\partial u/\partial y = 0$ at $y = 0$. In the chosen axis system the continuity and momentum equations are

$$\frac{\partial u}{\partial x} + \frac{\partial u}{\partial y} + \frac{v}{y} = 0$$

$$u\frac{\partial u}{\partial x} + v\frac{\partial u}{\partial y} = \nu \frac{1}{y}\frac{\partial}{\partial y}\left(y\frac{\partial u}{\partial y}\right)$$

    a. After introducing an axisymmetric stream function such that

$$u = \frac{1}{y}\frac{\partial \psi}{\partial y} \qquad v = -\frac{1}{y}\frac{\partial \psi}{\partial x}$$

assume a transformation of variables of the form

$$\psi(x,y) = Ax^a f(\eta) \qquad \eta = Bx^b y^c$$

and show that the governing differential equations reduce to the form

$$C_1\left(f'' - \frac{f'}{\eta}\right)' + C_2\left(\frac{ff'}{\eta}\right)' = 0$$

where the C_n are constants and the prime denotes differentiation with respect to η.

b. Show that this equation has a solution of the form

$$f(\epsilon) = \frac{\epsilon^2}{C_3 + C_4 \epsilon^2}$$

where $\epsilon = \alpha \eta$ with α a constant.

c. Use the physical condition that the rate of momentum transfer across any plane normal to the x-axis is a constant to evaluate the constant α.

d. Plot the velocity profiles for $u(x,y)$ and $v(x,y)$ for several positions along the x-axis, say at $x = 1$, 5 and 10 feet. Assume that the momentum transfer across any plane normal to the x-axis is 10 slugs ft/sec^2, the kinematic viscosity is 1×10^{-5} ft^2/sec and the density is 2 slugs/ft^3.

e. Determine the transverse velocity at the edge of the jet.

f. Determine the volume rate of discharge through the jet.

g. Compare the results with those for the solution of the two-dimensional jet (see Problem 3–8).

3–10 Consider the analogous problem to that described in Problem 2–12. Here we consider the viscous incompressible steady flow without body forces of a constant velocity revolving flow over a fixed plate. Due to the action of viscosity, a boundary layer forms on the plate. For rotationally symmetric flow, the steady viscous incompressible boundary layer equations in cylindrical polar coordinates reduce to

$$\frac{\partial P}{\partial z} = 0$$

$$u_r \frac{\partial u_r}{\partial r} + u_z \frac{\partial u_r}{\partial z} - \frac{u_\theta^2}{r} = -\frac{1}{\rho}\frac{\partial P}{\partial r} + \nu \frac{\partial^2 u_r}{\partial z^2}$$

$$u_r \frac{\partial u_\theta}{\partial r} + u_z \frac{\partial u_\theta}{\partial z} + \frac{u_r u_\theta}{r} = \nu \frac{\partial^2 u_\theta}{\partial z^2}$$

with the continuity equation given by

$$\frac{\partial u_r}{\partial r} + \frac{u_r}{r} + \frac{\partial u_z}{\partial z} = 0$$

a. Assuming that the no-slip condition holds at the surface ($z = 0$) and that the boundary layer flow smoothly transitions into the inviscid flow at some large distance above the plate, determine the boundary conditions.

b. Using transformations of the form

$$u_r = A r^a \omega^m F(\eta)$$
$$u_\theta = B r^b \omega^n G(\eta)$$
$$u_z = C r^c \omega^q H(\eta)$$
and
$$\eta = E r^e \omega^t z^i$$

show that the governing equations reduce to the system of simultaneous ordinary differential equations

$$C_1 F + C_2 H' = 0 \tag{1}$$

$$C_3 F^2 + C_4 H F' + C_5 G^2 + C_6 F'' + C_7 = 0 \tag{2}$$

$$C_8 FG + C_9 H G' + C_{10} G'' = 0 \tag{3}$$

with boundary conditions

$$\eta = 0 \qquad F = G = H = 0$$

and

$$\eta \to \infty \qquad F = 0 \qquad G = 1$$

where the C_n are integer constants generally equal to ± 1 or ± 2, and the prime denotes differentiation with respect to η. Note: Be careful in choosing a, b, ... etc. not to eliminate one of the independent variables.

c. Using Eq. (1), show that the solution is reduced to the integration of a pair of coupled ordinary differential equations of the form

$$C_{11} H''' + C_{12} H H'' + C_{13} G^2 + C_{14} H'^2 = 0$$

and

$$C_{15} G'' + C_{16} H G' - C_{17} H' G = 0$$

with appropriate boundary conditions. The solution for $F(\eta)$ is then obtained sequentially.

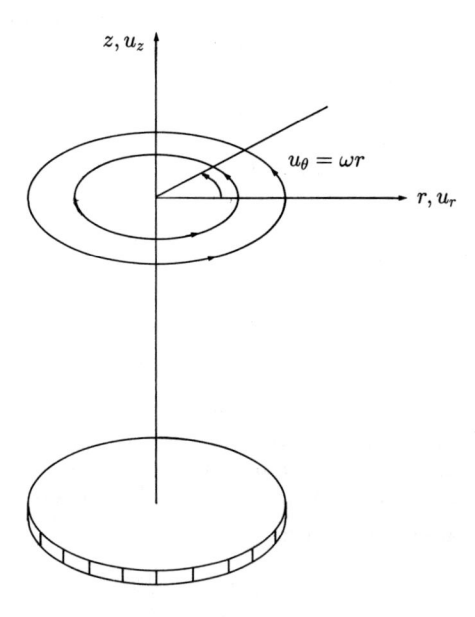

This problem is of a form similar to that governing the laminar free convection boundary layer discussed in Chapter 5. Its numerical integration is addressed in Problem 5–4.

3–11 Consider the three-dimensional incompressible viscous boundary layer on a yawed cylinder such as shown below.

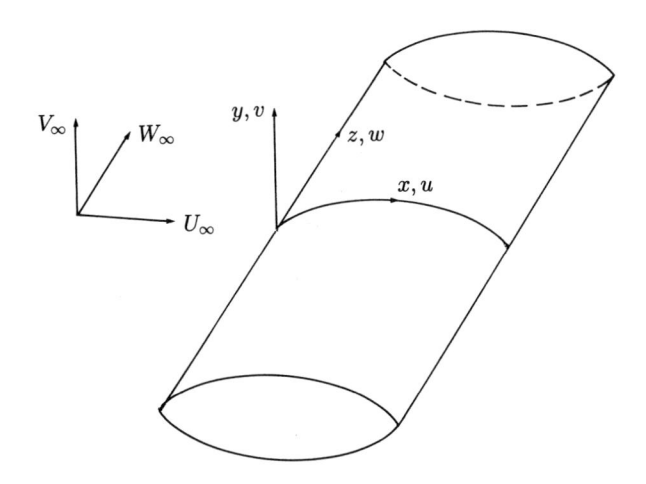

a. Using an order-of-magnitude analysis based on the thin layer concept, show that the continuity and incompressible Navier-Stokes (momentum) equations reduce to

$$\frac{\partial u}{\partial x} + \frac{\partial v}{\partial y} + \frac{\partial w}{\partial z} = 0$$

$$u\frac{\partial u}{\partial x} + v\frac{\partial u}{\partial y} + w\frac{\partial u}{\partial z} = -\frac{1}{\rho}\frac{\partial P}{\partial x} + \nu\frac{\partial^2 u}{\partial y^2}$$

$$\frac{\partial P}{\partial y} = 0$$

$$u\frac{\partial w}{\partial x} + v\frac{\partial w}{\partial y} + w\frac{\partial w}{\partial z} = -\frac{1}{\rho}\frac{\partial P}{\partial z} + \nu\frac{\partial^2 w}{\partial y^2}$$

with boundary conditions

$$y = 0 \quad u = v = w = 0$$
$$y \to \infty \quad u \to U \quad w \to W$$

b. Show that when the potential flow at the edge of the boundary layer is a function of x only, i.e., when

$$U = U(x) \quad W = W(x)$$

the three-dimensional incompressible boundary layer equations for an infinite yawed cylinder reduce to

$$\frac{\partial u}{\partial x} + \frac{\partial v}{\partial y} = 0$$

$$u\frac{\partial u}{\partial x} + v\frac{\partial u}{\partial y} = U\frac{\partial U}{\partial x} + \nu\frac{\partial^2 u}{\partial y^2}$$

$$u\frac{\partial w}{\partial x} + v\frac{\partial w}{\partial y} = \nu\frac{\partial^2 w}{\partial y^2}$$

396   Laminar Flow Analysis

with boundary conditions

$$y = 0 \quad u = v = w = 0$$
$$y \to \infty \quad u \to U \quad w \to W$$

c. The boundary layer flow for an infinite yawed wedge at zero angle of attack also satisfies these equations. For this case the potential velocity is given by

$$U = \text{constant } U_\infty x^m \quad W = \text{constant}$$

Using a transformation of the form

$$\psi = Ax^a y^b f(\eta)$$
$$v = Bg(\eta)$$
$$\eta = Cx^c y^d$$

show that the x momentum equation reduces to the Falkner-Skan equation

$$f''' + ff'' + \beta(1 - f'^2) = 0$$

and the z momentum equation to an ordinary differential equation of the form

$$C_1 g'' + C_2 fg' = 0$$

where the C_n are constants.

Further show that the boundary conditions are

$$\eta = 0 \quad f = f' = g = 0$$
$$\eta \to \infty \quad f' \to 1 \quad g \to 1$$

d. (c) Note that the x and z momentum equations are uncoupled. Using the **stag2d** program of Chapter 2 as a model, develop the appropriate main program for the general boundary layer driver program of Appendix D. Obtain numerical solutions for $\beta = \beta_{\text{sep}}$, -0.1, 0, 0.2, 0.5, 1.0, 1.6, 2.0, with $f''(0) \geq 0$. Attempt a solution for $\beta = 0.1$ and $f''(0) < 0$. Plot the nondimensional velocity profiles.

e. (c) Further modify the program to calculate the displacement and momentum thickness for both the x and z directions. Obtain results for $\beta = -0.1$, 0, 0.2, 0.5, 1.0, 1.6 and 2.0, and plot the results.

3–12   Derive the three-equation model boundary conditions for the locally nonsimilar method with mass transfer at the surface.

Modify the **faskan** program to calculate the results. Determine the shearing stress at the surface as a function of the mass transfer rate. Consider both a flat plate and a flow for which $U(x) = U_\infty x^m$ with uniform mass transfer. Plot the velocity profiles, and show the effect of mass transfer on them. Discuss your results.

Compare your results with those of the two-equation model.

Problems 397

3–13 A well-known boundary layer flow is Howarth's retarded flow. In this flow the inviscid velocity varies as

$$U(x) = U_\infty \left(1 - \frac{x}{\ell}\right)$$

a. Using the dependent and independent variable transformations

$$\psi = (\nu U x)^{1/2} f(\xi, \eta)$$

$$\xi = \frac{x}{\ell} \qquad \eta = y \left(\frac{U}{\nu x}\right)^{1/2}$$

show that the two-dimensional steady incompressible constant property boundary layer equation (see Eq. 3–43) and the associated no-slip, no-mass-transfer boundary conditions (see Eqs. 3–44) reduce to

$$f''' + \left(\frac{\Omega + 1}{2}\right) f f'' + \Omega(1 - f'^2) = \xi(f' f'_\xi - f'' f_\xi)$$

$$f(\xi, 0) = f'(\xi, 0) = 0$$

$$f(\xi, \eta \to \infty) \to 1$$

b. Letting $S = f_\xi$, derive the two-equation locally nonsimilar model.

c. Develop the appropriate main program and the **eqmot** routine for the **gblpdriv** routine, and numerically integrate the resulting system of equations. Obtain the shearing stress as a function of distance along the surface. Plot the velocity profiles for several positions along the surface. Discuss your results.

d. Letting $h = f_{\xi\xi} = S_\xi$, derive the three-equation locally nonsimilar model.

e. Do part (c) for the three-equation model.

f. Compare the results for the two- and three-equation model.

3–14(c) Develop the main program and the **eqmot** routine for the flow past a circular cylinder, using the two-equation locally nonsimilar method described by Eqs. (3–109) and (3–112), with the boundary conditions given by Eq. (3–113).

3–15(c) Develop the main program and the **eqmot** routine for the flow past a circular cylinder, using the three-equation locally nonsimilar method described by Eqs. (3–109), (3–116) and (3–117), with the boundary conditions given by Eq. (3–118).

3–16 The text discussed nonsimilarities caused by either the form of the external inviscid flow or by the surface boundary conditions. Nonsimilarities also result from the inclusion of additional terms in the governing equations. Typical of these additional terms are those associated with the transverse curvature of an axisymmetric body. In order to illustrate this effect, consider a long slender cylinder with its axis parallel to the flow direction. Assume that the flow is axisymmetric, steady and incompressible, with constant properties. The governing boundary layer equations in x, r cylindrical coordinates are then

continuity

$$\frac{\partial}{\partial x}(ru) + \frac{\partial}{\partial r}(rv) = 0$$

398 Laminar Flow Analysis

momentum

$$u\frac{\partial u}{\partial x} + v\frac{\partial u}{\partial r} = \frac{\nu}{r}\frac{\partial}{\partial r}\left(r\frac{\partial u}{\partial r}\right)$$

where the transverse curvature term is included in the term on the RHS of the momentum equation. The no-slip no-mass-transfer boundary conditions are

$$u(x, R) = 0 \qquad v(x, R) = 0$$

and

$$u(x, r \to \infty) \to U_e$$

where R is the radius of the cylinder and U_e is constant.

a. Introduce dependent and independent variable transformations

$$\psi = R(\nu U_e x)^{1/2} f(\xi, \eta)$$

$$\xi = \left(\frac{4}{R}\right)\left(\frac{\nu x}{U_e}\right)^{1/2}$$

$$\eta = \left(\frac{U_e}{\nu x}\right)^{1/2} \frac{(r^2 - R^2)}{4R}$$

and show that the governing equations and boundary conditions reduce to

$$(1 + \eta\xi)f''' + ff'' + \xi f'' = \xi(f' f'_\xi - f'' f_\xi)$$

with

$$f(\xi, 0) = f'(\xi, 0) = 0$$

and

$$f'(\xi, \eta \to \infty) \to 2$$

b. Letting $S = f_\xi$, derive the two-equation, locally nonsimilar model.

c. Develop the appropriate main program and the **eqmot** routine for the **gblpdriv** routine, and numerically integrate the resulting system of equations. Obtain the shearing stress as a function of distance along the surface. Plot the velocity profiles for several positions along the surface. Discuss your results.

d. Letting $h = f_{\xi\xi} = S_\xi$, derive the three-equation locally nonsimilar model.

e. Do part (c) for the three-equation model.

f. Compare the results for the two- and three-equation model.

CHAPTER 4

4–1 Derive Eqs. (4–40a) and (4–40b).

4–2(c) Modify the **faskan** program to generate the nondimensional temperature distributions shown in Figures 4–9 and 4–11.

4–3 Obtain solutions for the energy equation for the forced convection boundary layer on an isothermal surface without mass transfer ($f_0 = 0$, $n = 0$) for $\beta < 0$, $f''(0) < 0$, i.e. , for the reverse flow solutions.

4–4(c) Modify the **faskan** program to calculate $\alpha_0(\text{Pr})$ where

$$\alpha_0(\text{Pr}) = \frac{1}{\displaystyle\int_0^\infty [f''(\xi)]^{\text{Pr}} d\xi} \qquad (4-87)$$

Investigate the effect of $\eta_{\max}$ on these results. Use these results to determine the nondimensional temperature gradients, $\theta_1'(0)$, shown in Figure 4–10.

4–5 Using the method of variation of a parameter, derive Eq. (4–94), i.e.

$$\theta_2(\eta) = \text{Pr} \int_{\xi=\eta}^\infty [f''(\xi)]^{\text{Pr}} \left(\int_0^\xi [f''(\tau)]^{2-\text{Pr}} d\tau \right) d\xi$$

Show that for the special case of unit Prandtl number this reduces to

$$\theta_2(\eta) = \frac{1}{2}(1 - f'^2(\eta))$$

4–6 Show that the heat transfer is

$$q = k \left[\frac{1}{(2-\beta)} \frac{U}{\nu x} \right]^{1/2} \overline{T_w - T_e} \theta'(0)$$

for forced convection from a nonisothermal wall with a pressure gradient.

4–7 Show that plotting $\theta'(0)/(\text{Pr})^a$ vs n for forced convection from a nonisothermal surface with a pressure gradient reduces the parametric representation of $\theta'(0)$ vs n with Pr to almost one curve. What is the value of a?

4–8(c) Determine the effect of nonunit Prandtl number on the nondimensional heat transfer parameter, $\theta'(0)$, for forced convection from a nonisothermal surface without viscous dissipation. Consider various values of β.

4–9(c) Consider the forced convection boundary layer for parallel flow past a flat plate.

a. Show that making the substitution $\alpha = \theta_1(\eta)/\theta_1'(0)$ reduces Eqs. (4–82) and (4–83) to an initial value problem.

b. Describe the method you would use to numerically integrate the resulting boundary value problem.

c. Use the **forcb** program to demonstrate this method for Pr $= 1$, $\beta = 0$. Compare your result for $\theta_1'(0)$ with that determined using Eq. (4–88).

d. Is this method applicable to Eqs. (4–84) and (4–85)?

4–10 Derive the three-equation model for the locally nonsimilar boundary layer method.

4–11(c) Consider the inviscid flow past a circular cylinder with its axis perpendicular to the flow (see Section 3–11). Assume that the inviscid flow temperature is constant and that the cylinder has an isothermal wall.

a. Modify the `forcb` program to obtain solutions for the two-equation model for the locally nonsimilar boundary layer method.

b. Obtain the heat transfer rate (e.g., in terms of the Nusselt number) as a function of distance along the surface. Plot your results.

c. Obtain and plot the nondimensional temperature profiles along the surface. Discuss your results.

d. Modify the `forcb` program to determine the adiabatic wall temperature as a function of distance along the surface. Plot and discuss your results.

4–12(c) Consider Howarth's retarded flow.

a. Using the transformations given in Problem 3–13, derive the two-equation model for the locally nonsimilar method. Use

$$\theta = \frac{T - T_w}{T_e - T_w}$$

for the nondimensional temperature ratio, and $s = f_\xi$, $t = \theta_\xi$ for the momentum and energy auxiliary functions.

b. Modify the `forcb` program to obtain solutions for a constant external inviscid flow temperature with an isothermal wall temperature. Obtain the heat transfer rate (e.g., in terms of the Nusselt number) as a function of distance along the surface. Plot your results.

c. Obtain and plot the nondimensional temperature profiles along the surface. Discuss your results.

d. Modify the `forcb` program to determine the adiabatic wall temperature as a function of distance along the surface. Plot and discuss your results.

4–13(c) Using the results of Problem 4–9, modify the `forcb` program to obtain solutions for the three-equation locally nonsimilar boundary layer method. Compare your results with those for the two-equation model for either the circular cylinder or Howarth's retarded flow.

CHAPTER 5

5–1(c) Recall that the viscous incompressible steady flow about a flat disc rotating in its own plane with constant angular velocity was reduced to the solution of a pair of coupled ordinary differential equations of the form (see Problem 2–12)

$$C_{13}H''' + C_{14}HH'' + C_{15}G^2 + C_{16}H'^2 = 0$$

and

$$C_{17}G'' + C_{18}(HG)' = 0$$

with boundary conditions

$$\eta = 0 \quad H = 0 \quad H' = 0 \quad G = 1$$
$$\eta \to \infty \quad H' \to 0 \quad G \to 0$$

a. Modify the `freeb` program to obtain the solution of the above boundary value problem. Also obtain the solution for $P(\eta)$ and $F(\eta)$.

b. Plot the radial, tangential and axial velocity functions $F(\eta)$, $G(\eta)$ and $H(\eta)$, as well as the pressure function, $P(\eta)$, and their derivatives versus η.

c. Determine the tangential and radial components of the shearing stress at the surface.

d. Determine the displacement and momentum thicknesses.

5–2 The differential equations describing laminar free convection from the outer surface of a vertical circular cylinder for incompressible steady constant property flow are

$$\frac{\partial}{\partial z}(ru) + \frac{\partial}{\partial r}(rv) = 0$$

$$u\frac{\partial u}{\partial z} + v\frac{\partial u}{\partial r} = g\bar{\beta}(T - T_\infty) + \nu\left(\frac{\partial^2 u}{\partial r^2} + \frac{1}{r}\frac{\partial u}{\partial r}\right)$$

$$u\frac{\partial T}{\partial z} + v\frac{\partial T}{\partial r} = \frac{\nu}{\Pr}\left(\frac{\partial^2 T}{\partial r^2} + \frac{1}{r}\frac{\partial T}{\partial r}\right)$$

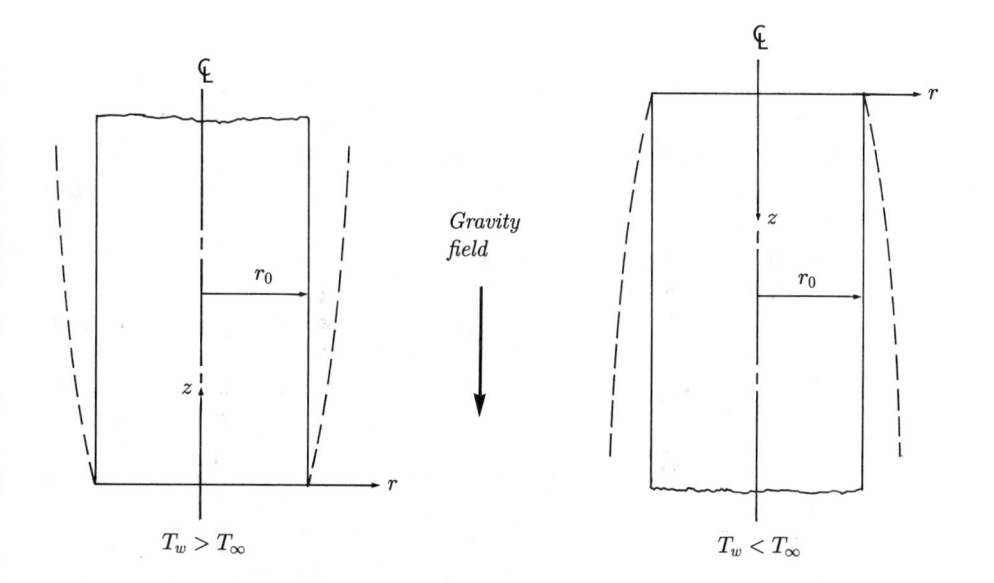

The corresponding boundary conditions are

$$r = r_0 \quad u = v = 0 \quad T = T_w(z) \quad \text{or} \quad \frac{\partial T}{\partial r} = -Q(z)$$

$$r \to \infty \quad u = 0 \quad T = T_\infty$$

a. Satisfy the continuity equation by introducing an axisymmetric stream function of the form

$$u = \frac{1}{r^a}\frac{\partial \psi}{\partial r} \qquad v = -\frac{1}{r^a}\frac{\partial \psi}{\partial z}$$

What is a?

b. Using an appropriate nondimensional form of the radial coordinate as the similarity variable, η, e.g.

$$\eta = \frac{r}{r_0^b}$$

introduce dependent variable transformations of the form

$$F(\eta) = A\frac{\psi(z,r)}{g(z)} \quad \text{and} \quad \theta(\eta) = B\frac{T(z,r) - T_\infty}{T_w(z) - T_\infty}$$

and show that a similar solution of the governing boundary value problem is of the form

$$C_1\eta^2 F''' + C_2\eta(g'F' - 1)F'' + C_3\eta(g'F + 1)F' + C_4(T_w - T_\infty)\frac{\eta^3}{g}\theta = 0$$

$$C_5\eta\theta'' + C_6(\Pr g'F + 1)\theta' - C_7\Pr(g'F')\theta = 0$$

with boundary conditions

$$F(1) = F'(1) = 0 \qquad F'(\eta \to \infty) \to 0$$
$$\theta(1) = 1 \qquad \theta(\eta \to \infty) \to 0$$

where the C_n are constants and the prime denotes differentiation with respect to the appropriate argument.

c. What are the conditions on $g(z)$ and $T_w(z)$ which yield a similar solution?

d. What are the specific functional forms of $g(z)$ and $T_w(z)$ which yield a similar solution?

e. Modify the **freeb** program to obtain a numerical solution of the resulting similar boundary value problem.

1) Determine the effect of Prandtl number on the nondimensional shearing stress and heat transfer at the cylinder surface.

2) Calculate and plot several nondimensional velocity and temperature profiles. Discuss these results.

5-3 Consider the problem of laminar free convection from a nonisothermal vertical cone shown below. The governing boundary layer equations for steady axisymmetric constant property free convection flow without viscous dissipation are

$$\frac{\partial}{\partial x}(ru) + \frac{\partial(rv)}{\partial y} = 0$$

$$u\frac{\partial u}{\partial x} + v\frac{\partial u}{\partial y} = g\bar{\beta}\cos\gamma(T - T_\infty) + \nu\frac{\partial^2 u}{\partial y^2}$$

$$u\frac{\partial T}{\partial x} + v\frac{\partial T}{\partial y} = \frac{\nu}{\Pr}\frac{\partial^2 T}{\partial y^2}$$

Note that here we assumed the boundary layer is thin with respect to the local cone radius and hence replaced the local radius to a point in the boundary layer with the local radius to the cone surface. Note that this condition is not satisfied in the neighborhood of the cone tip. The appropriate boundary conditions are

$$y = 0 \qquad u = v = 0 \qquad T = T_w(x)$$
$$y \to \infty \qquad u \to 0 \qquad T \to T_\infty$$

a. Introduce nondimensional variables such that

$$\bar{x} = \frac{x}{\ell} \qquad \bar{y} = \frac{y}{\ell} \qquad \bar{r}(\bar{x}) = \bar{x}\sin\gamma$$

$$\bar{u} = \frac{u\ell}{\nu} \qquad \bar{v} = \frac{v\ell}{\nu} \qquad G = \frac{g\bar{\beta}\cos\gamma(T-T_\infty)\ell^3}{\nu^2}$$

b. Satisfy the continuity equation by introducing the axisymmetric stream function.
c. Seek similar solutions to the resulting boundary value problem by introducing the similarity transformations

$$\eta = \frac{\bar{y}}{g(\bar{x})} \qquad F(\eta) = \frac{\psi(\bar{x},\bar{y})}{\bar{r}f(\bar{x})}$$

$$\theta(\eta) = \frac{G(\bar{x},\bar{y})}{G_w(\bar{x})} = \frac{T-T_\infty}{T_w(x)-T_\infty}$$

Show that the governing equations reduce to the form

$$F''' + \frac{g}{\bar{x}}(f\bar{x})' FF'' - g^2\left(\frac{f}{g}\right)' F'^2 + G_w \frac{g^3}{f}\theta = 0$$

$$\theta'' + \Pr\left[\frac{g}{\bar{x}}(f\bar{x})' F\theta' - \frac{fg}{G_w} G'_w F'\theta\right] = 0$$

with boundary conditions

$$\eta = 0 \qquad F'(0) = F(0) = 0 \qquad G(0) = 1$$
$$\eta \to \infty \qquad F'(\eta \to \infty) \to 0 \qquad \theta(\eta \to \infty) \to 0$$

where the prime denotes differentiation with respect to the appropriate argument.
d. Under what conditions do similar solutions to these equations exist, i.e., under what conditions are these equations ordinary differential equations?

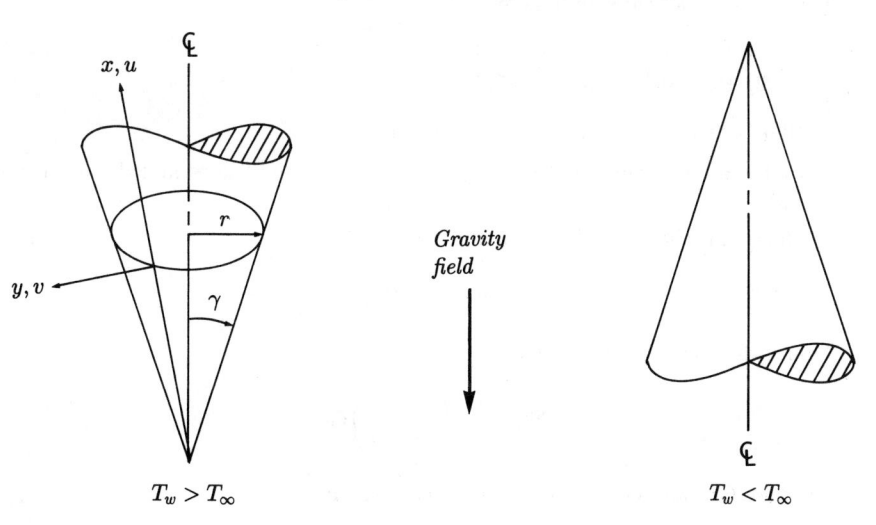

e. Determine $f(\bar{x})$, $g(\bar{x})$ and $G_w(\bar{x})$. Hint: Note that there are more conditions than the number of unknowns. Hence, the conditions are not all independent. Postulate that
$$T_w - T_\infty = T_1 x^n$$
and use this fact to determine $f(\bar{x})$, $g(\bar{x})$ from the first two conditions and thence $G_w(\bar{x})$. What are η, $F(\eta)$, $\theta(\eta)$, $\bar{u}$, $\bar{v}$?

f. Show that for an isothermal cone ($n = 0$) the governing boundary value problem is
$$F''' + \frac{7}{4}FF'' - \frac{1}{2}F'^2 + \theta = 0$$
$$\theta'' + \frac{7}{4}\Pr F\theta' = 0$$

with
$$F(0) = F'(0) = 0 \qquad \theta(0) = 1$$
$$F'(\eta \to \infty) \to 0 \qquad \theta(\eta \to \infty) \to 0$$

What are η, $F(\eta)$ and $\theta(\eta)$ in this case?

g. Show that for a linear surface temperature distribution ($n = 1$) the governing boundary value problem is
$$F''' + 2FF'' - F'^2 + \theta = 0$$
$$\theta'' + \Pr(2F\theta' - F'\theta) = 0$$

with
$$F(0) = F'(0) = 0 \qquad \theta(0) = 1$$
$$F'(\eta \to \infty) \to 0 \qquad \theta(\eta \to \infty) \to 0$$

What are η, $F(\eta)$ and $\theta(\eta)$ in this case?

h. Modify the **freeb** program to obtain numerical solutions for the generalized form of the governing equations.

i. Obtain numerical solutions for

  1) $\Pr = 1$ $\qquad n = 0, 0.5, 1$
  2) $\Pr = 0.7$ $\qquad n = 0, 0.5, 1$

Hint: $F''(0) = O(1)$ and $-\theta'(0) = O(1)$.

Plot the nondimensional velocity and temperature profiles and discuss the results.

j. Show that the local Nusselt number is
$$\mathrm{Nu}_x = -\mathrm{Gr}_x^{1/4}\,\theta'(0)$$
and the mean Nusselt number
$$\bar{\mathrm{Nu}}_\ell = -\left(\frac{8}{5n+7}\right)\mathrm{Gr}_\ell^{1/4}\,\theta'(0)$$
where Gr_x and Gr_ℓ are local and mean Grashof numbers.

5–4 Recall that the viscous incompressible steady flow of a rotating fluid over a fixed plate was reduced to the solution of a pair of coupled ordinary differential equations of the form (see Problem 3–10)

$$C_{11}H''' + C_{12}HH'' + C_{13}G^2 + C_{14}H'^2 = 0$$

and

$$C_{15}G'' + C_{16}HG' - C_{17}H'G = 0$$

with boundary conditions

$$\eta = 0 \qquad G = H = H' = 0$$
$$\eta \to \infty \qquad H' \to 0 \quad G \to 0$$

a. Modify the **freed** program to obtain the solution of the above boundary value problem. Also obtain the solution of $F(\eta)$.

b. Determine the appropriate displacement and momentum thicknesses, and modify the program to calculate them.

c. Plot the radial, tangential and axial velocity functions $F(\eta)$, $G(\eta)$ and $H(\eta)$ and their derivatives versus η.

d. Determine the tangential and radial components of the shearing stress at the surface.

5–5 Complete the details of the derivation of the two-equation model for the locally nonsimilar method as applied to the free convection boundary layer, i.e., complete the details of the derivation of Eqs. (5–32) to (5–35) and the boundary conditions given by Eqs. (5–27) and (5–36).

5–6 Derive the three-equation model for the locally nonsimilar method as applied to the free convection boundary layer with mass transfer.

5–7 Consider the nonsimilar boundary layer which occurs on an isothermal vertical flat plate with *constant* mass transfer at the surface. Consider both suction and injection.

a. Determine and plot the nondimensional velocity and temperature profile for several locations along the plate.

b. Determine and plot the variation of shearing stress and nondimensional heat transfer rate (Nusselt number) along the plate surface.

c. Compare your results with those from the locally similar approach.

5–8 Complete the details of the derivation of Eqs. (5–46), (5–47) and (5–48).

5–9 Show that use of the free convection similarity variables

$$\eta = \left(\frac{g\bar{\beta}T_1}{4\nu^2}\right)^{1/4} yx^{\frac{n-1}{4}}$$

$$\psi = 4\nu\left(\frac{g\bar{\beta}T_1}{4\nu^2}\right)^{1/4} x^{\frac{n+3}{4}} f(\eta)$$

$$\theta = \frac{T - T_\infty}{T_w - T_\infty}$$

406 Laminar Flow Analysis

also yields a similar solution of the governing equations for combined forced and free convection boundary layer flow.

5-10 Obtain numerical results for a uniform heat flux distribution for combined forced and free convection flows with $n = 1/5$, $m = 3/5$. Show that the separation value $f''(0) = 0$ occurs for $Gr_x/Re_x^2 = -1.15$. Explain this result.

5-11 Derive the three-equation model for the locally nonsimilar method for the combined forced and free convection boundary layer with mass transfer.

5-12 Consider the nonsimilar combined forced and free convection boundary layer which occurs on an isothermal surface with mass transfer at the surface when the external inviscid velocity is given by

$$\frac{U(x)}{U_\infty} = \frac{1-x}{\ell}$$

Assume that the temperature of the external inviscid flow is also constant. Determine and plot the nondimensional velocity and temperature profiles for several locations along the plate. Determine and plot the nondimensional shearing stress and heat transfer rate (Nusselt number) along the plate surface. Compare your results with those obtained using a locally similar approach.

CHAPTER 6

6-1 Obtain Eqs. (6-29) and (6-30) from Eq. (6-24).

6-2 Consider compressible Couette flow between rotating concentric circular cylinders.

a. Assuming that all the variables are functions of r only, show that the only nonzero velocity component is u_θ and that the two-dimensional compressible Navier-Stokes equations in polar coordinates without body forces, i.e.

continuity

$$\frac{\partial}{\partial r}(r u_r) + \frac{\partial u_\theta}{\partial \theta} = 0 \tag{1}$$

r momentum

$$\rho\left(\frac{Du_r}{Dt} - \frac{u_\theta^2}{r}\right) = -\frac{\partial P}{\partial r} + \frac{\partial}{\partial r}\left[2\mu\frac{\partial u_r}{\partial r} - \frac{2}{3}\mu\left(\frac{\partial u_r}{\partial r} + \frac{u_r}{r} + \frac{1}{r}\frac{\partial u_\theta}{\partial \theta}\right)\right]$$

$$+ \frac{1}{r}\frac{\partial}{\partial \theta}\left[\mu\left(\frac{1}{r}\frac{\partial u_r}{\partial \theta} + \frac{\partial u_\theta}{\partial r} - \frac{u_\theta}{r}\right)\right]$$

$$+ \frac{2\mu}{r}\left(\frac{\partial u_r}{\partial r} - \frac{1}{r}\frac{\partial u_\theta}{\partial \theta} - \frac{u_r}{r}\right) \tag{2}$$

θ momentum

$$\rho\left(\frac{Du_\theta}{Dt} + \frac{u_r u_\theta}{r}\right) = -\frac{1}{r}\frac{\partial P}{\partial \theta}$$

$$+ \frac{1}{r}\frac{\partial}{\partial \theta}\left[\frac{2\mu}{r}\frac{\partial u_\theta}{\partial \theta} - \frac{2}{3}\mu\left(\frac{\partial u_r}{\partial r} + \frac{u_r}{r} + \frac{1}{r}\frac{\partial u_\theta}{\partial \theta}\right)\right]$$

$$+ \frac{\partial}{\partial r}\left[\mu\left(\frac{1}{r}\frac{\partial u_r}{\partial \theta} + \frac{\partial u_\theta}{\partial r} - \frac{u_\theta}{r}\right)\right]$$

$$+ \frac{2\mu}{r}\left(\frac{1}{r}\frac{\partial u_r}{\partial \theta} + \frac{\partial u_\theta}{\partial r} - \frac{u_\theta}{r}\right) \qquad (3)$$

energy

$$\rho\frac{Dh}{Dt} = \frac{DP}{Dt} + \frac{1}{r}\frac{\partial}{\partial r}\left(kr\frac{\partial T}{\partial r}\right) + \frac{1}{r^2}\frac{\partial}{\partial \theta}\left(k\frac{\partial T}{\partial \theta}\right)$$

$$+ \mu\left[\left(\frac{\partial u_r}{\partial r}\right)^2 + 2\left(\frac{1}{r}\frac{\partial u_\theta}{\partial \theta} + \frac{u_r}{r}\right)^2\right.$$

$$\left. + \left(\frac{1}{r}\frac{\partial u_r}{\partial \theta} + \frac{\partial u_\theta}{\partial r} - \frac{u_\theta}{r}\right)^2\right]$$

$$- \frac{2}{3}\mu\left(\frac{\partial u_r}{\partial r} + \frac{1}{r}\frac{\partial u_\theta}{\partial \theta} + \frac{u_r}{r}\right)^2 \qquad (4)$$

where

$$\frac{D}{Dt} = u_r\frac{\partial}{\partial r} + \frac{u_\theta}{r}\frac{\partial}{\partial \theta}$$

reduce to

$$\frac{dP}{dr} - \rho\frac{u_\theta^2}{r} = 0 \qquad (5)$$

$$\frac{d}{dr}\left[r^2\mu\left(\frac{du_\theta}{dr} - \frac{u_\theta}{r}\right)\right] = 0 \qquad (6)$$

$$\frac{1}{\Pr}\frac{1}{r}\frac{d}{dr}\left(r\mu\frac{dH}{dr}\right) + \mu\left(\frac{du_\theta}{dr} - \frac{u_\theta}{r}\right)^2 = 0 \qquad (7)$$

Write the appropriate boundary conditions.

b. Introduce an angular velocity $\omega(r)$ such that $u_\theta = \omega r$, and show that Eqs. (6) and (7) become

$$\mu r^3 \frac{d\omega}{dr} = \text{constant} = A \tag{8}$$

and

$$\mu r \frac{dH}{dr} + \text{Pr}\, A\omega = \text{constant} = B \tag{9}$$

c. (c) Introduce the appropriate nondimensional variables (see Section 6–3), and make the analogous assumption to $\bar{\mu} = g$ of Section 6–3 (see Eq. 6–18). Integrate the resulting equations. (Hint: a numerical integration may be necessary.) Plot and discuss your results.

d. Solve for $P(r)$, $\rho(r)$ and $T(r)$, using Eq. (5) and the ideal gas equation of state, $P = \rho RT$.

e. Compare these results with those for plane Couette flow (see Section 6–3).

6–3 Derive the von Mises transformation for incompressible flow.

6–4 Derive the boundary layer equations in von Mises coordinates for axisymmetric compressible two-dimensional flow. Show that the continuity equation is automatically satisfied.

6–5 Develop the solution for compressible Couette flow using Crocco variables.

6–6 Show that when $\text{Pr} = 1$ and $g_w = \text{constant}$ then $g(\eta) = 1$ is a solution of the energy equation, and that $g_{aw} = 1$, i.e., derive the Crocco energy integral.

6–7 Derive Eq. (6–82) from Eq. (6–81).

a. State the conditions under which the result is applicable.

b. Determine an explicit equation for the physical dimension, y, for this case.

c. Modify the `forcb` program of Chapter 4 to integrate the $M_e = \text{constant}$, $\text{Pr} = \text{constant}$, $\hat{\beta} = 0$ problem, and calculate the physical dimension, y, for each value of η (see Section 6–7).

6–8 Show that Eq. (6–91) is a solution of the boundary value problem given by Eqs. (6–89) and (6–90).

6–9 Show that $\hat{\beta} = 1$ for compressible two-dimensional stagnation point flow.

6–10 Show that $\hat{\beta} = 1/2$ for compressible axisymmetric stagnation point flow.

6–11 Using the numerical results of Appendix C, show that for the unit Prandtl number solutions of the compressible boundary layer the physical thickness of the boundary layer, y, decreases as g_w increases, rather than increases as is indicated by η.

6–12(c) Using the numerical results given in Chapter 5, calculate and plot the heat transfer parameter as a function of Mach number, M_e, and wall-to-stagnation temperature ratio, g_w, for a flat plate in compressible flow.

6–13(c) For the similar compressible boundary layer, the use of constant Mach number, M_e, is strictly valid only for the flat plate, i.e., $\hat{\beta} = 0$ (see Section 6–5). However, it may be valid to assume the external Mach number constant in the energy equation for small values of $\hat{\beta}$. Using the `comb` program, determine the effect of Mach number (say $\bar{\sigma} = 0$, 0.5, 1, 1.5, 2) on the heat transfer rate and shearing stress at the surface for $\hat{\beta} = -0.1, 0, 0.1$ and Prandtl number $= 0.723$. Plot and discuss the results, including the velocity and enthalpy profiles.

6–14 Using the appropriate solutions of the Blasius equation with mass transfer at the surface, investigate the effects of surface mass transfer on the heat transfer at the surface for a flat plate in compressible flow. Discuss the physical effect on the enthalpy profiles and heat transfer rate at the surface. Hint: use the method of linear superposition to solve the energy equation.

6–15 Using the numerical results in Appendix C, compare the shearing stress and heat transfer rate for the compressible boundary layer when $M_e = 0$ and when $M_e \neq 0$ but $\Pr = 1$.

6–16(c) By plotting g vs f' for the unit Prandtl number solutions, show that for favorable pressure gradients the thermal boundary layer is thicker than the velocity boundary layer, with the reverse being true for adverse pressure gradients. Use the comb program to generate the required profiles.

6–17(c) Using the comb program, obtain solutions for the nonunit Prandtl number low speed similar compressible boundary layer ($\bar{\sigma} = 0$) for mass transfer at the surface ($f_0 \neq 0$). Investigate the effects of pressure gradient, heat and mass transfer at the surface. Plot and discuss your results.

6–18(c) For the compressible boundary layer with reverse flow, show that $f(\eta) = 0$ at some value of η in the boundary layer. (Reason this out physically, and then show it numerically). Also, show that the value of η for which $f(\eta) = 0$ increases monotonically and approaches 'infinity' as the singularity at $\hat{\beta} = 0^-$, $g'(0) = 0^{\pm}$.

6–19(c) For the unit Prandtl number compressible boundary layer, use the comb program to calculate the velocity and enthalpy profiles for 'large' blowing rates, say $f_0 = -1, -1.5, -2, -3$. Discuss the results physically. Explain the physical mechanism which allows mathematical solutions for values of the mass transfer parameter, f_0, for larger magnitudes than that for the flat plate blow-off value of -0.875745 (see Section 3–10). Plot your numerical results, and show that near the surface the boundary layer fluid consists mostly of injected fluid. What is the dominant heat transfer mechanism in the region?

6–20(c) Using the numerical results given in Appendix C, estimate the adiabatic wall-to-stagnation temperature ratio, g_{aw}, for a favorable pressure gradient of $\hat{\beta} = 0.5$ and a nonunit Prandtl number, $\Pr = 0.723$. Then, using the comb program, confirm your estimates of g_{aw} and estimate your errors. Would the errors be significant in determining the heat transfer rate? Plot your results as a function of η, and discuss them physically.

6–21 Modify the comb program in order to calculate the values of $\hat{\beta}$ and $g(0) = g_{aw}$, which yield the adiabatic wall separation solution, i.e., $f''(0) = 0$, $g'(0) = 0$. Calculate the value for $\hat{\beta} = 2/3$, $\Pr = 0.723$.

6–22 Derive the three-equation model for the locally nonsimilar boundary layer method for the compressible boundary layer.

6–23(c) Consider an isothermal flat plate with uniform mass transfer at the surface immersed in a compressible viscous fluid. Consider both suction and injection. For injection, assume that the injected fluid is the same as that in the external stream, that it has the same temperature as the surface and is injected normal to the surface. Using the two-equation model for the locally nonsimilar boundary layer method, obtain a solution. Modify the **comb** program to perform the required numerical integration. Determine and plot nondimensional velocity, shear, enthalpy and heat transfer profiles at several locations along the plate. Obtain and plot the nondimensional shearing stress and heat transfer rate along the surface.

6–24   Using the two-equation model for the locally nonsimilar method, obtain the solution for the compressible viscous boundary layer on an isothermal surface when the external velocity field varies as

$$\frac{U_e(x)}{U_\infty} = 1 - \frac{x}{\ell}$$

Assume that the external stagnation enthalpy field is constant. Modify the **comb** program to perform the numerical integration. Obtain and plot the nondimensional velocity and enthalpy profiles for several locations along the surface. Obtain and plot the nondimensional shearing stress and heat transfer rate along the surface. Compare your results with a local similarity analysis.

INDEX

Acceleration
    convective, 4
    local, 4
    substantial, 4
    total, 4
Adams, J.A., 173, 353
Adiabatic wall, 21, 115, 118, 119, 132, 144, 213
    compressible boundary layer
        flat plate, 225, 228
        hypersonic, 257, 261
            nonunit Prandtl number, 233
        low speed, 224, 225
        mass transfer, 269
        unit Prandtl number, 231
    Couette flow, simple, 112, 114
        compressible, 203, 204, 206
    temperature, 21, 122
Adverse pressure gradient, 77, 92, 95, 185, 187
    compressible boundary layer, 244, 246–249, 264, 265
        hypersonic, 257
    Couette flow, nonsimple, 115
    velocity profile in, 78
Asymptotic boundary conditions, satisfaction of, 299
Asymptotic boundary value problem, numerical integration of, 299
Asymptotic convergence, 40
Axisymmetric
    boundary layer, 84–86
        on thin needle, 88
    Couette flow, 199
    potential flow, 87
    stagnation point flow, 44–51, 72, 88, 89
        solution for (table), 50

Ball, K.O.W., 263
Baxter, D.C., 95
Beckmann, W., 168
Beckwith, Z.E., 262
Bernoulli equation, 36
Bertram, M.H., 275
Blasius equation, 64, 65, 70, 72, 91, 128, 130, 131, 161, 162, 214, 215, 219, 224–226, 237, 238, 257
    group properties, 69
    results for (table), 71
    series expansion for, 94
    solution of, 60, 74, 79, 88
Blasius, H., 64, 66
Blow off, 91
    compressible boundary layer, 264
Blowing, 90, 148, 177, 178, 223, 261–263, 265
    See also injection; mass transfer
Body force, 107, 199
Boerner, C.J., 96–98, 100, 103
Boundary conditions
    adiabatic wall, 20
    asymptotic, 98
    axisymmetric stagnation point flow, 45
    compressible boundary layer, 215
        edge, 209
        mass transfer, 262
        nonsimilar, 279
    Crocco coordinates, 213
    field, 20, 21
    infinity, 47
    mass transfer, nonsimilar, forced convection, 158
    no-mass-transfer, 45, 60, 128, 157, 158, 160, 179, 180, 185, 200
    no-slip, 20, 25, 27, 35, 45, 53, 54, 60, 128, 185, 200, 209

412   Index

Boundary conditions (Cont.):
    no-temperature-jump, 20, 53, 54, 200
    suddenly accelerated plane wall, 30, 33
    surface, 20
    two-dimensional stagnation point flow, 37, 39
    von Mises coordinates, 211
    zero heat transfer, 20
Boundary layer
    combined forced and free convection, 183–193
        nonsimilar 180–193
        numerical integration of, 186
        See also combined forced and free convection;
        energy equation;
        heat transfer;
        momentum equation;
        shearing stress;
        temperature profile;
        three-equation locally nonsimilar method;
        two-equation locally nonsimilar method;
        velocity profile
    compressible, 60, 95, 195, 208–281
        axisymmetric nonsimilar, 277
        flat plate, 225–230
        hypersonic, 257
        nonunit Prandtl number, 232, 233, 257–261
        low speed, 223–225, 237–241
        mass transfer 261–269
        nonsimilar, 276–281
            mass transfer, 277
        numerical integration of, 236, 237
        shock-boundary layer interaction, 270–276
        transformation of, 210–223
        unit Prandtl number, 230–232, 241–256
            equations for, 230
        See also adiabatic wall;
        adverse pressure gradient;
        blow off;

Boundary layer, compressible (Cont.):
    boundary conditions;
    continuity equation;
    displacement thickness;
    enthalpy;
    energy equation;
    eqmot;
    eqmotcb;
    favorable pressure gradient;
    heat transfer;
    injection;
    iteration;
    locally nonsimilar method;
    mass transfer;
    momentum equation;
    momentum thickness;
    Nachtsheim-Swigert iteration;
    reverse flow;
    shearing stress;
    similarity;
    suction;
    three-equation locally nonsimilar method;
    two-equation locally nonsimilar method;
    viscous dissipation;
    velocity profile
    derivation of equations for, 54–60
    forced convection, 105, 127–147 162, 164
        mass transfer, 147–154, 175
        See also boundary conditions;
        continuity equation;
        energy equation;
        forced convection;
        heat transfer;
        locally nonsimilar method;
        mass transfer;
        momentum equation;
        nonsimilar boundary layer;
        three-equation locally nonsimilar method;
        two-equation locally nonsimilar method;
        velocity profile
    free convection, 105, 159–183
        isothermal vertical flat plate, 164–170

Boundary layer, free convection (Cont.):
  level curves of error, 310
  nonisothermal vertical
    flat plate, 170–178
  nonsimilar, 182
  numerical integration of, 164
    mass transfer, 173,
      175–178
    See also energy equation;
      eqmot;
      eqmotcv;
      free convection;
      heat transfer;
      iteration ;
      locally nonsimilar method;
      mass transfer;
      momentum equation;
      Nachtsheim-Swigert
        iteration;
      nonsimilar boundary layer;
      shearing stress;
      temperature profile;
      three-equation locally
        nonsimilar method;
      two-equation locally
        nonsimilar method;
      velocity profile
  general program, 353
  general two-dimensional
    incompressible, 65
  Görtler series expansion for, 94
  incompressible, on flat plate, 60–64
  mass transfer in, 90, 92–94
  nonsimilar, 93
    free convection, 178
      mass transfer, 182
  Prandtl's, 53, 54
  pressure gradient, 65
  program for, 375–378
  results for (table), 340–342, 344
    mass transfer (table),
      344–350
    See also Appendix C
  thermal, 54
  velocity, 54
Boundary value problem, two-point
  asymptotic, 47
Brown, W.B., 240, 262
Bulk heat addition, 106, 123, 200

Buoyancy force, 105, 107, 108, 128,
  177, 195
Buoyancy term, 123, 184, 185
Busemann integral, 224, 226
Bussmann, K., 91

Catherall, D., 103
Cauchy's equations, 5
Channel flow 26, 68
Chapman, D.R., 197
Chapman-Rubesin viscosity law, 197,
  198, 202, 214, 215, 218, 222, 272,
  273, 277
Chen, K., 262
Circular cylinder
  nonsimilar boundary layer on, 99
  shearing stress on, 100
  skin friction coefficient for, 100
Clutter, D.W., 95
Cohen, C.B., 231, 240, 243
comb, 237–239, 246, 247, 257, 262,
  276
Combined forced and free convection,
  124
  See also boundary layer, combined
    forced and free convection
combl, 376–378
Compressible boundary layer *See*
  boundary layer, compressible
Compressive work, 208
Continuity equation, 3, 19, 22, 24, 25,
  30, 45, 55, 106, 123, 124, 183, 199
  axisymmetric, 84
  axisymmetric stagnation point flow,
    44
  boundary layer, 59
  compressible boundary layer, 208,
    210
    Crocco coordinates for, 212
  Couette flow, simple, compressible,
    200
  cylindrical polar coordinates, 44
  flat plate boundary layer, 60
  forced convection, 128, 137
    mass transfer, 147
    nonsimilar, 154
  incompressible, 21
  two-dimensional stagnation point
    flow, 36

414   *Index*

convbl, 372–374
Convergence asymptotic, 40
Couette flow, 24, 26
    adiabatic wall in, 113, 114, 203, 204, 206
    adverse pressure gradient in, 115
    axisymmetric, 199
    compressible, 195, 199
    continuity equation for, 200
    energy equation for, 200
    enthalpy in, 203, 204
    heat transfer in, 203, 205, 207
    momentum equation for, 200
    nonsimple, 27, 28
    Reynolds analogy for, 208
        temperature effects in, 115–119
    simple, 27
        compressible, 202
        temperature effects in, 110–115
    shearing stress in, 200, 207
    skin friction coefficient for, 207, 208
    temperature profile for, 117, 118
    velocity profile for, 203, 206
    viscous dissipation in 203-206
Crocco coordinates *See* energy equation, enthalpy, momentum equation
Crocco equations, 212, 218
    physical coordinate in, 213
Crocco integral, 224, 226
Crocco, L., 228
Crocco transformation, 210, 211,
    See also shearing stress
Cylindrical polar coordinates, 44

Derivative, Eulerian, 4
Dewey, C.F., 95
Diffusion coefficient momentum, 125
Diffusivity, thermal, 160
Dilatation, 12
Dilatational viscosity, 12
Displacement thickness, 77, 79
    compressible boundary layer, 234, 236, 237, 252, 253, 255, 264
    Falkner-Skan solution, 80, 81
    shock-wave boundary layer interaction, 270–276

Displacement thickness (Cont.):
    two-dimensional stagnation point flow, 80
Donoughe, P.L., 240, 262
Dynamic viscosity, 57

Eckert number, 57, 108–110, 121, 129, 136, 139, 141, 156, 184, 195
Eichhorn, R., 176, 186–188
Elzy, E., 143, 148
Emmons, H.W., 91
Energy, conservation of, 1, 13, 14
Energy equation, 13, 18, 22, 54, 56, 183, 200
    boundary layer, 59, 122–124, 208, 210, 221
        compressible, 215
        Crocco coordinates, 212
        derivation of, 58, 59
        flat plate, 225
        nonsimilar, 278
    combined forced and free convection, nonsimilar, 191
    Couette flow, simple, compressible, 200
    forced convection, 128, 129, 137, 219
        nonsimilar, 155–157
    free convection, 162, 164
        nonsimilar, 179, 180
    mass transfer results for (table), 332–336
    Poiseuille flow, 119
    results for (table), 331
    thermal boundary layer, 125, 126
    thermal flows, 106, 108, 109
    two-dimensional stagnation point flow, 39
Enthalpy, 19
    Crocco coordinates, at surface, 213
    overshoot, compressible boundary layer, hypersonic, 257
    profile
        boundary layer
            compressible, 240, 241, 244, 248, 250
                hypersonic, 257, 258

Enthalpy, profile, boundary layer, compressible (Cont.):
    mass transfer, effect of, 265
    $M_e \to 0$, 204
    Mach number, effect of, 228
  Couette flow, simple
    adiabatic wall, 203
    compressible, 203, 206
eqmot, 100, 370
  compressible boundary layer, 237
  free convection boundary layer, 186
eqmot2d, 48, 366
eqmotcb, 378
eqmotcv, 374
eqmotfs, 80, 369
Equation, continuity *See* continuity equation
Equation of state, 20, 21, 107
Euler's equation, 53, 65, 85, 273, 277
Eulerian derivative, 4
Evans, H.L., 84, 86, 88, 92, 95, 143, 154
Experimental data, free convection, 168, 169
explain, 355
explain2d, 366
explaincb, 379
explaincv, 275
explainfs, 370

Falkner, V.M., 67
Falkner-Skan equation, 65–68, 70–77, 82, 137, 139, 141, 184, 222, 225, 233, 238, 301, 303
  eqmotfs, 369
  main program, 366–369
  mass transfer, 92
  Newton-Raphson iteration, 299
  numerical integration of, 70–77
  physical interpretation, 81–84
  results (table), 322
    mass transfer (table), 323–329
  reverse flow solutions, 76–78
Falkner-Skan pressure gradient parameter, 224
Falkner-Skan solution, 70–79, 84, 94, 234

Falkner-Skan solution (Cont.):
  axisymmetric flow, 86
  displacement thickness, 80
  momentum thickness, 80
  potential flows for, 83
Fanucci, J.B., 262
faskan, 70, 72, 75, 76, 80, 91, 100, 103, 130–132, 139, 140, 164, 367–369
Favorable pressure gradient
  compressible boundary layer, 250, 252, 255, 263–266
    hypersonic, 257, 261
    mass transfer, 267–269
Field, 1
Finite difference-differential method, 94, 95
  results for circular cylinder, 100
Flat plate, shock-wave boundary layer interaction on, 271
Flow
  axisymmetric
    conical, 86, 87
    Couette flow, 199
    stagnation point, 64, 72, 88
    See also axisymmetric
  between parallel plates, 199
  channel, 26
    with parallel walls, 119
  circular, compressible, 199
  Couette, 24, 26
    compressible, 195
    nonsimple, 27, 28
      temperature effects in, 115–119
    simple, 27
      temperature effects in, 110–115
    See also Couette flow
  flat plate
    Blasius solution, 60–64
    forced convection on, 127
  free convection, 159, 161
  Hagen-Poiseuille, 24
  hypersonic, nonunit Prandtl number, 232
  inviscid
    axisymmetric stagnation point, 45

Flow, inviscid (Cont.):
  two-dimensional stagnation
    point, 35
  parallel, 23
  separation, 77
  stagnation point, 66
  suddenly accelerated plane wall, 28
  two-dimensional stagnation point,
    35, 46, 48, 49, 51, 64, 82,
    145, 243
  wedge, 82
Fluid
  compressible, 20
  continuous, 2–4, 13, 20
  deformation, 8, 9
  element
    deformation of, 9
    rotation of, 9
    translation of, 9
  heat conducting, 20
  homogeneous, 2, 3, 20
  isotropic, 7, 20
  Stokesian, 20
  viscosity, 12
  viscous, 20
Flügge-Lotz, I., 95
forcb, 140, 141, 143, 148, 164
Force
  body, 5
  surface, 5
Forced convection, 123, 187–189
  boundary layer
    flat plate on, 127
    heat transfer (table), 331
    main program for, 370
    mass transfer in, 175
    nonisothermal surface, 136
    nonsimilar, 154
    numerical integration of, 139
    pressure gradient, 136
    results for (table), 330
    See also boundary layer, forced
      convection
Fourier's law, 18, 207
Free convection, 124, 159, 187–189,
  237
  boundary layer
    level curves of error, 310
    main program for, 371–374

Free convection, boundary layer (Cont.):
    results (table), 336–339
Free parameter method, 65
freeb, 164, 170, 176, 186, 193, 237,
  308

Gallo, W.F., 95
gblpdriv, 70, 100, 353–355, 370
General boundary layer program,
  353–380
Gnos, U.S., 95
Goldstein, S., 64, 86
Grashof number, 108, 109, 123, 159,
  166, 169, 182, 185
Gregg, J.L., 162, 170, 186–188
Gross, J.F., 95
Görtler, H., 96
Görtler series expansion for boundary
  layer, 94

Hagen-Poiseuille flow, 24
Hansen, A.G., 61, 65, 84
Hartnett, J.P., 148
Hartree, D.R., 67, 75, 76
Hayes, W.D., 271, 273, 276
header, 357
Heat addition, bulk rate of, 14
Heat conduction
  coefficient of, 17
  diffusion of, 30
Heat conductivity, 21
Heat transfer, 126, 127, 135
  boundary layer
    combined forced and free
      convection, 187
    compressible, 237, 241, 242,
      247, 248, 252, 264
      flat plate, 225, 229
      hypersonic, 257, 260
        shock-boundary layer
          interaction, 274
      low speed, 225
      mass transfer in, 267
      unit Prandtl number, 232
      von Mises
        coordinates, 211
    forced convection, 132, 143,
      144
      mass transfer, 149–152

Index

Heat transfer, boundary layer (Cont.):
    free convection,
        nonisothermal wall, 170
    convective, 105
    Couette flow, simple, compressible, 203, 205, 207
        conduction in, 207
Heat transfer parameter, 146
    boundary layer
        forced convection, effect of viscous dissipation on 146, 147
Homann, F., 49
Howarth, L., 64, 213
Howarth transformation for compressible boundary layer, 213, 214
Hypersonic flow, 271
Hypersonic interaction parameter, 272
Hypersonic shock-wave boundary layer interaction, 269
Hypersonic strong interaction, 273, 275, 276
    results for (table), 351

Illingworth, C.R., 199, 220
Illingworth-Levy transformation, 220
Initial, 356
Injection, 148, 223, 261
    compressible boundary layer, 268
Inner and outer expansion, 64
Integral solution, 94
Inviscid flow,
    axisymmetric stagnation point, 45
Iteration
    Nachtsheim-Swigert, 40, 42, 43, 49, 71, 73, 76, 98, 142, 299–319, 353
        boundary layer
            compressible, 311, 312
                zero heat transfer, 313
                zero shearing stress, 313
        energy equation, 306
            adiabatic wall, 307
        Falkner-Skan equation, 302–304
            reverse flow, 304
            zero shearing stress, 305
        free convection, 308

        adiabatic wall, 310
        solution determined parameter 314
        systems of equations, 316
    Newton-Raphson, 72, 299
        Falkner-Skan equation, 300, 301

Kinematic viscosity, 125
Koh, J.C.Y., 148, 151
Kuethe, A.M., 99

Laplace's equation, 86
Lees, L., 95, 275
Legendre's equation, 86
Leigh, D., 91
Levy, S., 220, 231, 234, 240, 262
Lew, H.G., 262
Li, T.Y., 215, 221, 231, 275
Libby, P.A., 76, 262
Liepman, H.W., 270–272
Local similarity, 95, 156
Locally nonsimilar method, 96, 97, 237
    boundary layer, 156–158
        compressible, 279
        forced convection, 156
        free convection, 179
    three-equation model, 98, 99
        mass transfer, 102, 103
    two-equation model, 98, 100
        mass transfer, 102, 103
Lock, R.C., 91
Low, G.M., 240, 262
Lowell, R.L., 173

Mach number, 57, 201, 207, 221–223, 227, 229, 231
Mangler transformation, 84, 86
Mangler, W., 84–86
Marvin, J.G., 95
Mass, conservation of, 1, 2
Mass flow rate, 21
Mass transfer
    boundary layer, 90
        compressible, 222, 223, 261, 264
        forced convection, 147

Mass transfer, boundary layer (Cont.):
    free convection
        isothermal wall
            temperature profile for, 176
        nonisothermal on vertical flat plate, 173, 175
        nonsimilar, 182
    nonsimilar, 101
    shearing stress, 92
    velocity profile for, 94
    See also blowing; injection; suction
Matched asymptotic expansion, 64
McCroskey, W.S., 270
Meksyn, D., 95, 96
Meksyn-Merk method, 95, 154
Merk, H.J., 95
Momentum equation, 24, 25, 123, 124, 183, 199
    axisymmetric, 84
    axisymmetric stagnation point flow, 44, 46
    boundary layer, 59
        combined forced and free convection
            nonsimilar, 191
        compressible, 208, 210, 215
            Crocco coordinates, 212
        derivation of, 58
        flat plate, 60, 225
        nonsimilar, 278
        forced convection, 128, 137
            mass transfer, 147
            nonsimilar, 155
        free convection, 162, 164
            nonsimilar, 179, 180
        thermal, 125, 126
    Couette flow, simple, compressible, 200
    cylindrical polar coordinates, 44
    nondimensional, 55
    thermal flows, 106–109
    two-dimensional stagnation point flow, 36, 39
Momentum thickness, 77, 79, 80
    compressible boundary layer, 234, 236, 238, 253, 255
    Falkner-Skan solution, 80, 81
    two-dimensional stagnation point flow, 80
Monatomic gas, 195, 274
Moore, F.K., 75, 95

Nachtsheim, P.R., 301, 303, 304, 308, 310
Nachtsheim-Swigert, 40, 42, 193
Nachtsheim-Swigert iteration, 49, 71, 73, 76, 98, 142, 164, 237, 299–319, 353, 359
    boundary layer
        compressible, 238, 311, 312
            zero heat transfer, 313
            zero shearing stress, 313
    energy equation, 306
        adiabatic wall, 307
    Falkner-Skan equation, 302–304
        reverse flow, 304
        zero shearing stress, 305
    free convection, 308
        adiabatic wall, 310
    solution determined parameter, 314
    systems of equations, 316
    See also iteration
Nagakura, T., 276
Nagamatsu, H.T., 215, 221, 231, 275
Naruse, H., 276
Natural convection, 124, 159
Navier-Stokes equation, 1–19, 21, 22, 53, 54, 60, 68, 105–107
    compressible, 13, 23, 195, 199
    solutions
        analytical, 23–35
        approximate, 23
        exact, 23, 35–51
    suddenly accelerated flat plate, 28–35
Newton's law of friction, 26
Newton's second law, 1, 3–5
Newton-Raphson iteration, 40, 72, 299
    Falkner-Skan equation, 300–302
Nikuradse, J., 72
Nonsimilar boundary layer
    forced convection, 154
    free convection, 178
        with mass transfer, 182
    on circular cylinder, 99

Nonsimilar boundary layer (Cont.):
  See also locally nonsimilar method;
   three-equation locally
     nonsimilar method;
   two-equation locally
     nonsimilar method
ns, 70, 359–361
Numerical integration, free convection, 164
Numerical solution, two-dimensional stagnation point flow, 41
  table, 42
Nusselt number, 126, 127, 136, 152, 153, 166, 181, 207, 230, 232
  compressible boundary layer, 250

Order-of-magnitude, 54–58, 130
Ostrach, S., 160, 166
output, 362

parameter, 355, 356
Perfect gas thermal, 21
Pohlhausen, E., 130
Poiseuelle flow, 119
Prandtl, L., 53, 54, 64
Prandtl number, 56–58, 107–109, 112, 113, 124, 125, 127, 131–136, 144, 148, 164, 166, 168, 170, 172, 186, 189, 198, 207, 219, 222–226, 229, 231, 232, 237, 240, 241, 262, 276
Prandtl-Eckert product, 113, 114, 117, 118, 122, 134
Pressure gradient, adverse, 77
Probstein, R.F., 271, 273, 276
Pun, W.M., 154
Punnis, B., 143

Quack, H., 96–98, 100, 103

Rate-of-dilatation volume, 12
Rate-of-shearing-deformation, 11
Rate-of-strain, 1, 8
Rate-of-strain tensor, 11
Recovery enthalpy, Couette flow, simple, compressible, 203
Recovery temperature Couette flow, simple, compressible, 204

Reeves, B.L., 95
Reshotko, E., 240, 243
Retarded flow, compressible boundary layer, 240, 249, 250
Reverse flow
  boundary layer
   compressible, 240, 244, 246, 250, 263, 266
   hypersonic, 261
   mass transfer in, 267
  for energy equation, 143
Reverse flow solutions, 76–78, 93
Reynolds analogy, 126, 127, 208, 230
  compressible
   boundary layer, 251, 254
   Couette flow simple, 208
Reynolds number, 57, 100, 107–109, 110, 123–125, 134, 135, 152–154, 170, 183, 185, 201, 207
rk, 70, 358, 359, 370
Rogers, D.F., 76, 231, 240, 257, 261, 353
Rosenhead, L., 55, 81, 84
Roshko, A., 270–272
Runge-Kutta integration, 289–297
  equivalent first-order equation set, 289
  example, 296
  first-order differential equation, 290
  fourth-order, 293, 353
   graphical explanation, 293, 294
   simultaneous equations, 295
  recursion formulas, 290
  routine, 358
  second-order, 291, 293
  third-order, 293

Schlichting, H., 73, 91, 243
Schmidt, E., 168
Schönauer, W., 100
Separated flow, 77
Separation, 84, 101, 185
  boundary layer
   compressible, 240, 247, 248, 251, 265
    mass transfer, 267
  mass transfer, effect on, 93
Separation of variables, 30, 36

Separation of variables (Cont.):
   axisymmetric two-dimensional stagnation point flow, 46
Separation point, 77
Series expansion, 94
Shearing force, viscous, 15
Shearing stress, 16, 17, 26–28, 78, 93
   boundary layer
      compressible, 237, 241, 246, 248, 263, 264
         hypersonic, 257, 259
         mass transfer, 200, 209
         unit Prandtl number, 231
      combined forced and free convection, 187
      flat plate with mass transfer, 102
      free convection, 166
         on a nonisothermal wall, 170
      nonsimilar, 100
   circular cylinder, 100, 101
   Couette flow, simple, compressible, 200, 207
   Crocco transformation, 211
   hypersonic shock-boundary layer interaction, 274
   mass transfer, effect on, 92
   tensor, 5
   two-dimensional stagnation point flow, 51
   von Mises coordinates, 211
Shock wave structure, 199
Shock-wave boundary layer interaction, 269, 270
Shooting method, 299
Similarity
   analysis, suddenly accelerated plane wall, 31
   conditions, compressible boundary layer, 219, 221, 222
   requirements, compressible boundary layer mass transfer, 262
   solution, suddenly accelerated plane wall, 32
Sinclair, R.A., 95
Sisson, R.M., 143, 148

Skin friction, 77
   in Crocco coordinates, 213
Skin friction coefficient, 78, 79, 126, 127
   boundary layer
      compressible, 240, 250
         unit Prandtl number, 231, 232
   Couette flow, simple, compressible, 207, 208
   circular cylinder, 100
Smith, A.M.O., 95
Spalding, D.B., 92, 154
Sparrow, E.M., 96–98, 100, 103, 162, 170, 186–188
Specific heat at constant pressure, 195, 198
stag2d, 41, 43, 48, 364–366
Stagnation point flow,
   axisymmetric, 44, 45, 49, 51, 64
      solution (table), 50
   two-dimensional, 64, 243
Stanton number, 127, 134, 136, 152, 153, 207
Stewartson, K., 54, 76, 243
Stokes approximation, 12, 17
Streeter, L., 197
Stress tensor, 5, 8
Stress-rate-of-strain relation, 7, 8, 16
Suction, 148, 177, 178, 223, 261, 262
   in boundary layer, 90
      compressible, 263, 268
   See also mass transfer
Suddenly accelerated plane wall, 28, 34, 62
Sutherland equation, 196, 197
Sutherland viscosity law, 218, 228
Swigert, P., 301, 303, 304, 308, 310

Tangent wedge approximation, 271
Taylor series expansion, 303, 305, 306, 308, 315
   Runge-Kutta integration, 290
      in two variables, 292
      second-order, 291
Temperature profile
   Couette flow
      simple, adiabatic wall, 117
      nonsimple, 117, 118

Temperature profile (Cont.):
  combined forced and free
      convection, 189
  forced convection, 131, 133–135,
      168, 171
  free convection (table), 148, 167
  mass transfer
      isothermal wall, 176
      nonisothermal wall, 177
  Poiseuille flow, 121
  two-dimensional stagnation point
      flow, 145, 148
      mass transfer, 153, 154
Terrill, R.M., 100
Thermal conductivity, 195, 198, 198
Thermal diffusivity, 160
Thermally perfect fluid, 125
Three-equation locally nonsimilar model
  combined forced and free
      convection, 193
  compressible, 280
  forced convection, 158
  free convection, 180
Trace, rate-of-strain tensor, 11
Transformation, Mangler, 85
Transport properties, 195, 198
Transverse curvature, 277
Tsien, H.S., 210
Two-dimensional stagnation point flow,
    35–44, 46, 49, 51, 82, 145, 243
  displacement thickness, 80
  main program, 363–366
  momentum thickness, 80
  numerical solution (table), 42
  temperature distribution for, 148
  viscous dissipation effect of, 147
Two-equation locally nonsimilar model
  combined forced and free
      convection, 192
  compressible boundary layer, 280
  forced convection, 156, 157
  free convection, 180
      mass transfer, 182
  results for circular cylinder
      (table), 330
Töpfer, C., 64, 69

Van Dyke M., 64
Velocity, angular, 8, 10

Velocity overshoot in compressible
    boundary layer, 244
Velocity profile
  adverse pressure gradient, 78
  blowing, effect of, 93
  combined forced and free
      convection, 188
  compressible boundary layer, 241,
      243–245, 264
      adiabatic flat plate, 229
      $M_e \to 0$, 204
  Couette flow, simple, compressible,
      203, 206
      adiabatic wall, 203
  Falkner-Skan, 75
  forced convection boundary layer
      (table), 140
  free convection, 168, 171
  mass transfer
      isothermal wall, 176
      nonisothermal wall, 177
      table, 167
  mass transfer, effect of, 94
  nonsimilar, 100
  suction, effect of, 93
Viscosity, 12, 21, 195
  bulk, 12
  Chapman-Rubesin law, 197, 198,
      202, 214, 215, 218, 222, 272,
      273, 277
  dilatational, 12
  dynamic, 12, 13, 57, 272
  kinematic, 125
  Sutherland equation, 196, 197,
      218, 228
  variation with temperature, 196
Viscous dissipation, 18, 54, 108, 125,
    143, 145–147, 148, 160, 179
  compressible boundary layer, 208
  Couette flow, simple, compressible,
      203–206
  heat transfer, effects of (table), 332
  importance in compressible flow,
      210
  unit Prandtl number, 231
Viscous heating, 195
  simple Couette flow, 112, 113
Volume coefficient of thermal expansion,
    106

Volume rate-of-dilatation, 12
von Driest, E.R., 228, 229
von Karman, T, 210
von Karman integral technique, 94
von Mises equations, 218
von Mises transformation, 210

Vorticity function, 10

Weast, R.C., 197
Work, 14, 15, 16, 17

Yang, K.T., 170